Bioenergy Production by Anaerobic Digestion

Interest in anaerobic digestion (AD), the process of energy production through the production of biogas, has increased rapidly in recent years. It comes as the need to seek alternative renewable energy sources to fossil fuels, as well as reduce landfill waste and greenhouse gases, has accentuated. Agricultural and other organic waste are important substrates that can be treated by AD.

This book is one of the first to provide a broad introduction to anaerobic digestion and its potential to turn agricultural crops or crop residues, animal and other organic waste, into biomethane. The substrates used can include any non-woody materials, including grass and maize silage, seaweeds, municipal and industrial wastes. These are all systematically reviewed in terms of their suitability from a biological, technical and economic perspective. In the past the technical competence and high capital investment required for industrial-scale anaerobic digesters has limited their uptake, but the authors show that recent advances have made smaller-scale systems more viable through a greater understanding of optimising bacterial metabolism and productivity. Broader issues such as life cycle assessment and energy policies to promote AD are also discussed.

Nicholas E. Korres is a consultant in sustainable agriculture and biofuels/renewable energy.

Padraig O'Kiely works at Teagasc, the Irish Agricultural Authority, Ireland.

John A.H. Benzie is a Professor in the Environmental Research Institute at University College Cork, Ireland.

Jonathan S. West is a senior scientist at Rothamsted Research, Harpenden, UK.

Bioenergy Production by Anaerobic Digestion

Using agricultural biomass and organic wastes

Edited by Nicholas E. Korres, Padraig O'Kiely, John A.H. Benzie and Jonathan S. West

First published 2013
by Routledge

2 Park Square, Milton Park, Abingdon, Oxfordshire OX14 4RN
711 Third Avenue, New York, NY 10017

Routledge is an imprint of the Taylor & Francis Group, an informa business

First issued in paperback 2018

© 2013 Nicholas E. Korres, Padraig O'Kiely, John A.H. Benzie and Jonathan S. West selection and editorial material; individual chapters, the contributors

The right of the editors to be identified as the authors of the editorial material, and of the authors for their individual chapters, has been asserted in accordance with sections 77 and 78 of the Copyright, Designs and Patents Act 1988.

All rights reserved. No part of this book may be reprinted or reproduced or utilised in any form or by any electronic, mechanical, or other means, now known or hereafter invented, including photocopying and recording, or in any information storage or retrieval system, without permission in writing from the publishers.

Trademark notice: Product or corporate names may be trademarks or registered trademarks, and are used only for identification and explanation without intent to infringe.

British Library Cataloguing in Publication Data
A catalogue record for this book is available from the British Library

Library of Congress Cataloging-in-Publication Data
Bioenergy production by anaerobic digestion : using agricultural biomass and organic wastes / edited by Nicholas E. Korres, Padraig O'Kiely, John A.H. Benzie and Jonathan S. West.
    pages cm
  Includes bibliographical references and index.
  1. Sewage – Purification – Anaerobic treatment. 2. Biogas. 3. Agricultural wastes as fuel. I. Korres, Nicholas E., editor of compilation. II. O'Kiely, Padraig, editor of compilation. III. Benzie, John A. H., editor of compilation. IV. West, Jon (Jon S.), editor of compilation.
  TD756.45.B54 2013
  628´.746–dc23   2013006456

ISBN13: 978-0-415-69840-5 (hbk)
ISBN13: 978-1-138-36410-3 (pbk)

Typeset in Goudy
by HWA Text and Data Management, London

Printed in the United Kingdom
by Henry Ling Limited

A large part of this book was completed in a hospital room where, for the past 3 years, I was keeping company to my mother Sophia Korres while she was fighting against her illness. I dedicate this book to her memory because she taught me the value of hard and honest work but above all how to pursue my dreams with dignity.

Nicholas E. Korres

Contents

List of figures   x
List of tables   xiii
List of reviewers   xvi
Preface   xxi
Acknowledgements   xxv
Acronyms and abbreviations   xxvii

PART I
Legislation and energy policy   1

1. Sustainable agriculture and greenhouse gas emissions   3
   JONATHAN S. WEST

2. Energy and agricultural policy in relation to biomethane, with particular reference to the transport sector   8
   BEATRICE SMYTH

3. Biomethane production with reference to land-use change   30
   SURAJBHAN SEVDA, DEEPAK PANT AND ANOOP SINGH

PART II
Feedstocks   47

4. Grass and grass silage: agronomical characteristics and biogas production   49
   JOSEPH MCENIRY, NICHOLAS E. KORRES AND PADRAIG O'KIELY

5. Maize and maize silage for biomethane production   66
   MARKUS NEUREITER

6. Suitability of microalgae and seaweeds for biomethane production   82
   JOHN A.H. BENZIE AND STEPHEN HYNES

7	Organic wastes for biomethane production ANOOP SINGH	99
8	Industrial residues for biomethane production MARKUS ORTNER, BERNHARD DROSG, ELITZA STOYANOVA AND GÜNTHER BOCHMANN	111

PART III
Anaerobic digestion technology   137

9	Anaerobic digesters: perspectives and challenges ABDUL-SATTAR NIZAMI, BRADLEY A. SAVILLE AND HEATHER L. MACLEAN	139
10	Biogas upgrading and compression PRASAD KAPARAJU, SAIJA RASI AND JUKKA RINTALA	152
11	Storage and distribution of biomethane NICHOLAS E. KORRES	183
12	Variation in anaerobic digestion: need for process monitoring NICHOLAS E. KORRES AND ABDUL SATAR NIZAMI	194
13	General principles of data warehouse and data mining in anaerobic digestion NICHOLAS E. KORRES, ANASTASIOS DEKAZOS, DIMITRIS N. ARGYROPOULOS, AMMAR AHMED AND PAUL STACK	231

PART IV
Molecular biology and population dynamics   259

14	Microbial communities and their dynamics in biomethane production DIMITRIS N. ARGYROPOULOS, THEODOROS H. VARZAKAS AND JOHN A.H. BENZIE	261
15	The role of molecular biology in optimizing anaerobic digestion and biomethane production DIMITRIS N. ARGYROPOULOS, CHAROULA C. PSALLIDA AND JOHN A.H. BENZIE	290

PART V
Sustainability in anaerobic digestion   315

16	Life cycle assessment as a tool for assessing biomethane production sustainability NICHOLAS E. KORRES	317

17	The use of digestate as a substitute for manufactured fertilizer BRIAN J. CHAMBERS AND MATTHEW TAYLOR	359
18	The sustainability of small-scale anaerobic digesters at farm scale PHILLIP HOBBS AND ALLAN BUTLER	375
19	Biogas technology for developing countries: an approach to sustainable development M.S. DHANYA, S. PRASAD AND ANOOP SINGH	397
20	Concluding remarks	422
	Appendix A: Anaerobic digestion application in a typical Cambodia family farm – a case study ANDREA SALIMBENI AND GIULIANO GRASSI	427
	Index	432

Figures

1.1	GHG emissions (CO$_2$ equivalents per hectare) associated with the production of a typical winter oilseed rape crop in the UK	5
1.2	Effects of changes in farming practices on yields of winter wheat grown in the Broadbalk wheat experiment at Rothamsted Research, UK, since 1843	6
2.1	Policy, industry structure and stakeholders in the bioCNG industry	19
2.2	Roadmap for the development of a bioCNG industry	23
4.1	Schematic representation of changes in the chemical composition of grass with advancing maturity	52
5.1	Harvest of whole crop maize	71
5.2	Ensiling of whole crop maize	71
6.1	Major processing pathways for energy production from algae	85
6.2	Representative results for the range in microalgae productivity observed in a number of types of production systems	86
6.3	Representative results for the range in macroalgae productivity observed over a range of types of production systems	89
6.4	The classic two-stage anaerobic digestion process used for macroalgae	91
6.5	Present costs of producing biomass for energy from microalgae using photobioreactors and commercial raceways	93
6.6	Present costs and alternative uses of algae	94
6.7	The total world production of algae in 2009	95
8.1	Process scheme of the AD-plant	115
8.2	State-of-the-art stillage treatment process in dry-grind bioethanol plants	119
8.3	Overview of the accumulation of by-products in sugar beet processing	128
9.1	Anaerobic digestion as a means to produce energy and bioproducts	140
9.2	Types of high rate digesters	143
10.1	Flow chart of water scrubbing technology	163
10.2	Flow chart of chemical absorption process	165
10.3	Pressure-swing adsorption schematic process	166
10.4	Relative permeation rate of various biogas components	167
10.5	Flow chart of membrane biogas purification process	169
10.6	Flow chart of cryogenic biogas purification process	169
11.1	Floating tank cover	187
11.2	Biogas low-pressure gas bag storage facility	188
11.3	Weather-proof outer layer of flexible membrane cover	189
11.4	Cross-section of flexible membrane covers	190
12.1	Biogas production chain	195
12.2	Schematic representation of methane production by AD	195

12.3	Dendrogram based on hierarchical cluster analysis for the characterisation of the similarity of various physico-chemical characteristics of grass silages	203
12.4	Process control system of a digester	215
13.1	Data warehouse and data mining in the knowledge discovery process	231
13.2	Gathering data from various sources and transforming them into useful information available for decision-making procedures	234
13.3	De-normalized multi-dimensional star schema for AD performance data management	235
13.4	Representation of a data cube for AD	237
13.5	Pivot table	239
13.6	Performance of the selected attributes for each stage of AD	239
13.7	The data mining process	240
13.8	Categorization of DMN operations and techniques	243
13.9	Classification of biogas yield data according to volatile solids concentration	248
13.10	Summary of a training process	248
13.11	Regression curve showing data points and the prediction line	250
13.12	Relationship between maize dry matter (dry tonnes/ha) and methane yield	252
14.1	A simplified conceptual illustration of AD for I, II, and III stage digesters	263
14.2	The four major stages of chemical transformation: hydrolysis (H), acidogenesis (Ac), acetogenesis (At) and methanogenesis (M)	263
14.3	An example of the marked changes in bacterial diversity over time	271
14.4	Different relative abundances of 16S rRNA gene sequences	273
14.5	Principal component analysis	276
14.6	Sigmoidal curve of bacterial growth during batch culture	278
14.7	Generic scheme for development of models of the dynamics in a digester	279
15.1	The main stages and major representative chemical reactions during methanogenesis	294
15.2	A summary of three major pathways for methanogenesis.	296
15.3	Universal tree of life based on small-subunit ribosomal (ssu) rRNA sequences	302
15.4	Bio-molecular strategies for improved biomethane production	305
15.5	Illustrative example of possible paths in directed evolution of microbial communities for biomethane production combined with molecular methods	306
16.1	Generalised representation of a product's life cycle	319
16.2	Life cycle assessment framework and its direct applications	319
16.3	LCA of feedstock for biogas production and system boundaries	321
16.4	Schematic illustration after LCA application between biogas and a fossil fuel (a reference system) derived electricity	327
16.5	A generic step-by-step LCI development plan	328
16.6	Life cycle inventory and determination of system boundaries	329
16.7	Flow chart for the calculation of GHG emissions for bioenergy and fossil fuels	338
16.8	A supplementary flow chart to that in Figure 16.7 in which unit processes of the agricultural biomass production and AD is represented in detail	339
16.9	Energy flows (and related greenhouse gas emissions) in AD processing	340
16.10	Possible nutrient losses from manure management	344
16.11	Estimation of direct and indirect N_2O emissions due to run-off of the slurry produced by outdoor fed dry cattle	344
16.12	Integration of biotic, abiotic and factors influencing soil nitrogen dynamics	345
16.13	Multifunctional process with several input products and resources consumed accompanied by various wastes and emissions along with two co-products as output	346
16.14	Representation of midpoint and damage or endpoint impact categories	348

17.1	Readily available N (RAN) contents of food-based digestate in comparison with typical values for pig and cattle slurries	362
17.2	Quantofix meter for rapid on-site ammonium-N measurement	363
17.3	Digestate N use efficiency (percentage of total N applied)	364
17.4	Bandspread application to arable land	368
17.5	Shallow injection application to grassland	368
17.6	Application type and effect on ammonia (NH_3) losses	369
18.1	A conceptual framework for the sustainability of small-scale anaerobic digestion	379
18.2	Complex interplay in developing sustainable small-scale anaerobic digestion	388
18.3	Shadow price of carbon based on the EU ETS between 2008 and 2050	389
18.4	Marginal carbon cost curve per kW electricity produced on a dairy farm	390
19.1	Biogas production process	401
A.1	Life cycle of biogas production	429

Tables

2.1	GHGs and other sustainability issues of selected biofuels	12
3.1	General characteristics of biogas	30
3.2	Energy-yielding reactions of methanogens	33
3.3	Greenhouse gas emissions from land use	38
3.4	Land-use intensity of petroleum fuels and biofuels	39
3.5	Summary of the various initiatives proposed or developing proposals for measures to mitigate indirect impacts from biofuels	41
4.1	Some adaptation and management characteristics of perennial temperate grasses	51
4.2	Perennial grasses and red clover as feedstocks for anaerobic digestion	51
4.3	Effect of growth stage and harvest date on specific CH_4 yields from grassland	53
4.4	Summary of the potential dry matter (DM) losses that can occur during ensilage and at feedout	58
4.5	Fermentation efficiencies of the main fermentation pathways employed by silage microorganisms	59
5.1	Yields and composition of whole crop maize	68
5.2	Composition and properties of digestate from energy crop digesters	74
6.1	Major characteristics distinguishing the nature, cultivation and harvest of microalgae and macroalgae	83
6.2	Representative data illustrating the major differences in composition and productivity between microalgae, macroalgae and two examples of terrestrial crops	85
6.3	Representative figures for yield and CH_4 content of biogas from anaerobic digestion of microalgae	87
6.4	Representative figures for yield and CH_4 content of biogas from anaerobic digestion of macroalgae and some data on H_2S production characteristic of some species of macroalgae	90
7.1	Average collectable slurry production during housing period from different animals	100
7.2	Biodegradable municipal waste generation in different countries	101
7.3	The world crop residue production in 1991 and 2001	102
7.4	Different pre-treatments, their effects and pros and cons	104
7.5	Effect of pre-treatments on the methane production	105
7.6	Biomethane production potential of organic waste	106
7.7	Country-wise biomethane potential of biological municipal waste (BMW)	107
7.8	World-wide biomethane potential of some agricultural residues	107

8.1	Overview of estimated amounts of animal by-products (ABP) deriving from slaughter	113
8.2	Chemical characterisation of slaughterhouse waste processed in the biogas plant	115
8.3	Examples of increase in bioethanol production capacities from 2008 to 2010	117
8.4	Mean values of standard parameters of the stillage fractions in a dry-grind bioethanol plant	119
8.5	Potential for energy recovery per stillage fraction in a large-scale bioethanol plant	121
8.6	Amounts of brewery residues and biogas yield	122
8.7	Composition of brewers' spent grains	123
8.8	Chemical characterisation of the waste from olive oil production residues	125
8.9	Chemical characterisation of the wastes from sugar production from sugar beet	129
8.10	Composition of sugar beet	129
9.1	Various types of anaerobic digesters	141
9.2	Comparison of advantages and disadvantages of various digester types	142
9.3	Comparison of different digester characteristics	143
9.4	Factors affecting anaerobic digester inputs and outputs	144
9.5	Comparison of different anaerobic digesters applied for high solid substrates	145
9.6	Challenges of the AD industry	146
9.7	Conditions of a stabilized anaerobic digester	146
10.1	Composition of biogas from dedicated AD plants, sewage treatment plants and landfills and as a reference composition of natural gas in the Netherlands	153
10.2	Theoretical methane percentage in biogas calculated for different major substrates using Buswell's equation	155
10.3	Biogas contaminants and possible impact on the downstream application	156
10.4	Some common siloxanes occurring in biogas	159
10.5	Solubility of CO_2 in water at different temperatures and pressures	162
10.6	Technical information along with the advantages and disadvantages of different biogas upgrading techniques	171
10.7	Summary of energetic requirements of biogas upgrading technologies	172
10.8	Investment and maintenance costs for the five different upgrading technologies	174
10.9	Cost estimates of upgrading biogas to biomethane from studies undertaken during 2007–2009.	175
10.10	Summary of CH_4 losses associated with biogas upgrade technologies	176
10.11	Most commonly used biogas and biomethane storage systems	178
11.1	On-farm storage options for biogas and biomethane	186
12.1	Typical composition of biogas	196
12.2	Biowastes suitable for biological treatment according to the European Waste Catalogue	198
12.3	Important economic and technical considerations for the selection of the feedstock in AD	199
12.4	Variation of chemical characteristics, retention time and biogas production for various feedstocks	200
12.5	Examples of biomethane yield (m^3 kg^{-1} VS) from digestion of various plant species and plant material.	201
12.6	Effects of grassland type on biogas and methane production	202
12.7	Range of process parameters affecting optimum AD performance	208
12.8	Comparison of the performance of mesophilic and thermophilic anaerobic digestion	209

12.9	Variables in AD and their measuring techniques	217
13.1	Differences between DWH and OLTP	232
13.2	Training data of a hypothetical methane production in relation to process parameters temperature, pH and VFA	245
13.3	Estimation of probability frequencies	246
14.1	Examples of different microorganisms shown to be involved in different stages of AD	264
14.2	Two examples demonstrating how digestate communities can be very diverse	266
14.3	Main patterns of bacterial and archaeal diversity in anaerobic digesters	267
14.4	Environmental variables whose effects on population dynamics of microbial communities in AD are being increasingly investigated	268
15.1	Major substrates used in anaerobic digestion and examples of the principal types of enzymes or enzyme groups used in their breakdown to fermentable compounds	292
15.2	Summary of the main techniques for molecular genetic analysis	298
16.1	Example of a renewable/biofuel data collection sheet	330
16.2	Relating biogas data to unit process	333
16.3	Financial activity data in winter barley production	334
16.4	Decision-making process for selecting an allocation method	347
16.5	Various LCA impact categories	350
17.1	BSI PAS100 quality standards	361
17.2	Mean nutrient concentrations in food-based digestate and livestock slurries	362
17.3	Mean digestate and livestock slurry heavy metal concentrations	365
17.4	Financial value of food-based digestate	366
17.5	Displaced carbon footprint benefit of food-based digestate	366
17.6	Proposed Biofertiliser Matrix (PAS110/ADQP input materials): agriculture and field horticulture	370
17.7	Biofertiliser Matrix: crop categories	370
18.1	Some very small digesters available commercially	378
18.2	Principal incentives that are used to promote energy generation from biogas	381
18.3	Range in values assumed for DM, ODM and potential biogas yield	384
18.4	Improving biogas production by adding high energy feedstocks	384
18.5	Discounting the shadow price of carbon (SPC) for SSAD plant scenarios	389
18.6	Viable SSAD according to 2012 costs	392
19.1	Composition and properties of biogas	401
19.2	Potential biogas production from different feedstocks	402
19.3	Composition of biogas from alternative feedstocks	403
19.4	Energy efficiency of cooking fuels	404
19.5	Pollution reductions due to use of biogas plant	405
19.6	Energy content of biogas compared with other fuels	405
19.7	Pollution reductions due to biogas used as vehicle fuel	406
19.8	Health benefits of biogas over traditional fuel wood usage	407
A.1	Parameters for methane estimation from rice crop residuals	430
A.2	Parameters for methane estimation from pig manure	430
A.3	Parameters for methane estimation from buffalo manure	430
A.4	Total methane yield	431
A.5	Annual energy consumed for the satisfaction of the family farm basic needs	431

Reviewers

Prof. Dr Nuri Azbar
Ege University
Engineering Faculty, Bioengineering Department
Bornova, Izmir 35100, Turkey
Email: nuri.azbar@ege.edu.tr or nuriazbar@yahoo.com

Dr Guenther Bochmann
Institute for Environmental Biotechnology, BOKU
University of Natural Resources and Life Sciences
Vienna, Austria
Email: guenther.bochmann@boku.ac.at

Dr Wojciech Budzianowski
Wroclaw University of Technology
ul. Wybrzeze Wyspianskiego 27
50-370 Wroclaw, Poland
Email: wojciech.budzianowski@pwr.wroc.pl
Web: http://works.bepress.com/wojciech_budzianowski

Dr Wim Corre
Plant Research International
Wageningen University and Research Centre
The Netherlands
Email: wim.corre@wur.nl

Dr Ruth Delzeit
Kiel Institute for the Wold Economy
Hindenburgufer 66
D-24105 Kiel, Germany
Email: ruth.delzeit@ifw-kiel.de
Web: www.ifw-kiel.de

Dr Antonio Domínguez
Biogas Fuel Cell, S.A.
Parque Científico Tecnológico
c/ Ada Byron, nº 107, 1º izq., 33203
Gijón, Asturias, Spain
Email: a.dominguez@grupobfc.com
Web: www.grupobfc.com

Dr Alessandro Flammini
Energy Programme
Food and Agriculture Organization of the United Nations
Climate, Energy and Tenure Division (NRC)
Viale delle Terme di Caracalla Rome, Italy
Email: alessandro.flammini@fao.org
Web: http://www.fao.org/energy

Prof. Dr Uwe R. Fritsche
IINAS-International Institute for Sustainability Analysis and Strategy
Heidelberger Str. 129 1/2
D-64285 Darmstadt, Germany
Email: uf@iinas.org
Web: www.iinas.org

Dr Makarand Ghangrekar
Department of Civil Engineering
Indian Institute of Technology
Kharagpur-721, India
Email: ghangrekar@civil.iitkgp.ernet.in; makarand@ghangrekar.com; m_ghangrekar@rediffmail.com
Web-page: www.Ghangrekar.com

Dr Thomas Gitsopoulos
Hellenic Agricultural Organisation-Demeter
Plant Protection Institute of Thessaloniki
PO Box 60324, Thermi Thessaloniki, Greece
Email: gitsopoulos@yahoo.gr, gitsopoulos@nagref.gr

Dr Giuliano Grassi
EUBIA, European Biomass Industry Association
Rue d'Arlon 63–65
B-1040 Brussels, Belgium
Email: eubia@eubia.org
Web: www.eubia.org

List of reviewers

Mr Sjors van Iersel (MSc)
SQ Consult B.V.
P.O. Box 8239
3503 RE Utrecht, The Netherlands
Email: s.vaniersel@SQConsult.com

Dr Bodik Igor
Oddelenie environmentálneho inžinierstva
(Dept. of Environmental Engineering)
Fakulta chemickej a potravinárskej technológie STU
(Faculty of Chemical and Food Technology SUT)
Radlinského 9, 812 37 Bratislava, Slovak Republic
Email: igor.bodik@stuba.sk

Prof. Dr Heriber Insam
Institut für Mikrobiologie
Technikerstraße 25d
A-6020 Innsbruck, Austria
Email: heribert.insam@uibk.ac.at
Web: http://uni.microbial-ecology.at
Web: http://www.biotreat.eu
Web: http://www.bio4gas.at
Chief Editor of Applied Soil Ecology

Prof. Dr Gunnur Kocar
Solar Energy Institute
Ege University, Izmir, Turkey
Email: gkocar@gmail.com
Web: http://eusolar.ege.edu.tr/en/

Prof. Ken Krich
California Institute for Energy & Environment
University of California
2087 Addison Street
Berkeley, CA 94704, USA
Email: ken.krich@uc-ciee.org

Dr Clare Lukehurst
Director, Task 37 (UK)
52 Broadstairs Road
Broadstairs, Kent CT10 2RJ, UK
Email: clare.lukehurst@green-ways.eclipse.co.uk

Dr Didier Marchal
Belgian Alternate Team Leader, IEA Bioenergy Task 40
Walloon Forest Service, Belgium
Email: Didier.MARCHAL@spw.wallonie.be

Dr David McIlveen-Wright
Centre for Sustainable Technologies (CST)
School of the Built Environment
Faculty of Art, Design and the Built Environment
University of Ulster
Shore Road, Newtownabbey BT37 0QB
Northern Ireland, UK
Email: david@mcilveen-wright.com or Drmcilveen-wright@ulster.ac.uk

Dr Richard E. Muck
USDA, Agricultural Research Service
US Dairy Forage Research Center
1925 Linden Drive West
Madison, WI 53706, USA
Email: remuck@wisc.edu

Dr Philip Owende
School of Biosystems Engineering
University College Dublin
Belfield, Dublin 4, Ireland
Email: philip.owende@ucd.ie

Dr Tim Patterson
Wales Centre of Excellence for Anaerobic Digestion
Sustainable Environment Research Centre (SERC)
Faculty of Health, Sport & Science
University of Glamorgan
Pontypridd, CF37 4AT, Wales, UK
Email: tlpatter@glam.ac.uk
Web: http://www.walesadcentre.org.uk
Web: http://www.serc.research.glam.ac.uk
Web: http://www.glam.ac.uk

Prof. Annette Prochnow
Leibniz-Institute for Agricultural Engineering Potsdam-Bornim
Department of Technology Assessment
Max-Eyth-Allee 100
14469 Potsdam, Germany
Email: aprochnow@atb-potsdam.de
Web: www.atb-potsdam.de

Dr A. Gangagni Rao
Bioengineering and Environmental Centre (BEEC)
Indian Institute of Chemical Technology (IICT)
Tarnaka, Hyderabad-500007, India
Email: gangagnirao@yahoo.com

Dr Andrea Salimbeni
EUBIA, European Biomass Industry Association
Rue d'Arlon 63–65
B-1040 Brussels, Belgium
Email: andrea.salimbeni@eubia.org or eubia@eubia.org
Web: www.eubia.org

Prof. Paul C. Struik
Centre for Crop Systems Analysis
Wageningen UR, Droevendaalsesteeg 1 (Building 107)
6708 PB Wageningen, The Netherlands
P.O. Box 430, 6700 AK Wageningen
The Netherlands
Email: paul.struik@wur.nl
Web: http://www.csa.wur.nl/

Dr Han Vervaeren
Hogeschool West-Vlaanderen – Campus GKG
Associatie Universiteit Gent
Graaf Karel de Goedelaan 5 – B-8500 Kortrijk, Belgium
Email: han.vervaeren@howest.be
Web: http://www.howest.be

Prof. Peter Weiland
Federal Research Institute for Rural Areas, Forestry and Fisheries
Institute of Agricultural Technology and Biosystems Engineering
Bundesallee 50
D-38116 Braunschweig, Germany
Email: peter.weiland@vti.bund.de

Preface

Lignocellulosic biomass and organic wastes are the most investigated type of feedstock used for biomethane production because they are some of the most abundant resources with wide availability in most countries. It has been projected that a major part of the European renewable energy production, for example, will originate from farming and forestry and at least 25 per cent of all bioenergy in the future will originate from biogas, produced from wet organic materials such as animal manure, whole crop silages, wet food and feed wastes. In addition, while electricity and heat can be produced by a variety of renewable sources, the only alternative to fossil resources for production of liquid and gas fuels and chemicals is biomass. Anaerobic digestion is the most appropriate technology to convert the available biomass and other organic wastes to biomethane. The frame under which modern agriculture operates including the significant role agriculture can hold in energy production and in environmental pollution reduction under the need for agriculture diversification and sustainability is discussed in Chapter 1. The potential and significant role of agriculture in the energy sector is enhanced by the current trends and changes in the legal framework of many countries which have triggered the emergence of energy crop-based biogas digesters. These issues are discussed extensively in Chapter 2. Land use and land use change is one of the main topics in the debate of food production vs. biofuels that merits further consideration. Some initial thoughts concerning this important subject are discussed in Chapter 3. These chapters form Part I of this book which focuses on legislation and energy policy matters.

Second-generation biofuels, i.e. these originated from lignocellulosic biomass such as grass and grass silage, maize silage, as well as micro- and macro-algae but also non-food resources such as organic solid and industrial wastes, have started to gain pace in the race of biofuel research. Part II of the book deals with the production and/or processing of selected feedstocks and their utilisation for optimum biomethane production. Chapters 4 to 8 discuss holistically and analytically the important issues on the production of second-generation biofuels by anaerobic digestion by taking grass and maize silage, micro- and macro-algae, organic and industrial wastes as model feedstocks. More particularly, Chapter 4 analyses the suitability of grass species as a potential feedstock in anaerobic digestion in terms of their fitness, ecophysiology and husbandry. Maize as a sole energy crop or as part of a mixed substrate, consisting usually of manure and other agricultural residues, has been proven to be an invaluable resource for biomethane production. An extensive analysis is performed in Chapter 5. Microalgae and seaweeds could not be excluded from the list of highly desirable lignocellulosic feedstocks for the production of second-generation biomethane. Macro- and micro-algae production systems, productivity rates and biogas production potential along with economic production issues and future trends are comprehensively discussed in Chapter 6. The potential use of the organic and industrial wastes as feedstock for biomethane production through anaerobic digestion is discussed in Chapters 7 and 8. More specifically, Chapter 7 focuses on organic origin wastes (e.g. household wastes, human and animal excreta, biodegradable wastes, agricultural refuse and the organic fraction of municipal solid waste) and summarises general aspects of their anaerobic digestion and potential. Chapter 8 focuses on industrial wastes (e.g. abattoir wastes, slaughterhouse wastes, bioethanol production wastes, brewery residues, olive oil production wastes and wastes from sugarbeet processing facilities) and analyses in detail their availability, potential, possible bottlenecks and economics along with production processing.

Part III of this book consists of chapters related to anaerobic digestion technology. Technological issues such as digester design and configurations, biogas upgrading and biogas/biomethane storage are discussed in detail in Chapters 9 to 11. As such, Chapter 9 discusses digester design, substrate properties, variable biogas yields from similar digester types and biomethane potential assays, suggesting future research needs on factors that may affect anaerobic digester design. Chapter 10 discusses technologies of biogas upgrading for biomethane production with emphasis on CO_2 removal. More particularly, it refers to biogas contaminants and their treatment followed by an extensive report on CO_2 and various technologies for its removal. The comparison of various biogas upgrading techniques in terms of technical availability, energetic performance, economic assessment, investment and maintenance cost occupies a large part of this chapter. Environmental pollution of upgrading and biomethane compression and storage complete this chapter. Chapter 11 discusses various biogas storage options. Storage prerequisites and more common low pressure storage options such as floating cover, gas bags, rigid digester cover and flexible membrane cover are explained. Medium and high pressure storage options accompanied with biogas distribution and transportation conclude this chapter. The cumulative variation in biomethane production due to numerous factors and the need for monitoring in order to increase biomethane yield potential is discussed in Chapter 12. More particularly, the inter- and intra-feedstock variation, the pre-treatment (i.e. physical, chemical and biological) and the variation resulting due to process parameters (i.e. temperature, pH, C/N ratio, alkalinity, loading rate, retention time, volatile fatty acids etc.) is analysed and discussed comprehensively. The need for process monitoring is clearly established and process control systems (i.e. instrumentation, programmable logic controller, human machine interface and supervisory control and data acquisition programmes) are described. It is widely known that constraints exist in the exploitation of current knowledge and available information in anaerobic digestion. A consequence of this is that decision-making processes often lack scientific support. Chapter 13 attempts to highlight the benefits that biogas production stakeholders could gather by the incorporation, into the existing monitoring and decision-making system, of two of the most important techniques of knowledge development from databases – namely data warehouse and data mining techniques. Conceptual examples in data warehousing as these of multidimensional data modelling (i.e. star schema and data cubes) have been employed to highlight the usefulness of the data warehouse technique. Additionally, data mining techniques, particularly these under classification (i.e. naïve Bayes, time series, decision trees, neural networks etc.) and regression (linear and non-linear) categories support the integration of these data analytics into the anaerobic digestion system. Hierarchical clustering, a descriptive data mining technique, is also analysed in detail and completes the third part of the book.

In the last few decades, technological progress in molecular biology has become of major significance in the study of the physiology of gene function of microorganisms. Therefore, it is highly appropriate to discuss these subjects like population dynamics, molecular biology and molecular genetics of the anaerobic bacteria as described in Chapter 14 and Chapter 15 respectively. These topics form Part IV of the book. Chapter 14 deals with the variety of microorganisms present in anaerobic digestion and the dynamic changes that occur in these populations over time process. The kinetics and modelling of methanogenesis and future trends concerning the development of accurate microbial population dynamics models for better process understanding complete this chapter. The biochemistry of anaerobic digestion in relation to various substrates and the description of the biochemical pathways in anaerobic digestion is provided in the first half of Chapter 15. The second half of this chapter summarises the techniques for molecular genetic analysis applied to anaerobic digestion and the impact these have had on understanding these systems. Principal among those recognised are the ability to identify and enumerate the biological community involved in anaerobic digestion, the dissection of their metabolic process and the growing capability to genetically engineer organisms for more efficient gas production.

Part V, the final part of this book, analyses sustainable biogas/biomethane production issues along with methods to investigate them. Biomethane production from agricultural biomass and organic residues can be an efficient technique to minimise emissions from energy production. This feature is most probably the reason for the wide applicability of life cycle assessment (LCA) in the renewable/biogas production sector, particularly when it is used as transport fuel. This is analysed in Chapter 16 where the general working protocol for LCA application supported by examples based on various feedstocks for biogas/biomethane production are discussed. The recycling of digestate, a residue of anaerobic digestion, to land is regarded as the best practicable environmental option in most circumstances, completing both natural nutrient and carbon cycles. Chapter 17 discusses the use of the digestate as a substitute for manufactured fertilisers and examines how valuable this can be proved in relation to sustainability of the anaerobic digestion and biomethane production. More specifically, the quality of the digestate and standards in relation to microbial pathogens, heavy metals, stability and physical contaminants is discussed in detail. Digestate properties, i.e. nutrient content, organic matter, heavy metal concentration etc. along with its financial values, carbon footprint, land application controls and its integration with manufactured fertilisers conclude this chapter. Chapter 18 discusses in detail the sustainability of small-scale anaerobic digesters and their contribution to climate change mitigation in relation to national and international policy incentives that support anaerobic digestion. The factors influencing the development of small-scale anaerobic digestion and the effect of quality of feedstock on their revenue flow is discussed along with process monitoring controls. An important part of this chapter is based on valuing the social benefits of greenhouse gas emissions reduction from small-scale anaerobic digesters based on the shadow price of carbon in a social cost-benefit analysis. The final part of this book closes with Chapter 19 in which the benefits of anaerobic digestion in developing countries are extensively discussed. The negative impacts of conventional energy use in comparison with biomethane are analysed. The production of biogas from local resources along with various socio-economic benefits for developing countries is mentioned. Finally, an extensive reference is made to biogas development for a number of representative developing countries. The book closes with the final conclusions and future needs for a sustainable biogas and/or biomethane production.

The editors are grateful to the chapter authors and publishers in bringing this collection of key information relating to bioenergy production into the format of a book.

Despite the great effort that editors have invested in this work and the extensive checks conducted by many experts in the field of anaerobic digestion and biogas/biomethane production, mistakes may have been made. We would like to highlight that any comments or suggested changes to improve and update the book contents for future editions are welcomed.

Nicholas E. Korres, Padraig O'Kiely, John A.H. Benzie, Jonathan S. West

nkorres@yahoo.co.uk (Nicholas E. Korres)
Padraig.OKiely@teagasc.i.e. (Padraig O'Kiely)
j.benzie@ucc.i.e. (John A.H. Benzie)
jon.west@rothamsted.ac.uk (Jonathan S. West)

Acknowledgements

The editors and authors would like to express their gratitude for their invaluable contribution into this effort to the following:

- Prof. John A. Lucas, Rothamsted Research, UK for manuscript review (Chapter 1)
- Dr Christoph Resch, Universität für Bodenkultur Wien, Department IFA-Tulln, Institute für Umweltbiotechnologie for copyrights permission (Figures 5.1 & 5.2)
- Dr Ulrike Schmid-Staiger, Fraunhofer Institute for Interfacial Engineering and Biotechnology IGB, Department of Environmental Biotechnology and Biochemical Engineering for copyright permission (Figure 6.6)
- Dipl.-Ing. (FH) Michael Beil, Fraunhofer Institute for Wind Energy and Energy System Technology, Germany for suggestions and comments (Chapter 10)
- Dr Alfons Schulte-Schulze Berndt, Schmack Carbotech GmbH, Germany for suggestions and comments (Chapter 10)
- Dr Arthur Wellinger, IEA Bioenergy, for the data in Table 10.1 adopted from IEA Bioenergy Task 37, 2009, 'Biogas upgrading technologies-developments and innovations' (Chapter 10)
- Dr Anneli Petersson, IEA Bioenergy, for the data in Table 10.1 adopted from IEA Bioenergy Task 37, 2009, 'Biogas upgrading technologies-developments and innovations' (Chapter 10)
- Dr Mona Arnold, VTT Technical Research Centre, Finland for the data in Table 10.4 adopted from Arnold, M. 2009, Research Notes 2496 (Chapter 10)
- Dr Alfons Schulte-Schulze Berndt, Schmack Carbotech GmbH, Germany for the data in Table 10.7 (Chapter 10)
- Dipl.-Ing. (FH) Michael Beil, Fraunhofer Institute for Wind Energy and Energy System Technology, Germany for reproducing Figures 10.1–10.3 and 10.6 from his presentation Beil, M., 2009, 'Over view on biogas upgrading technologies' at European Biomethane Fuel Conference, Goteborg, Sweden (Chapter 10)
- Dr M. Miltner, Dr A. Makaruk and Prof. M. Harasek of Institute of Chemical Engineering, Vienna University of Technology and ETA-Florence Renewable Energies, Italy for reproducing Figures 10.4 and 10.5 from the conference paper (Miltner et al, 2008, 16th European Biomass Conference and Exhibition, Valencia, Spain (Chapter 10)
- Dr Christoph Grieder, ETH Zürich Institut für Agrarwissenschaften, LFW A4, Gruppe Prof. A. Walter, Universitätsstrasse 2, 8092 Zuerich, Switzerland for providing the data used to develop Figure 13.12 (Chapter 13)

- National Development Plan, through the Research Stimulus Fund (#RSF 07 557), administered by the Department of Agriculture, Food & Marine, Ireland for providing financial support (Chapter 4).
- Siemens Industry Inc. for copyright permission (Figure 11.4)
- ADAS colleagues for their contribution to the digestate characterisation and field experimental work (Chapter 17).
- Waste and Resources Action Programme (WRAP) for providing financial support (Chapter 17)
- Department for Environment Food and Rural Affairs (Defra) for providing financial support (Chapter 17)
- Elsiever Ltd for the data in Table 10.1 adopted from Rasi et al, 2007, *Energy*, vol. 32, pp1375–1380 (Chapter 10)
- Elsiever Ltd for the data in Table 10.2 adopted from Angelidaki and Sanders, 2004, *Re/Views in Environmental Science and Bio/Technology*, vol, 3, no 2, pp117–129 (Chapter 10)
- Elsiever Ltd for the data in Table 10.3 adopted from Ryckebosch et al, 2011, *Biomass and Bioenergy*, vol. 35, pp1633–1645 (Chapter 10)
- Elsiever Ltd for the data in Table 10.4 adopted from Rasi et al, 2007, *Energy*, vol. 32, pp1375–1380 (Chapter 10)
- Elsiever Ltd for the data in Table 10.5 adopted from Appels et al, 2008, *Energy Conversion and Management*, vol 49, no 10, pp2859–2864 (Chapter 10)
- Elsiever Ltd for the data in Table 10.6 adopted from Patterson et al, 2011, *Energy Policy*, vol. 39(3), no. 18, pp1806–1816 (Chapter 10)
- Elsiever Ltd for the data in Table 10.8 adopted from Patterson et al, 2011, *Energy Policy*, vol 39(3), no. 18, pp1806–1816 (Chapter 10)
- Agency for Renewable Resources, Germany for the data in Table 10.6 adopted from *Biomethan*. Gülzow, 2012 (Chapter 10)
- ADAS UK Ltd for Figure 17.5 (Chapter 17)

Last, but not least, our sincere gratitude goes to our commissioning editor, Tim Hardwick, for his patience, help and professionalism particularly in turning points of this project. Ashley Wright, our editorial assistant, deserves a sincere vote of appreciation for her efforts in securing high-quality outcomes and providing all kinds of support.

Acronyms and abbreviations

16S RNA	16S ribosomal RNA
ABP	animal by-products
AD	anaerobic digestion
ADAM	anaerobic digestion analytical model
ADF	acid detergent fibres
ADL	acid detergent lipids
ADQP	anaerobic digestate quality protocol
AFLP	amplified fragment length polymorphism
BC	buffering capacity
BCR	benefit-cost ratio
BDTC	Biogas Development and Training Centres
bioCNG	blend of biomethane and compressed natural gas
BMP	biomethane potential
BMW	biodegradable municipal waste
BOD	biological oxygen demand
BSE	bovine spongiform encephalopathy
BSG	brewers' spent grains
BSP	Biogas Support Programme
BSS	biogas spent sludge
CAD	centralised anaerobic digestion
CAP	Common Agricultural Policy
CBA	cost-benefit analysis
CBM	compressed biomethane
CBP	community biogas plants
CCM	corn cob mix
CDM	clean development mechanism
CERs	certified emission reductions
CH_4	methane
CHP	combined heat and power
CMO	Common Market Organisation
CNG	compressed natural gas
CO_2	carbon dioxide
CO_2e	CO_2-equivalent
CoA	coenzyme A
COD	chemical oxygen demand
CP	crude protein
CSTR	continuously stirred tank reactor

DDBJ	DNA Data Bank of Japan
DDGS	distillers × dried grains with solubles
DGGE	denaturing gradient gel electrophoresis
dH	German degrees
DM	dry matter
DMI	dry matter intake
DMN	data mining
DMo	desugared molasses
DNA	deoxyribose nucleic acid
DS	dry solids
DSS	decision support system
DWH	data warehouse
E	exa- ($\times 10^{18}$)
EDP	ecosystem damage potential
EEA	European Environment Agency
EEIOEF	environmentally extended input–output emission factor
EEPROM	electrically erasable programmable read-only memory
EF	emission factor(s)
EMBL	European Molecular Biology Laboratory
EMP	Emden–Meyerhof–Parnas pathway
EPROM	erasable programmable read-only memory
ETL	extract, transform and load
EU	European Union
FAS	Farm Accounts Survey
FAS	farm assurance schemes
FBS	Farm Business Survey
FERC	Federal Energy Regulatory Commission
FIRR	financial internal rates of return
FISH	fluorescent in-situ hybridization
FIT	feed-in-tariff
FLC	fuzzy logic control
FM	fresh matter
GAEC	good agricultural and environmental condition
GDP	gross domestic product
GHG	greenhouse gas
GOLD	Genomes OnLine Database
GWh	gigawatt hour
GWP	global warming potential
H_2S	hydrogen sulphide
ha	hectare
HACCP	hazard analysis and critical control point
HANPP	human appropriation of net primary production
HDPE	high-density polyethylene
HHs	households
HHV	higher heating value
hl	hectolitre
HMI	human–machine interface
HPWS	high pressure water scrubbing
HRT	hydraulic retention time
IBP	institutional biogas plants
IC	internal combustion
IEA	International Energy Agency

ILUC	indirect land use change
INC	initial national communication
IPCC	Intergovernmental Panel on Climate Change
IRG	Italian ryegrass (*Lolium multiflorum*)
ISO	International Standards Organisation
ITDG	Intermediate Technology Development Group
J	joule
k	kilo- ($\times 10^3$)
KDD	knowledge discovery in databases
ktoe	thousand tonnes of oil equivalent
KVIC	Khadi and Village Industries Commission
kWh	kilowatt hour
LBM	liquefied biomethane
LCA	life cycle assessment
LCFA	long chain fatty acids
LCI	life cycle inventory
LCIA	life cycle impact assessment
LDC	least developed countries
LDPE	low-density polyethylene
LHV	lower heating value
LNG	liquefied natural gas
LUC	land use change
MDGs	Millennium Development Goals
MFIT	micro-feed-in-tariff
MJ	megajoule
MMTCO$_2$e	million metric tons of carbon dioxide equivalents
MPS	massively parallel sequencing
mRNA	messenger RNA
MSW	municipal solid waste
MT	megatonne
NBMMP	National Biogas and Manure Management Programme
NBP	night-soil based biogas plants
NDF	neutral detergent fibre
NDP	natural degradation potential
NGO	non-governmental organisation
NGV	natural gas vehicle
NIRS	near-infrared spectroscopy
Nm^3	normal cubic metres
NTG	*N*-methyl-*N'*-nitro-*N*-nitrosoguanidine
NVZ	nitrate vulnerable zones
ODM	organic dry matter
OFMSW	organic fraction of municipal solid waste
OLAP	online analytical process
OLR	organic loading rate
OLTP	online transaction processing
OM	organic matter
OMAFRA	Ontario Ministry of Agriculture, Food and Rural Affairs
OMSW	olive mill solid waste
OMW	olive mill wastewater
OPA	Ontario Power Authority
OUT	operational taxonomic unit
PCF	product carbon footprint

PCR	polymerase chain reaction
PHA	polyhydroxyalkanoates
PLC	programmable logic controller
ppm	parts per million
PRG	perennial ryegrass (*Lolium perenne*)
PSA	pressure swing adsorption
RAN	readily available nitrogen
RAPD	random amplification of DNA
RBP	residual biogas potential
RD	rational protein design
RECs	Renewable Energy Certificates
RETs	renewable energy technologies
RFLP	restriction fragment length polymorphism
RHI	Renewable Heat Initiative
ROCs	Renewable Obligations Certificates
RPS	Renewable Portfolio Standard
RT-PCR	real-time PCR
SBP	sugar beet pulp
SCADA	supervisory control and data acquisition
SCF	standard cubic feet
SDM	site-directed mutagenesis
SGIP	small generator interconnection process
SNV	Netherlands Development Organisation
SOM	soil organic matter
SPC	shadow price of carbon
SRI	Silage maize Ripeness Index
SSADs	small-scale anaerobic digesters
SSM	soft system methodology
SSU rRNA	small subunit ribosomal RNA
STP	standard temperature and pressure
SVM	support vector machine
SW	slaughterhouse wastes
T	tera- ($\times 10^{12}$)
t	tonne
TA	total acids
TKN	total Kjeldahl nitrogen
TOC	total organic carbon
toe	tonnes of oil equivalent
TS	total solids
TSE	transmissible spongiform encephalopathy
UASB	upflow anaerobic sludge blanket
UK	United Kingdom
UNDP	United Nations Development Programme
US	United States of America
UV	ultraviolet
VFA	volatile fatty acid
VFR	vertical flow reactor
VS	volatile solids
Wh	watt hour
WI	Wobbe index
WSC	water-soluble carbohydrates
WWTP	wastewater treatment plant

Part I
Legislation and energy policy

Chapter 1

Sustainable agriculture and greenhouse gas emissions

Jonathan S. West

Bawden Plant Pathology Lab, Plant Biology and Crop Science Department
Rothamsted Research, Harpenden AL5 2JQ, UK
Email: jon.west@rothamsted.ac.uk

Introduction

At a time when oil reserves are running out and carbon emissions from human activities are widely acknowledged to be causing environmental change, there has been an increasing emphasis on alternative 'clean' energy sources. Coupled with this is the challenge to increase food production to feed the world's population. Currently over 1 billion people do not have enough to eat (www.fao.org), but the population is predicted to increase by over 35 per cent in the next 40 years, from 7 billion now to over 9 billion by 2050 (Beddington, 2010; Anon., 2011). Already, a reduction in global food stocks has occurred in recent years due to increased demand and decreased yields as a result of environment change (principally insufficient rainfall) and changes to diet (increased meat consumption, associated with increasing affluence) (Anon., 2011). Additionally, the use of some potential food materials such as maize grain for bioenergy production has led to instability in food prices with food price spikes in 2007–8 associated with food export bans in some countries and even riots (Anon., 2012). Without new developments in science and technology, the problem can only worsen as the world's population increases, water sources continue to be overexploited (currently 70 per cent of water is used for agriculture, much extracted from rivers and aquifers) and particularly if a sub-set of food-crops such as oilseeds, maize or wheat grain is used for biofuel production, since there will then be less food available. This may not seem a problem where food is plentiful or for countries with enough wealth to import food, but the problem is passed on to other regions, usually in tropical climates. As a result, more land in tropical countries is being converted from forest to agriculture, often by burning areas of forest. This land use change causes a release of carbon from the burnt vegetation and also from carbon that was stored in the soil. This, together with the loss of productive forest area (which efficiently sequesters carbon) more than cancels out the benefits of biofuel production in the temperate areas. As such, when considered globally, certain biofuels are not 'carbon neutral' as claimed.

Fortunately, new scientific advances show great promise in delivering sustainable production of both food (through genetic improvement of crops such as wheat, rice and oilseed rape) and bioenergy. Of course renewable forms of energy are available from wind, wave or tidal action, hydroelectric or geothermal sources but these each produce electricity, rather than a liquid or gas that can be used in conventional engines for transport. Liquid bioethanol, produced from sugarcane (effectively a non-food crop) in Brazil is available for this purpose. However, other liquid fuels such as bio-diesel (produced from oilseed rape/

canola) and ethanol (from cereals and particularly maize grain) have the disadvantage of using a potential food source, vegetable oil or carbohydrates, as their respective starting materials (Parry and Hawkesford, 2010; Parry and Jing, 2011). Therefore it is desirable to use non-food crops or waste materials from crops as a feedstock for production of biofuels such as biomethane and many new biofuel crop species are currently being investigated and genetic improvements are being made both for production of liquid and solid biofuels (Karp et al., 2011; Mariani et al., 2010). The EU 2003 biofuels directive targets an increase in biofuel transport energy from 5.75 per cent in 2010 to 10 per cent by 2020 (Anon., 2007). In addition to this, biomass derived liquid or gaseous fuels could substitute current transport fuels and natural gas used for domestic and industrial purposes. Conversion of lignin, cellulose and other carbohydrates in plant cell walls is a potential approach to produce biofuel from non-food and perennial crops or waste-products.

Agriculture and carbon emissions

In addition to producing fuels from renewable biological sources, it is also desirable to reduce the carbon footprint of all agricultural activities associated with food production. Agriculture currently contributes a significant proportion of global carbon emissions. Globally, greenhouse gas (GHG) emissions from agriculture are estimated to amount to 10–12 per cent of all emissions (Smith et al., 2007). For example, GHG emissions from the UK agricultural sector amounted to 7 per cent of the UK total in 2007 (43.3 Mt CO_2 eq out of 618.6 Mt CO_2 eq) (National Atmospheric Emissions Inventory; www.naei.org.uk). This is similar to other parts of Western Europe and the UK is committed to reducing agricultural GHG emissions in England by 3 Mt CO_2 eq by 2020 (UK Committee on Climate Change; www.theccc.org.uk/sectors/non-co2-gases/agriculture). Much of the agricultural GHG emissions in northwestern Europe are associated with animal production (particularly as methane) and new research on diets, breeds and species of animals is in progress to produce animal products with much lower GHG emissions (Smith et al., 2007). For arable crops, the largest contribution to GHG emissions is by the manufacture and use of fertilisers; for example over 79 per cent of emissions associated with the production of a typical hectare of winter oilseed rape is associated with the manufacture of nitrogen-containing fertiliser (1433 kg CO_2 eq/ha) and a further 1242 kg CO_2 eq/ha is associated with the breakdown of a proportion of the applied nitrogen-containing fertiliser into N_2O, which is a powerful greenhouse gas (Figure 1.1; Mahmuti et al. 2009).

In comparison, only 9.41 kg CO_2 eq/ha or 0.3 per cent of emissions were associated with the manufacture of the pesticides (herbicides, insecticides and fungicides) typically used. Yet fungicides alone were found to increase yields of winter oilseed rape by an average of 12.7 per cent having contributed to 0.04 per cent of GHG emissions in their use (Mahmuti et al. 2009). In the UK, fungicide treatment is estimated to have reduced GHG emissions by 1.64 Mt CO_2 for four major UK arable crops (winter barley, spring barley, winter wheat, and winter oilseed rape) in 2009 compared with releases calculated to have occurred by producing the same yield on the necessarily increased land area but without fungicide-based crop protection (Hughes et al. 2011). Globally, diseases are associated with losses of 16 per cent of crops and more generally losses to pests, weeds and diseases amount to 40 per cent of annual yields (Oerke, 2006). Climate change may itself alter the severity of crop disease epidemics (Evans et al. 2008; Madgwick et al. 2011). Recent studies by Berry et al. (2008), Mahmuti et al. (2009) and Hughes et al. (2011) illustrate that disease control measures

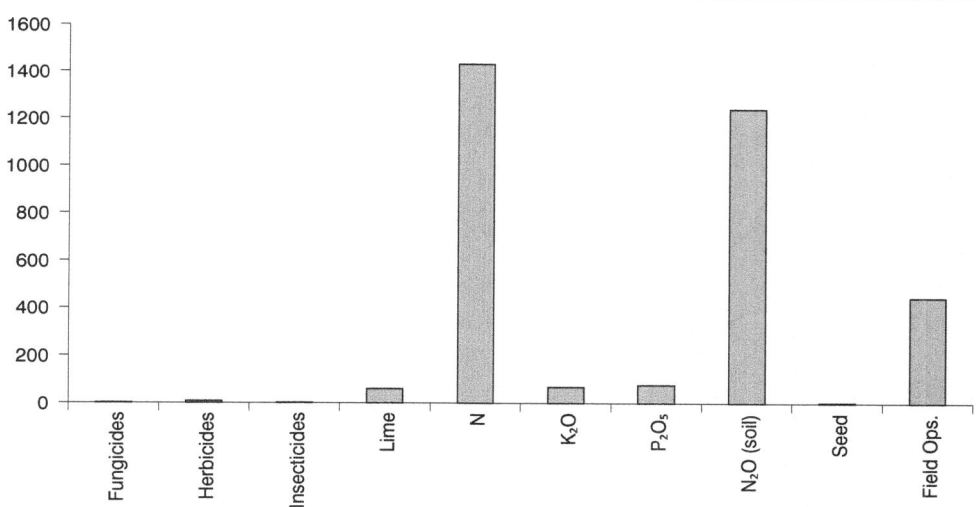

Figure 1.1 GHG emissions (CO_2 equivalents per hectare) associated with the production of a typical winter oilseed rape crop in the UK (data from Mahmuti et al., 2009). 'Field ops.' describes GHG emissions associated with mechanical equipment such as tractors and combine harvesters.

can not only reduce crop losses but also reduce the carbon footprint of crop production per tonne of grain produced and play a substantial part as a strategy to reduce agricultural GHG emissions by producing food efficiently on a smaller land area. A substantial reduction in GHG emissions is therefore possible by optimising and even increasing crop protection and by breeding crops that use nutrients more efficiently so that less nitrogen and other fertilisers need be applied. Good crop protection alongside effective application of nutrients and improved plant varieties has delivered substantial increases in yields over the last 60 years in particular (Figure 1.2).

Land use and carbon sequestration

An additional benefit to crop protection and GHG emissions has been realised recently by Berry et al. (2010) in research that has shown that growing arable crops efficiently using good crop protection products, elite cultivars and optimised fertiliser inputs not only increases yield per hectare and reduces the carbon footprint per tonne of grain produced but also means that less land area is required for this food production. This releases land for additional food production and/or for perennial biofuel crops, permanent grassland or woodland, which each sequester CO_2 into their soils to reach a steady state in which a larger amount of CO_2 is stored than in soils of arable crops. Less efficient crop production would require a larger land area to be cropped and Berry et al. (2010) show that this land use change (from pasture to arable crops) will lead to the release of CO_2 stored in converted grassland soils. In terms of GHG emissions associated not only with food production but also with land use, sustainable intensive arable crop production can therefore be considered as a climate-smart, environmentally conscious form of farming, using integrated pest management to reduce the carbon footprint of food production.

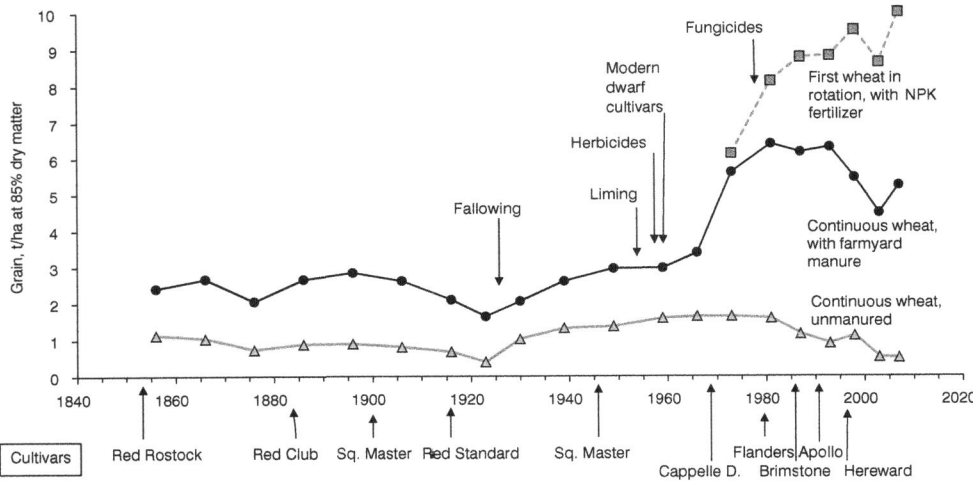

Figure 1.2 Effects of changes in farming practices on yields of winter wheat grown in the Broadbalk wheat experiment at Rothamsted Research, UK, since 1843. This dataset is part of the Long-Term Experiments National Capability at Rothamsted Research, funded by the UK Biotechnology and Biological Sciences Research Council and the Lawes Agricultural Trust. NPK = nutrients: Nitrogen, Phosphorus and Potassium, respectively.

Conclusions

To quote Sir John Beddington, Chief Scientific Advisor to the UK Government, "Food production must increase through climate-smart sustainable intensive arable crop production and this will need new scientific advancements, including use of some biotechnology approaches, improved crop varieties and species, and enhanced crop protection to produce more food with decreased associated GHG emissions". Simultaneously, the policy of the EU and some national governments towards the choice of biofuels must place a strong emphasis on the use of grasslands and (non-food) waste products, rather than grains and oilseeds, since grasslands serve a dual purpose in carbon sequestration in soil and production of a clean form of energy – biomethane – without decreasing food production. Biomethane uses the principle of anaerobic digestion for its production and this is discussed in more detail in later chapters. Financial incentives must be made available to encourage the uptake of this technology. Advances in microbiology, molecular and cellular biology, biochemistry, synthetic biology and bioengineering offer potential solutions towards biofuel production as part of a sustainable form of agriculture that minimises GHG emissions. These solutions are discussed in subsequent chapters along with methods of biomass production.

References

Anon. (2007) *Biofuels Progress Report: Report on the progress made in the use of biofuels and other renewable fuels in the Member States of the European Union.* http://ec.europa.eu/energy/energy_policy/doc/07_biofuels_progress_report_en.pdf

Anon. (2011) Executive Summary. *Foresight. The future of food and farming.* London: The Government Office for Science

Anon. (2012) 'FAO Food Price Index'. http://www.fao.org/worldfoodsituation/wfs-home/foodprices index/en/

Beddington, J. (2010). 'Food security: contributions from science to a new and greener revolution'. *Philosophical Transactions of the Royal Society B*, 365, 61–71.

Berry, P.M., Kindred, D.R. and Paveley, N.D. (2008) 'Quantifying the effects of fungicides and disease resistance on greenhouse gas emissions associated with wheat production', *Plant Pathology*, 57, 1000–1008.

Berry, P.M., Kindred, D.R., Olesen, J.E., Jorgensen, L.N. and Paveley, N.D. (2010) 'Quantifying the effect of interactions between disease control, nitrogen supply and land use change on the greenhouse gas emissions associated with wheat production', *Plant Pathology*, 59, 753–763.

Evans, N., Baierl, A., Semenov, M.A., Gladders, P. and Fitt, B.D.L. (2008) 'Range and severity of plant disease increased by global warming', *Journal of the Royal Society Interface*, 5, 525–553.

Hughes, D.J., West, J.S., Atkins, S.D., Gladders, P., Jeger, M.J. and Fitt, B.D.L. (2011) 'Effects of disease control by fungicides on greenhouse gas emissions by UK arable crop production'. *Pest Management Science* 67, 1082–1092.

Karp, A., Hanley, S.J., Trybush, S.O., Macalpine, W., Pei, M. and Shield, I. (2011) 'Genetic improvement of willow for bioenergy and biofuels', *Journal of Integrated Plant Biology* 53, 151–165.

Madgwick, J., West, J.S., White, R., Semenov, M., Townsend, J.A., Turner, J.A. and Fitt, B.D.L. (2011) 'Future threat; direct impact of climate change on wheat fusarium ear blight in the UK' *European Journal of Plant Pathology*, 130, 117–131.

Mahmuti, M., West, J.S., Watts, J., Gladders, P. and Fitt, B.D.L. (2009) 'Controlling crop disease contributes to both food security and climate change mitigation'. *International Journal of Agricultural Sustainability* 7, 189–202.

Mariani, C., Cabrini, R., Danin, A., Piffanelli, P., Fricano, A., Gomarasca, S., Dicandilo, M., Grassi, F. and Soave, C. (2010) 'Origin, diffusion and reproduction of the giant reed (*Arundo donax* L.): A promising weedy energy crop', *Annals of Applied Biology*, 157, 191–202.

Oerke, E.C. (2006) 'Crop losses to pests', *Journal of Agricultural Sciences*, 144, 31–43.

Parry, M.A.J. and Hawkesford, M. (2010) 'Food security: increasing yield and improving resource use efficiency', *Proceedings of the Nutrition Society* 69, 1–9.

Parry, M.A.J. and Jing, H.-C., 2011 'Bioenergy plants: hopes, concerns and prospectives' *Journal of Integrative Plant Biology*, 53, 94–95.

Smith, P., Martino, D., Cai, Z., Gwary, D., Janzen, H., Kumar, P., McCarl, B., Ogle, S., O'Mara, F., Rice, C., Scholes, B. and Sirotenko, O. (2007) 'Agriculture'. In B. Metz, O.R. Davidson, P.R. Bosch, R. Dave, L.A. Meyer (eds) *Climate Change 2007: Mitigation. Contribution of Working Group III to the Fourth Assessment Report of the Intergovernmental Panel on Climate Change*. Cambridge and New York: Cambridge University Press.

Chapter 2

Energy and agricultural policy in relation to biomethane, with particular reference to the transport sector

Beatrice Smyth

Clean Energies Research Cluster, School of Mechanical and Aerospace Engineering, Queen's University Belfast, Ashby Building, Stranmillis Road, BT9 5AH, Belfast, Northern Ireland
E-mail: beatrice.smyth@qub.ac.uk

Introduction

The use of biomethane for transport is an area of growing interest, with ongoing research into feedstock suitability, anaerobic digestion, gas upgrading, grid injection techniques, and vehicle and filling station technology (e.g. Bordelanne et al., 2011; Gerin et al., 2008; Hagen et al., 2001; Lehtomäki et al., 2008; Nizami and Murphy, 2010; Petersson and Wellinger, 2009). Due to the complex nature of the biomethane-for-transport industry, which is comprised of the areas mentioned above, a wide range of existing policies, particularly those in energy and agriculture, have both direct and indirect impacts on the sector.

Despite the importance of policy for the biomethane-for-transport industry, there exists limited information on the impact of policy on the industry. The aim of this chapter is to start fill that knowledge gap and to explore energy and agricultural policy in relation to biomethane. The various policies and policy instruments are discussed and a number of case studies are presented, showing experience of both successful and unsuccessful industries. Drawing on the case studies and policy discussion, a policy roadmap for a successful industry is developed. The focus of the chapter is on biomethane injected into the gas grid for use as a transport fuel, although many of the aspects discussed also pertain to biomethane used (on- or off-site) for heat or electricity generation.

Background to biomethane

Biogas, biomethane and bioCNG

Biomethane is biogas that has been upgraded to the same standard as natural gas, and is typically composed of 97 per cent methane (CH_4), 3 per cent carbon dioxide (CO_2) and some minor constituents. Biogas is produced through the anaerobic digestion (AD) of organic materials and typically consists of 55–70 per cent CH_4 (but this can be higher), 30–45 per cent CO_2 and some minor constituents, e.g. hydrogen sulphide (H_2S) and water. Many feedstocks are suitable for anaerobic digestion, including crops such as grass and maize, agricultural wastes such as animal slurries and slaughterhouse waste, industrial wastes, the organic fraction of municipal solid waste (OFMSW) and sewage sludge. Biogas composition depends on the feedstock, e.g. biogas from grass contains around 55 per cent CH_4 (Smyth et al., 2009), whereas biogas from chicken slurry contains 60–80 per cent CH_4 (Steffen et al., 1998). The digested feedstock that remains at the end of the AD process is known as digestate and can be used as a substitute for conventional manufactured fertiliser (Chapter 17 of this book).

Biomethane is mixable and interchangeable with natural gas and can be used in all applications intended for natural gas. It can be used directly for heat and/or electricity generation, or can be compressed for use in natural gas vehicles (NGVs). Compressed biomethane mixed (in any proportion) with compressed natural gas (CNG) is known as bioCNG. There is a considerable existing market for gas as a transport fuel, with over 14.7 million NGVs worldwide (NGV, 2012). Iran, Pakistan and Argentina have the largest fleets globally, each with over two million NGVs. Within Europe, the largest NGV fleet is in Italy (approximately 760,000), while Germany, Bulgaria and Sweden also have significant fleets, with around 95,000, 61,000 and 36,000 NGVs respectively (NGV, 2012). The use of biomethane as a vehicle fuel is increasing; biomethane accounts for over 60 per cent of fuel used in Swedish NGVs (Petersson, 2011), and Austria aims to replace 20 per cent of natural gas used in the transport sector with biomethane (Jönsson, 2006).

The potential biomethane resource is large. Using European Environment Agency (EEA) data, Åhman (2010) reported that the potential biogas supply in 2030 in the EU from 'wet' manure, sewage sludge and food processing residues is 0.8 EJ, and a further 2.15 EJ of environmentally compatible biogas (i.e. that which can be produced in line with environmental policies and assuming that there are no additional pressures on biodiversity, soil and water resources compared with the business-as-usual situation) is estimated to be available from agricultural crops. This compares to 0.25 EJ of biogas production in 2007. If all this biogas were converted to biomethane, it would account for 20 per cent of transport fuel in 2030 (business as usual) or 31 per cent under an energy-efficient scenario (Åhman, 2010). A 1998 US study concluded that it is feasible to capture and use over a third of the biogas potential of animal waste, sewage sludge and landfill in the country. If all this biogas were used for transport, it would displace 38 billion litres (10 billion gallons) of petrol equivalent per year (USDOE, 2011). Total US petrol (gasoline) consumption in 2011 was about 507 billion litres (134 billion gallons) (USEIA, 2012).

Biomethane policy in the literature

Progress in the renewable energy sector is inextricably linked to policy, but there is no "one size fits all" and different types of policies are needed for different technologies and applications (Gross et al., 2003). Very few countries have long-term policy experience of a mature biofuel market (Worldwatch Institute, 2007), let alone a mature bioCNG market, and there exists no step-by-step guide on how to promote biofuels (Bomb et al., 2007) or, indeed, biomethane.

There is limited information relating specifically to policies for biomethane or bioCNG for transport, although research on general biofuel policies and on particular aspects of the biomethane-for-transport industry (e.g. NGVs, AD) is more prevalent and is helpful in discussing the effect of policy on a biomethane-for-transport industry.

The research most relevant to biomethane policy is an article by Thamsiriroj et al. (2011), which developed a country-specific roadmap for a bioCNG industry in Ireland and recommended a range of supports, including an obligation for a minimum percentage of biomethane in gaseous transport fuel and subsidies for biomethane facilities. The paper was specific to the Irish context and had a strong focus on national legislation and resources. Work by Patterson et al. (2011) evaluated the policy and techno-economic factors affecting the use of biomethane as a transport fuel in the UK (a country with very low penetration of NGVs), and discussed in detail available biogas upgrading technologies and the economic viability of upgrading biogas for use as a vehicle fuel. Although not focused on biomethane, Yeh (2007) conducted an empirical analysis of the adoption of NGVs in eight countries and discussed policies and other factors influencing the industry.

More generally, Wisenthal et al. (2009) analysed the strengths and weaknesses of biofuel support policies implemented in the EU, while Bomb et al. (2007) investigated the biofuel industries in Germany and the UK, and discussed policy issues with a focus on the early stages of a biofuel industry. Also focusing on the initial stages of the industry, van der Laak et al. (2007) analysed various projects in the Netherlands and put forward guidelines for policy development in the Dutch biofuel sector. Silvestrini et al. (2010) looked at experience with biofuels in the EU cities of Berlin, London, Milan and Helsinki, and highlighted the importance of cities as test cases for policies that may in future be implemented at national level.

The findings of these studies with relevance to policy in the biomethane-for-transport industry are discussed in the following sections.

Energy and agricultural policy in the biomethane industry

Energy policy

Renewable energy

Arising from concerns over climate change, increasing energy prices, dwindling fossil fuel supplies and security of energy supply, policies and targets have been put in place to promote renewable sources of energy. Although renewable energy targets have been set in all three energy sectors (heat, electricity and transport), transport lags behind in terms of the penetration of renewable resources. In the EU, where there is a target for 10 per cent renewable transport energy by 2020 (EC, 2009a), progress has been relatively slow and in 2010 there was less than 5 per cent biofuels penetration, which is below the 2003 Biofuels Directive target of 5.75 per cent for 2010 (EurObserv'ER, 2011). The relatively slow development of renewable energy in the transport sector means that considerable growth is required if targets are to be met. If renewable energy policies are to drive the biomethane-for-transport industry, they must be translated into specific energy, transport and biofuel policies with a direct impact on AD and biomethane production and use.

Anaerobic digestion, biogas and biomethane

Renewable energy policy can encourage the production of biogas through, for example, grants for the construction of anaerobic digestion plants. The use of that biogas, i.e. for heat, electricity or transport, is then also dictated by policy. In the EU, biogas energy is mainly recovered in the form of electricity. In 2009, primary biogas energy output in the EU was 8346 ktoe, and gross biogas electricity output was 2164 ktoe. Biogas heat output in the EU in the same year was 174 ktoe (EurObserv'ER, 2010). The quantity used as vehicle fuel for trains, buses and other vehicles is relatively minor, although it is growing (REN21, 2011).

For biogas to be used as a vehicle fuel, it must first be upgraded to biomethane standard and then delivered to the point of use. The most efficient means of transporting biomethane is through use of the existing natural gas pipeline network (where available), although transport by pressurised container is also possible. For biomethane to be injected in to the grid and used as a vehicle fuel, plant operators, the gas network operator, and vehicle and filling station providers, need to be involved in the development of policies to ensure that biomethane meets quality and safety standards.

In the EU, Directive 2009/73/EC on the natural gas market (EC, 2009b) states that biogas should be granted non-discriminatory access to the gas system. Several countries, e.g. Sweden, Germany and Switzerland, have national policies for biomethane injection to the gas grid and/or use in vehicles, but the EU Directive has yet to be implemented through national policy in all countries in the EU, and despite considerable discussion there is as yet no European-wide standard for biomethane injected into the gas grid.

Energy in transport and biofuels

Targets for renewable transport energy can be met through different sources, including electric vehicles, hydrogen and biofuels. Depending on the specifics of the renewable transport energy policy that is in place, biomethane may or may not be supported by the policy. Brazil, for example, set a target for 5 per cent biodiesel by 2013 and for ethanol to account for 20–25 per cent of current petrol usage (Worldwatch Institute, 2007). Previous targets in the US focused on ethanol (e.g. 2.8 billion litres by 2012), as did targets in a number of Canadian states and in China (Worldwatch Institute, 2007). EU policy (Directive 2009/28/EC) includes support for biomethane, but demands that certain sustainability criteria are met in order for the renewable transport energy source to be counted towards meeting the target for renewable energy in transport. The Directive stipulates that biofuels (including biomethane) must effect greenhouse gas (GHG) savings of 35 per cent compared with the fossil fuel replaced, rising to 60 per cent for new facilities in 2018 (EC, 2009a). GHG savings are calculated from GHG balances, which, along with energy balances, are critical for assessing the sustainability of bioenergy systems (Buchholz et al., 2009). For an energy balance, a cradle-to-grave life-cycle assessment is carried out and the parasitic energy demands of the system, e.g. energy use in agriculture, are subtracted from the gross energy of the feedstock to determine the net energy of the biofuel. A GHG balance is conducted in a similar manner by comparing the emissions saved through fossil fuel replacement with the net emissions from biofuel production. Biofuels with higher net energy values and higher GHG savings are considered preferable to those with poor energy and GHG balances.

Biomethane can be produced from many different feedstocks, including crops (such as grass and maize) and wastes (such as agricultural slurries and the organic fraction of municipal solid waste), and the GHG savings of biomethane are heavily dependent on the feedstock used. Typical GHG savings for biomethane from different feedstock are presented in Table 2.1, along with values for ethanol and biodiesel. The GHG savings of biomethane compare well to conventional temperate biofuels, and even when mixed with natural gas to form bioCNG considerable emissions savings can still be achieved. Dedicated CNG vehicles running on natural gas have emissions that are around 17 per cent lower than petrol vehicles (Bordelanne et al., 2011). In addition to the associated GHG benefits, transport policies promoting CNG prior to the introduction of bioCNG and biomethane are beneficial to the biomethane industry (Silvestrini et al., 2010). Countries with successful biomethane-for-transport industries, e.g. Sweden and Austria, began by promoting compressed natural gas (CNG) as a transport fuel, followed by the introduction of biomethane to the market (by

Table 2.1. GHGs and other sustainability issues of selected biofuels (adapted from Smyth et al. 2010b).

Feedstock and fuel	GHG savings with no LUC[a] (%)			GHG savings with LUC[b] (%)		Other environmental and sustainability issues
	Process/ process fuel	Typical	Default		Previous land use	
Grass biomethane		75 (range 22–150)[c]				Existing grassland is used for livestock and the expansion of a grass biomethane industry may result in competition with existing agriculture (and food production).[d]
Maize/corn biomethane		60[e]				Maize is generally grown in rotation, so additional land is required under contract. Arable land is needed for maize; its use for energy crops may result in competition with food supplies. There are considerable environmental concerns[f] when maize is grown as a monoculture.
Municipal waste (OFMSW) biomethane		80	73			Compared with composting, AD of OFMSW leads to reduced emissions of GHGs; AD of OFMSW reduces waste volume.[g] Digestate needs to be dealt with carefully.
Manure biomethane		84–86	81–82			Slurry has a relatively low biogas yield per unit volume (due to high water content), leading to higher production costs per unit energy.[h] Digested slurry contains nutrients in a form more easily absorbed by plants.[i] Digestate needs to be dealt with carefully.
Slaughter waste biomethane		>100[j]				Animal by-products regulations restrict the use of certain slaughter wastes for AD. Digestate needs to be dealt with carefully.
Palm oil biodiesel	Methane capture at oil mill	68	65	−135	Forestland in Malaysia	Excellent energy balance, but there are concerns over land use change from the expansion of palm oil plantations.[k]
				−185	Forestland in Indonesia	
				−12	Grassland in Malaysia	
				−84	Grassland in Indonesia	

Rapeseed biodiesel		45	38	–569	Forestland in UK	Generally poor energy balance, although the use of by-products, such as straw and rape cake, improves the efficiency of the system.[k]
				–123	Grassland in UK	
Sugarcane ethanol		71	71			Excellent energy balance, but there are concerns over land use change and displacement effects.[k]
Wheat ethanol	Lignite (CHP plant)	32	16			Poor energy balance, although the use of by-products such as straw improves the efficiency of the system.[k] Wheat is generally grown in rotation, so additional land is required under contract. Arable land is needed; its use for energy crops may result in competition with food supplies.
	Natural gas (CHP plant)	53	47			
	Straw (CHP plant)	69	69			
Corn/maize ethanol	Natural gas	56	49	–93	Worldwide displacement	Poor energy balance, although use of by-products such as straw improves the efficiency of the system.[k] See notes on maize biomethane for further issues concerning the crop.
Sugar beet ethanol		61	52			Good energy balance.[k] Sugar beet is generally grown in rotation, so additional land is required under contract. Arable land is needed; its use for energy crops may result in competition with food supplies.

AD = anaerobic digestion; CHP = combined heat and power; GHG = greenhouse gas; LUC = land use change; OFMSW = organic fraction of municipal solid waste

[a] Values taken from EC (2009a), unless otherwise stated

[b] Searchinger et al., 2008; Upham et al., 2009

[c] Korres et al., 2010

[d] Smyth and Murphy, 2011

[e] Baxter, 2010

f Uekoetter, 2011

[g] Borjesson and Mattiasson, 2007

[h] Smyth et al., 2010a

[i] Yiridoe et al., 2009

[j] Singh and Murphy, 2009

[k] Smyth et al., 2010b

blending with CNG to form bioCNG) and increasing the percentage in the blend as the industry developed.

Returning to sustainability, biofuel policies often impose further requirements. The EU Renewable Energy Directive requires the exclusion of biofuels from peatlands and land with high biodiversity value or high carbon stock, as well as the assessment of social sustainability, food prices and other development issues.

Agricultural policy

Energy crops and land use

Agricultural policy can drive the availability of energy crops by encouraging farmers to grow crops for energy purposes. The Common Agricultural Policy (CAP) in the EU is a system of subsidies and support programmes for agriculture and is the main vehicle used to deliver agricultural policy. Council Regulation EC No. 1782/2003 (EC, 2003) established the Single Payment Scheme and introduced a payment of €45 per hectare per annum for areas under energy crops. Any agricultural raw material may be grown under the scheme, provided that the crops are intended primarily for energy purposes, i.e. for biofuels (including biogas), electric or thermal energy. It should be highlighted, however, that not all energy crops policies support biogas production. Also under the CAP are mechanisms to promote the cultivation of oilseed crops on set-aside land but only if contracted solely for the production of biodiesel or other industrial products (Schnepf, 2006).

Where there is general support for energy crops for biogas production, a decision must be made on which crop to grow. Although many crops are technically suitable for anaerobic digestion, there are numerous other factors and related policies which influence energy crop choice. Crop yield, energy balance and GHG savings (Table 2.1) are all important, as is the type of land needed for the crop.

Agricultural policies may place restrictions on land use change (LUC) or on the use of certain land types for energy crops. Direct LUC can be said to occur when land is converted from a previous use to bioenergy crop production; indirect land use change (ILUC) occurs when, for example, grassland or forest is converted to cropland to meet the demand for commodities which have been displaced by the production of biofuel feedstock elsewhere (Plevin et al., 2010).

In the EU, cross-compliance regulations, which are part of the CAP, require that the ratio of the area of permanent pasture to the total agricultural area of each member state must not decrease by 10 per cent or more from the 2003 reference ratio (EC, 2004), thus limiting the conversion of grassland to arable cropping. The expansion of the cultivation of energy crops that require arable land may therefore have an impact on existing arable crops grown for food, fibre, feed or energy purposes. In Germany, there has been considerable development in the use of maize for biogas production over the last few years. Between 2008 and 2009, the area under energy maize increased by 29 per cent, and by a further 40 per cent between 2009 and 2010 (BMELV, 2011). Maize in Germany has traditionally been grown for animal fodder and the increase in the area under energy maize has mainly been at the expense of this fodder maize, i.e. there is direct competition between food and biofuels. According to government, this direct change in land use has negatively affected the landscape and biodiversity in some areas, as well as leading to rent increases, and as a result the German government has changed its support schemes for certain energy crops (BMELV,

2011; Strauch and Krassowki, 2012). Widespread use of monocultures and the resultant homogeneous land use is a serious concern for both long-term agricultural productivity and the environment (Uekoetter, 2011).

Although generally not yet covered by agricultural or biofuel policy, the issue of indirect effects, particularly ILUC, is an area of growing concern. The "corn connection" in the US is one such example; subsidies for corn bioethanol have led to a move from soy to corn, resulting in decreased soy output and increased soy and beef prices (soy is used for animal feed). This in turn has led to increased production of soy and beef in South America and is strongly linked to deforestation (Laurance, 2007). It is argued (Liska and Perrin, 2009) that the emissions from ILUC caused by biofuel production should, if significant, be considered when calculating the GHG impact of the biofuel. However, measuring the emissions from ILUC is considerably more difficult than measuring those from direct LUC (Liska and Perrin, 2009), as the causal effects assumed in such calculations are open to interpretation. The assumptions and data inputs used to calculate the magnitude of potential LUC associated with biofuels have been questioned (CBES, 2009). It has been argued that the relationships between biofuels, commodity prices, trade and land-cover changes that are assumed by current modelling approaches are not consistent with historic data for initial land conversion and expansion. Empirical verification of ILUC due to recent expansion of the biofuel industry is problematic because those expansions constitute a very small driver relative to global LUC, so the biofuel impact is likely to be overshadowed by other causes (Liska and Perrin, 2009).

Agricultural wastes

Agri-environmental policies can direct improved management and treatment of agricultural wastes, such as agricultural slurries and slaughterhouse waste, which can in turn promote anaerobic digestion as a method for treating these wastes. In Denmark in the 1980s, concerns over nitrate leaching from agriculture led to tightening of agri-environmental legislation, including a requirement for farmers to have sufficient capacity for 6–9 months manure storage. Farmers were only permitted to spread manure when the risk of nitrate leaching was low and had to store manure for the remainder of the year. Arising from this, farmers began to participate in centralised anaerobic digestion (CAD) plants, which managed the transportation, storage and distribution of manure and digestate (Raven and Gregersen, 2007).

Similar policies were introduced in Sweden in the 1980s to reduce nitrogen leakage from the agricultural sector. Measures, such as increased seasonal manure storage, exclusion periods for manure spreading and the construction of AD plants, were implemented. These measures have resulted in significant environmental benefits, for example, the Laholm biogas plant in west Sweden, which was built in 1992 due to environmental concerns, has substantially reduced regional eutrophication and nitrogen leakage in to the Laholm Bay area (IEA Bioenergy, 2005). The plant treats 28,000 t of animal manure and 20,000 t of other wastes (including vegetable and slaughterhouse waste) per year, and recycles the digestate to 17 farms in the surrounding area (IEA Bioenergy, 2005).

The use of AD for the treatment of agricultural wastes can reduce GHG emissions from the agricultural sector, and so assist in complying with GHG reduction policies. Agricultural slurry applied directly to land results in uncontrolled GHG emissions, whereas if the slurry is treated in an anaerobic digester, methane is captured in the form of biogas. The biogas can then be used for energy purposes, including transport, thus replacing fossil fuel and reducing GHG emissions from the energy sector. Taking into account the reduction in GHG emissions from both slurry treatment and fossil fuel replacement, GHG savings of 82 per cent have been reported for cattle slurry biomethane compared with fossil diesel (Singh and Murphy, 2009).

Digestate

Following the production of biogas, the material that remains at the end of the AD process is known as digestate. Digestate consists of a solid and a liquid fraction, and can be used as a fertiliser; its use as a fertiliser is controlled by agri-environmental policy. Like for conventional fertiliser, the spreading of digestate on agricultural land must follow regulations designed to prevent nutrient leaching and protect ground and surface water. Depending on the feedstock, further policies relating to the use of animal by-products (ABP) may also apply. The purpose of these policies is to safeguard human and animal health, and to minimise the risk of contaminants entering the food chain. In the EU, the AD of animal wastes and the use of digestate as fertiliser are controlled by the Animal By-Products Regulations (EC, 2009c). These regulations lay down requirements for the collection, transport, storage, handling, processing and use or disposal of all ABP. Requirements include specific hygienisation steps during the processing of the ABP, and restrictions on grazing and on the type of crops that can be grown on land fertilised with digestate.

The interpretation of the ABP Regulations varies between countries, with stricter interpretation in some countries than in others. In Ireland, for example, past food and animal health scares, the importance (and reputation) of agriculture in the export economy, and limited experience of AD have led to strict interpretation of the regulations. Certain slaughter wastes, such as blood, are not permitted in Irish AD plants under the national regulations (DAFF, 2009), even though they are allowed under EU regulations and are used in AD plants throughout Europe. Very strict interpretation can act as a barrier to the industry, through, for example, stringent controls on the processing of feedstock (which can add considerable cost) and restrictions on the type of feedstock that can be used for AD.

While there is no doubt that regulations concerning ABP are necessary, it is also recognised that very strict interpretation of these regulations can cause problems for the development of the AD industry (Farrar, 2009). Policy-makers can look to successful AD industries, such as those in Germany and Sweden, and use the experience gained there to develop policies that both regulate the safe use of ABP and facilitate AD.

Other policies

Waste policy

In many cases it is advantageous for AD plants to co-digest wastes with energy crops due to increased methane yields (Uzodinma and Ofoefule, 2009) and the gate fees that can be charged for treating wastes (Smyth et al., 2010a). Policy relating to waste can therefore have a significant impact on the AD and biomethane industries. Increased landfill tariffs or policies demanding treatment of organic wastes can be an indirect driver of anaerobic digestion. Waste-related energy policies can also drive biogas and biomethane production, such as the EU Renewable Energy Directive (EC, 2009a), which offers double credits for renewable transport fuels produced from wastes. On the downside, other waste-related policies can pose barriers to AD, such as policies relating to digestate disposal, transport and handling of wastes, and planning permission for waste treatment facilities.

Environmental policy

At the biogas production end, the use of AD as part of a waste treatment strategy can help to achieve targets set by environmental policy, as GHG emissions and pollution from poor waste management are reduced (Börjesson and Mattiasson, 2007; Yiridoe et al., 2009).

The use of digestate as a replacement for conventional fertiliser, as long as it is applied following best practice guidelines, can bring benefits of reduced pollution. This is because digestate contains nutrients in a form more easily absorbed by plants (Yiridoe et al., 2009), thus reducing the risk of run-off into water sources, as well as the risk of nitrogen losses by ammonia emissions (DCMNR and SEI, 2004). In addition, the pathogen content of animal slurries (which are commonly spread as fertiliser in their raw form) can be reduced by AD as a result of the temperatures reached during the digestion process. A specific pasteurisation step can also be added to the process (Lukehurst et al., 2010).

When it comes to the use of the fuel, biomethane can assist in compliance with targets for improved air quality. In terms of local pollutants, methane (whether it is biomethane or natural gas) is much cleaner burning than many other fuels and, when used as a replacement for oil, can benefit air quality, especially in urban areas, leading to improved public health and associated cost reductions in the health sector (Goyal and Sidhartha, 2003; Mediavilla-Sahagún and ApSimon, 2003; Rabl, 2002).

The reduction in pollution from agricultural wastes is promoted by numerous policies, including, in the EU, the Nitrates Directive, the Water Framework Directive, Biodiversity Action Plan and national agri-environmental schemes.

Policy instruments for promotion of biomethane

This section discusses different policy instruments that can be used to promote biomethane. As experience of biomethane (and bioCNG) markets is limited, general biofuel policies are also included.

Regulatory and economic instruments

Tax exemptions

Tax exemptions have been found to be very successful in promoting biofuels, both in the EU and in the US (Wisenthal et al., 2009). There is a strong relationship between the level of tax reduction and the penetration of biofuels; a full tax exemption in Germany resulted in a 3.75 per cent biofuels share in 2005, compared with no tax exemption and no biofuels penetration in Finland (Silvestrini et al., 2010). There is a solid argument for permanent tax breaks for biofuels because of the associated environmental benefits (Bomb et al., 2007) and the fact that conventional technologies are currently subsidised, with fossil fuels receiving very little or no penalty for their negative environmental impacts (Silveira, 2005).

Maintaining the price of bioCNG below that of petrol and diesel can be an effective method of developing the market. A study of existing NGV markets by Yeh (2007) found that keeping natural gas fuel prices 40–50 per cent below petrol and diesel prices, along with a payback period of 3–4 years or lower, was very important for the development of a mainstream NGV market.

An advantage of tax exemptions over other means of promoting biofuels is that it is a low-cost method that can make use of the existing administrative and collection system

(Ryan et al., 2006). On the downside, tax exemptions can result in a loss of revenue for the government as the biofuel market grows. In Belgium, this has been compensated for by a simultaneous increase in the tax on fossil fuel, making the policy budget-neutral. Other EU member states have switched from tax exemptions to an obligation or mixed system to reduce losses to the exchequer (Wisenthal et al., 2009).

Obligation systems

Compared with tax exemptions, which can be revised every year, an obligation system provides stability to the market by setting a long-term framework for biofuels (Wisenthal et al., 2009). An advantage of obligation systems over tax exemptions is that, with an obligation system, the government can control the quantity of biofuel produced and/or used. A popular obligation system is to set a requirement for blended fuels, but, while low-level blending is straightforward and relatively cheap to implement, it may not be sufficient to achieve significant penetration of biofuels (Bomb et al., 2007). Obligation systems may well increase overall biofuel consumption, but they are relatively ineffective at promoting particular biofuels (Wisenthal et al., 2009) and are unlikely to result in a step-change in behaviour.

Subsidies

Subsidies can play an important part in the development of a bioCNG industry, especially in the early stages. Subsidies for specific crops or for growing crops for a particular purpose, e.g. transport fuel, can assist in feedstock availability for AD plants. Subsidies are also of importance to consumers. Research has shown that customers purchase "cheap rather than green" (Bomb et al., 2007). The additional cost of purchasing and maintaining NGVs is a barrier to the industry; a study in the UK (where there is no mainstream NGV market) found that support in this area could result in the rapid expansion of the industry (Patterson et al., 2011). Once the industry has been established, the level and availability of subsidies should be reviewed.

Information instruments

Demonstration projects

Demonstration projects have a number of valuable roles; they bring new technologies into the public eye and can help garner public acceptance, as well as providing opportunity for research and development, and the dissemination of results and information. In the AD and upgrading industry, demonstration projects are particularly important in countries where there are limited existing plants. Farm visits can also be arranged to showcase novel energy crops and improved techniques for growing existing crops.

For the wider public acceptance of bioCNG vehicles, demonstration projects involving fleet vehicles are common practice; a review of biofuels in the Netherlands describes such projects as having an exemplary role (van der Laak et al., 2007). The development of biomethane for transport in other countries has often been based on an existing CNG market, and the development of CNG has often begun with the introduction of CNG to captive fleets (e.g. buses, waste collection lorries, taxis) followed by private cars (Smyth et al., 2010a). As fleet vehicles return to a depot each day, an advantage, especially in the early stages of market development, is that only one filling station is required. Such filling stations can also be open to the public. However, while niche markets, such as fleet vehicles, can be an important part of the early stages of an alternative vehicle market, they are insufficient to develop the market into mainstream (McNutt and Rodgers, 2004, cited in Yeh, 2007).

Stakeholder involvement

For a biomethane (i.e. upgraded biogas) industry, there must be a biogas/AD industry, and, for a biomethane-for-transport industry, it is beneficial if a conventional CNG transport industry exists. Stakeholders in a biogas/biomethane industry include farmers, the waste sector, and AD and upgrading plant operators (Thamsiriroj et al., 2011). Stakeholders in an NGV industry include the fuel suppliers, suppliers of natural gas industry equipment (e.g. fuelling stations and vehicles) and consumers (Yeh, 2007). The general public, all levels of government, research institutes and NGOs (non-governmental organisations), such as environmental groups, are important stakeholders at all stages of a bioCNG industry (Figure 2.1).

Regular meetings should be held between stakeholders (van der Laak et al., 2007). Communication between the industry and the general public is important, particularly when it comes to the benefits of the industry (e.g. improved air quality), as this can help create a demand pull. The demand pull and the technology push (through technology- or fuel-based regulation) work together with consumer and producer incentives to promote the adoption of alternative fuels (Yeh, 2007).

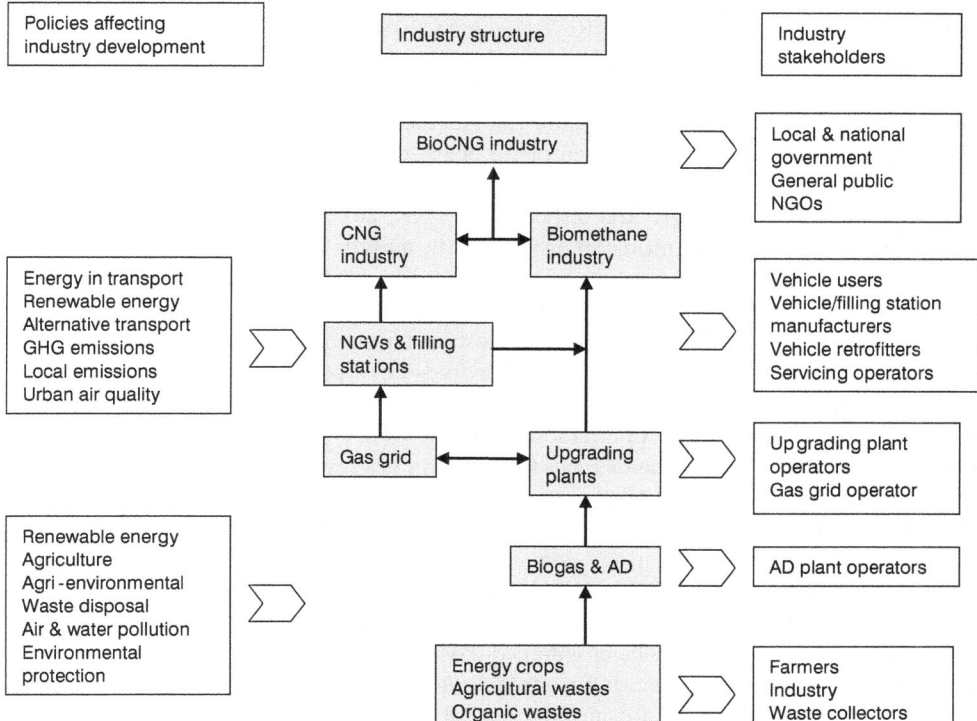

Figure 2.1 Policy, industry structure and stakeholders in the bioCNG industry.

Experience with biomethane and CNG for transport

Two success stories: Sweden and Germany

Sweden

In 2009, Sweden had 230 biogas plants (including 12 farm plants, 21 co-digestion plants and 136 municipal sewage treatment plants) and 48 upgrading plants (Petersson, 2011). Annual biogas production in the same year was 1363 TWh and 26 per cent of this was used as vehicle fuel (Petersson, 2011). The NGV market began in Sweden in the mid-1990s and there are now around 32,000 NGVs and almost 170 filling stations. In 1999, there were about 1400 NGVs served by 22 filling stations (NGV, 2011).

Initially based solely on CNG, the amount of biomethane in the bioCNG vehicle fuel mix has been increasing year-on-year and there is now over 60 per cent biomethane infiltration in the market (Petersson, 2011). A suite of measures has been put in place to promote the use of biomethane as a vehicle fuel. Taxation measures include energy tax, CO_2 tax and CO_2 differentiated vehicle tax. Market-based mechanisms include carbon emissions trading, while support systems are in place that offer investment grants, agricultural grants, support for filling stations and green cars, including reduced vehicle tax and free parking. Regulations also demand the availability of renewable fuel at filling stations (Petersson, 2011).

Germany

Germany, a country with only around 100 biogas plants in the early 1990s, had an estimated 7200 agricultural plants in 2011 (FNR, 2012; Linke, 2011). There were about 50 upgrading plants in operation in 2010, rising to 83 at the end of 2011 (FNR, 2012). Biomass crops account for 49 per cent of substrates in biogas plants by mass and over 70 per cent by energy content (FNR, 2012; Linke, 2011). Of the 12,000 kha of agricultural crop land in Germany, 650 kha (5 per cent) were under cultivation for biogas in 2010, and the area was predicted to rise to 800 kha by the end of 2011. The areas under cultivation for biodiesel and bioethanol in 2010 were 940 kha and 240 kha respectively (FNR, 2012). Animal wastes are another major substrate for AD, accounting for 43 per cent by mass or 11 per cent by energy content; the majority of German biogas plants use more than 30 per cent manure by mass content in their daily substrate blend (FNR, 2012; Linke, 2011; Weiland, 2010).

The German AD and biogas upgrading industries have benefited from an effective tariff structure based on graded tariffs, which depend on feedstock type, plant size and AD technology type, among other factors. The produced biogas and biomethane is used for heat and electricity as well as for vehicle fuel. The International Energy Agency (IEA) has stated that the high investor security provided by the German feed-in-tariff has resulted in a rapid deployment of renewables, the entrance of many new actors to the market and a subsequent reduction in costs (IEA and OECD, 2007). It should be noted, however, that government support schemes are regularly revised. For example, arising from concerns over the expansion of maize cultivation for biogas production, the 2012 Renewable Energy Act limits the proportion of maize and other cereal grains in the biogas substrate mix, while biogas production from municipal biowaste, other residues and pastureland is being enforced (BMELV, 2011; Strauch and Krassowki, 2012).

Coupled with policies promoting a strong agricultural biogas sector in Germany are policies leading to the successful introduction of biofuels, including biomethane, to the market (Silvestrini et al., 2010). Also key to the development of the biomethane in the transport sector was full tax exemption on biofuels from the outset, receptive agricultural and automotive sectors and the promotion of natural gas as a transport fuel (Silvestrini et al., 2010). There has been significant growth in the CNG sector since the 1990s; in 2010, there were almost 92,000 NGVs being served by 900 filling stations, rising from only 50 filling stations and 3245 NGVs in 1999 (NGV, 2011).

Policy barriers and lessons learned

Ireland – conflicting policies

Despite the many benefits of biomethane and its success in some countries, it is still a relatively niche area. A recurring theme in faltering biomethane industries is a lack of consistent and cohesive policy. Policy relating to energy and agriculture, as well as to waste, GHG emissions, transport, and air and water quality all affect the development of a bioCNG industry and, if there is no joined-up thinking, the industry will struggle (Smyth et al., 2010b; Thamsiriroj et al., 2011; Yeh, 2007). For example, while waste legislation may support the treatment of organic wastes by AD and agri-environmental policy promotes reducing fossil fertiliser usage, the spreading of digestate can be severely restricted by ABP regulations, presenting challenges for the AD sector. The strict ABP controls in Ireland are largely due to the importance of the agricultural sector to the economy and concerns over the impact that problems from poor digestate management could have on the sector, particularly on the export market. Uncertainty in the waste collection sector can also limit the development of AD plants, since a secure supply of feedstock is required. There has been much discussion in Ireland over waste ownership and preferred waste treatment options, with little consensus from the various government parties and planning authorities. Several legal actions have arisen and are ongoing in the courts. Both ABP regulations and uncertainty in the waste sector have hampered the AD industry in Ireland, where, despite the introduction of capital grants for AD plants and tariffs for energy from biogas, the industry has failed to get off the ground (Smyth et al., 2010a). The small size of the tariffs is also a contributory factor.

Changing policies in Denmark and New Zealand

Changing policies can also pose a barrier to development, as the introduction of sustainable technologies is often a long-term process that requires stability (Raven and Gregersen, 2007). The detrimental effect of policy transition can be seen in the AD sector in Denmark. Although the Danish AD sector is well established, with 20 CAD and over 35 farm-scale plants, there have been no new CAD plants constructed since 1998 (Raven and Gregersen, 2007). A change in government led to the withdrawal of many grants and support schemes, and a shift in policy direction. These changes, along with uncertainty over tariffs, deterred new operators from entering the market (Raven and Gregersen, 2007).

In New Zealand, changing policies resulted in the collapse of the NGV market (Yeh, 2007). Initially strongly supported by government through incentives and targets to promote the adoption of NGVs, the country had over 10 per cent penetration of NGVs by the mid-1980s. However, the withdrawal of incentives in 1985 resulted in a rapid decline in the number of NGVs and the failure of the NGV market (Yeh, 2007). There were only 200 NGVs in New Zealand in 2010 (NGV, 2011).

A stable and consistent policy framework is essential for the development of renewable energy technologies (Foxon et al., 2005).

Developing a policy roadmap for a biomethane industry

Roadmap

The main purpose of roadmaps, as discussed in a review by Amer and Daim (2010), is to forecast future market directions and developments in technology, and to assist in decision-making. Put simply, a roadmap is generally used to answer three basic questions: what is our current situation, where do we want to go, and how are we going to get there (Amer and Daim, 2010).

The roadmap in Figure 2.2 draws together knowledge gained from the case studies and the discussion of biomethane policy to present a strategy for the development of a bioCNG transport industry. The roadmap draws on other work in the literature, with particular reference to Thamsiriroj et al. (2011). The roadmap is based on a 15-year period, which is a typical time horizon for renewable energy roadmaps (Amer and Daim, 2010). The key points of the roadmap are outlined in the following paragraphs.

Identify stakeholders and get them on board

The first step in the development of a biomethane-for-transport industry is to identify the industry structure and stakeholders (Figure 2.1). The success of a bioCNG industry is based on numerous symbiotic relationships between these stakeholders (Thamsiriroj et al., 2011) and good communication between stakeholders during the development of new policies is important (van der Laak et al., 2007). Stakeholders should be brought on board in the early stages of industry development and involved in policy formation from the outset.

Align existing and develop new policies and targets to ensure a cohesive policy framework

Existing policies and targets

Many existing policies and targets already directly and indirectly support biomethane/bioCNG (e.g. renewable energy in the transport sector, the use of alternative fuels, energy crops, treatment of organic wastes, reducing pollution from agricultural wastes and improving urban air quality) and the ability of biomethane to contribute to these policies and targets is a benefit that should be highlighted to stakeholders. However, while existing policies and targets can lend support to a bioCNG industry, they can also pose barriers to the development of the industry. It is recommended that a government working group be set up to liaise between different government departments and stakeholders in order to align existing policies relating to biogas, biomethane and bioCNG, and to put in place new policies and targets to drive the development of the industry.

New policies and targets

A review of existing resources and resource potential should inform policy, and targets should take account of existing renewable energy and GHG emissions policies and targets as well as other related targets, such as those for local air quality and waste treatment. Ambitious targets are beneficial as they give a clear signal of intent from government and can provide the security needed for private sector firms to invest in the technology. It is important that targets are feasible and that there are policies in place to enable targets to be achieved.

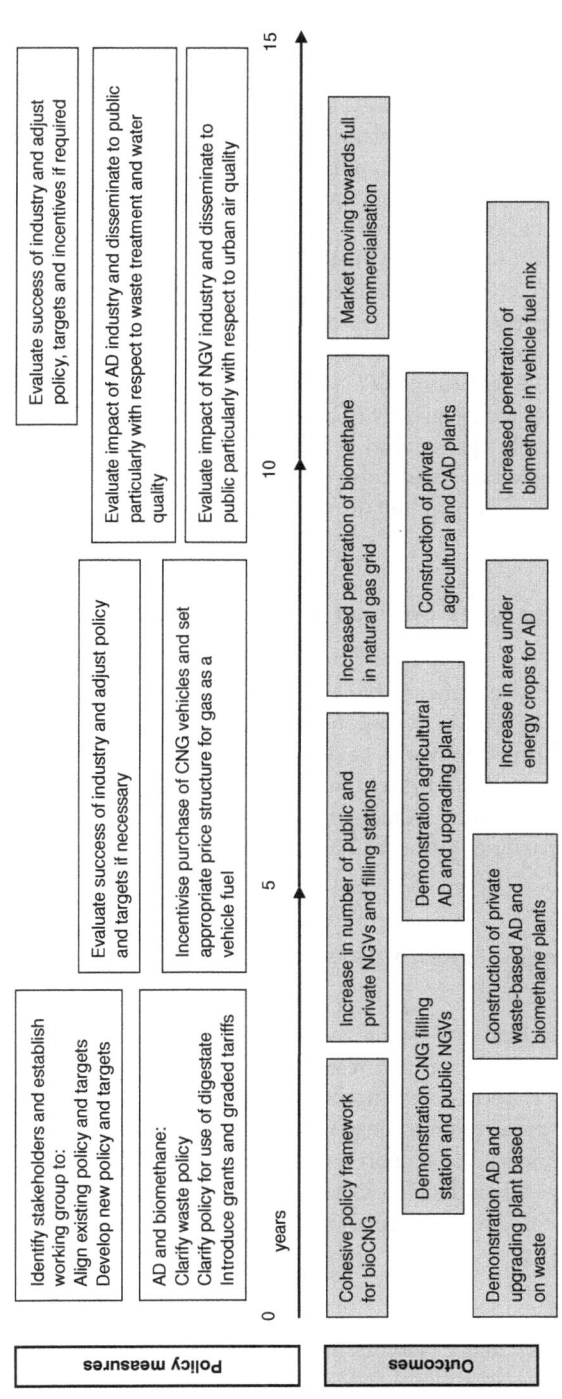

Figure 2.2 Roadmap for the development of a bioCNG industry (adapted from Thamsiriroj et al., 2011)

Energy policy has an impact on all stages of a bioCNG industry (Figure 2.1), as it sets targets for and regulates the use of renewable energy in the transport sector, the use of alternative fuels and the development of AD and upgrading plants. Agricultural policy influences the availability of both energy crops and agri-wastes for AD. Agricultural and agri-environmental policies largely determine the actions of farmers, e.g. through monetary incentives to grow energy crops or treat agricultural wastes. The use of agri-wastes for AD can help meet agri-environmental policies to reduce nutrient leaching. A key to the success of biofuels industries in Europe is biofuel policies that support local farmers (Worldwatch Institute, 2007), and development of agriculture and rural employment has been shown to be crucial for transport biofuels (di Lucia and Nilsson, 2007).

For the industry to develop, it is important to make the link between policy objectives and anticipated outcomes (Jablonski et al., 2010). Whether targets are voluntary or mandatory will have an impact on the development of the industry. European experience suggests that it can prove difficult to meet voluntary targets, especially if effective support programmes and incentives are not in place (Worldwatch Institute, 2007). Targets can be set in both the public and private sectors. As governments are frequently one of the largest energy users in a country, they have an important role to play in generating a large and consistent market (Worldwatch Institute, 2007), contributing to targets and also in fostering public acceptance. Strong government commitment has a large positive influence on the biofuel industry (Bomb et al., 2007).

Provide incentives for reaching targets and get the right policy instruments in place

Once policies across various government departments have been aligned to consistently promote bioCNG, governments must then look at policy implementation. To ensure the expansion of a bioCNG (or, indeed, any biofuel) industry, governments require "an effective 'toolbox' of wide-ranging policy strategies" (Worldwatch Institute, 2007), encompassing regulatory, economic and information instruments. Demonstration projects in key industry areas (i.e. AD, biogas upgrading, CNG refuelling stations, CNG vehicles), play an important role in the early stages of the industry.

Facilitate communication between stakeholders and build flexibility into the programme

It is important that visions do not become rigid and that regular meetings are held between stakeholders (farmers, policy makers, technology firms and research institutes) (van der Laak et al., 2007). Policies and targets should be flexible and should be regularly assessed to take account of the success or otherwise of the industry (and the industry's impacts), and amended if necessary. The type of policy instrument needed to support an industry varies as the industry develops (Foxon et al., 2005) and, as both needs and technologies change with time, a roadmap should be seen as a "living document" (Amer and Daim, 2010) rather than as static legislation.

Summary and conclusions

A biomethane-for-transport industry brings many benefits: it can reduce GHG emissions and local pollutants, while providing waste treatment and a source of renewable energy.

Anaerobic digestion, upgrading and NGV technology are all well established, and many sustainable feedstocks are suitable for biomethane production. A biomethane industry complies with many existing policies and can help meet existing targets. It is these advantages that are driving the biomethane and bioCNG transport industry in a number of countries, e.g. Sweden, Germany and Austria; however, while growing, the use of biomethane for transport is still a relatively niche area. The structure of the bioCNG industry means that there is a complicated web of policy and stakeholders, and a recurring theme in faltering industries is a lack of consistent legislation across the policy areas. The alignment of policy to form a cohesive framework for bioCNG, addressing production, distribution and use, is a key factor in the development of the industry. Stakeholders need to be brought on board at an early stage and involved in policy formation, and there should be regular communication between the various parties as the industry develops. Policy implementation should be via a toolbox of measures, including regulation, economic instruments and the provision of information. A roadmap for the development of a bioCNG industry shows significant progress over a 15-year period, led by specific energy and agricultural policies. Flexibility and communication are key; assessment and reassessment of policy are essential for the success of the industry.

References

Åhman, M. (2010) 'Biomethane in the transport sector – An appraisal of the forgotten option', *Energy Policy*, vol 38, no 1, pp208–217

Amer, M. and Daim, T.U. (2010) 'Application of technology roadmaps for renewable energy sector', *Technological Forecasting & Social Change*, vol 77, no 8, pp1355–1370

Baxter, D. (2010) 'European Commission JRC – Institute for Energy', IEA Bioenergy Task 37, Copenhagen, Denmark, 26–28 May 2010, available at http://www.iea-biogas.net/_content/publications/member-country-reports.html, accessed 21 July 2012

BMELV (2011) *Press release no. 045 from 17.02.11 Federal Minister Aigner advocates adjustment of biogas support*, Federal Ministry of Food, Agriculture and Consumer Protection (Germany), http://www.bmelv.de/SharedDocs/Pressemitteilungen/EN/2011/045-AI-EEG-Kongres, accessed 15 May 2012

Bomb, C., McCormick, K., Deurwaarder, E. and Kåberger, T. (2007) 'Biofuels for transport in Europe: Lessons from Germany and the UK', *Energy Policy*, vol 35, no 4, pp2256–2267

Bordelanne O., Montero, M., Frédérique, B., Prieur-Vernat, A., Oliveti-Selmi O., Pierre, H., Papadopoulo, M. and Muller, T. (2011) 'Biomethane CNG hybrid : A reduction by more than 80% of the greenhouse gas emissions compared to gasoline', *Journal of Natural Gas Science and Engineering*, vol 3, no 5, pp617–624

Börjesson, P. and Mattiasson, B. (2007) 'Biogas as a resource-efficient vehicle fuel', *Trends in Biotechnology*, vol 26, no 1, pp7–13

Buchholz, T., Luzadis, V.A. and Volk, T.A. (2009) 'Sustainability criteria for bioenergy systems: results from an expert survey', *Journal of Cleaner Production*, vol 17, supplement 1, ppS86–S98

CBES (2009) 'Land-use change and bioenergy: Report from the 2009 workshop', ORNL/CBES-001, Center for BioEnergy Sustainability (CBES), U.S. Department of Energy, Office of Energy Efficiency and Renewable Energy and Oak Ridge National Laboratory, http://www.ornl.gov/sci/ees/cbes/workshop.shtml, accessed 30 July 2012

DAFF (2009) *Conditions for approval and operating of biogas plants treating animal by-products in Ireland*, Department of Agriculture, Fisheries and Food (DAFF), Dublin, Ireland, http://www.agriculture.ie/agrifoodindustry/animalbyproducts/applicationformsconditionsforabpprocessingoperations/, accessed 12 May 2009

DCMNR and SEI (2004) *Bioenergy in Ireland*, Bioenergy Strategy Group, Department of Communications, Marine and Natural Resources (DCMNR) and Sustainable Energy Ireland (SEI), Dublin, http://www.seai.ie/About_Energy/Energy_Policy/National_Policy_Drivers/, accessed 4 February 2009

Di Lucia, L. and Nilsson, L.J. (2007) 'Transport biofuels in the European Union: The state of play', *Transport Policy*, vol 14, no 6, pp533–543

EC (2003) 'Council Regulation (EC) No 1782/2003 of 29 September 2003 establishing common rules for direct support schemes under the common agricultural policy and establishing certain support schemes for farmers and amending Regulations (EEC) No 2019/93, (EC) No 1452/2001, (EC) No 1453/2001, (EC) No 1454/2001, (EC) No 1868/94, (EC) No 1251/1999, (EC) No 1254/1999, (EC) No 1673/2000, (EEC) No 2358/71 and (EC) No 2529/2001', *Official Journal of the European Union*, L270/1–69

EC (2004) 'Commission Regulation 796/2004/EC of 21 April 2004 laying down detailed rules for the implementation of cross-compliance, modulation and the integrated administration and control system provided for in Council Regulation (EC) No 1782/2003 establishing common rules for direct support schemes under the common agricultural policy and establishing certain support schemes for farmers', *Official Journal of the European Union*, L141/18–58

EC (2009a) 'Directive 2009/28/EC of the European Parliament and of the Council of 23 April 2009 on the promotion of the use of energy from renewable sources and amending and subsequently repealing Directives 2001/77/EC and 2003/30/EC', *Official Journal of the European Union*, L140/16–62

EC (2009b) 'Directive 2009/73/EC of the European Parliament and of the Council of 13 July 2009 concerning common rules for the internal market in natural gas and repealing Directive 2003/55/EC', *Official Journal of the European Union*, L211/94–136

EC (2009c) 'Regulation (EC) No 1069/2009 of the European Parliament and of the Council of 21 October 2009 laying down health rules as regards animal by-products and derived products not intended for human consumption and repealing Regulation (EC) No 1774/2002 (Animal by-products Regulation)', *Official Journal of the European Union*, L300/1–33

EurObserv'ER (2010) 'Biogas barometer', November 2010, www.eurobserv-er.org/downloads.asp, accessed 15 January 2012

EurObserv'ER (2011) 'Biofuels barometer', July 2011, www.eurobserv-er.org/downloads.asp, accessed 22 January 2012

Farrar, M. (2009) 'Regulatory update…Animal by-products (ABP) as a feedstock in anaerobic digestion (AD)', 26 May 2009, Department of Agriculture, Fisheries and Food (DAFF), Dublin, Ireland, http://www.sei.ie/Renewables/Bioenergy/Anaerobic_Digestion/animal_byproducts/, accessed 30 June 2009

FNR (2012) 'Facts and figures (Daten und Fakten)', Fachagentur Nachwachsende Rohstoffe (Agency for Renewable Resources), Germany, http://mediathek.fnr.de/grafiken/daten-und-fakten.html, accessed 22 July 2012

Foxon, T.J., Gross, R., Chase, A., Howes, J., Arnall, A. and Anderson, D. (2005) 'UK innovation systems for new and renewable energy technologies: drivers, barriers and systems failures', *Energy Policy*, vol 33, no 16, pp2123–2137

Gerin, P.A., Vliegen, F. and Jossart, J.-M. (2008) 'Energy and CO_2 balance of maize and grass as energy crops for anaerobic digestion', *Bioresource Technology*, vol 99, no 7, pp2620–2627

Goyal, P. and Sidhartha (2003) 'Present scenario of air quality in Delhi: a case study of CNG implementation', *Atmospheric Environment*, vol 37, no 38, pp5423–5431

Gross, R., Leach, M. and Bauen, A. (2003) 'Progress in renewable energy', *Environment International*, vol 29, no 1, pp105–122

Hagen, M., Polman, E., Jensen, J., Myken, A., Jönsson, O. and Dahl, A. (2001) *Adding gas from biomass to the gas grid* (Report SGC 118), Swedish Gas Centre (SGC), Malmö, Sweden, http://www.sgc.se/display.asp?ID=184&Typ=Rapport&Menu=Rapporter, accessed 9 January 2008

IEA Bioenergy (2005) *Injection of biogas into the natural gas grid in Laholm, Sweden*, IEA Bioenergy Task 37, http://www.iea-biogas.net/_content/case-studies/succes-stories.html, accessed 9 January 2008

IEA and OECD (2007) *Energy policies of IEA countries, Germany 2007 Review*, International Energy Agency (IEA) and Organisation for Economic Co-operation and Development (OECD), Paris, France, www.iea.org/publications/free_new_Desc.asp?PUBS_ID=1922, accessed 15 June 2009

Jablonski, S., Strachan, N., Brand, C. and Bauen, A. (2010) 'The role of bioenergy in the UK's future formulation and modelling of long-term UK bioenergy scenarios', *Energy Policy*, vol 38, no 10, pp5799–5816

Jönsson, O. (2006) *Market development for biogas as a vehicle fuel in Europe – STATUS 2006*, Swedish Gas Centre, Malmö, Sweden, www.sgc.se/nyhetfiler/Status2006.pdf, accessed 5 May 2009

Korres, N.E., Singh, A., Nizami, A.S. and Murphy, J.D. (2010) 'Is grass biomethane a sustainable transport biofuel?', *Biofuels, Bioproducts and Biorefining*, vol 4, no 3, pp310–325

Laurance, W.F. (2007) 'Switch to corn promotes Amazon deforestation', *Science*, vol 318, no 5857, p1721

Lehtomäki, A., Viinikainen, T.A. and Rintala, J.A. (2008) 'Screening boreal energy crops and crop residues for methane biofuel production', *Biomass and Bioenergy*, vol 32, no 6, pp541–550

Linke, B. (2011) 'Country report, Germany', IEA Bioenergy Task 37, Istanbul, Turkey, 13–15 April 2011, www.iea-biogas.net/_content/publications/member-country-reports.html, accessed 17 October 2011

Liska, A.J. and Perrin, R.K. (2009) 'Indirect land use emissions in the life cycle of biofuels: regulations vs science', *Biofuels Bioproducts & Biorefining*, vol 3, no 3, pp318–328

Lukehurst C., Frost P. and Al Seidi T. (2010) *Utilisation of digestate from biogas plants as biofertiliser*, IEA Bioenergy, http://www.iea-biogas.net/_content/publications/publications.php, accessed 22 July 2012

McNutt, B. and Rodgers, D. (2004) 'Lessons learned from 15 years of alternative fuels experience—1988–2003', in Sperling, D. and Cannon, J. (eds) *The hydrogen energy transition: moving toward the post petroleum age in transportation*, Elsevier Academic Press, Burlington, MA

Mediavilla-Sahagún, A. and ApSimon, H.M. (2003) 'Urban scale integrated assessment of options to reduce PM10 in London towards attainment of air quality objectives', *Atmospheric Environment*, vol 37, no 33, pp4651–4665

NGV (2011) 'Worldwide NGV statistics', *The GVR Gas Vehicles Report*, vol 10#9, no 118, November 2011, NGV Communications Group, www.ngvjournal.com/en/statistics, accessed 3 December 2011

NGV (2012) 'Worldwide NGV statistics', www.ngvjournal.dreamhosters.com/en/statistics/item/911-worldwide-ngv-statistics, accessed 22 January 2012

Nizami, A.-S. and Murphy, J.D. (2010) 'What type of digester configurations should be employed to produce biomethane from grass silage?', *Renewable and Sustainable Energy Reviews*, vol 14, no 6, pp1558–1568

Patterson, T., Esteves, S., Dinsdale, R. and Guwy, A. (2011) 'An evaluation of the policy and techno-economic factors affecting the potential for biogas upgrading for transport fuel use in the UK', *Energy Policy*, vol 39, no 3, pp1806–1816

Petersson, A. (2011) 'Country report Sweden', IEA Bioenergy Task 37, Istanbul, Turkey, 13–15 April 2011, www.iea-biogas.net/_content/publications/member-country-reports.html, accessed 17 October 2011

Petersson, A. and Wellinger, A. (2009) *Biogas upgrading technologies – developments and innovations*, IEA Bioenergy Task 37, http://www.iea-biogas.net/publicationspublic.htm, accessed January 2010

Plevin, R.J., Hare, M., Jones, A., Torn, M. and Gibbs, H.K. (2010) 'Greenhouse gas emissions from biofuels' indirect land use change are uncertain but may be much greater than previously estimated', *Environmental Science and Technology*, vol 44, no 21, pp8015–8021

Rabl, A. (2002) 'Environmental benefits of natural gas for buses', *Transportation Research Part D: Transport and Environment*, vol 7, no 6, pp391–405

Raven, R.P.J.M. and Gregersen, K.H. (2007) 'Biogas plants in Denmark: successes and setbacks', *Renewable and Sustainable Energy Reviews*, vol 11, no 1, pp116–132

REN21 (2011) *Renewables 2011 global status report*, REN21 Secretariat, Paris, France, http://www.ren21.net/default.aspx?tabid=5434, accessed 26 June 2011

Ryan, L., Convery, F. and Ferreira, S. (2006) 'Stimulating the use of biofuels in the European Union: Implications for climate change policy', *Energy Policy*, vol 34, no 17, pp3184–3194

Schnepf, R. (2006) 'European Union biofuels policy and agriculture: An overview', CRS Report for Congress, http://www.usembassy.it/pdf/other/RS22404.pdf, accessed 4 June 2009

Searchinger, T., Heimlich, R., Houghton, R.A., Dong, F., Elobeid, A., Fabiosa, J., Tokgoz, S., Hayes, D. and Yu, T.-H. (2008) 'Use of U.S. croplands for biofuels increases greenhouse gases through emissions from land-use change'. *Science*, vol 319, no 5867, pp1238–1240

Silveira, S. (2005) 'Bioenergy – realizing the potential now!' in S. Silveira (ed.) *Bioenergy: Realizing the potential*, Elsevier, Oxford, England

Silvestrini A., Monni, S., Pregernig, M., Barbato, A., Dallemand, J.-F., Croci, E. and Raes, F. (2010) 'The role of cities in achieving the EU targets on biofuels for transportation: The cases of Berlin, London, Milan and Helsinki', *Transportation Research Part A*, vol 44, no 6, pp403–417

Singh, A. and Murphy, J.D. (2009) 'Biomethane from animal waste and grass for clean vehicular biofuel in Ireland', *Twelfth International Waste Management and Landfill Symposium*, 5–9 October 2009, Padova, Italy, International Waste Working Group (IWWG)

Smyth, B.M. and Murphy, J.D. (2011) 'The indirect effects of biofuels and what to do about them: the case of grass biomethane and its impact on livestock', *Biofuels, Bioproducts and Biorefining*, vol 13, no 5, pp165–184

Smyth, B.M., Murphy, J.D. and O'Brien, C.M. (2009) 'What is the energy balance of grass biomethane in Ireland and other temperate northern European climates?' *Renewable and Sustainable Energy Reviews*, vol 13, no 9, pp2349–2360

Smyth, B.M., Smyth, H. and Murphy, J.D. (2010a) 'Can grass biomethane be an economically viable biofuel for the farmer and the consumer?', *Biofuels, Bioproducts and Biorefining*, vol 4, no 5, pp519–537

Smyth, B.M., Ó Gallachóir, B.P., Korres, N.E. and Murphy, J.D. (2010b) 'Can we meet targets for biofuels and renewable energy in transport given the constraints imposed by policy in agriculture and energy?', *Journal of Cleaner Production*, vol 18, no 16–17, pp1681–1685

Steffen, R., Szolar, O. and Braun, R. (1998) 'Feedstocks for anaerobic digestion', Institute for Agrobiotechnology, Tulln, and University of Agricultural Sciences, Vienna, Austria, www.adnett.org/dl_feedstocks.pdf, accessed 7 July 2007

Strauch S. and Krassowki J. (2012) 'Overview of biomethane markets and regulations in partner countries', Green Gas Grids, Intelligent Energy Europe, http://www.greengasgrids.eu/sites/default/files/files/120325_D2_2_Overview_of_biomethane_markets_final.pdf, accessed 22 July 2012

Thamsiriroj, T., Smyth, H. and Murphy, J.D. (2011) 'A roadmap for the introduction of gaseous transport fuel: A case study for renewable natural gas in Ireland', *Renewable and Sustainable Energy Reviews*, vol 15, no 9, pp 4641–4651

Uekoetter, F. (2011) 'The magic of one reflections on the pathologies of monoculture', *Perspectives*, Issue 2/2011, RCC (Rachel Carson Centre), Munich, Germany, http://www.carsoncenter.uni-muenchen.de/publications/new_perspectives/index.html, accessed 15 May 2012

Upham, P., Thornley, P., Tomei, J. and Boucher, P. (2009) 'Substitutable biodiesel feedstocks for the UK: a review of sustainability issues with reference to the UK RTFO', *Journal of Cleaner Production*, vol 17, supplement 1, ppS37–S45

USDOE (2011) 'Biogas production', United States Department of Energy Alternative and Advanced Vehicles Data Center (USDOE), www.afdc.energy.gov/afdc/fuels/emerging_biogas_production.html, accessed 23 January 2011

USEIA (2012) 'Frequently asked questions: How much gasoline does the United States consume?', United States Energy Information Administration (USEIA), http://www.eia.gov/tools/faqs/faq.cfm?id=23&t=10, accessed 21 July 2012

Uzodinma, E.O. and Ofoefule, A.U. (2009) 'Biogas production from blends of field grass (*Panicum maximum*) with some animal wastes', *International Journal of Physical Sciences*, vol 4, no 2, pp91–95

van der Laak, W.W.M., Raven, R.P.J.M. and Verbong, G.P.J. (2007) 'Strategic niche management for biofuels: Analysing past experiments for developing new biofuel policies', *Energy Policy*, vol 35, no 6, pp3213–3225

Weiland, P. (2010) 'Country report, Germany', IEA Bioenergy Task 37, Copenhagen, Denmark, 26–28 May 2010, www.iea-biogas.net/publicationsreports.htm, accessed 30 August 2010

Wisenthal, T., Leduc, G., Christidis, P., Schade, B., Pelkmans, L., Govaerts, L. and Georgopoulos, P. (2009), 'Biofuel support policies in Europe: Lessons learnt for the long way ahead', *Renewable and Sustainable Energy Reviews,* vol 13, no 4, pp789–800

Worldwatch Institute (2007) *Biofuels for transport global potential and implications for sustainable energy and agriculture,* Earthscan, London

Yeh, S. (2007) 'An empirical analysis on the adoption of alternative fuel vehicles: The case of natural gas vehicles', *Energy Policy,* vol 35, no 11, pp5865–5875

Yiridoe, E.K., Gordon, R. and Brown, B.B. (2009) 'Nonmarket cobenefits and economic feasibility of on-farm biogas energy production', *Energy Policy,* vol 37, no 3, pp1170–1179

Chapter 3

Biomethane production with reference to land-use change

Surajbhan Sevda,[1,2] Deepak Pant[2] and Anoop Singh[3*]

[1] Waste Treatment Lab, Department of Biochemical Engineering and Biotechnology, Indian Institute of Technology Delhi, New Delhi, India 110016; [2] Separation and Conversion Technology, VITO – Flemish Institute for Technological Research, Boeretang 200, Mol 2400, Belgium; [3] Department of Scientific and Industrial Research (DSIR), Ministry of Science and Technology, Technology Bhawan, New Mehrauli Road, New Delhi 110016 India *Corresponding author email: apsinghenv@gmail.com

Introduction

Escalating energy costs and energy shortages in recent years have become problems of national significance and have prompted the search for new sources of energy. Methane, biogas and biomethane are all names for similar products, which each differ slightly in composition. Methane is the main component of natural gas and biogas. It is a natural hydrocarbon consisting of one carbon atom and four hydrogen atoms (CH_4). The heat content of methane is approximately 1000 Btu/scf (standard cubic feet). Methane is a greenhouse gas with 21 times the global warming potential of carbon dioxide on a weight basis. Biogas is a naturally occurring gas formed as a by-product of the breakdown of organic waste materials in a low-oxygen (e.g. anaerobic) environment (Jeffery et al., 1981). Biogas is composed primarily of methane (typically 55–70 per cent by volume) and carbon dioxide (typically 30–45 per cent). Biogas may also include smaller amounts of hydrogen sulphide (typically 50–2000 parts per million [ppm]) (Table 3.1), water vapour, oxygen, and various trace hydrocarbons (Braun, 2007). Due to its lower methane content (and therefore lower heating value) compared with natural gas, biogas use is generally limited to engine-generator sets and boilers adapted to

Table 3.1 General characteristics of biogas.

Composition	55–70% methane (CH_4)
	30–45% carbon dioxide (CO_2)
	Traces of other gases
Energy content	6.0–6.5 kWh m^{-3}
Fuel equivalent	0.60–0.65 L oil/m³ biogas
Ignition temperature	650–750 °C (with the above-mentioned methane content)
Critical pressure	75–89 bar
Critical temperature	−82.5 °C
Normal density	1.2 kg m^{-3}
Molar mass	16.043 kg kmol^{-1}

Source: Vij, 2011

combust biogas as fuel. Biomethane is produced from biogas by removing hydrogen sulphide, moisture, and carbon dioxide to enhance its quality as a fuel.

A theoretical or maximal methane yield (YCH_4, m³ STP/kg substrate converted) can be calculated from the elemental composition of a substrate, $C_cH_hO_xN_nS_s$, as shown in Eq. 3.1: with 22.4 as the molar volume of any ideal gas (L STP/mol).

$$Y_{CH4} = \frac{22.4\left(\frac{c}{2}+\frac{h}{8}-\frac{x}{4}-\frac{3n}{8}-\frac{s}{4}\right)}{12c+h+16x+14n+16s} \quad \text{(Eq. 3.1)}$$

Application of Eq. 3.1 gives 0.37 m³ STP/kg carbohydrates (CH_2O), 0.51 m³ STP/kg proteins ($C_{106}H_{168}O_{34}N_{28}S$), 1 m³ STP/kg fat ($C_8H_{15}O$), and 0.48 m³ STP/kg biomass (based on the formula of Roels (1983), $C_5H_9O_{2.5}NS_{0.025}$). In practice, however, the methane yield of biomass does not often exceed 60 per cent of the theoretical value (Moletta, 2008), because it contains compounds that are poorly or not biodegradable (e.g., lignin, peptidoglycan, membrane-associated proteins), or compounds the solubilisation of which might be limited by the hydrolytic deficiency of the actual components (cellulose, hemicellulose, proteins).

By removing hydrogen sulphide, moisture, and carbon dioxide, biogas can be upgraded to biomethane, a product equivalent to natural gas, which typically contains more than 95 per cent methane. Biomethane can be used interchangeably with natural gas, whether for electricity generation, heating, cooling, pumping, or as a vehicle fuel. Biomethane can also be pumped into the natural gas supply pipeline.

High pressures can be used to store and transport biomethane as compressed biomethane (CBM), which is analogous to compressed natural gas (CNG), or very low temperatures can be used to produce liquefied biomethane (LBM), which is analogous to liquefied natural gas (LNG). From a functional point of view, biomethane is extremely similar to natural gas except that it comes from renewable sources.

Direct and indirect land carbon dynamics

Land is one quarter of earth's surface and it holds three times as much carbon as the atmosphere does. About 1600 billion tons of this carbon is in the soil as organic matter and some 540–610 billion tons is in living vegetation. Surface carbon moves from the atmosphere to the land and back. Plants use carbon dioxide (CO_2) from the atmosphere to grow and produce food and resources that sustain the rest of the biota. When these organisms breathe, grow, die, and eventually decompose, carbon is released to the atmosphere and the soil. Indeed, life depends on this harmonised movement of carbon from one sink to another. Large-scale disruption or changes on land drastically alter the dynamic movement of carbon (Scherr and Sthapit, 2009a).

The emphasis is now very much on sustainable biofuel production; biofuels from wastes and lignocellulosic material are now seen as good sustainable biofuels that affect significantly better greenhouse gas balances as compared with first generation biofuels (Korres et al., 2010; Nigam and Singh, 2011). The use of agricultural crops for energy purposes (e.g. biomethane production) has, however, other non-climate-related environmental impacts such as eutrophication and acidification due to fertiliser runoff.

Overall, land-use and land-use changes account for around 31 per cent of total human-induced greenhouse gas (GHG) emissions into the atmosphere (Santilli et al., 2005).

Growing plants can remove huge amounts of carbon from the atmosphere and store it in vegetation and soils in ways that not only mitigate climate change but also benefit food and fibre production and the environment. Extensive action to influence land-use is also going to be essential to sustain food and forest production in the face of climate change.

Land-use associated impacts, such as habitat degradation, fragmentation and destruction, are identified as the main threats to biodiversity (IUCN, 2005). However, sustainability studies of bioenergy have to a large extent focused on energy and greenhouse gas balances (Cherubini et al., 2009) and in some cases on chemical pollution (Hill et al., 2006). Indirect land-use change, caused by increased demand on agricultural crop land, has been identified as a critical issue, but the focus has been on climate change (Cherubini et al., 2009) instead of biodiversity. Other impacts to ecosystems (e.g. soil compaction, loss of biodiversity, reduced productivity, salination) caused by land occupation have been ignored to a large extent. Partially this has been caused by the lack of a consensus about which indicators to use for impact assessment for land use (Milá i Canals et al., 2007). Compared with other impacts, land use is a multifaceted environmental issue, as it can be seen as habitat degradation, resource competition or even as an alteration of biogeochemical cycles (Guinee et al., 2006).

Microbial consortia and fermentation aspects of biomethane production

Methane fermentation is a versatile biotechnology capable of converting almost all types of polymeric materials to methane and carbon dioxide under anaerobic conditions. This is achieved as a result of the consecutive biochemical breakdown of polymers to methane and carbon dioxide in an environment in which a variety of microorganisms which include fermentative microbes (acidogens); hydrogen-producing, acetate-forming microbes (acetogens); and methane-producing microbes (methanogens) harmoniously grow and produce reduced end-products (McCarty, 1982). Anaerobes play important roles in establishing a stable environment at various stages of methane fermentation.

Hydrolysis, acidogenesis and methanogenesis

Polymeric materials such as lipids, proteins, and carbohydrates are primarily hydrolysed by extracellular hydrolases, excreted by microbes. Hydrolytic enzymes (lipases, proteases, cellulases, amylases, etc.) hydrolyse their respective polymers into smaller molecules, primarily monomeric units, which are then used by microbes.

Proteins are generally hydrolysed to amino acids by proteases. The amino acids produced are then degraded to fatty acids such as acetate, propionate, and butyrate, and to ammonia (Zeikus, 1980). Polysaccharides such as cellulose, starch, and pectin are hydrolysed by cellulases, amylases, and pectinases, to produce glucose. Microbial hydrolysis of raw starch to glucose requires amylolytic activity, which consists of five amylase species. Pectins are degraded by pectinases, including pectinesterases and depolymerases. Xylans are degraded with α^2-endo-xylanase and α^2-xylosidase to produce xylose. Hexoses and pentoses are generally converted to C_2 and C_3 intermediates and to reduced electron carriers (e.g. NADH) via common pathways. Most anaerobic bacteria undergo hexose metabolism via the Embden–Meyerhof–Parnas pathway (EMP) which produces pyruvate as an intermediate along with NADH. The pyruvate and NADH thus generated are transformed into fermentation endo-products by other enzymatic activities which vary highly with microbial species.

Table 3.2 Energy-yielding reactions of methanogens.

Reaction	$-G°$ (kJ/mol substrate)
$CO_2 + 4 H_2 \rightarrow CH_4 + 2 H_2O$	−130.7
$HCO_3^- + 4 H_2 + H^+ \rightarrow CH_4 + 3 H_2O$	−135.5
$CH_3COO^- + H^+ \rightarrow H_4 + CO_2$	−37.0
$CH_3COO^- + H_2O \rightarrow CH_4 + HCO_3^-$	−32.3
$HCOO^- + H^+ \rightarrow 0.25 CH_4 + 0.75 CO_2 + 0.5 H_2O$	−36.1
$CO + 0.5 H_2O \rightarrow 0.25 CH_4 + 0.75 CO_2$	−52.7
$CH_3OH \rightarrow 0.75 CH_4 + 0.25 CO_2 + 0.5 H_2O$	−79.9
$CH_3NH_3^+ + 0.5 H_2O \rightarrow 0.75 CH_4 + 0.25 CO_2 + NH_4^+$	−57.4
$(CH_3)_2NH_2^+ + H_2O \rightarrow 1.5 CH_4 + 0.5 CO_2 + NH_4^+$	−112.2
$(CH_3)_2NCH_2CH_3H^+ + H_2O \rightarrow 1.5 CH_4 + 0.5 CO_2 + {}^+H_3NCH_2CH_3$	−105.0
$(CH_3)_3NH + 1.5 H_2O \rightarrow 2.25 CH_4 + 0.75 CO_2 + NH_4^+$	−170.8

Source: Thauer et al., 1977

Thus, in hydrolysis and acidogenesis, sugars, amino acids, and fatty acids produced by microbial degradation of biopolymers are successively metabolised by fermentation endo-products such as lactate, propionate, acetate, and ethanol by other enzymatic activities, which vary powerfully with microbial species and the type of feedstock used. Methane and carbon dioxide are produced as the end-products of fermentation. The complex organic (cellulose, starch, lignin, wastewater, spent wash) compounds are used as the feedstock.

Although some acetate (20 per cent) and H_2 (4 per cent) are directly produced by acidogenic fermentation of sugars and amino acids, both products are primarily derived from the acetogenesis and dehydrogenation of higher volatile fatty acids (McCarty, 1982).

Methanogens are physiologically united as methane producers in anaerobic digestion. Although acetate and H_2/CO_2 are the main substrates available in the natural environment, formate, methanol, methylamines, and CO are also converted to CH_4 (Table 3.2).

How do we convert agricultural waste/residues to biomethane?

Although the concept of biorefinery is mostly applied to the production of ethanol and biodiesel as of now, an anaerobic-digestion-based biorefinery would deserve to be more carefully evaluated for the conversion of crops, as it would generate potentially more renewable energy, as methane. For instance ca. 60 per cent of the energy from sugarcane that has been used for the production of bioethanol for decades in Brazil can be converted to biogas while only 38 per cent of the cane energy is converted into alcohol (van Haandel, 2005). With wheat or maize, up to three times more net energy yield can be obtained per hectare by making methane instead of biodiesel or bioethanol (Börjesson and Mattiasson, 2008; De Baere, 2007).

In effect, the food versus fuel debate is making it hard to justify the diversion of lands traditionally harvested for human feed and convert them into lands harvested for transportation. These first-generation biorefineries, operating from first-generation crops, were probably a necessary step in the evolution towards more sustainable practices for renewable energy production. A second generation of biorefineries is under way, with the production of biofuels from lignocellulosic material, crops, agricultural wastes, or forestry feedstocks.

Anaerobic digesters can be built more locally, and a variety of feedstock can be used for biomethanation (more versatile). Also, there is flexibility on the type of energy produced, where feedstock can be transformed into heat, combined heat and power (electricity), or purified and used as compressed natural gas for use as vehicle fuel. In effect, anaerobic digestion is one of the most energy efficient, as well as environmentally benign ways to produce vehicle biofuel (LBS 2002; Singh et al., 2011). Biogas production from energy crops represents a more thermodynamically efficient option than converting plant matter into liquid fuels (Samson et al., 2008). Moreover, biomethane obtained from anaerobic digestion is the most efficient, clean-burning biofuel available today (Rutz and Janssen, 2007). Although biogas may contain siloxanes (0–50 mg/m^3) and dust particles, the combustion of methane reduces emissions of NO_x, CO, particulate matter and unburned hydrocarbons, by 80, 50, 98, and 80 per cent, respectively, as compared with petroleum-derived diesel (Braun, 2007).

Anaerobic digestion

Anaerobic digestion (AD) is an old technology used for stabilising waste and wastewaters, and more recently for energy production. The process of anaerobic digestion also occurs in nature when organic matter degrades and decays, e.g. the cow's digestive system, marshes and swamps, landfills etc. Biogas, the major end-product of the AD process, is either produced naturally or artificially in airtight vessels known as anaerobic digesters (Salminen and Rintala, 2002).

The process of biogas production is not efficient unless carried out in a controlled environment within an anaerobic digester. The digester technology should be designed so as to optimise the conversion of the specific organic material to gaseous products (Demirbas and Ozturk, 2005). A range of digester types and configurations may be utilised. The configuration chosen must be based on various process parameters such as: solids content of the feedstock; number of phases or stages of digestion activities; operating temperature; method of feeding substrate; retention time in the digester and organic loading rate (Nizami and Murphy, 2010; Karagiannidis and Perkoulidis, 2009).

The application of anaerobic digestion technology covers a wide range of uses and substrates, e.g. farm waste, wastewater, industrial organic waste, municipal solid waste, agricultural residues, crops, crop residues, grass and grass silage (Vandevivere, 1999).

Land-use change for biomethane production

Bioenergy to replace fossil fuels can be generated from agricultural feedstocks, including by-products of agricultural production, and dedicated energy crops. The production of biofuels has a different dynamic in relation to fossil fuels, being based on agricultural products, where the main input is land (Rathmann et al., 2010). The main effects of the competition for land between biofuel crops and food crops will be less food and higher food prices. Indeed, if the competition between biofuel crop production and food crop production is extensive and severe enough, it is possible that the consequent increases in agricultural prices will cause some people to go hungry and even starve (Runge and Senauer, 2007); this is why production of bioenergy from non-food products, such as grass, is important. Biofuel crop harvesting practices can affect soil erosion, nutrient and organic content of the soil (Reijnders and Huijbregts, 2009) and can also affect soil microbial populations and their activity. Large-scale bioenergy production has met concerns about sustainability; it is important to note that bioenergy will play an important role in a decarbonised economy. Bioenergy pathways that are not directly linked to land-use competition and land-use changes are those which utilise biogenic wastes and agricultural residues. Other interconnections between land-use systems and energy systems based on agricultural residues exist, yet wastes and residues, such as those from agricultural processing, represent a still largely untapped energy potential worldwide. Utilising the energy potential of available agricultural residues could be a strategy for farms and companies in the agricultural sector to cope with the enduring power supply problems (Bringezu et al., 2009).

Current bioenergy production focuses on well-established first-generation technologies, such as fermentation of agricultural crops to produce ethanol and combustion of biomass to produce heat and power. Most of today's biomass agriculture production uses "classical" food crops, such as maize and oil seeds, and production requirements are very similar compared with when used for feed and food. This can lead to increased environmental pressures if land-use intensity or area is increased. However, on the other hand, energy crops may provide some opportunities for reducing soil erosion and nutrient leaching risks from agricultural land use. When assessing the environmental impacts of bioenergy cropping, it is necessary to take into account three main issues:

- potential land-use changes as a result of the increasing bioenergy demand;
- the different input (e.g. water and nutrients) requirements of the crops and the related management practices;
- the impacts of production processes – biogas as a transport fuel or for the generation of heat or power is more efficient than many of the production paths of bioethanol and biodiesel.

The relationship between bioenergy production and its impact on the environment is very complex. The removal of crop residues from agricultural fields for the production of biofuels might result in soil erosion and losses of nutrients and organic matter returned back to the field and additional fertiliser may be required to balance such losses that will result in additional environmental impacts. However, this can be reduced or avoided completely if non-food crops are harvested at the end of the growing season after nutrients have been translocated down to the roots or rhizomes for recycling into the next season's growth, as occurs with *Miscanthus* and some perennial grasses.

Residues from agriculture

The energy production and GHG mitigation potentials depends on yield/product ratios, and the total agricultural land area as well as type of production system. Less intensive management systems require re-use of residues for maintaining soil fertility. Intensively managed systems allow for higher utilisation rates of residues, but also usually deploy crops with higher crop to residue ratios. Estimates of energy production potential from agricultural residues vary between 15 and 70 EJ/year. Dried dung can also be used as an energy feedstock. The total estimated contribution could be 5 to 55 EJ/year worldwide, with the range defined by current global use at the low end, and technical potential at the high end. Organic wastes and residues together could supply 20–125 EJ/year by 2050, with organic wastes making a significant contribution.

Dedicated energy crops

The energy production and GHG mitigation potentials of dedicated energy crops depends on availability of land, which must also meet demands for food as well as for nature protection, sustainable management of soils and water reserves, and other sustainability criteria. Because future biomass resource availability for energy and materials depends on these and other factors, an accurate estimate is difficult to obtain.

Various studies have arrived at differing figures for the potential contribution of biomass to future global energy supplies, ranging from below 100 EJ/yr to above 400 EJ/yr in 2050. Agricultural systems have developed during a time of relatively predictable local weather patterns. The choice of crops and varieties, the timing of input application, vulnerability to pests and diseases, the timing of management practices, are all closely linked to temperature and rainfall.

Carbon dioxide (77 per cent), nitrous oxide (8 per cent), and methane (14 per cent) are the three main greenhouse gases that trap infrared radiation and contribute to climate change. The GHG emissions during the cultivation stage due to various inputs are dominated by emissions for the production of nitrogen fertiliser and the use of nitrogen fertiliser provokes the emissions of N_2O from the soil (Edwards et al., 2010).

The effects of bioenergy production and consumption can be divided into direct effects and indirect effects.

Direct effects

The direct effects of production and consumption of product (biogas) are a result of the activities involved in the production process. Land-use change (LUC) is one of the main direct effects and occurs mainly when new areas are converted to produce the biomass. The LUC could have both positive and negative significances on biodiversity, carbon stocks and livelihoods of the local habitants. The development of voluntary certification schemes such as the Roundtable on Sustainable Palm Oil and the Round Table on Responsible Soy aim to prevent negative direct effects from crop cultivation. For fully controlled LUC effects, there is a need to establish such certification schemes for worldwide production of energy crops and to enforce land-use planning or alternative control mechanism effectively (Dehue et al., 2011).

Indirect effects

In the present scenario most of the biomass is sourced from existing plantations for the production of bioenergy, which has resulted in a hike on the commodity price and affects the food security and the additional demand for the feedstock triggers additional yield increases (Cornelissen et al., 2009; Dehue et al., 2011).

Indirect land-use change (ILUC) can occur when existing plantations are replaced to cultivate the biomass to meet the demand of additional biofuel production. This displacement can cause an expansion of the land-use for biomass production to new areas, if the previous users of the feedstock do not reduce their feedstock demand and any demand-induced yield increases are insufficient to produce the additional demand. The ILUC is uncertain and is out of the control of the bioenergy sector, wherever it takes place (Dehue et al., 2011).

The unwanted impact of ILUC from bioenergy is evident through unwanted direct LUC for the production of agricultural products for other sectors such as the food and feed sector. If biofuels are to meet policy goals such GHG savings, intermediate solutions will need to be enforced, which recognise the lack of sustainability control in other biomass consuming sectors (Dehue et al., 2011). Displacement effects act across borders because some raw materials are traded globally. Achieving effective national land-use planning in some producing countries should therefore not be taken as full protection against indirect effects.

Dehue et al. (2011) described three mitigation measures that can be implemented to prevent or minimise unwanted indirect effects from a bioenergy production system. The first measure is to prevent unwanted direct LUC, globally and for all sectors, which can be achieved through better land-use planning and corresponding enforcement, and would thus eliminate unwanted ILUC altogether. They also emphasised that because of the international characteristics of ILUC and the competition for land between different sectors, this mitigation measure requires global implementation for all land-based sectors to be effective. The second measure is to reduce pressure on land from the agricultural sector as a whole by increasing yields, supply chain efficiencies and/or a reduction in consumption. The third measure is implementation of practical production models that prevent indirect impacts at a project level.

However, as long as not all worldwide production is controlled by such certification schemes, effective enforcement of land-use planning, or alternative control mechanisms, such mechanisms are not able to fully control indirect effects.

Dehue et al. (2011) summarises three main solutions from different studies (Klessmann et al., 2007; Blok et al., 2008; Cornelissen and Dehue, 2009; Dehue and van de Staaij, 2009) that could be put forward for producers to expand biomass usage for energy purposes without causing unwanted indirect effects.

a   Biomass production on "unused land". Clearly, expanding production on unused land does lead to a direct LUC. The big advantage is that direct LUC is controllable and can be limited to those areas where effects are acceptable, while the effects of indirect LUC are largely uncontrollable.
b   Introducing energy crop cultivation without displacing the original land use through increased land productivity or integration models.
c   Bioenergy production from residues. Current functions and utilisation of these residues must be well understood; otherwise displacement and the associated indirect effects may not be avoided.

Land-use changes contribute to the release of all three greenhouse gases (Table 3.3) – of the total annual human-induced GHG emissions in 2004 of 49 billion tons of carbon dioxide equivalent, roughly 31 per cent (15 billion tons) was from land use. By comparison, fossil fuel burning accounts for 27.7 billion tons of CO_2 emissions annually.

Increasing energy consumption has exerted great pressure on natural resources; this has led to a move towards sustainable energy resources to improve security of supply and to reduce greenhouse gas emissions. Climate change may in particular affect agro-ecosystems and is currently thought to have positive as well as negative effects on yields in different regions of the world (Auvinen et al., 2007). The major sources of carbon dioxide pollution

Table 3.3 Greenhouse gas emissions from land use.

Land-use	Annual emissions (million tons CO_2e)	Greenhouse gas emitted
Soil fertilisation (inorganic fertilisers and applied manure)	2,100	Nitrous oxide
Gases from food digestion in cattle (enteric fermentation in rumens)	1,800	Methane
Biomass burning	700	Methane, nitrous oxide
Paddy (flooded) rice production (anaerobic decomposition)	600	Methane
Livestock manure	400	Methane, nitrous oxide
Other (e.g., delivery of irrigation water)	900	Carbon dioxide, nitrous oxide

Source: Scherr and Sthapit, 2009b

in the atmosphere are burning fossil fuels and land-use change including vegetation fires (Uherek, 2004), which make CO_2 a bigger contributor to global warming than methane.

Biomethane can be made from a variety of sources within the forestry, agricultural, municipal and industrial waste streams. Making biomethane from biomass prevents incineration or the creation of landfills, which are also significant sources of atmospheric methane. Biomethane can be used to achieve environmental improvements through:

- reduction of traffic induced CO_2 emissions;
- replacement of fossil fuels;
- improvement of municipal air quality;
- sustainable waste management and soil conservation.

Economics inevitably play an important role in the decision to start up a biomethane business within a municipality or region. The economic objectives typically include:

- a commercial approach to fuel change management;
- the need for an economical and economically sustainable vehicle fuel solution;
- the development of new markets and/or job creation;
- the creation of stable supply and demand on the local, regional or national vehicle fuel market.

Political considerations also play a part in the decision to implement the biomethane solution.

The political decision will most likely be based on the environmental and economic aspects of the biomethane project, but some additional objectives are:

- image-building as a progressive municipality or region;
- positive partnerships with new, community-based stakeholders;
- development of efficient "best-practices" related to solving different municipal problems.

The use of biomass for energy production as a substitute for fossil energy is often seen as an attractive option to reduce fossil-fuel dependency and help reduce GHG emissions

Table 3.4 Land-use intensity* of petroleum fuels and biofuels.

Fuel	Land-use intensity (km^2/TWh/yr)
Biodiesel from soy	890
Ethanol from corn	350
Ethanol from cellulose	460
Ethanol from sugarcane	286
Electricity from biomass	543
Maize biogas	617
Petroleum	45

Sources: McDonald et al., 2009; Smil, 2008, 2010; Graebig et al., 2010
* The land requirement per unit of delivered biofuel can be calculated simply as the product of the yield (crop output per unit area), the production intensity (energy per unit crop), and a factor that accounts for the land-use impacts of any co-products of the production process (Delucchi, 2010).

(Rassi et al., 2001). The interrelations between food and bioenergy depend on a host of factors, including economic factors (e.g., prices and trade), agricultural technology (e.g., crop yields, conversion efficiencies), changes in demand (e.g., diets, population numbers), as well as patterns and trajectories of global land-use.

The land requirement for biofuel production, even expressed relative to some measure of available land (Table 3.4), is just a rough indicator of other land-use impacts that society cares about, such as soil erosion, dust and smoke from agricultural activities, loss of habitat, biodiversity, and ecosystem services, and the effects of competition for land on the prices of commodities and services produced by land (Delucchi, 2010). The use of monocultural feedstocks for biofuel production purposes can reduce biological diversity and the associated biocontrol services in agricultural landscapes (UNEP, 2009; Reijnders and Huijbregts, 2009). Each biofuel production system can severely degrade natural habitats. Monocultures could be replaced by "natural, diversified, multifunctional vegetation, residues from crops and forest to mitigate these effects that could meet the broad demand for goods and other resource functions in a sustainable manner" (Klay, 2000).

An agricultural landscape should simultaneously provide food and fibre, meet the needs of nature and biodiversity, and support viable livelihoods for people who live there. In terms of climate change, landscape and farming systems should actively absorb and store carbon in vegetation and soils, reduce emissions of methane from rice production, livestock, reduce emissions from burning, and reduce nitrous oxide emissions from inorganic fertilisers. Examples of situations where landscape and farming systems could help mitigate climate change are important to increase the resilience of production systems and ecosystem services to climate change.

Biomass is the most common basis for biomethane production. It is defined by the European Union as the organic products, by-products and waste streams from forestry and agriculture (including animal husbandry) as well as municipal and industrial waste streams (EC, 2005). The use of biomass for fuel production has several advantages. Natural materials are plentiful and multiple types of biomass materials can be found easily in many different locations or at the same location. The most popular types of biomass for biomethane production are: sorted biodegradable municipal waste, sewage sludge, manure and leftover agricultural and agro-industrial products. Agricultural feedstock sources include energy crops, agricultural residues and by-products and can include leguminous plants, grasses, and residues of harvests and food production. An issue apart is the use of energy crops. Debates are ongoing about whether or not dedicated land should be used for the production of such feedstock. Wood can be used as a feedstock in a separate process called thermal gasification. Thermal gasification is achieved by reacting the material at high temperature (>700 °C), without combustion, with a controlled amount of oxygen/steam. The resulting gas mixture is called syngas. In addition to biogas, landfill gas can also be used as a starting point for biomethane production. Landfill gas is the gaseous by-product of waste deposits in landfills, which would otherwise be flared off.

Fischer et al. (2010), in a study on biofuel production potential in Europe, specifically with respect to land-use scenarios, pointed out that the maximum amount of crop residues that can be removed from the field without significant effect on soil fertility is under debate. A few researchers have discussed crop residues as unused waste and make a strong case for the use of crop residue as raw material for biofuel production (Somerville, 2006; Prasad et al., 2007), while others perceive that crop residues is a valuable resource that provides irreplaceable environmental sustainable services and defended the removal of crop residues and highlighted that it would exacerbate risks of soil erosion by water and wind, deplete soil organic matter, degrade soil quality, increase non-point source pollution, decrease agronomic productivity and reduce crop yields (Lal, 2005; Smil, 1999; Lal and Pimentel, 2007; Pimentel and Lal, 2007). Additional fertiliser may be required to balance the nutrient loss, which will result in additional environmental impacts (Delucchi, 2010). Lynd et al. (2002) suggested that removal of up to 50 per cent of crop residues can be considered without any significant impact on soil erosion and fertility. In a recent study, Fischer et al. (2010) demonstrate that significant agricultural land reserves can be freed up for bioenergy production. A total of 53 million hectares of cultivated land and about 19 million hectares of pastures could become available by 2030 for bioenergy feedstocks without compromising Europe's food and feed sectors. Significant scope for yield improvements in the food and feed sector combined with modest increases in domestic demand, create opportunities for relatively large extents of land to be freed up.

Methane produced in the rumen (the first stomach of cattle, sheep, and goats and other species that chew the cud) accounts for about 1.8 billion tons of CO_2-equivalent emissions (Scherr and Sthapit, 2009a). Nutrient supplements and innovative feed mixes have been developed that can reduce methane production by 20 per cent, though these are not yet commercially viable for most farmers. Some feed additives can make diets easier for animals to digest their food and reduce methane emissions. These require fairly sophisticated management, so they are mainly useful in larger-scale livestock operations (which are, in any case, the main sources of methane emissions). Manure is a major source of methane; it is responsible for some 400 million tons of CO_2-equivalent, and poor manure management is a leading source of water pollution. But it is also an opportunity for an alternative fuel that reduces a farm's reliance on fossil fuels. By using appropriate technologies like an anaerobic biogas digester, farmers can profit from their farm waste while helping the climate. Some communities in developing countries are already using manure to produce cooking fuel. By installing anaerobic digesters, a large pile of manure can be used to produce biogas as well as fertiliser for farms. Even collecting the methane and burning it to convert it to carbon dioxide will be an improvement, and generated heat can be used to produce electricity. By thinking creatively, previously undervalued and dangerous wastes can be converted into new sources of energy, saving costs, and even providing income. Biogas digesters involve an initial cash investment that often needs to be advanced for low-income producers, but lifetime benefits far outweigh costs. This technology could be extended to millions of farmers with

Table 3.5 Summary of the various initiatives proposed or developing proposals for measures to mitigate indirect impacts from biofuels.

Initiative	Measure	Scope
US Renewable Fuels Standard	GHG-factor	GHG
Californian Low Carbon Fuel Standard	GHG-factor	GHG
Low Indirect Impact Biofuels (Dehue et al., 2010)	Preventing displacement by expanding on land without provisioning services	GHG Biodiversity Land rights Food consumption
	Preventing displacement through agricultural intensification	GHG Biodiversity Land rights Food consumption
	Preventing displacement through using wastes or residues	GHG Biodiversity Land rights Food consumption
EU Renewable Energy Directive	Various policy options are being considered	

Adopted from Dehue et al., 2011

benefits for the climate as well as for human well-being through expanded access to energy (Scherr and Sthapit, 2009b).

Table 3.5 shows a summary of the various initiatives that have proposed or are developing proposals for measures to mitigate indirect effects of biofuels. For each initiative, the main measure and its scope are also included. The number of mitigation measures that currently exist is small and most are not yet fully operational. Most of the mitigation measures focus only on GHG effects of biofuels by incorporating an ILUC factor in the general life cycle analysis of feedstock-based biofuel pathways. The Low Indirect Impact Biofuels initiative is the only initiative to work on pragmatic solutions for biofuel feedstock production that has a minimised risk of indirect effects by preventing displacement effects from occurring at the project level (Dehue et al., 2011).

Several methods for transforming land occupation to improve environmental impacts have been proposed, ranging from use of natural bioproductivity (Wackernagel, 1994) to exergy (available energy) retention in ecosystems (Wagendorp et al., 2006) and to landscape naturalness (Brentrup et al., 2002). Due to the lack of consensus for a single indicator, three impact assessment methods were used: eco footprint biocapacity (Wiedmann et al., 2009), human appropriation of net primary production (HANPP) (Haberl et al., 2007) and habitat loss (Koellner and Scholz, 2006). These represent the influences of land use to human resource use, ecological life support functions and to biodiversity, respectively. The ecological footprint describes the amount of productive land needed to produce the renewable raw materials for society and to absorb the biological wastes produced by society. The land-use intensity metric is not a very effective indicator of these impacts because it does not reflect the impact of the land use on habitat integrity, wildlife corridors, and interactions at the "edges" of the affected area (Delucchi, 2010). Brentrup et al. (2002) discussed a more direct indicator of the impacts of land use on habitat and biodiversity, the natural degradation potential (NDP), which ranged between 0 and 1 on the basis of influence of human activity. Koellner and Scholz (2007) propose a similar measure of land-use impacts, the ecosystem damage potential (EDP), based on the extent of the degradation of the "ecological quality" of the occupied land. Dubreuil et al. (2007) suggest a variety of indicators for impacts on biodiversity (e.g., measures of species lost), biotic production potential (e.g., energy required to restore productive potential of the soil), and soil quality (e.g., measures of soil erosion). Lindeijer (2000) proposes free net primary biomass production as an indicator for the "life support" function of an ecosystem, and species number as an indicator of biodiversity. A land-use intensity index gives only a sense of the potential scale of the problems (Delucchi, 2010).

Delucchi (2010) suggested that policies for sustainable biofuel production should promote those biofuels programs that have very low inputs of fossil fuels and chemicals, can rely on rainfall or abundant groundwater, and utilise land with little or no economic or ecological value in alternative uses. Physical parameters (e.g. reflectivity and evapotranspiration rate) can be changed with the changes in land-use and vegetation, which directly affect the absorption and disposition of energy at the earth surface and changes in local and regional temperatures (Bala et al., 2007; Feddema et al., 2005; Lobell et al., 2006).

Conclusions

Selection of the proper digester design for biomethane production is an important management/design decision that merits further investigation. The wet continuous two-stage system, the leach bed system with UASB, the dry continuous system and batch digesters all have potential for biomethanation of agricultural residues. There is need to compare the potential of various pre-treatment options to increase the process efficiency. Upgrading of biogas by removing CO_2, H_2S and other possible pollutants produces biomethane, which can either be used directly on the site as a transport fuel or after grid injection, may be used off site where better energy efficiencies and financial returns may be achieved. The removal of biomass from land could affect the soil quality as it can increase erosion, losses in nutrients and organic carbon and decrease in microbial population and activity. To overcome all these effects, additional fertiliser could be needed that again creates further emissions. The shifting of agricultural land to energy crops could also raise the price of food grains and their products.

References

Auvinen, A., Hilden, M., Toivonen, H., Primmer, E., Niemela, J. and Aapala, K. (2007) 'Evaluation of the Finnish national biodiversity action plan 1997–2005', *Monogr Boreal Environ Res*, vol 29.

Bala, G., Caldeira, K., Wickett, M., Phillips, T.J., Lobell, D.B., Delire, C. and Mirin, A. (2007) 'Combined climate and carbon-cycle effects of large-scale deforestation', *Proc Natl Acad Sci USA*, vol 104, pp. 6550–6555.

Blok, K., van Breevoort, P., Roes, L., Coenraads, R. and Müller, N. (2008) *Global status report on energy efficiency 2008*, REEEP Global Status Report on Energy Efficiency.

Börjesson, P. and Mattiasson, B. (2008) 'Biogas as a resource-efficient vehicle fuel', *Trends Biotechnology*, vol. 26, pp. 7–13.

Braun, R. (2007) 'Anaerobic digestion: a multi-facetted process for energy, environmental management and rural development', in R. Paolo (ed.)*Improvement of crop plants for industrial uses*, pp. 335–416. Dordrecht: Springer.

Brentrup, F., Kusters, J., Lammel, J. and Kuhlmann, H. (2002) 'Life cycle impact assessment of land-use based on the Hemeroby concept'. *Int. J. Life Cycle Assess*. 20: 247–264.

Bringezu, S., Schütz, H., O'Brien, M., Kauppi, L., Howarth, H.W., and McNeely, J. (2009) *Towards sustainable production and use of resources—assessing biofuels*, Nairobi: United Nations Environment Programme.

Cherubini, F., Bird, N.D., Cowie, A., Jungmeier, G, Schlamadinger, B. and Woess-Gallasch, S. (2009) 'Energy- and greenhouse gas-based LCA of biofuel and bioenergy systems: key issues, ranges and recommendations', *Resource Conservation Recycling*, vol. 53, pp. 434–447.

Cornelissen, S. and Dehue, B. (2009) *Summary of approaches to accounting for indirect impacts of biofuel production*, Utrecht: Ecofys International BV.

Cornelissen, S., Dehue, B. and Wonink, S. (2009) *Summary of approaches to account for and monitor indirect impacts of biofuel production*, Utrecht: Ecofys International BV..

De Baere, L. (2007) 'Dry continuous anaerobic digestion of energy crops', 11th IWA World congress on Anaerobic Digestion; Brisbane, Australia.

Dehue, B., van de Staaij, J. (2009) *Mitigating indirect effects of biofuel production*, Utrecht: Ecofys International BV.

Dehue, B., Meyer, S. and van de Staaij, J. (2010) *Responsible cultivation areas – Identification and certification of feedstock production with a low risk of indirect effects*, Utrecht: Ecofys International BV.

Dehue, B., Cornelissen, S., and Peters, D. (2011) *Indirect effects of biofuel production: Overview prepared for GBEP*, Utrecht: Ecofys International BV.

Delucchi, M.A. (2010) 'Impacts of biofuels on climate change, water use, and land-use', *Annals of the New York Academy of Sciences*, Issue: The Year in Ecology and Conservation Biology, 1195, pp. 28–45.

Demirbas, A. and Ozturk, T. (2005) 'Anaerobic digestion of agricultural solid residues', *International Journal of Green Energy*, vol.1, pp. 483–494.

Dubreuil, A., Freiermuth Knuchel, R., and Gaillard, G. (2007) 'Key elements in a framework for land-use impact assessment in LCA', *Internation Journal Life Cycle Assessesmt* 12:5–15.

EC (European Commission) (2005) *Biomass, green energy for Europe*. Brussels: European Commission, DG Research – Unit 3 Sustainable Energy Systems.

Edwards, R., Mulligan, D. and Marelli, L. (2010) *Indirect land-use change from increased biofuels demand: Comparison of models and results for marginal biofuels production from different feedstocks*. Ispra, Italy: European Commission, Joint Research Centre, Institute for Energy.

Feddema, J.J., Oleson, K.W., Bonan, G.B., Mearns, L.O., Buja, L.E., Meehl, G.A., and Washington, W.M. (2005) 'The importance of land-cover change in simulating future climates', *Science* vol. 310, pp. 1674–1678.

Fischer, G., Prieler, S., van Velthuizen, H., Berndes, G., Faaij, A., Londo, M. and de Wit, M. (2010) 'Biofuel production potentials in Europe: Sustainable use of cultivated land and pastures, Part II: Land-use scenarios'. *Biomass and Bioenergy*, vol. 34, pp. 173–187.

Graebig, M., Bringezu, S. and Fenner, R. (2010) 'Comparative analysis of environmental impacts of maize-biogas and photovoltaics on a land-use basis', *Solar Energy*, vol. 84, no 7, pp. 1255–1263.

Guinee, J., van Oers, L., de Koning, A., and Tamis, W. (2006) *Life cycle approaches for conservation agriculture*, Leiden: CML.

Haberl, H., Erb, K.H., Krausmann, F., Gaube, V., Bondeau, A., Plutzar, C., Gingrich, S., Lucht,W. and Fischer-Kowalski, M. (2007) 'Quantifying and mapping the human appropriation of net primary production in earth's terrestrial ecosystems', *Proceedings of the National Academy of Sciences of the USA* 104, 12942–12947.

Hill, J., Nelson, E., Tilman, D., Polasky, S. and Tiffany, D. (2006) 'Environmental, economic, and energetic costs and benefits of biodiesel and ethanol biofuels', *PNAS*, vol. 103, pp. 11206–11210.

IUCN (2005) *IUCN Red list of threatened species. A global species assessment*. Gland, Switzerland: IUCN World Conservation Union.

Jeffery, A.C., Peter, J.V., William, J.J.B.R. and James, M.G. (1981) 'Predicting methane fermentation biodegradability', *Biotechnology and Bioengineering Symposium*, vol.11, pp. 93–117.

Karagiannidis, A. and Perkoulidis, G. (2009) 'A multi-criteria ranking of different technologies for the anaerobic digestion for energy recovery of the organic fraction of municipal solid wastes', *Bioresource Technology*, vol.100, pp. 2355–2360.

Klay, A. (2000) 'The Kyoto Protocol and the carbon debate', Development and Environment Report No. 18, Center for Development and Environment, University of Berne, Berne, Switzerland. www.cde.unibe.ch/info/pdf/waldco2.pdf.

Klessmann, C., Graus, W., Harmelink, M. and Geurts, F. (2007) *Making energy-efficiency happen: from potential to reality*, Utrecht: Ecofys International BV.

Koellner, T., and Scholz, R.W. (2006) 'Assessment of land-use impacts on the natural environment. Part 2: Generic characterization factors for local species diversity in Central Europe'. *Int J Life Cycle Ass*, vol. 20 pp. 1–20.

Koellner, T. and Scholz, W. (2007) 'Assessment of land-use impacts on the natural environment, Part 1: an analytical framework for pure land occupation and land-use change', *Int J Life Cycle Ass*, vol. 12, pp. 16–33.

Korres, N.E., Singh, A., Nizami, A.S. and Murphy, J.D. (2010) 'Is grass biomethane a sustainable transport biofuel?', *Biofuels, Bioproducts and Biorefining*, vol. 4, no 3, pp. 310–325.

Lal, R. (2005) 'World crop residues production and implications of its use as a biofuel', 603 *Environ. Int.*, vol. 31, pp. 575–584.

Lal, R., and Pimentel, D. (2007) 'Bio-fuels from crop residues'. *Soil and Tillage Research*, 93(2): 237–238.

LBS (2002) *GM well-to wheel analysis of energy use and greenhouse gas emission of advanced fuel/vehicle system – a European study*. Ottobrumn: L-B-Systemtechnik GmbH.

Lindeijer, E. (2000) 'Biodiversity and life support impacts of land-use in LCA', *Journal of Cleaner Production*, vol. 8, pp. 313–319.

Lobell, D.B., Bala, G. and Duffy, P.B. (2006) 'Biogeophysical impacts of cropland management changes on climate', *Geophys. Res. Lett.* vol. 33, L06708.

Lynd, L.R., Haiming, J., Joseph, G.M., Charles, E.W. and Bruce, D. (2002) 'Bioenergy: background, potential, and policy', Policy briefing prepared for the Centre for Strategic and International Studies. Available at: http://i-farmtools.iastate.edu/ref/Lynd_et_al_2002.pdf [accessed July 2012].

McCarty, P.L. (1982) 'One hundred years of anaerobic treatment', in D.E. Hughes, D.A. Stafford, B.I. Wheatley, W. Baader, G. Lettinga, E.J. Nyns, W. Verstraete, and R.L. Wentworth (eds.), *Anaerobic Digestion, 1981*, pp. 3–21, Amsterdam: Elsevier Biomedical Press B.V.

McDonald, R.I., Fargione, J., Kiesecker, J., Miller, W.M. and Powell, J. (2009) 'Energy sprawl or energy efficiency: climate policy impacts on natural habitat for the United States of America', *PLoS ONE*, vol. 4, no 8, e6802, http://dx.doi.org/10.1371/journal.pone.0006802 [accessed July 2012].

Milà i Canals, L., Romanyà and Cowell, S.J. (2007) 'Method for assessing impacts on life support functions (LSF) related to the use of 'fertile land' in life cycle assessment (LCA)', *Journal of Cleaner Production*, vol 15, pp. 1426–1440.

Moletta, R. (2008) 'Technologies de la méthanisation de la biomasse – Déchets ménagers et agricoles', in R. Moletta (ed.) *La méthanisation*, pp. 181–204. Paris: Lavoisier, Editions Tec & Doc.

Nigam, P.S. and Singh, A. (2011) 'Production of liquid biofuels from renewable resources', *Progress in Energy and Combustion Science*, vol. 37, pp. 52–68.

Nizami, A.S. and Murphy, J.D. (2010) 'What is the optimal digester configuration for producing grass biomethane?', *Renewable and Sustainable Energy Reviews*, vol.14, pp. 1558–1568.

Pimentel, D. and Lal, R. (2007) 'Biofuels and the environment', *Science*, vol.317, p. 897.

Prasad, S., Singh, A. and Joshi, H.C. (2007) 'Ethanol as an alternative fuel from agricultural, industrial and urban residues', *Resources, Conservation and Recycling*, vol. 50, pp. 1–39.

Rassi, P., Alanen, A., Kanerva, T. and Mannerkoski, I. (2001) *Endangerment of Finnish species 2000*, Helsinki: Finnish Ministry of Environment.

Rathmann, R., Szklo, A. and Schaeffer, R. (2010) 'Land-use competition for production of food and liquid biofuels: An analysis of the arguments in the current debate', *Renewable Energy*, vol. 35, pp. 14–22.

Reijnders, L. and Huijbregts, M.A.J. (2009) *Biofuels for road transport, a seed to wheel perspective*, Green Energy and Technology Series. London: Springer-Verlag.

Roels, J.A. (1983) *Energetics and kinetics in biotechnology*, Amsterdam: Elsevier Biomedical Press.

Runge, C.F. and Senauer, B. (2007) 'How biofuels could starve the poor', *Foreign Affairs*, vol. 86, pp. 41–53.

Rutz, D. and Janssen, R. (2007) *Biofuel technology handbook*. Munich: WIP Renewable Energies.

Salminen, E. and Rintala, J. (2002) 'Anaerobic digestion of organic solid poultry slaughterhouse waste: a review', *Bioresource Technology*, vol. 83 , pp. 13–26.

Samson, R., Lem, C.H, Bailey Stamler, S. and Dooper, J. (2008) 'Developing energy crops for thermal applications. Optimizing fuel quality, energy security and GHG mitigation', in D. Pimentel (ed.) *Biofuels, solar and wind as renewable energy systems*, pp. 395–423. Dordrecht: Springer.

Santilli, M., Moutinho, P., Schwartzman, S., Nepstad, D., Curran, L., and Nobre, C. (2005) 'Tropical deforestation and the Kyoto Protocol: an editorial essay', *Climatic Change*, vol. 71, pp. 267–276.

Scherr, S.J. and Sthapit, S. (2009a) 'Farming and land-use to cool the planet', *State of the World 2009 confronting climate change*, pp. 30–49. London: Earthscan.

Scherr, S.J. and Sthapit, S. (2009b) 'Mitigating climate changes through food and land use', Worldwatch Report 179. Washington, DC: Worldwatch Institute.

Singh, A., Nizami, A.S., Korres, N.E. and Murphy, J.D. (2011) 'The effect of reactor design on the sustainability of grass biomethane', *Renewable and Sustainable Energy Reviews*, vol.15, no 3, pp. 1567–1574.

Smil, V. (1999) 'Crop residues: agriculture's largest harvest', *BioScience*, vol. 49, no 4, pp. 299–308.

Smil, V. (2008) *Energy in nature and society: general energetics of complex systems*, Cambridge, MA: The MIT Press.

Somerville, C. (2006) 'The billion-ton bio fuels vision', *Science*, vol.312, p. 1277.

Thauer, Rudolf, K., Jungermann, K. and Decker, K. (1977) 'Energy conservation in chemotrophic anaerobic bacteria', *Bacteriological Reviews*, vol. 41, no. 1, pp. 100–180.

Uherek, E. (2004) 'The greenhouse gases – carbon dioxide and methane', Environmental Science Published for Everybody Round the Earth (ESPE) website. http://www.atmosphere.mpg.de/enid/04 6524c2e25d1ef90dfcf9176209675e,0/2__Radiation___greenhouse_gases/-_CO2__CH4_253.html [accessed July 2012].

UNEP (United Nations Environment Programme) (2009) 'Working Group on Biofuels of the International Panel for Sustainable Resource Management: Towards sustainable production and use of resources – assessing biofuels'. http://www.unep.org/PDF/Assessing_Biofuels.pdf

van Haandel, A.C. (2005) 'Integrated energy production and reduction of the environmental impact at alcohol distillery plants', *Wat Sci Tech*, vol. 52, nos (1–2), pp. 49–58.

Vandevivere, P. (1999) 'New and broad applications of anaerobic digestion' *Critical Reviews in Environmental Science and Technology*, vol.29, pp. 151–173.

Wackernagel, M. (1994) 'Ecological footprint and appropriated carrying capacity: A tool for planning toward sustainability' (PhD thesis). Vancouve School of Community and Regional Planning. The University of British Columbia.

Wagendorp,T., Gulinck,H., Coppin,P. and Muys, B. (2006) 'Land use impact evaluation in life cycle assessment based on ecosystem thermodynamics', *Energy*, vol. 31, 112–125.

Wiedmann, T. (2009) 'Carbon footprint and input-output analysis – an introduction', *Economic Systems Research*, vol. 21, pp.175–186.

Vij, S. (2011) 'Biogas production from kitchen waste', Department of Biotechnology and Medical Engineering National Institute of Technology, Rourkela. E-thesis. http://ethesis.nitrkl.ac.in/2547/1/THESIS_FINAL_REPORT.pdf (accessed 15 June 2012).

Zeikus, J.G. (1980) 'Chemical and fuel production by anaerobic bacteria', *Ann. Rev. Microbiol*, vol. 34, pp. 423–464.

Part II
Feedstocks

Chapter 4

Grass and grass silage

Agronomical characteristics and biogas production

Joseph McEniry,[1] Nicholas E. Korres[2] and Padraig O'Kiely[1]*

[1]Grassland Science Research Department, Animal & Grassland Research and Innovation Centre, Teagasc, Grange, Dunsany, Co. Meath, Ireland; [2]26 Grigoroviou St, Patisia, GR–11141, Athens, Greece
*Corresponding author email: padraig.okiely@teagasc.ie

Introduction

Grasslands play a major role in global agriculture, accounting for approximately 70 per cent of the world agricultural land area and 26 per cent of total land area (FAO, 2012). They vary greatly in their degree and intensity of management, ranging from extensively managed rangelands and savannahs to intensively managed continuous pasture and hay land (Verchot et al., 2006).

Grasslands are predominantly used for animal production, particularly as a principal source of food for ruminants, but they also have an important multifunctional role in maintaining floral and faunal diversity (Isselstein et al., 2005), providing water catchment protection and soil erosion control (Cerdan et al., 2010), reducing the impact of global warming through carbon sequestration (Peeters and Hopkins, 2010) and providing landscape and amenity value (Gibon, 2005). More recently, grass biomass has been considered for the production of renewable energy (Prochnow et al., 2009a, b) and other materials (Mandl, 2010).

Methane-rich biogas can be produced from a wide range of feedstocks (e.g. agricultural crops, animal manure, organic wastes, etc.) through anaerobic digestion (AD) and can be used to generate heat and electricity (Al Seadi et al., 2008) or can be upgraded to biomethane and used as a transport fuel (Korres et al., 2010). In recent years, dedicated energy crops (e.g. maize, grass, sugar beet) have been grown specifically for biogas production (Murphy et al., 2011). In temperate regions, grassland in particular represents a significant biomass resource for biogas production.

The aim when providing a feedstock for biogas production is to achieve a high CH_4 yield per area of land (Prochnow et al., 2009a). Thus, similar to producing high-quality conserved feed for ruminants, biomass yield, digestible energy content and ensilability are important factors for consideration. These factors are determined by the botanical composition of the sward, environmental factors (e.g. soil fertility, rainfall and temperature) and specific management factors (e.g. harvest date, nutrient management). This chapter outlines these agronomic factors and their potential impact on biogas production.

Grass species

The genome of each grass species imposes differences in adaptability, productivity and chemical composition, and this impacts on their CH_4 production potential. For example, there are distinct differences between the photosynthetic pathways of C_3 (i.e. cool season or temperate) and C_4 (i.e. warm season or tropical) grass species which affect light, water and nitrogen use efficiencies (Lattanzi, 2010). In warm regions, C_4 grasses can outyield C_3 grasses due to their more efficient photosynthetic pathway. However, the lower temperatures and shorter growing seasons in northern Europe limit the growth of C_4 grasses (Lewandowski et al., 2003). Furthermore, C_4 grasses generally have a lower dry matter (DM) (i.e. total solids) digestibility (Reid et al., 1988) than C_3 grasses and this is accentuated by the increased lignification of forages grown under high temperatures (Buxton and Fales, 1994). Thus, perennial C_4 grasses are usually less suitable for AD.

Several perennial grasses have been identified as promising energy crops, including the C_4 grasses miscanthus (*Miscanthus* × *giganteus*; Clifton-Brown et al., 2004) and switchgrass (*Panicum virgatum*; McLaughlin and Kszos, 2005) and the C_3 grasses reed canary grass (*Phalaris arundinacea*; Burvall, 1997) and giant reed (*Arundo donax*; Angelini et al., 2005). High biomass yields and their suitability for a single delayed harvest are some of the advantages of these grass species (Lewandowski et al., 2003). However, their relatively low digestibility makes them unsuitable for biogas production and research has primarily focused on the potential of these grasses for combustion (Prochnow et al., 2009b).

Much of the work on grass for biogas production has focused on temperate grasslands which are currently (or have recently been) used for animal production. Intensively managed temperate grasslands are dominated by a small number of grass species including perennial ryegrass (PRG; *Lolium perenne*), Italian ryegrass (IRG; *Lolium multiflorum*), timothy (*Phleum pratense*), cocksfoot (*Dactylis glomerata*) and tall fescue (*Festuca arundinacea*) (Plantureux et al., 2005; Hopkins and Holz, 2006). These grasses may be present as indigenous species in permanent pastures, part of swards reseeded many years previously or they may be resown every few years as part of a crop rotation (Buxton and O'Kiely, 2003). The majority of reseeded temperate grassland in Europe is now dominated by PRG due to its high digestibility when harvested at the appropriate growth stage, high yield in response to nitrogen fertiliser application and ease of preservation as silage due to its relatively high water-soluble carbohydrate (WSC) content (Whitehead, 1995). However, in situations where ryegrasses are limited by winter survival there is a continuing role for timothy, while in drier temperate regions there is a role for more drought tolerant species such as cocksfoot and tall fescue (Table 4.1; Hopkins and Wilkins, 2006).

Legumes such as white clover (*Trifolium repens*), lucerne (*Medicago sativa*) and red clover (*Trifolium pratense*) can also represent a prominent component of temperate grasslands (Peeters et al., 2006). These legumes can be grown alone or in combination with compatible grasses and they can reduce the requirement for fertiliser N through fixation of atmospheric N, while also increasing herbage yield and quality (Peyraud et al., 2010). However, legumes generally have a lower concentration of WSC and a higher buffering capacity (BC) than grasses, making them more difficult to preserve as silage (Buxton and O'Kiely, 2003).

The aim with providing a feedstock for biogas production is to achieve the highest CH_4 yield per area of land (i.e. m^3 CH_4 ha^{-1}; Prochnow et al., 2009a). The two criteria decisive in choosing a grass species are: (1) the specific CH_4 yield per unit of biomass and (2) the biomass yield per hectare (Taube et al., 2006). The specific CH_4 yield of a feedstock will be determined

Table 4.1 Some adaptation and management characteristics of perennial temperate grasses.

Common name	Winter hardiness	Drought tolerance	Heading dates		Primary use	Forage quality
			Relative	Range		
Perennial ryegrass	Fair	Poor	Early	Wide	Pasture	Excellent
Timothy	Excellent	Poor	Late	Wide	Hay or silage	Very good
Tall fescue	Good	Very good	Early	Wide	Pasture or hay	Poor to good
Cocksfoot	Good	Good	Early	Wide	Hay or silage	Good

Source: Balasko and Nelson (2006)

Table 4.2 Perennial grasses and red clover as feedstocks for anaerobic digestion.

Common name	Botanical name	Photo-synthetic pathway	Biomass yield[1] (t DM ha^{-1} a^{-1})	Specific CH$_4$ yield[2] (m^3 t^{-1} VS)	Area specific CH$_4$ yield[3] (m^3 ha^{-1})
Perennial ryegrass	Lolium perenne	C$_3$	9–20	198–410	1,639–7,544
Timothy	Phleum pratense	C$_3$	9–18	308–365	2,550–6,044
Tall fescue	Festuca arundinacea	C$_3$	8–14	296–394	2,179–5,075
Cocksfoot	Dactylis glomerata	C$_3$	8–10	308–382	2,267–3,514
Reed canary grass	Phalaris arundinacea	C$_3$	7–13	340–430	2,190–5,143
Giant reed	Arundo donax	C$_3$	3–37	–	–
Miscanthus	Miscanthus x giganteus	C$_4$	5–44	179–218	823–8,825
Switchgrass	Panicum virgatum	C$_4$	5–23	191–309	879–6,538
Red clover	Trifolium pratense	C$_3$	7–13	300–350	1,932–4,186

[1] DM = dry matter; data adapted from Lewandowski et al. (2003) and Peeters and Kopec (1996) Peeters et al. (2006).
[2] VS = volatile solids; data adapted from Murphy et al. (2011), Masse et al. (2010), Seppala et al. (2009) and Kaiser and Gronauer (2007).
[3] Assuming 0.92 VS in grasses and red clover; not accounting for any field, harvesting or storage losses.

by its chemical composition and different species can differ markedly in terms of quality. However, within the C$_3$ grass species only small differences in quality are observed when they are harvested at the same developmental growth stage (Balasko and Nelson, 2006) and numerous authors have reported that these species have similar specific CH$_4$ yields, ranging from 300 to 400 L CH$_4$ kg^{-1} volatile solids (VS = organic matter), and that the area specific CH$_4$ yield is more dependent on biomass yield (Table 4.2; Prochnow et al., 2009a).

A limited number of studies have provided information on the specific CH$_4$ yield of different temperate grass species grown under similar management conditions. In addition, only limited information on the chemical composition of these feedstocks has generally been provided. Mahnert et al. (2005) used fresh grass in batch digestion tests and reported that PRG (830 L biogas kg^{-1} VS) produced higher biogas yields (specific CH$_4$ yields not reported) than cocksfoot (740 L kg^{-1} VS) and meadow foxtail (720 L kg^{-1} VS). In a larger study, Kaiser and Gronauer (2007) reported specific CH$_4$ yields (L CH$_4$ kg^{-1} VS) for five ryegrass varieties

(198–330), three smooth meadow grass varieties (260–310), two meadow fescue varieties (310–320), cocksfoot (315) and timothy (345). Furthermore, Seppala et al. (2009) reported that there was no significant difference between the specific CH_4 yields (L CH_4 kg^{-1} VS) of cocksfoot (318), tall fescue (314) and timothy (311) grown under boreal conditions.

Information on the biomass yields of different grassland species in Europe is difficult to compile due to the diverse conditions in which experiments were carried out. In an extensive study comparing the productivity of PRG and timothy across 32 European sites, yields of PRG varied from almost 2 (Vila Real, Portugal) to 20 t DM ha^{-1} (Kiel, Germany) and were higher than timothy in most cases (Peeters and Kopec, 1996). Green et al. (1973) also reported that varieties of ryegrasses outyielded other common grass species at a given digestibility.

Growth stage and harvest date

The most important factor influencing the chemical composition of a specific herbage is the growth stage at harvest (Buxton, 1996). As a plant matures, the proportion of cell wall components increases, while the proportion of cell contents decreases (Figure 4.1; Stefanon et al., 1996).

This reflects the general decrease in the plant leaf to stem ratio and the increasing cell wall content within the stems (Buxton, 1996). Since this process is accompanied by increasing lignification within the cell wall fraction, there is an overall reduction in digestibility (Ugherughe, 1986). Thus, advancing maturity from the vegetative to the inflorescence growth stage is characterised by an increase in fibre components (Buxton, 1996) and a decrease in digestibility (Ballard et al., 1990), BC (Muck et al., 1991) and ash concentrations (Buxton, 1996). A much faster rate of decline in herbage quality is observed during the primary reproductive spring growth than during subsequent more vegetative regrowths (Balasko and Nelson, 2006).

Delaying the harvest date also has an important negative influence on CH_4 production from AD (Table 4.3; Gunaseelan, 1997; Prochnow et al., 2009a).

Most studies demonstrate a decreasing CH_4 yield from grasses and legumes with a more fibrous growth stage and this has been attributed to the increase in lignin concentration with advancing maturity, which is difficult to degrade under anaerobic conditions (Shiralipour and Smith, 1984).

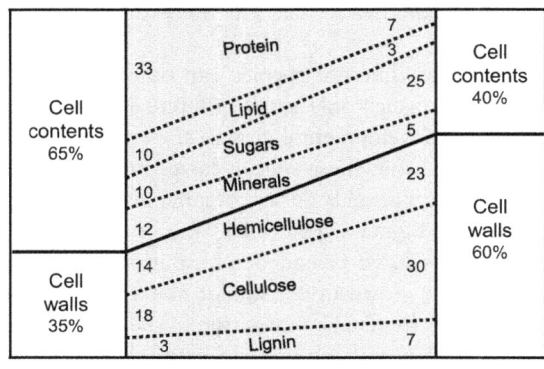

Figure 4.1 Schematic representation of changes in the chemical composition of grass with advancing maturity (adapted from Holmes, 1980)

As well as influencing herbage quality, harvest date also has a significant effect on herbage DM yield. Keating and O'Kiely (2000a) reported that delaying harvest date generally increased DM yield, but it simultaneously decreased the DM digestibility of old permanent grassland, PRG and IRG swards. The timing of the first silage cut is also of key importance in determining the total biomass yield (and thus CH_4 yield per hectare). In general, taking the first silage cut between the boot and early anthesis growth stages provides a good

Table 4.3 Effect of growth stage and harvest date on specific CH_4 yields from grassland.

Grassland management conditions	Specific CH_4 yield (L CH_4 kg^{-1} VS)	Digestion conditions	Reference
Clover: Vegetative growth stage Flowering stage	210 140	Laboratory/ 2 L batch/ 35°C/ 155 days	Kaparaju et al. (2002)
Landscape management grass, harvested monthly from June to February over 3-year period: Cut in June Cut in September Cut in February	 298 229 155	Laboratory/ 2 L batch/ 35°C/ 28 days	Prochnow et al. (2005)
Extensively managed hill site: 3 cuts per year, 1st cut inflorescence emerged 2 cuts per year, 1st cut anthesis 1 late cut in August Intensively managed valley site: 4 cuts per year, 1st cut stem elongation 3 cuts per year, 1st cut inflorescence emerged 3 cuts per year, 1st cut inflorescence emerged-anthesis	Cut 1/ 2/ 3/ 4 221/ 167/ 152 171/ 128 153 257/ 392/ 317/ 351 315/ 272/ 190 362/ 192/ 243	Laboratory/ 1 L batch/ 38°C/ 42 days	Amon et al. (2007)
Timothy-clover grass: 1st cut, vegetative growth stage 2nd cut, silage stage Red clover: 1st cut, vegetative growth stage 2nd cut, flowering stage	 370 380 300 280	Laboratory/ 2 L batch/ 35°C/ 124–146 days	Lehtomaki et al. (2008)
Cocksfoot: 1st cut, early flowering stage 2nd cut, vegetative regrowth Tall fescue: 1st cut, early flowering stage 2nd cut, vegetative regrowth Reed canary grass: 1st cut, early flowering stage 2nd cut, late flowering stage	 361 332 358 348 316 258	Laboratory/ 1 L batch/ 35°C/ 75–95 days	Seppala et al. (2009)
Switchgrass (ensiled): 1st cut mid-summer 2nd cut early fall	 266–309 269–276	Laboratory/ 20 L batch/ 35°C/ 70 days	Masse et al. (2010)

VS = volatile solids

compromise for yield and persistence of most temperate grass species, while also providing a high-quality feedstock. Cutting later in the primary growth gives a lower forage quality and slower subsequent regrowth, while an earlier harvest date can result in better quality forage but yield may be compromised (Gilliland et al., 1995; Balasko and Nelson, 2006). Furthermore, the cost of producing a grass feedstock for AD will decrease as the biomass yield increases (McEniry et al., 2011), but there must be an economic equilibrium between yield and digestibility. Gunnarsson et al. (2008) reported that compared with harvesting dates for milk production, it may be optimal to harvest grass for AD at a later date, since the lower biogas yield would be compensated for by higher biomass yields.

Most on-farm anaerobic digesters are designed to co-digest energy crops with liquid manure (Murphy et al., 2011). However, the tendency for grass particles to float can prove problematic for the mechanical mixing systems in these digesters (Thamsiriroj and Murphy, 2010). The scale of this problem may increase for herbage harvested at a more advanced growth stage, and a finer chop length and a more robust mixing system may be required to ensure proper mixing. Furthermore, the slower digestion rate of more mature (i.e. fibrous) material will result in a longer retention time in the digester if the CH_4 yield (per unit VS incubated) is to be maintained.

Environmental factors

Temperature and sunlight

Temperature is the driving force behind most physiological processes that occur in the plant including leaf expansion rate and floral initiation (Peacock, 1976), with considerable variation observed between grass species and varieties. In general, a rise in temperature increases the rate of plant development and reduces DM digestibility (Buxton, 1996). The optimum temperature range for growth of C_3 and C_4 grasses is 18–24°C and 30–35°C, respectively (Volenec and Nelson, 2003), and consequently C_4 plants dominate warmer climatic regions (Sage et al., 1999).

The length of the growing season is also strongly determined by temperature. As light intensity and duration increase in spring, soil and air temperatures begin to rise (Volenec and Nelson, 2003). The threshold soil temperature for grass growth in temperate regions is 6°C (Frame and Laidlaw, 2011), with little herbage being produced below this threshold. As spring temperature rises, the rate of grass growth accelerates. In Nordic countries, such as Finland, the growing season varies from north to south and lasts for 5 to 6 months (Seppala et al., 2009), while in more temperate countries such as Ireland it varies from north-east to south-west and lasts for 8 to 10 months (Frame and Laidlaw, 2011). In general, the longer the growing season the greater the annual grass production.

Solar radiation is closely associated with air temperature and provides the energy via photosynthesis that fuels grass growth. The impact of solar radiation depends on the combination of day length and the intensity of solar radiation, the latter varying with latitude, altitude and climate (Frame and Laidlaw, 2011). Differences in sunlight are thought to be less significant than other factors affecting herbage chemical composition, but can still be reflected in diurnal variations in plant chemical composition. For example, in temperate regions, the concentration of WSC typically increases from morning to evening with changes in temperature and sunlight (Buxton, 1996), and may provide an opportunity for matching herbage chemical composition to silage fermentation requirements via timing of harvest (Griggs et al., 2007).

Water

Adequate soil moisture is essential for plant growth and can have a substantial impact on herbage production potential (Hopkins and Wilkins, 2006). For example, Brereton and Keane (1982) estimated that DM yield reductions of 1.4 to 4.0 t/ha could be lost for intensively managed grassland in the drier south-east of Ireland due to limiting soil moisture availability. Soil moisture conditions are dependent on both weather conditions and soil physical characteristics and, as a result, exhibit large variation between soil types, regions, seasons and years (Schulte et al., 2005). Annual precipitation accounts for most of the variance in the primary production of grassland systems, with Morrison et al. (1980) and Sala et al. (1988) reporting that the year-to-year variation in grassland yield was determined by both annual precipitation and soil moisture availability.

Low soil moisture content, during the summer months for example, can lead to water stress and suboptimal grass growth (Allen et al., 1998). Water stress will affect many processes involved in herbage growth, from cell growth and division to the photosynthetic capacity of the leaves (Hsiao and Acevedo, 1974). Transpiration and nutrient uptake by the sward are also reduced, ultimately reducing plant growth. Furthermore, grass under water stress may decrease its leaf content (Laidlaw, 2009) and may seek to produce seed heads, thereby suffering a more rapid decrease in herbage DM digestibility (Han et al., 2003).

However, in some grass growing regions, excess rather than a lack of soil moisture can often be a problem. For example, prolonged waterlogging of poorly drained, heavy soils in high rainfall climates rapidly depletes soil oxygen, alters plant metabolic processes (Kozlowski, 1984) and reduces the uptake of nutrients (Huang et al., 1995), thereby limiting plant growth. Thomasson (1979) reported a 20–50 per cent reduction in herbage production, depending on the severity and duration of excess soil moisture. It should be noted however, that the influence of water stress on plant growth is also dependent on grass species (Table 4.1).

Soil fertility

Apart from climatic factors such as water and temperature, the availability of soil nutrients is the most likely factor limiting the productivity of temperate grasslands. Plants obtain their C, H and O (and N to some extent) requirements from air and water, while other essential plant nutrients (e.g. N, P, K, Ca, Mg and S) are supplied from the soil or from fertiliser inputs (Coulter and Lalor, 2008). Of these nutrients, N is the most limiting element for grass production, while P and K are required to maintain these production levels (Balasko and Nelson, 2003). Soil fertility is the capacity of the soil to provide these nutrients and is determined by soil chemical composition (i.e. nutrient status), soil physical structure (e.g. soil texture, organic matter, water supply) and soil biology (e.g. nutrient cycling, soil biota) (Barker and Collins, 2006).

Carbon dioxide

The increase in the concentration of atmospheric CO_2 and associated global warming has stimulated research into plant response to elevated CO_2. In general, elevated CO_2 is expected to increase grassland plant yield, root mass and leaf area, and to alter plant chemical composition (Campbell et al., 1997). For example, Casella and Soussana (1997) reported that elevated CO_2 resulted in a significant increase in WSC concentration in PRG. However, there is considerable variation in the direction and magnitude of this response and this can be influenced by plant development, species and availability of primary resources (Poorter and Perez-Soba, 2001). For example, a greater increase in DM production is expected for C_3 than C_4 plants (Newton, 1991) as C_4 plants already have a built-in mechanism to maintain an enriched CO_2 atmosphere in the bundle sheath (Newman et al. 2001). Furthermore, this growth response to elevated CO_2 is dependent on soil water availability (Newman et al. 2001), nutrient availability (Newman et al. 2006), and temperature and ozone concentration (Poorter and Perez-Soba, 2001).

Grassland management

Nutrient management

Fertilisation of grassland, in particular the application of N, is employed primarily to ensure that economically viable yields are available for harvesting at a time of adequate herbage quality (Keating and O'Kiely, 2000b). Nitrogen is an essential element for plant growth and development, and facilitates many functions including photosynthesis and enzyme synthesis (Addiscot, 2005). While the yield response is greatest for N application, P, K and regular applications of lime, are also required to maintain grassland productivity (Balasko and Nelson, 2003).

The main inputs of N to grassland are as animal manures, imported inorganic N fertiliser and fixation of atmospheric N by *Rhizobia* bacteria in the root nodules of legumes. Animal manure can be an effective source of plant nutrients, especially P and K, and application to grassland can be a safe method of use (Sanderson and Jones, 1997). However, a generally poorer crop response to manure, compared with inorganic N fertiliser, is reported due to NH_3 volatilisation, slow availability of organic N and the negative effect of particulate material adhering to grass leaves (Whitehead, 1995). On farms with an AD facility, manure is generally utilised as a feedstock for biogas production, with the resulting digestate reported to have an improved fertiliser value (Lukehurst et al., 2010). Forage legumes can also be incorporated into grassland swards to reduce the requirement for inorganic N fertiliser (Peeters et al., 2006) and yield benefits of grass–clover mixtures can be equivalent to fertiliser N inputs of 150–350 kg N ha^{-1} (Peyraud et al., 2009).

Most grass yield response curves show an almost linear increase in biomass yield up to application rates of 200 and 400 kg N ha^{-1}, beyond which the response declines (Sparrow, 1979; Whitehead, 1995). In addition to increasing biomass yields, N fertiliser is also considered an important tool for efficient management. For example, the increased growth rate following N application can reduce the growth period between grazing or silage cuts and can result in more harvests per year.

Nitrogen fertiliser can also impact on herbage quality and ensilability. Increasing the rate of N fertiliser application generally increases herbage CP concentration (Keady and O'Kiely, 1998) and BC (O'Kiely et al., 1997), and reduces herbage DM (Whitehead, 1995) and WSC concentrations (Keating and O'Kiely, 2000b). This increase in herbage BC will make the grass more difficult to preserve as silage than indicated by the decrease in WSC concentration alone (Keating and O'Kiely, 2000b). Although, some studies have reported moderate increases in grass fibre components in response to N fertiliser (Peyraud et al., 1997; Nordheim-Viken and Volden, 2009), this is generally considered to have little effect on digestibility (Keating and O'Kiely, 2000b; Buxton and O'Kiely, 2003). In contrast to N fertiliser, P and K applications have little direct effect on herbage quality and ensilability (O'Kiely and Tunney, 1997; Keady and O'Kiely, 1998).

Although inorganic N fertiliser can increase herbage yields, there are financial and environmental issues associated with very high rates of application. For example, the use of N fertiliser in agricultural systems is one of the biggest contributors to greenhouse gas emissions through fertiliser manufacture and N_2O emissions from soils (Dillon, 2010). The valuable nutrient content of animal manures and/or digestate, and the incorporation of legumes into grassland, all have the potential to reduce the requirement for inorganic N fertiliser resulting in positive financial and environmental gains (Gerin et al., 2008; Peyraud et al., 2009; Smyth et al., 2010; McEniry et al., 2011).

Ensilage

In order to ensure a predictable quality and a constant supply of feedstock to an AD facility, grass usually needs to be harvested and stored as silage. Preservation is achieved by the combination of an anaerobic environment and the bacterial fermentation of sugar, the lactic acid produced from the latter process lowering the pH and preventing the proliferation of spoilage microorganisms (Muck, 1988). The main objective of ensilage is the preservation of the crop at an optimum growth stage for later use during seasons when the fresh crop is unavailable (McDonald et al., 1991). Ideally crops for preservation as silage should have an adequate content of fermentable substrate in the form of WSC, a relatively low BC and a DM content above 200 g kg^{-1} (Buxton and O'Kiely, 2003). Among the temperate grasses, typically the higher concentration of WSC of the ryegrasses make them more suitable substrate for ensiling (Buxton and O'Kiely, 2003). Wilson and Collins (1980), in a study investigating the chemical composition of silages made from different grass genera, reported that 0.97 of the IRG samples preserved satisfactorily compared with 0.72 for PRG and with other genera being significantly lower. The more successful the silage preservation process, the greater the energy content of the feedstock conserved for CH_4 production.

Several authors have reported that the gross energy value of well-preserved silage is higher than that of the fresh parent material (Alderman et al., 1971; McDonald and Edwards, 1976). The specific CH_4 yield of some silages has also been reported to be higher than for the original parent material due to the formation of fermentation products (e.g. ethanol, 1,2-propanediol) with a higher potential CH_4 yield than the original fermentation substrates (Buswell and Mueller, 1952; Pakarinen et al., 2008; Herrmann et al., 2011). It has also been suggested that ensiling increases the rate of CH_4 formation (Heiermann et al., 2002) as some of the fermentation products produced act as precursors to CH_4 formation. However, the losses occurring during ensilage must also be taken into account. Potential losses during ensilage which could impact on CH_4 production potential include field losses, effluent production, fermentation losses in the silo and aerobic deterioration during storage and at feedout (Table 4.4; McDonald et al., 1991). Some of these losses are inevitable but the extent can be minimised by good farm management practices.

Plants maintain metabolic activity for extended periods after cutting, most notably the continued aerobic respiration of sugars to CO_2 and water (Moser, 1980). Plant proteolytic activity is also responsible for the hydrolysis of plant proteins to peptides, free amino acids and amines (Carpintero et al., 1979). The length of this aerobic field phase can have a significant impact on DM losses and in particular, the amount of fermentable substrate available for preservation. If crops are directly harvested and ensiled, these losses can be

Table 4.4 Summary of the potential dry matter (DM) losses that can occur during ensilage and at feedout.

Process	Causative factor(s)	Potential losses (g kg^{-1} DM harvested)	Classification
Respiration	Plant enzymes Aerobic microorganisms	10–20	Unavoidable
Fermentation	Lactic acid bacteria	20–40	Unavoidable
Effluent production or Field losses due to wilting	Low DM concentration or Weather Wilting technique	50–70 or 20–40	Mutually unavoidable
Secondary fermentation	Clostridia	0–50	Avoidable
Aerobic deterioration during storage	Aerobic microorganisms Filling time Sealing and compaction Crop ensilability	0–100	Avoidable
Aerobic deterioration at feedout	Aerobic microorganisms Silage quality DM concentration	0–100	Avoidable

Source: Woolford (1984); McDonald et al. (1991).

kept to a minimum (McDonald et al., 1991). However, if a period of wilting is employed, considerable DM losses can occur if weather conditions are unfavourable (Kormos and Chestnutt, 1968a, b). Wilting is often preferred as it increases herbage DM concentration, and this discourages an undesirable clostridial fermentation and reduces or eliminates the discharge of effluent (Woolford, 1990). In general, respiration and proteolysis can be minimised with a rapid wilt under dry conditions.

After the grass is harvested and sealed in a silo, air is still present between the plant particles enabling continued plant and microbial respiration (Weinberg and Muck, 1996). Under good storage conditions this stage is confined to a few hours and is characterised by the steadily diminishing effect of oxygen, as plant respiration quickly consumes the trapped oxygen with only a minor loss of sugars (Pahlow et al., 2003). The subsequent anaerobic environment created terminates plant respiration, inhibits the activity of aerobic microorganisms and creates conditions conducive to the growth and development of the epiphytic lactic acid bacteria (Pitt et al., 1985). However, if the silo is filled slowly or imperfectly sealed, excessive respiration can result in further energy losses and can delay the required onset of pH decline (Muck, 1988). Pakarinen et al. (2008) reported that under extremely suboptimal silage storage conditions more than half of the potential CH_4 yield can be lost.

Dry matter and energy losses arising from fermentation depend on the nutrients fermented and the microorganisms responsible (McDonald et al., 1991). Under good ensiling conditions, lactic acid bacteria become the dominant microbial population, producing mainly lactic acid as a fermentation product with a consequent decrease in pH (Weinberg and Muck, 1996). In general, silage which has undergone a desirable fermentation is characterised by a low pH, high lactic acid content and low concentrations of butyric acid and ammonia-N (Haigh and Parker, 1985). McDonald et al. (1973) reported that ensiled energy is almost completely recoverable in a closed lactic acid dominant fermentation system (Table 4.5).

Table 4.5 Fermentation efficiencies of the main fermentation pathways employed by silage microorganisms.

Microorganism	Pathway	Substrate	Product	% Recovery	
				Energy	DM
Lactic acid bacteria	Homofermentative	Glucose	2 lactate	96.9	100
Lactic acid bacteria	Heterofermentative	Glucose	1 lactate + 1 acetate	79.6	83
Lactic acid bacteria	Heterofermentative	Glucose	1 lactate + 1 ethanol	97.2	83
Clostridia		Glucose	1 butyrate	77.9	66
Enterobacteria		2 glucose	2 lactate + 1 acetate + 1 ethanol	88.9	83
Yeast		Glucose	2 ethanol	97.4	51

Source: McDonald et al. (1973)

Considering the amount of acid produced (i.e. indication of conservation efficiency) and the energy losses incurred, homofermentative lactic acid bacteria are the most efficient silage microorganisms producing two molecules of lactic acid from one molecule of glucose with minimal energy loss. In contrast, and despite the negligible loss of energy, the production of ethanol by yeast during fermentation is undesirable because no acidification occurs (Pahlow et al., 2003). Similarly, clostridia have a low acidification potential and cause large energy losses. Under sub-optimal ensiling conditions a secondary clostridial fermentation may lead to considerable DM and energy losses due to extensive production of CO_2 and H_2 from the fermentation of lactate and hexose sugars (McDonald et al., 1991).

The ensiling of wet crops (< 250 g DM kg^{-1}; Bastiman, 1976) can result in the production of large volumes of effluent and DM losses can exceed 100 g kg^{-1} (Weissbach and Peters, 1983). The degree of compaction in the silo and crop pretreatments can also influence effluent flow. Silage effluent is composed of a range of highly digestible components including carbohydrates and organic acids, and is also a major source of mineral elements (Patterson and Walker, 1979; Kemppainen, 1987). Effluent represents an excellent feedstock for AD (Barry and Colleran, 1982; Abu-Dahrieh et al., 2011), and thus losses associated with effluent production should be minimised by directing the silage effluent stream to the digester (McEniry et al., 2011).

Once silage is exposed to air at feedout, microorganisms which were dormant under anaerobic conditions can be reactivated, causing spoilage (McDonald et al., 1991). These microorganisms multiply in the presence of air and contribute to heating and to major chemical changes within the silage, indicated at its simplest by a reduction of lactic acid, a corresponding rise in pH and a substantial decrease in available energy value (Pahlow et al., 2003). Aerobic deterioration of silage is undesirable and energy losses can amount to 30–50 g DM kg^{-1} within one day's exposure to air (Honig and Woolford, 1980). Baserga and Egger (1997) reported that biogas yields from baled grass silage decreased from 500 L kg^{-1} VS at opening, to 370 and 250 L kg^{-1} VS after 5 and 30 days aerobic exposure, respectively.

The application of silage additives during ensiling is sometimes used to encourage beneficial microbial activity and/or inhibit detrimental microbial activity (Whittenbury,

1968; Kung et al., 2003). Silage additives can be employed to enhance silage fermentation and increase aerobic stability, thus reducing ensilage losses. Studies assessing the impact of silage additives on preservation and their ability to enhance CH_4 yields have proved inconclusive (Lehtomaki, 2006; Pakarinen et al., 2008; Plochl et al., 2009; Herrmann et al., 2011). In general, the potential economic benefits of increased CH_4 yield need to be considered in relation to the additional costs of additive application.

Concluding comments

The aim when providing a feedstock for biogas production is to achieve a high CH_4 yield per area of land. The two essential criteria are the biomass yield per hectare and the specific CH_4 yield per unit of biomass. The specific CH_4 yield will be determined by the available energy content of the herbage, and this will determine the rate of biogas production and the retention time of the feedstock in the digester.

The principles controlling conservation of grass as silage for biogas production are the same as those for producing high-quality conserved feed for ruminants. The more strictly these principles are employed and the more successful the silage conservation process, the greater the energy content of the feedstock conserved for biogas production.

References

Abu-Dahrieh, J., Orozco, A., Groom, E. and Rooney, D. (2011) 'Batch and continuous biogas production from grass silage liquor', *Bioresource Technology*, doi:10.1016/j.biortech.2011.09.072.

Addiscot, T.M. (2005) *Nitrate, agriculture and the environment*, Wallingford: CABI Publishing.

Alderman, G., Collins, F.C. and Dougall, H.W. (1971) 'Laboratory methods of predicting feeding value of silage', *Grass and Forage Science*, 26, 109–111.

Allen, R.G., Pereira, L.S., Raes, D. and Smith, M. (1998) 'Crop evapotranspiration. Guidelines for computing crop water requirements', FAO Irrigation and Drainage Paper 56, Rome: FAO.

Al Seadi, T., Rutz, D., Prassl, H., Kottner, M., Finsterwalder, T., Volk, S. and Janssen, R. (2008) *The biogas handbook*, Esbjerg: University of Southern Denmark.

Amon, T., Amon, B., Kryvoruchko, V., Machmuller, A., Hopfner-Sixt, K., Bodiroza, V., Hrbek, R., Friedel, J., Potsch, E., Wagentristl, H., Schreiner, M. and Zollitsch, W. (2007) 'Methane production through anaerobic digestion of various energy crops grown in sustainable crop rotations', *Bioresource Technology*, 98, 3204–3212.

Angelini, L.G., Ceccarini. L. and Bonari. E. (2005) 'Biomass yield and energy balance of giant reed (*Arundo donax* L.) cropped in central Italy as related to different management practices', *European Journal of Agronomy*, 22, 375–389.

Balasko, J.A. and Nelson, C.J. (2003) 'Grasses for northern areas', in R.F. Barnes, C.J. Nelson, M. Collins and K.J. Moore (eds) *Forages – an introduction to grassland agriculture*, Ames, IA: Blackwell Publishing.

Ballard, R.A., Simpson, R.J. and Pearce, G.R. (1990) 'Losses of the digestible components of annual ryegrass (*Lolium rigidum gaudin*) during senescence', *Australian Journal of Agricultural Research*, 41, 719–731.

Barker, D.J. and Collins, M. (2006) 'Forage fertilization and nutrient management', in R.F. Barnes, C.J. Nelson, M. Collins and K.J. Moore (eds) *Forages – an introduction to grassland agriculture*, Ames, IA: Blackwell Publishing.

Barry, M. and Colleran, E. (1982) 'Anaerobic digestion of silage effluent using an upflow fixed bed reactor', *Agricultural Wastes*, 4, 231–239.

Baserga, U. and Egger, K. (1997) *Anaerobic digestion of energy grass for biogas production*, Tonikon: Bundesamt für Energiewirtschaft, Forschungsprogramme Biomasse.

Bastiman, B. (1976) 'Factors affecting silage effluent production', *Experimental Husbandry*, 31, 40–46.
Brereton, A.J. and Keane, T. (1982) 'The effect of water on grassland productivity in Ireland', *Irish Journal of Agricultural Research*, 21, 227–248.
Burvall, J. (1997) 'Influence of harvest time and soil type on fuel quality in reed canary grass (*Phalaris arundinacea* L.)', *Biomass and Bioenergy*, 12, 149–154.
Buswell, A.M. and Mueller, H.F. (1952) 'Mechanism of methane fermentation', *Industrial & Engineering Chemistry*, 44, 550–552.
Buxton, D.R. (1996) 'Quality-related characteristics of forages as influenced by plant environment and agronomic factors', *Animal Feed Science and Technology*, 59, 37–49.
Buxton, D.R. and Fales, S.L. (1994) 'Plant environment and quality', in G.C. Fahey (ed.) *Forage quality, evaluation and utilization*, Madison, WI: American Society of Agronomy.
Buxton, D.R. and O'Kiely, P. (2003) 'Preharvest plant factors affecting ensiling', in D.R. Buxton, R.E. Muck and J.H. Harrison (eds) *Silage science and technology*, Madison, WI: American Society of Agronomy.
Campbell, B.D., Stafford Smith, D.M. and McKeon, G.M. (1997) 'Elevated CO_2 and water supply interactions in grasslands: A pastures and rangelands management perspective', *Global Change Biology*, 3, 177–187.
Carpintero, C.M., Henderson, A.R. and McDonald, P. (1979) 'The effect of some pre treatments on proteolysis during the ensiling of herbage', *Grass and Forage Science*, 34, 311–315.
Casella, E. and Soussana, J.F. (1997) 'Long-term effects of CO_2 enrichment and temperature increase on the carbon balance of a temperate grass sward', *Journal of Experimental Botany*, 48, 1309–1321.
Cerdan, O., Govers, G., Le Bissonnais, Y., Van Oost, K., Poesen, J., Saby, N., Gobin, A., Vacca, A., Quinton, J. and Auerswald, K. (2010) 'Rates and spatial variations of soil erosion in Europe: A study based on erosion plot data', *Geomorphology*, 122, 167–177.
Clifton-Brown, J.C., Stampfl, P.F. and Jones, M.B. (2004) 'Miscanthus biomass production for energy in Europe and its potential contribution to decreasing fossil fuel carbon emissions', *Global Change Biology*, 10, 509–518.
Coulter, B.S. and Lalor, S. (2008) *Major and micronutrient advice for productive agricultural crops*, Johnstone Castle: Teagasc.
Dillon, P. (2010) 'Managing European grasslands to increase the sustainability and competitiveness of livestock production systems', MULTISWARD, EU Seventh Framework Programme – Food, Agriculture and Fisheries, Biotechnology, <http://www.multisward.eu/multisward_eng/Output-deliverables>, accessed 13 January 2012.
FAO (2012) 'FAOSTAT' < http://faostat.fao.org/site/291/default.aspx>
Frame, J. and Laidlaw, A.S. (2011) *Improved grassland management*, Marlborough: Crowood Press.
Gerin, P.A., Vliegen, F. and Jossart, J.M. (2008) 'Energy and CO_2 balance of maize and grass as energy crops for anaerobic digestion', *Bioresource Technology*, 99, 2620–2627.
Gibon, A. (2005) 'Managing grassland for production, the environment and the landscape: Challenges at the farm and the landscape level', *Livestock Production Science*, 96, 11–31.
Gilliland, T.J., Camlin, M.S. and Johnston, J. (1995) 'Effect of harvest date and cultivar maturity on perennial ryegrass (*Lolium perenne* L.) yield and digestibility'. *Irish Journal of Agricultural and Food Research*, 34, 133–142.
Green, J.O., Corrall, A.J. and Terry, R.A. (1973) *Grass species and varieties – relationship between stage of growth, yield and forage quality*, Hurley: Grassland Research Institute.
Griggs, T.C., MacAdam, J.W., Mayland, H.F. and Burns, J.C. (2007) 'Temporal and vertical distribution of nonstructural carbohydrate, fiber, protein, and digestibility levels in orchardgrass swards', *Agronomy Journal*, 99, 755–763.
Gunaseelan, V.N. (1997) 'Anaerobic digestion of biomass for methane production: A review', *Biomass and Bioenergy*, 13, 83–114.
Gunnarsson, C., Vagstrom, L. and Hansson, P.A. (2008) 'Logistics for forage harvest to biogas production – timeliness, capacities and costs in a Swedish case study', *Biomass and Bioenergy*, 32, 1263–1273.

Haigh, P.M. and Parker, J.W.G. (1985) 'Effect of silage additives and wilting on silage fermentation, digestibility and intake, and on liveweight change of young cattle', *Grass and Forage Science*, 40, 429–436.

Han, D., O'Kiely, P. and Wen Sun, D. (2003) 'Application of water-stress models to estimate the herbage dry matter yield of a permanent grassland pasture sward regrowth', *Biosystems Engineering*, 84, 101–111.

Heiermann, M., Plochl, M., Linke, B. and Schelle, H. (2002) 'Preliminary evaluation of some cereals as energy crops for biogas production', in A.A.M. Sayigh (ed.), *Proceedings of the World Renewable Energy Congress VII*, Cologne: Pergamon.

Herrmann, C., Heiermann, M. and Idler, C. (2011) 'Effects of ensiling, silage additives and storage period on methane formation of biogas crops', *Bioresource Technology*, 102, 5153–5161.

Holmes, W. (1980) *Grass: Its production and utilisation*, Oxford: Blackwell Scientific Publications.

Honig, H. and Woolford, M.K. (1980) 'Changes in silage on exposure to air', in C. Thomas (ed.) *Forage conservation in the 80s*, Occasional Symposium No. 11, Hurley: British Grassland Society.

Hopkins, A. and Holz, B. (2006) 'Grassland for agriculture and nature conservation: Production, quality and multi-functionality', *Agronomy Research*, 4, 3–20.

Hopkins, A. and Wilkins, R.J. (2006) 'Temperate grassland: key developments in the last century and future perspectives', *Journal of Agricultural Science*, 144, 503–523.

Hsiao, T.C. and Acevedo, E. (1974) 'Plant responses to water deficits, water-use efficiency, and drought resistance', *Agricultural Meteorology*, 14, 59–84.

Huang, B., Johnson, J.W., Nesmith, D.S. and Bridges, D.C. (1995) 'Nutrient accumulation and distribution of wheat genotypes in response to waterlogging and nutrient supply', *Plant and Soil*, 173, 47–54.

Isselstein, J., Jeangros, B. and Pavlu, V. (2005) 'Agronomic aspects of biodiversity targeted management of temperate grasslands in Europe – a review', *Agronomy Research*, 3, 139–151.

Kaiser, F. and Gronauer, A. (2007) 'Methane potential of renewable resources in biogas plants', Freising: Bayerische Landesanstalt fur Landwirtschaft.

Kaparaju, P., Luostarinen, S., Kalmari, J. and Rintala, J. (2002) 'Co-digestion of energy crops and industrial confectionery by-products with cow manure: Batch-scale and farm-scale evaluation', *Water Science and Technology*, 45, 275–280.

Keady, T.W. and O'Kiely, P. (1998) 'An evaluation of potassium and nitrogen fertilization of grassland, and date of harvest, on fermentation, effluent production, dry-matter recovery and predicted feeding value of silage', *Grass and Forage Science*, 53, 326–337.

Keating, T. and O'Kiely, P. (2000a) 'Comparison of old permanent grassland, *Lolium perenne* and *Lolium multiflorum* swards grown for silage: 4. Effects of varying harvesting date', *Irish Journal of Agricultural and Food Research*, 39, 55–71.

Keating, T. and O'Kiely, P. (2000b) 'Comparison of old permanent grassland, *Lolium perenne* and *Lolium multiflorum* swards grown for silage: 3. Effects of varying fertiliser nitrogen application rate', *Irish Journal of Agricultural and Food Research*, 39, 35–53.

Kemppainen, E. (1987) 'Silage effluent: Nutrient content and capacity to reduce ammonia loss when mixed with urine or slurry', *Annales Agriculturae Finland*, 26, 95–105.

Kormos, J. and Chestnutt, D.M.B. (1968a) 'Measurement of dry matter losses in grass during the wilting period. (1) The effect of the length of the wilting period', *Record of Agricultural Research, Ministry of Agriculture for Northern Ireland*, 16, 145–150.

Kormos, J. and Chestnutt, D.M.B. (1968b) 'Measurement of dry matter losses in grass during the wilting period. (2) The effects of rain, mechanical treatment, maturity of grass and some other factors', *Record of Agricultural Research, Ministry of Agriculture for Northern Ireland*, 17, 59–65.

Korres, N.E., Singh, A., Nizami, A.S. and Murphy, J.D. (2010) 'Is grass biomethane a sustainable transport biofuel?', *Biofuels, Bioproducts and Biorefining*, 4, 310–325.

Kozlowski, T.T. (1984) 'Plant responses to flooding of soil', *BioScience*, 34, 162–167.

Kung, L., Stokes, M.R. and Lin, C.J. (2003) 'Silage additives', in D.R. Buxton, R.E. Muck and J.H. Harrison (eds) *Silage science and technology*, Madison, WI: American Society of Agronomy.

Laidlaw, A.S. (2009) 'The effect of soil moisture content on leaf extension rate and yield of perennial ryegrass', *Irish Journal of Agricultural and Food Research*, 48, 1–20.

Lattanzi, F.A. (2010) 'C_3/C_4 grasslands and climate change', *Grassland Science in Europe*, 15, 3–13.

Lehtomaki, A. (2006) 'Biogas production from energy crops and crop residues', PhD thesis, University of Jyvaskyla, Finland.

Lehtomaki, A., Huttunen, S., Lehtinen, T.M. and Rintala, J.A. (2008) 'Anaerobic digestion of grass silage in batch leach bed processes for methane production', *Bioresource Technology*, 99, 3267–3278.

Lewandowski, I., Scurlock, J.M.O., Lindvall, E. and Christou, M. (2003) 'The development and current status of perennial rhizomatous grasses as energy crops in the US and Europe', *Biomass and Bioenergy*, 25, 335–361.

Lukehurst, C., Frost, P. and Al Seadi, T. (2010) 'Utilisation of digestate from biogas plants as a biofertiliser', IEA Bioenergy – Task 37 Energy from Biogas.

Mahnert, P., Heiermann, M. and Linke, B. (2005) 'Batch and semi-continuous biogas production from different grass species', *Agricultural Engineering International: the CIGR Ejournal*, 7, 1–11.

Mandl, M.G. (2010) 'Status of green biorefining in Europe', *Biofuels, Bioproducts and Biorefining*, 4, 268–274.

Masse, D., Gilbert, Y., Savoie, P., Belanger, G., Parent, G. and Babineau, D. (2010) 'Methane yield from switchgrass harvested at different stages of development in eastern Canada', *Bioresource Technology*, 101, 9536–9541.

McDonald, P. and Edwards, R.A. (1976) 'The influence of conservation methods on digestion and utilization of forages by ruminants', *Proceedings of the Nutrition Society*, 35, 201–211.

McDonald, P., Henderson, A.R. and Ralton, I. (1973) 'Energy changes during ensilage', *Journal of the Science of Food and Agriculture*, 24, 827–834.

McDonald, P., Henderson, N. and Heron, S. (1991) *The biochemistry of silage*, Marlow: Chalchombe Publications.

McEniry, J., O'Kiely, P., Crosson, P., Groom, E. and Murphy, J.D. (2011) 'The effect of feedstock cost on biofuel cost as exemplified by biomethane production from grass silage', *Biofuels, Bioproducts and Biorefining*, DOI: 10.1002/bbb.322.

McLaughlin, S.B. and Kszos, L.A. (2005) 'Development of switchgrass (*Panicum virgatum*) as a bioenergy feedstock in the United States', *Biomass and Bioenergy*, 28, 515–535.

Morrison, J., Jackson, M.V. and Sparrow, P.E. (1980) *The response of perennial ryegrass to fertiliser nitrogen in relation to climate and soil*, Technical Report 27, Hurley: Grassland Research Institute

Moser, L.E. (1980) 'Quality of forage as affected by post-harvest storage and processing', in K.J. Moore and M.A. Peterson (eds) *Crop quality storage, and utilization*, Madison, WI: American Society of Agronomy.

Muck, R.E. (1988) 'Factors influencing silage quality and their implications for management', *Journal of Dairy Science*, 71, 2992–3002.

Muck, R.E., O'Kiely, P. and Wilson, R.K. (1991) 'Buffering capacities in permanent pasture grasses', *Irish Journal of Agricultural Research*, 30, 129–141.

Murphy, J.D., Braun, R., Weiland, P. and Wellinger, A. (2011) 'Biogas from crop digestion', IEA Bioenergy – Task 37 Energy from Biogas.

Newman, Y.C., Sollenberger, L.E., Boote, K.J., Allen, L.H., Thomas, J.M. and Littell, R.C. (2001) 'Carbon dioxide and temperature effects on forage dry matter production', *Agronomy Journal*, 41, 399–406.

Newman, Y.C., Sollenberger, L.E., Boote, K.J., Allen, L.H., Thomas, J.M. and Littell, R.C. (2006) 'Nitrogen fertilization affects bahiagrass responses to elevated atmospheric carbon dioxide', *Agronomy Journal*, 98, 382–387.

Newton, P.C. (1991) 'Direct effects of increasing carbon dioxide on pasture plants and communities', *New Zealand Journal of Agricultural Research*, 34, 1–24.

Nordheim-Viken, H. and Volden, H. (2009) 'Effect of maturity stage, nitrogen fertilization and seasonal variation on ruminal degradation characteristics of neutral detergent fibre in timothy (*Phleum pratense* L.)', *Animal Feed Science and Technology*, 149, 30–59.

O'Kiely, P. and Tunney, H. (1997) 'Silage conservation characteristics of grass that received a range of rates of phosphorus fertiliser', *Irish Journal of Agricultural and Food Research*, 36, 104.

O'Kiely, P., O'Riordan, E.G. and Maloney, A.P. (1997) 'Grass ensilability indices as affected by the form and rate of inorganic nitrogen fertiliser and the duration to harvesting', *Irish Journal of Agricultural and Food Research*, 36, 93.

Pahlow, G., Muck, R.E., Driehuis, F., Oude Elferink, S.J.W.H. and Spoelstra, S.F. (2003) 'Microbiology of ensiling', in D.R. Buxton, R.E. Muck and J.H. Harrison (eds) *Silage science and technology*, Madison, WI: American Society of Agronomy.

Pakarinen, O., Lehtomaki, A., Rissanen, S. and Rintala, J. (2008) 'Storing energy crops for methane production: Effects of solids content and biological additive', *Bioresource Technology*, 99, 7074–7082.

Patterson, D.C. and Walker, N. (1979) 'The use of effluent from grass silage in the diet of finishing pigs. I. Variation in composition of effluents', *Animal Feed Science and Technology*, 4, 263–274.

Peacock, J.M. (1976) 'Temperature and leaf growth in four grass species', *Journal of Applied Ecology*, 13, 225–232.

Peeters, A. and Hopkins, A. (2010) 'Climate change in European grasslands', *Grassland Science in Europe*, 15, 72–74.

Peeters, A., and Kopec, S. (1996) 'Production and productivity of cutting grasslands in temperate climates of Europe', *Grassland Science in Europe*, 1, 59–73.

Peeters, A., Parente, G. and Legall, A. (2006) 'Temperate legumes: Key species for sustainable temperate mixtures', *Grassland Science in Europe*, 11, 205–220.

Peyraud, J.L., Astigarraga, L. and Faverdin, P. (1997) 'Digestion of fresh perennial ryegrass fertilized at two levels of nitrogen by lactating dairy cows', *Animal Feed Science and Technology*, 64, 155–171.

Peyraud, J.L., Le Gall, A. and Luscher, A. (2009) 'Potential food production from forage legume-based systems in Europe: an overview', *Irish Journal of Agricultural and Food Research*, 48, 115–135.

Peyraud, J.L., Van Den Pol-Van Dasselaar, A., Dillon P. and Delaby, L. (2010) 'Producing milk from grazing to reconcile economic and environmental performances', *Grassland Science in Europe*, 15, 865–879.

Pitt, R.E., Muck, R.E. and Leibensperger, R.Y. (1985) 'A quantitative model of the ensilage process in lactate silages', *Grass and Forage Science*, 40, 279–303.

Plantureux, S., Peeters, A. and McCracken, D. (2005) 'Biodiversity in intensive grasslands: Effect of management, improvement and challenges,' *Agronomy Research*, 3, 153–164.

Plochl, M., Zacharias, H., Herrmann, C., Heiermann, M., Prochnow, A. and Potsdam, G. (2009) 'Influence of silage additives on methane yield and economic performance of selected feedstock', *Agricultural Engineering International: the CIGR Ejournal*, 11, 1–16.

Poorter, H. and Perez-Soba, M. (2001) 'The growth response of plants to elevated CO_2 under non-optimal environmental conditions', *Oecologia*, 129, 1–20.

Prochnow, A., Heiermann, M., Drenckhan, A. and Schelle, H. (2005) 'Seasonal pattern of biomethanisation of grass from landscape management', *Agricultural Engineering International: the CIGR Ejournal*, 7, 1–17.

Prochnow, A., Heiermann, M., Plochl, M., Linke, B., Idler, C., Amon, T. and Hobbs, P.J. (2009a) 'Bioenergy from permanent grassland – a review: 1. Biogas', *Bioresource Technology*, 100, 4931–4944.

Prochnow, A., Heiermann, M., Plochl, M., Amon, T. and Hobbs, P.J. (2009b) 'Bioenergy from permanent grassland – a review: 2. Combustion', *Bioresource Technology*, 100, 4945–4954.

Reid, R.L., Jung, G.A. and Thayne, W.V. (1988) 'Relationships between nutritive quality and fiber components of cool season and warm season forages: A retrospective study', *Journal of Animal Science*, 66, 1275–1291.

Sage, R.F., Wedin, D.A. and Li, M. (1999) 'The biogeography of C_4 photosynthesis, patterns and controlling factors', in R.F. Sage and R.K. Monson (eds) C_4 *plant biology*, Toronto: Academic Press.

Sala, O.E., Parton, W.J., Joyce, L.A. and Lauenroth, W.K. (1988) 'Primary production of the central grassland region of the United States', *Ecology*, 69, 40–45.

Sanderson, M.A. and Jones, R.M. (1997) 'Forage yields, nutrient uptake, soil chemical changes, and nitrogen volatilization from bermudagrass treated with dairy manure', *Journal of Production Agriculture*, 10, 266–271.

Schulte, R.P.O., Diamond, J., Finkele, K., Holden N.M. and Brereton, A.J. (2005) 'Predicting the soil moisture conditions of Irish grasslands', *Irish Journal of Agricultural and Food Research*, 44, 95–110.

Seppala, M., Paavola, T., Lehtomaki, I.A. and Rintala, J. (2009) 'Biogas production from boreal herbaceous grasses – specific methane yield and methane yield per hectare', *Bioresource Technology*, 100, 2952–2985.

Shiralipour, A. and Smith, P.H. (1984) 'Conversion of biomass into methane gas', *Biomass*, 6, 85–92.

Smyth, B.M., Smyth, H. and Murphy, J.D. (2010) 'Can grass biomethane be an economically viable biofuel for the farmer and the consumer?', *Biofuels, Bioproducts and Biorefining*, 4, 519–537.

Sparrow, P.E. (1979) 'The comparison of five response curves for representing the relationship between the annual dry-matter yield of grass herbage and fertilizer nitrogen', *Journal of Agricultural Science, Cambridge*, 93, 513–520.

Stefanon, B., Pell, A.N. and Schofield, P. (1996) 'Effect of maturity on digestion kinetics of water-soluble and water-insoluble fractions of alfalfa and brome hay', *Journal of Animal Science*, 74, 1104–1115.

Taube, F., Hermann, A. and Potsch, E.M. (2006) 'What are the consequences of producing energy crops in the European Union for grassland renovation and new forage production systems?', *Grassland Science in Europe*, 11, 463–471.

Thamsiriroj, T. and Murphy, J.D. (2010) 'The difficulties associated with mono-digestion of grass as exemplified by commissioning a pilot scale digester', *Energy and Fuels*, 24, 4459–4469.

Thomasson, A.J. (1979) 'The effect of drainage conditions on grassland production: Water control and grassland productivity', in *Proceedings of the Winter Meeting of the British Grassland Society*, Hurley: British Grassland Society.

Ugherughe, P.O. (1986) 'Relationship between digestibility of *Bromus inermis* plant parts', *Journal of Agronomy and Crop Science*, 157, 136–143.

Verchot, L., Krug, T., Lasco, R.D., Ogle, S. and Raison, J., Li. Y., Martino, D.L., McConkey, B.G. and Smith, P. (2006) 'Grassland', in *2006 IPCC Guidelines for national greenhouse gas inventories, Volume 4: Agriculture, forestry and other land use*, Geneva: IPCC.

Volenec, J.J. and Nelson, C.J. (2003) 'Environmental aspects of forage management', in R.F. Barnes, C.J. Nelson, M. Collins and K.J. Moore (eds) *Forages – an introduction to grassland agriculture*, Ames, IA: Blackwell Publishing.

Weinberg, Z.G. and Muck, R.E. (1996) 'New trends and opportunities in the development and use of inoculants for silage', *FEMS Microbiology Reviews*, 19, 53–68.

Weissbach, F. and Peters, G. (1983) 'Quantity, chemical composition and feed value of silage effluent,' *Feldwirtschaft*, 24, 78–81.

Whitehead, D.C. (1995) *Grassland nitrogen*, Wallingford: CAB International.

Whittenbury, R. (1968) 'Microbiology of grass silage', *Process Biochemistry*, 3, 27–31.

Wilson, R.K. and Collins, D.P. (1980) 'Chemical composition of silages made from different grass genera', *Irish Journal of Agricultural Research*, 19, 75–84.

Woolford, M.K. (1984) *The silage fermentation*, New York: Mercel Dekker.

Woolford, M.K. (1990) 'The detrimental effects of air on silage', *Journal of Applied Microbiology*, 68, 101–116.

Chapter 5

Maize and maize silage for biomethane production

Markus Neureiter

Institute for Environmental Biotechnology, BOKU – University of Natural Resources and Life Sciences, Vienna, Konrad Lorenz Str. 20, A-3430 Tulln an der Donau, Austria
E-Mail: markus.neureiter@boku.ac.at

Maize – an important and versatile crop

Maize (*Zea mays* L.) is one of the most intensively grown crops worldwide. According to FAOSTAT (2012) in 2010 maize was cultivated on an area of 162 million ha and the annual production (grains) was more than 818 Mt. Like other important crops such as sugar cane and sorghum, maize is a C_4-plant, which means that it uses a more efficient photosynthetic mechanism resulting in high biomass production (Byrt et al., 2011). Provided that conditions are optimal and suitable varieties are used, harvests of more than 10 t grains per ha (FAOSTAT, 2012) can be achieved. For whole crop biomass, as it is used to prepare silage for animal nutrition or biomethane production, yield values range from 30 to 78 t fresh material per ha (Amler, 2009). Advancements in breeding have led to a nearly fourfold increase in grain yield since the 1950s (Dhugga, 2007) and since the harvest index (ratio of grain yield to total aboveground biomass) for maize remained basically constant, it can be assumed that the total biomass yield increased by the same proportion (Lorenz et al., 2010).

Maize grains are a major source of food calories, especially in developing countries. Crop failure as well as increased use for livestock feed, as an energy crop and for industrial products recently led to increased prices and shortages (Shiferaw et al., 2011). In industrialized countries maize starch and its modifications are widely used as ingredients by the food industry. Glucose syrup or high fructose corn syrup, which are both produced by enzymatic hydrolysis of maize starch, are common sweeteners (Coker and Venkatasubramanian, 1985).

Starch and modified starches are also used for industrial purposes, for example the production of adhesives, agrochemicals, cosmetics, detergents, bioplastics and in the paper and textile industries (Ellis et al., 1998). Together with potatoes, wheat and tapioca, maize is one of the major crops for industrial starch production and also, in order to obtain the desired starch properties, special varieties which are not suitable for food purposes are grown. In addition, maize starch is the base material for glucose, which is not only a sweetener but also a major carbon source for the fermentation industry. Therefore many industrial fermentation products like bioethanol, lactic acid, citric acid, amino acids, antibiotics, etc. are currently based to a great extent on maize. Biorefinery concepts often include whole-crop wet-mill-based maize-processing facilities with the aim to produce starch for further processing. On the other hand, wet-milling technology also allows the use of grain germ for oil production and to recover proteins. Moreover the stems, leaves, stalks, and stover can be utilized either for heat and power generation or for further processing within a lignocellulosic biorefinery (Kamm et al., 2006).

Because of its advantages maize is a well-established commercial crop, which can be used for multiple purposes ranging from human nutrition and animal feed to industrial products and fuels. The importance of the crop creates intensive efforts in breeding and therefore nowadays varieties are available which are adapted for different regional and climatic requirements and exhibit variable properties.

The characteristics described above are also responsible for the fact that maize has been the first choice for energy crop production and has become the major feedstock for anaerobic digestion of energy crops. In addition to high yields and high energy content the pre-existing technology for cultivation, harvest and storage of the crop constitute the major factors for the popularity of maize as a feedstock for biogas facilities.

Maize for anaerobic digestion

Background

Until the late 1990s anaerobic digestion processes were basically used for the treatment of wastes like sewage sludge, industrial waste water, agricultural residues and solid biodegradable wastes from industry and municipalities (Braun, 2007). Although experiments on the anaerobic digestion of crops date back to the 1950s (Reinhold and Noak, 1956), the production of biomethane from energy crops on a larger scale was not common until the late 1990s. Before that time, maize and other crops were only occasionally used as a co-substrate, mostly in agricultural manure digesters, with the aim of enhancing gas yields and utilizing the installed capacity to a higher extent (Braun, 2007). With the background of diminishing fossil energy, rising energy prices and the need to produce renewable energy with respect to the reduction of greenhouse gas emissions in accordance with the Kyoto Protocol, the use of anaerobic digestion for the production of energy, and in particular electricity by the means of combined heat and power generation (CHP), gained importance. Changes in the legal framework in some countries triggered the emergence of energy-crop-based biogas digesters (Braun, 2007; Weiland, 2000). In particular, the Renewable Energy Sources Act in Germany (Erneuerbare-Energien-Gesetz, 2000) and the Green Electricity Act in Austria (Ökostromgesetz, 2002) provided attractive feed-in tariffs for electricity produced by anaerobic digestion of energy crops. Under this legislation tariffs were guaranteed for several years, which lowered the investment risks for the erection of new anaerobic digesters. Subsequently, the number of energy crop digesters considerably increased in these two countries and whole crop maize silage was very soon established as the major feedstock for this kind of facility (Weiland, 2006). In some cases, maize grains, grain silage and corn cob mix (CCM) were used because they contain more starch and less fibre, and therefore are more readily degraded during anaerobic digestion resulting in lower retention times. However, because of higher feedstock costs these products are predominately used for animal feed and only to a lesser extent as a substrate for anaerobic digestion. In addition, energy yields per ha are lower for these maize components compared with whole crop silages (Amon et al., 2007a).

The composition of whole crop maize as described in the literature is listed in Table 5.1. Variations are due to different varieties, growth conditions and harvest times. Usually total solids content as well as starch concentration will be considerably higher at later growth stages whereas the amount of water-soluble carbohydrates (WSC) will decline with ripeness of the crop.

Table 5.1 Yields and composition of whole crop maize. Variations in the range of the values are due to different growth conditions and harvest times (FM – fresh matter, TS – total solids, VS – volatile solids).

Parameter	Unit	Value	References
Harvest yield	t FM ha^{-1}	30.6–78.3	Amler (2009)
		39.9–55.5	Bruni et al. (2010a)
	t TS ha^{-1}	9.2–23.5	Amler (2009)
		10.2–17.5	Bruni et al. (2010a)
		15–30	Weiland (2006)
Total solids (TS)	g (100g FM)$^{-1}$	24.5–58.1	Amler (2009)
		21.8–40.5	Bruni et al. (2010a)
		23.6–33.5	McDonald et al. (1991)
		22–35	KTBL (2005)
Volatile solids (VS)	g (100g FM)$^{-1}$	20.8–39.0	Bruni et al. (2010a)
		22.5–31.7	McDonald et al. (1991)
		30.6–52.9	Amon et al. (2003)
		20.9–33.6	KTBL (2005)
Crude protein	g (100g TS)$^{-1}$	7.7–8.2	Bruni et al. (2010a)
		7.7–9.8	Amon et al. (2003)
Crude fibre	g (100g TS)$^{-1}$	9.9–26.8	Amler (2009)
		13.6–18.2	Bruni et al. (2010a)
		21.7–22.7	McDonald et al. (1991)
		16.0–21.8	Amon et al. (2003)
Crude fat	g (100g TS)$^{-1}$	2.2–3.8	Amon et al. (2003)
Starch	g (100g TS)$^{-1}$	1.7–50.1	Amler (2009)
		33.7–41.3	Bruni et al. (2010a)
		6.82–27.8	McDonald et al. (1991)
Water-soluble carbohydrates (WSC)	g (100g TS)$^{-1}$	8.0–30.7	McDonald et al. (1991)
Methane yield	l_N (kg VS)$^{-1}$	276–317	Amler (2009)
		247–375	Amon et al. (2003)
		300–380	Weiland (2006)
		296–312	KTBL (2005)

The net energy yield per hectare is a crucial factor affecting the economic viability of maize (Salter and Banks, 2009). Maize varieties and harvest time are key management parameters influencing this factor (Amon et al., 2003; Bruni et al., 2010a). Usually methane hectare yields tend to increase when maize is harvested close to maturity (Schittenhelm, 2008); however, depending on the ripening characteristics of the variety, higher yields can be achieved also at earlier growth stages (Amon et al., 2007b). A silage maize ripeness index (SRI) based on the ratio of total solids content of maize grains to total solids content of the residual parts of the crop has been proposed in order to determine the harvest time for ensiling. Thus, grain total solids content of 64 per cent, 33 per cent of starch in the total solids and an SRI between 2.55 and 2.9 are recommended in order to achieve maximum yields and optimal ensiling conditions (Amler, 2009).

Where the anaerobic digester is attached to a farm with animal husbandry the substrate is usually a mix of maize silage, other energy crops, manure and other agricultural residues, while facilities that are specifically designed to produce bioenergy also use monodigestion of maize larger than those commonly used for animal feed, and therefore a larger surface area is exposed to aerobic conditions after the silo has been opened. There is not much data available on mass balances of full-scale clamp silos for anaerobic digestion and it can be expected that anaerobic digester operators are often unaware of the material and energy losses during storage. Resch et al. (2007) found losses in total solids of 16.5 per cent when standard good practice was used. Improper conditions such as using unsuitable coverage of the silo resulted in total solids losses up to 27 per cent. In animal husbandry the use of additives in order to increase the aerobic stability of silages and to prevent spoilage and nutrient losses is often practised. Acetic acid has been shown to inhibit the growth of yeasts and fungi and therefore microbial additives often contain strains of *Lactobacillus buchneri*, which is able to convert lactic acid into acetic acid (Holzer et al., 2003). In a lab-scale study on the effect of chemical and microbial silage additives on various energy crops, Plöchl et al. (2009) assessed silage quality as determined by chemical parameters, storage losses and aerobic stability. In addition, methane yields per hectare and the costs and additional income related to the use of silage additives were considered. With regard to maize silage the use of additives resulted in income losses because the costs for these products could not be compensated by increased methane yields; however, when losses due to reduced aerobic stability were taken into account, chemical additives were found to significantly improve the economic returns.

While an improved aerobic stability will reduce storage losses, metabolites formed during the ensiling process may generally have an effect on methane formation during the anaerobic digestion process (Neureiter et al., 2005; Plöchl et al., 2009). Silage additives with special focus on energy crops are commercially available and two studies on the effect of silage additives on methane formation (Vervaeren et al., 2010; Herrmann et al., 2011) were published recently. The results of Vervaeren et al. (2010) suggest that additives composed of heterofermentative lactic acid bacteria or yeasts and enzymes are more suitable to improve methane yields because they may lead to the enhanced formation of intermediates for methanogenesis during the ensiling process. On the other hand, Herrmann et al. (2011) found in their survey that overall methane yields from treatments with different additives amount to similar values when storage losses are considered in the calculation.

Pre-treatment

Pre-treatment methods have been suggested in order to improve the anaerobic digestion of fibrous biomass (Hendriks and Zeeman, 2009; Bruni et al., 2010b). Reduction of particle size, steam treatment or pressure cooking with or without catalysts and the application of various chemicals have been mainly developed in order to enhance the enzymatic hydrolysis of lignocellulosic materials for bioethanol production (Kumar et al., 2009; Mosier et al., 2005; Wright, 1988). Lignocellulose basically consists of cellulose, hemicelluloses and lignin and because of its interwoven structure it is more recalcitrant to hydrolytic enzymes. The hydrolysis of these substances may become a rate limiting step during anaerobic digestion (Mata-Alvarez et al., 2000) and therefore pre-treatment has also been suggested for anaerobic digestion of wastes and biomass with high fibre content. Size reduction, thermal or chemical pre-treatments have been described to be effective with feedstocks like manure fibres (Angelidaki and Ahring, 2000; Bruni et al., 2010b) and fibrous herbages like grasses and sugarbeet tops (Seppälä et al., 2008). Treatments with alkali like NaOH or ammonia without additional feedstock. If operated without any additional feedstock, a typical medium size crop digestion facility with 500 kW$_{el}$ needs an annual input of approximately 9000 t maize silage, which requires an area of about 250 ha for cultivation (Resch et al., 2008).

During the anaerobic digestion of whole crop silage, digestible components of maize including the cellulose and hemicellulose fraction as well as fats and proteins are converted to biogas. The typical methane content of biogas from maize is 52 per cent (KTBL, 2005). Methane yields per ha whole crop maize have been specified with average values of 4570 m^3_N/ha (Amler, 2009) and 9886 m^3_N/ha (Braun, 2007), and up to maximum values of 9370 m^3_N/ha (Schittenhelm, 2008) and 12,390 m^3_N/ha (Amon et al., 2007b). This would be equivalent to a caloric value of 163,600–443,562 MJ per ha. The electrical efficiency of a CHP is in the range of 34–40 per cent (Scholwin et al., 2005) and therefore the energy efficiency of the overall process largely depends on whether useful applications for the secondary heat can be implemented. Especially in remote rural areas and during summer there is frequently not sufficient demand for the surplus heat.

While biogas from maize digestion is currently predominantly used to produce electricity and heat in a CHP, upgrading to methane for the use as gaseous vehicle fuel as well as a replacement for natural gas is generally seen as future development for the utilization of biogas (Weiland, 2010). Therefore these applications can also be expected to gain importance for the use of biogas from maize.

Storage and pre-treatment

Ensiling of maize for anaerobic digestion

Whole crop maize is harvested once a year (Figure 5.1) and therefore needs to be preserved in order to provide continuous operation of the anaerobic digester over the year. Ensiling (Figure 5.2) is a common method to preserve moist herbage for animal feed, and it is also preferred for the storage of moist energy crops. Compression of the material and an airtight sealing generate anaerobic conditions and promote the growth of autochthonous lactic acid bacteria, which convert free soluble carbohydrates into lactic acid. pH values around 4 and the anaerobic conditions prevent the growth of undesirable microorganisms like enterobacteria, clostridia, and moulds (McDonald et al., 1991). For large anaerobic digesters maize silage is usually stored in surface walled clamp silos sealed with plastic sheeting.

Storage losses in the ensiling process can occur during anaerobic conditions in the closed silo or due to aerobic deterioration as soon as the silo is opened to feed the material into the digester. Spoilage at anaerobic conditions is mainly due to the growth of saccharolytic *Clostridium* sp., which are able to convert lactic acid to butyric acid or proteolytic clostridia which release ammonia when degrading protein. Clostridial growth usually occurs in feedstocks with low total solids content (below 30 per cent) and low amounts of water soluble carbohydrates, which leads to insufficient acidification (McDonald et al., 1991).

Problems with the feedstock quality of maize are rare and therefore anaerobic losses are only of minor importance. Such clostridial losses may even go unnoticed, since the spoiled silage can result in higher methane yields per kg volatile solids than unspoiled maize (Bursche, 2011). However, this ignores the considerable loss of volatile solids that occurs during clostridial fermentation.

Aerobic spoilage due to the growth of yeasts and moulds is considered to have a large impact on the feedstock costs for anaerobic digesters, because the clamp silos are often

Figure 5.1 Harvest of whole crop maize (copyright: C. Resch)

Figure 5.2 Ensiling of whole crop maize; the material is filled into the silage clamp and compressed by the weight of the machinery (copyright: C. Resch)

were also found to improve methane production from maize straw or stover (Li et al., 2010; Zhong et al., 2011) but would interfere negatively with silage fermentation and therefore are not typically used on whole crop maize.

Whole crop maize silage total solids usually contains around 30 per cent starch and about 20 per cent fibres from stalks, leaves, cobs and husks (Table 5.1). Starch is readily hydrolysed in the anaerobic digestion process and therefore does not require any pre-treatment. Exposure to chemicals or high temperatures may even lead to feedstock losses and the formation of unwanted by-products like furanoids and other Maillard products. Although the utilization of the fibrous fraction could probably be improved, it is not common practice to pre-treat whole crop maize silages. For silage preparation the material is usually already chopped to a particle length < 25 mm during harvesting (McDonald et al., 1991). Bruni et al. (2010a) found that a further reduction of particle size to 2 mm resulted in an increase of the methane yield by 10 per cent. Results by Ellenrieder et al. (2010) indicate that a combination of size reduction and addition of amylase can reduce the viscosity of mashes from maize silages. This could help to decrease the energy consumption of stirrers and pumps. In addition to the investment costs it has, however, to be considered that size reduction can be an energy intensive process, which could impair the overall energy balance of the process. Cutting the material to a size of 3 mm using a knife mill would require an energy input of 20 kWh per t of silage (Kratky and Jirout, 2011). The effects of chopping length at feedstock harvest on harvest and transport costs, ensiling and silage quality and the anaerobic digestion process were assessed in detail by Herrmann (2010). According to these results a chopping length of 7–8 mm is considered to be optimal in most cases, while cutting to shorter sizes usually will not improve the overall process economy.

When ripe cereal grains are used as feedstock for anaerobic digestion, rolling or milling is recommended (Schumacher et al., 2006). Otherwise the hull will possibly remain intact and protect the easily digestible parts of the kernel and subsequently intact grains can be found in the digestate.

Frequently the pre-treatment methods for biogas feedstocks have been tested at a laboratory scale and often only results from batch fermentation tests are available to evaluate the effects. On the other hand, equipment for mechanical size reduction and disintegration is offered by several companies and such machines are already installed at full-scale anaerobic digesters in Germany in order to achieve higher biogas yields and to reduce the risk of scum formation.

Anaerobic digestion of maize – practical implementation and limitations

Anaerobic digester design and process parameters

Typical energy crop digestion facilities are in most cases designed as either a one- or two-step process with stirred tank reactors (Braun, 2007). In addition to the solid feedstock, either manure or process water, which is often recirculated within the process, is used to adjust a TS content of up to 10 per cent in the fermenter (Resch et al., 2008). For energy crop digestion, hydraulic retention times (HRT) of above 50 days for co-digestion with cattle manure and above 80 days for monodigestion or co-digestion with pig manure are recommended at maximum organic loading rates (OLR) of 3–4 kg VS m^{-3} d^{-1} (KTBL, 2005) and values from full-scale facilities using maize as feedstock are reported to be in that range (Laaber et al., 2005; Lindorfer et al., 2007; Resch et al., 2008). Relatively few publications discuss alternative digester concepts for maize, such as the use of leach bed reactors coupled with an anaerobic filter (Cysneiros et al., 2008); however, according to Gemmeke et al. (2009) several large-scale biogas facilities in Germany utilize solid-state fermentation processes with percolation technology for maize silage. One of the anaerobic digesters described in this report operates with seven leach bed reactors and maize silage as feedstock.

In general, maize silage is considered to be an unproblematic feedstock for anaerobic digestion. The composition of the substrate is in most cases favourable for the anaerobic digestion process. The C:N ratio can vary in the range 24–52 (Amon et al., 2007a) and usually this does not cause any difficulties. Compared with other energy crops like grass silages, major problems like blockage of feeding augers or the formation of scum layers are rarely reported. Still there are some peculiarities regarding the anaerobic digestion of maize, which will be addressed in the following paragraphs.

Temperature range and effects

Energy crop digestion processes are often conceived for the mesophilic temperature range (35–40°C). According to data collected in a survey of biogas facilities in Austria (Laaber et al., 2005), temperatures between 40 and 50°C are also common, especially for facilities that use maize grains in addition to whole crop silages. It has been shown that the anaerobic digestion of feedstocks with a high content of easily degradable carbohydrates at high organic loading rates can be exothermic and lead to a self-induced temperature increase in the digesters (Lindorfer et al., 2006). The effects of elevated temperature in the fermenter depend on whether the microbial population in the digester is able to adapt to the changed conditions. If no adaptation is possible, the methane production will stop completely and measurements like cooling and changes in substrate feed will be required in order to re-establish the process. In case of an adaptation of the microbial consortium, the anaerobic digestion process can be operated at elevated temperatures, however, in many cases this will be accompanied by the accumulation of propionic acid (Lindorfer et al., 2008).

Nutrient requirements

A problem associated with monodigestion of maize silage is the limitation of trace elements. Waste, manures and industrial waste waters usually contain plenty of metal ions and therefore with this kind of feedstocks a limitation of trace elements rarely occurs. Energy crop digestion – and monodigestion of maize in particular – may lead to a lack in elements like cobalt, nickel, molybdenum and iron as the concentrations of these elements in maize silage are below the recommended values for anaerobic digestion feedstocks (Hinken et al., 2008). Trace elements generally act as co-factors in enzymes that are important for the microbial metabolism. In the case of anaerobic digestion a lack of trace elements can therefore imply the malfunction of key enzymes of methanogenic archaea: e.g. the element cobalt is essential for the coenzyme M methyltransferase and nickel acts as cofactor for carbon monoxide dehydrogenase, methylreductase and other hydrogenases (Pobeheim et al., 2011; Somitsch, 2007; Schattauer et al., 2011). The absence of crucial co-factors may therefore result in complete cessation of methanogenesis, which will not only cause a decline in biogas formation but also lead to the acidification of the digester contents due to the accumulation of organic acids. Acidification induced by a lack of trace elements has been shown to change the microbial population during monodigestion of maize silage (Munk et al., 2010). In this study under-supply with Co (below 0.03 mg/L) and Na (below 10 mg/L) led to less diversity in the methanogenic population followed by acidification. While methanogens belonging to the family Methanosarcinaceae and the order Methanobacteriales are predominant during stable operation, after acidification the order Methanomicrobiales appears to be dominant. Negative effects due to trace element deficiency have been mainly reported for monodigestion of maize (Lebuhn et al., 2008) or model substrates that are similar to maize (Pobeheim et al., 2011). Anaerobic digestion facilities using a mixture of energy crops and/or manure are obviously less likely to undergo an under-supply with trace elements (Schattauer et al., 2011) and also grass silages contain significantly higher amounts of the critical elements and are therefore less prone to acidification (Munk et al., 2010). In order to overcome problems during monodigestion of maize therefore changes in the feedstock mix (e.g. the use of co-substrates) or the supplementation with trace elements is suggested.

Use and treatment of digestate

The best use of the digested material is as a fertilizer for crops. In contrast to waste digesting facilities there is usually no problem in distributing the digestate from crop digesters on surrounding fields, especially if the crops for the anaerobic digestion are supplied locally. In this case the reduction of mineral fertilizers will lower the energy input for the whole process and as a consequence the energy efficiency is enhanced while the negative ecological impacts are reduced. An overview on composition of digestate from energy crop digesters is given in Table 5.2.

It has to be considered that 1 t of feedstock (maize silage with 30 per cent total solids) will generate approximately 0.8–1.2 t of digestate with a content of total solids of only 6–9 per cent (Fuchs and Drosg, 2010). At larger facilities it is therefore not possible to distribute the complete digestate on surrounding fields because there are legal limits on the amounts of phosphorous and nitrogen that can be applied per ha within a certain time frame. According to Weiland (2008), transport costs for digestate beyond a distance of 10 km already exceed

Table 5.2 Composition and properties of digestate from energy crop digesters (Fuchs and Drosg, 2010).

Feedstock		Energy crops (including maize) with manure	Energy crops (including maize) without manure
pH		7–8	7–8
Chemical oxygen demand (COD)	kg t^{-1}	50–100	50–120
Total solids (TS)	%	6.0–9.0	6–9.5
Volatile solids (VS)	%	4.5–7	4.5–7.5
VS/TS	%	70–80	70–80
Total nitrogen (TN)	kg t^{-1}	4.5–10	3.5–6.5
TN/TS	kg (t TS)$^{-1}$	60–120	40–75
Ammonium-nitrogen (NH$_4$-N)	kg t^{-1}	2.5–6	1–4
NH$_4$-N/TS	kg (t TS)$^{-1}$	30–80	15–50
NH$_4$-N/TN	%	50–70	30–70
Volatile fatty acids (VFA)	mg L^{-1}	50–1,000	20–3,000

the fertilizer value in countries like Germany. Accordingly, the local utilization of the digestate is problematic for large energy crop digesters with electric power above 1 MW due to costs for storage and transportation. A possibility to overcome these problems is to dewater the digestate with, for example, a screw press in a first step. In crop digesters that do not use manure as co-substrate, a large part of the liquid phase may be recirculated as process liquid (Resch et al., 2008). Excessive nitrogen loads are a limiting factor for the application of digestates on fields. As the liquid phase contains a high amount of ammonia nitrogen, methods have been developed in order to concentrate ammonia or ammonium salts by means of evaporation or diverse membrane processes (Fuchs and Drosg, 2010). Concentrated ammonium salts from the liquid fraction of the digestate can be used instead of synthetic nitrogen fertilizers while the solid fraction is often dried and applied as fertilizer or soil conditioner. A general evaluation of methods for digestate treatment has been conducted by Fuchs and Drosg (2010) and also life cycle assessments for different digestate treatment methods are available (Rehl and Müller, 2011). Apart from the efficiency of the digestate treatment, the energy consumption and the reduction of emissions of ammonia or nitrous oxide during and after fertilization are major items that have to be considered.

Process evaluation

Environmental assessment and criticism

The main goal of encouraging anaerobic digestion of energy crops was to increase the production of environmentally friendly energy; however, it may be assumed that the strengthening of rural areas by creating an additional income for agriculture was a desired side-effect, since reduced revenues from animal husbandry and a general restructuring in agriculture entailed an increase in fallow land, which previously had been used as pastures. In the meantime the cultivation of energy crops has become an object of criticism. This is true for maize in particular, since the production of biomethane and bioethanol created an additional demand for maize in the recent years. Concerns are related to the energy input and greenhouse gas emissions during cultivation and due to land-use change (Kim et al., 2009; Barnett, 2010) and the reduction of biodiversity because of the increase of maize monocropping (Fletcher et al., 2011; Rösch and Skarka, 2008).

Compared with other crops, maize is a quite demanding crop with regard to requirements for fertilizer and herbicides. Fossil energy input for the production of fertilizer and direct or indirect nitrous oxide emissions from cropped soil are considered to impair the greenhouse gas balance of the overall process (Adler et al., 2007; van Groenigen et al., 2004). Results from a recent study (Meyer-Aurich et al., 2012) indicate that the production of electricity by anaerobic digestion of energy crops can nevertheless contribute to the mitigation of greenhouse gas emissions. Feedstock composition, utilization of heat energy, covering of the digestate storage and land-use change were identified as parameters that influence greenhouse gas emissions in anaerobic digesters. Facilities using manure as co-feedstock were found to be more beneficial because in this case GHG emissions from undigested manure can be reduced. According to this study, about 60 per cent of the GHG emissions of the biogas process are due to production and processing of maize and the total gross GHG emissions per hectare of maize production amounts to 3100 kg CO_2 equivalent per ha. For monodigestion facilities the effect on GHG emission may not be positive at all, if land-use change effects are in the range of 50 per cent of the area needed or above.

Monocropping of maize therefore appears to have a negative ecological impact and without crop rotation the system is also more prone to diseases and economic losses due to crop failures. In order to establish a more sustainable cultivation for maize, cropping systems with improved energy efficiency (Alluvione et al., 2011), intercrops with other energy crops like sunflower (Nassab et al., 2011) and crop rotation systems (Amon et al., 2007b) have been proposed.

In order to evaluate the energy efficiency of the overall biogas production process, all energy inputs and outputs have to be taken into account. As described above, anaerobic digestion of whole crop maize silage produces a high gross energy yield per ha due to high biomass yields and a good conversion efficiency in the anaerobic digestion process. On the other hand, the energy demand for feedstock production, transport, processing and conversion has to be considered as well (Salter and Banks, 2009). In the literature, estimates for energy input of 31 GJ/ha (Salter and Banks, 2009), 34 GJ/ha (Schumacher et al., 2010) and 70 GJ/ha (Braun, 2007) can be found. For energy efficiency it is also important to consider whether the digestate can be used for fertilization and whether the methane is used in a CHP or as a vehicle fuel (Salter and Banks, 2009; Schumacher et al., 2010). Pöschl et al. (2010) also consider factors like the scale of the facility, and transport and distribution distances, in their calculations. Depending on the scenario and the underlying estimations the output/input energy ratio for anaerobic digestion of whole crop maize is between 3 and 6.5, which makes the process more efficient compared with other selected energy crops (Braun, 2007) and other biomass conversion processes like bioethanol (Hoover and Abraham, 2009).

Economic considerations

A detailed economic assessment would exceed the scope of this chapter and most factors regarding the costs for anaerobic digestion of maize correspond with economic data for anaerobic digestion facilities in general and for energy crop digesters in particular. Therefore only selected aspects specifically regarding the economy of biogas production from maize are discussed below.

Feedstock supply is an important factor for the economic viability of energy crop digestion and the related costs are dependent on market prices for different crops as well as on the available agricultural crop land and harvest yields (Toews, 2009). Other relevant cost factors in connection with feedstock supply are harvesting and transport. Optimized systems for crop handling during harvest and transport can significantly reduce the costs for feedstock supply (Stürmer et al., 2011). While – compared with other energy crops – whole crop maize silage from monocultures appears to be the most cost effective feedstock, due to ecological reasons and with respect to sustainability, the implementation of crop rotation systems is increasingly considered and in some cases even legally required (Stürmer et al., 2011; Toews, 2009).

Crop usage and transport range also affect the costs or revenues for delivering digestate as a fertilizer. The possibilities for digestate utilization are also strongly linked to the mode of feedstock supply, such as local production and direct delivery from the field vs. external acquisition of energy crops (Walla and Schneeberger, 2008; Toews, 2009).

An important factor influencing the cost efficiency of energy crop digesters is the facility size. Fermenter size and the installed electric capacity affect transport costs for feedstock and digestate, investment costs and the efficiency of electric power generation (Walla and Schneeberger, 2008). Since the production of energy from crops is currently not competitive without governmental support, the optimal size of energy crop digesters is strongly dependent on the respective national regulations for feed-in tariffs and subsidies for the erection of new facilities (Toews, 2009; Walla and Schneeberger, 2008).

Summary and conclusions

Because of high biomass and methane yields per hectare and well-established techniques for cultivation, harvest and storage, maize is often a superior crop for the production of biomethane. This substrate allows stable operation of the anaerobic digester and problems are only reported for monodigestion and for high loading rates.

Since maize is also an important crop for human nutrition, animal feed and industrial use, intensified utilization for energy crops will contribute to an increased demand. Therefore shortages lead to increased prices and competition between the various uses. Maize monocultures have been criticized from an ecological point of view because of their impact on landscape and biodiversity and their increased susceptibility to pests. An important factor is the energy input for the cultivation and transport of maize. The ecological impact of an anaerobic digester can vary significantly depending on the facility size, transport distances and the utilization of the digestate.

It can be expected that maize will remain an important substrate or co-substrate for biomethane production. Efforts to improve the entire process are being undertaken in the areas of crop production, storage and pre-treatment, and fermentation, and in the use of digestate. The profitability of anaerobic digestion of maize, however, will also strongly depend on legal pre-conditions and in most cases the production of electricity by anaerobic digestion of energy crops will not be economically feasible without subsidies. In addition the development of prices due to increased competition between the different uses for the crop is expected to have an impact on the preconditions for the production of biomethane from maize.

References

Adler, P.R., Del Grosso, S.J. and Parton, W.J. (2007) 'Life-cycle assessment of net greenhouse-gas flux for bioenergy cropping systems', *Ecological Applications*, vol 17, no 3, pp 675–691.

Alluvione, F., Moretti, B., Sacco, D. and Grignani, C. (2011) 'EUE (energy use efficiency) of cropping systems for a sustainable agriculture', *Energy*, vol 36, pp 4468–4481.

Amler, R. (2009) 'Der Einfluss der Reifedifferenz von Korn zu Restpflanze auf den optimalen Erntezeitpunkt und die standortgerechte Sortenwahl von Energie- und Silomais', *Gesunde Pflanzen*, vol 61, pp 57–71.

Amon, T., Kryvoruchko, V., Amon, B., Buga, S., Mayer, K., Zollitsch, W. and Pötsch, E. (2003) 'Optimierung der Biogaserzeugung aus den Energiepflanzen Mais und Kleegras-/Feldfuttermischungen', http://umwelt.lebensministerium.at/article/articleview/50010/1/14986, accessed 26 October 2011.

Amon, T., Amon, B., Kryvoruchko, V., Zollitsch, W., Mayer, K. and Gruber, L. (2007a) 'Biogas production from maize and dairy cattle manure – Influence of biomass composition on the methane yield', *Agriculture, Ecosystems and Environment*, vol 118, pp 173–182.

Amon, T., Amon, B., Kryvoruchka, V., Machmüller, A., Hopfner-Sixt, K., Bodiroza, V., Hrbek, R., Friedel, J., Pötsch, E., Wagentristl, H., Schreiner, M. and Zollitsch, W. (2007b) 'Methane production through anaerobic digestion of various energy crops grown in sustainable crop rotations', *Bioresource Technology*, vol 98, pp 3204–3212.

Angelidaki, I. and Ahring, B. (2000) 'Methods for increasing the biogas potential from the recalcitrant organic matter contained in manure', *Water Science and Technology*, vol 41, pp 189–194.

Barnett, M.O. (2010) 'Biofuels and greenhouse gas emissions: green or red?', *Environmental Science and Technology*, vol 44, no 14, pp 5330–5331.

Braun, R. (2007) 'Anaerobic digestion: a multi-faceted process for energy, environmental management and rural development', in P. Ranalli (ed.) *Improvement of crop plants for industrial end uses*, pp 335–416. Dordrecht: Springer.

Bruni, E., Jensen, A.P., Pedersen, E.S. and Angelidaki, I. (2010a) 'Anaerobic digestion of maize focusing on variety, harvest time and pretreatment', *Applied Energy*, vol 87, pp 2212–2217.

Bruni, E., Jensen, A.P., Pedersen, E.S. and Angelidaki, I. (2010b) 'Comparative study of mechanical, hydrothermal, chemical and enzymatic treatments of digested biofibers to improve biogas production', *Bioresource Technology*, vol 101, pp 8713–8717.

Bursche, J. (2011) 'Silage preservation and biogas production. Effects of the addition of lactic acid bacteria and clostridium tyrobutyricum to energy crops', Doctoral thesis, University of Natural Resources and Life Sciences, Vienna, Austria.

Byrt, C.S., Grof, C.P.L. and Furbank, R.T. (2011) 'C_4 plants as biofuel feedstocks: Optimising biomass production and feedstock quality from a lignocellulosic perspective', *Journal of Integrative Plant Biology*, vol 53, no 2, pp 120–135.

Coker, L.E. and Venkatasubramanian (1985) 'Starch conversion processes', in M. Moo Young (ed.) *Comprehensive biotechnology Volume 3*, pp 777–787. Oxford: Pergamon Press, United Kingdom,

Cysneiros, D., Banks, C.J. and Heaven, S. (2008) 'Anaerobic digestion of maize in coupled leach-bed and anaerobic filter reactors', *Water Science and Technology*, vol 58, no 7, pp 1505–1511.

Dhugga, K.S. (2007) 'Maize biomass yield and composition for biofuels', *Crop Science*, vol 47, pp 2211–2227.

Ellenrieder, J., Schieder, D., Mayer, W. and Faulstich, M. (2010) 'Combined mechanical enzymatic pretreatment for an improved substrate conversion when fermenting biogenic resources', *Engineering in Life Sciences*, vol 10, no 6, pp 544–551.

Ellis, R.P., Cochrane, M.P., Dale, M.F.B., Duffus, C.M., Lynn, A., Morrison, I.M., Prentice, R.D., Swanston, J.S. and Tiller, S.A. (1998) 'Starch production and industrial use', *Journal of the Science in Food and Agriculture*, vol 77, pp 289–311.

Erneuerbare-Energien-Gesetz (2000) Gesetz für den Vorrang Erneuerbarer Energien vom 29.03.2000. German BGBl I: 305.

FAOSTAT (2012) 'Statistical databases and data-sets of the Food and Agriculture Organization of the United Nations', http://faostat.fao.org, accessed 11 May 2012.

Fletcher, R.J., Robertson, B.A., Evans, J., Doran, P.J., Alavalopati, J.R.R. and Schemske, D.W. (2011) 'Biodiversity conservation in the era of biofuels: risks and opportunities', *Frontiers in Ecology and the Environment*, vol 9, no 3, pp 161–168.

Fuchs, W. and Drosg, B. (2010) *Technologiebewertung von Gärrestbehandlungs- und Verwertungskonzepten*, Vienna: University of Natural Resources and Life Sciences.

Gemmeke, B., Rieger, C. and Weiland, P. (2009) *Biogas-Messprogramm II, 61 Biogasanlagen im Vergleich*, Gülzow: Fachagentur Nachwachsende Rohstoffe e.V.

Hendriks, A.T.W.M. and Zeeman, G. (2009) 'Pretreatments to enhance the digestibility of lignocellulosic biomass', *Bioresource Technology*, vol 100, pp 10–18.

Herrmann, C. (2010) *Ernte und Silierung pflanzlicher Substrate für die Biomethanisierung – Prozessgrundlagen und Bewertung*, Göttingen: Cuvillier Verlag.

Herrmann, C., Heiermann, M. and Idler, C. (2011) 'Effects of ensiling, silage additives and storage period on methane formation of biogas crops', *Bioresource Technology*, vol 102, pp 5153–5161.

Hinken, L., Urban, I., Haun, E., Urban I, Weichgrebe, D. and Rosenwinkel, K.-H. (2008) 'The valuation of malnutrition in the mono-digestion of maize silage by anaerobic batch tests', *Water Science and Technology*, vol 58, no 7, pp 1453–1459.

Holzer, M., Mayrhuber, E., Danner, H. and Braun, R. (2003) 'The role of *Lactobacillus buchneri* in forage preservation', *Trends in Biotechnology*, vol 21, no 6, pp 282–287.

Hoover, F.-A. and Abraham, J. (2009) 'A comparison of corn-based ethanol with cellulosic ethanol as replacements for petroleum-based fuels: a review', *International Journal of Sustainable Energy*, vol 28, no 4, pp 171–182.

Kamm, B., Kamm, M., Gruber, P.R. and Kromus, S. (2006) 'Biorefinery systems – an overview', in B. Kamm, P.R. Gruber and M. Kamm (eds) *Biorefineries: Industrial processes and products*, Weinheim: Wiley-VCH.

Kim, H., Kim, S. and Dale, B.E. (2009) 'Biofuels, land use change, and greenhouse gas emissions: Some unexplored variables', *Environmental Science and Technology*, vol 43, no 3, pp 961–967.

Kratky, L. and Jirout, T. (2011) 'Biomass size reduction machines for enhancing biogas production', *Chemical Engineering Technology*, vol 34, no 3, pp 391–399.

KTBL – Kuratorium für Technik und Bauwesen in der Landwirtschaft (ed.) (2005) *Gasaubeute in landwirtschaftlichen Biogasanlagen*, pp 11–15. Münster: Landwirtschaftsverlag.

Kumar, P., Barrett, D.M., Delwiche, M.J. and Stroeve, P. (2009) 'Methods for pretreatment of lignocellulosic biomass for efficient hydrolysis and biofuel production', *Industrial and Engineering Chemistry Research*, vol 48, no 8, pp 3713–3729.

Laaber, M., Kirchmayr, R., Madlener, R. and Braun, R. (2005) 'Development of an evaluation system for biogas plants', in B.K. Ahring and H. Hartmann (eds) *Conference proceedings 4th International Symposium on Anaerobic Digestion of Solid Waste*, vol 1, pp 631–635, Copenhagen, Denmark.

Lebuhn, M., Liu, F., Heuwinkel, H. and Gronauer, A. (2008) 'Biogas production from mono-digestion of maize silage – long-term process stability and requirements', *Water Science and Technology*, vol 58, no 8, pp 1645–1651.

Li, L., Yang, X., Li, X., Zheng, M., Chen, J. and Zhang, Z. (2010) 'The influence of inoculum sources on anaerobic biogasification of NaOH-treated corn stover', *Energy sources, Part A*, vol 33, no 2, pp 138–144.

Lindorfer, H., Braun, R. and Kirchmayr, R. (2006) 'Self-heating of anaerobic digesters using energy crops', *Water Science and Technology*, vol 53, no 8, pp 159–166.

Lindorfer, H., Pérez López, C., Resch, C, Braun, R. and Kirchmayr, R. (2007) 'The impact of increasing energy crop addition on process performance and residual methane potential in anaerobic digestion', *Water Science and Technology*, vol 56, no 10, pp 55–63.

Lindorfer, H., Waltenberger, R., Köllner, K., Braun, R. and Kirchmayr, R. (2008) 'New data on temperature optimum and temperature changes in energy crop digesters', *Bioresource Technology*, vol 99, pp 7011–7019.

Lorenz, A.J., Gustafson, T.J., Coors, J.G. and de Leon, N. (2010) 'Breeding maize for a bioeconomy: A literature survey examining harvest index and stover yield and their relationship to grain yield', *Crop Science*, vol 50, pp 1–12.

Mata-Alvarez, J., Macé, S. and Llabres, P. (2000) 'Anaerobic digestion of organic solid wastes. An overview of research achievements and perspectives', *Bioresource Technology*, vol 83, pp 37–46.

McDonald, P., Henderson, A.R. and Heron, S.J.E. (1991) *The biochemistry of silage*, 2nd edition, Marlow: Chalcombe Publications.

Meyer-Aurich, A., Schattauer, A., Hellebrand, H.J., Klauss, H., Plöchl. M. and Berg, W. (2012) 'Impact of uncertainties on greenhouse gas mitigation potential of biogas production from agricultural resources', *Renewable Energy*, vol 37, pp 277–284.

Mosier, N., Wyman, C., Dale, B., Elander, R., Lee, Y.Y., Holtzapple, M. and Ladisch, M. (2005) 'Features of promising technologies for pretreatment of lignocellulosic biomass', *Bioresource Technology*, vol 96, pp 673–686.

Munk, B., Bauer, C., Gronauer, A. and Lebuhn, M. (2010) 'Population dynamics of methanogens during acidification of biogas fermenters fed with maize silage', *Engineering in Life Sciences*, vol 10, no 6, pp 496–508.

Nassab, A.D.M., Amon, T. and Kaul, H.-P. (2011) 'Competition and yield in intercrops of maize and sunflower for biogas', *Industrial Crops and Products*, vol 34, pp 1203–1211.

Neureiter, M., dos Santos, J.T.P., Perez Lopez, C., Pichler, H., Kirchmayr, R. and Braun, R. (2005) 'Effect of silage preparation on methane yields form whole crop maize silages', in B.K. Ahring and H. Hartmann (eds) *Conference proceedings 4th International Symposium on Anaerobic Digestion of Solid Waste*, vol 1, pp 109–115, Copenhagen, Denmark.

Ökostromgesetz (2002) Austrian BGBl. I Nr. 244/2002.

Plöchl, M., Zacharias, H., Herrmann, C, Heiermann, M. and Prochnow, A. (2009) 'Influence of silage additives on methane yield and economic performance of selected feedstock', *Agricultural Engineering International: CIGR Journal*, vol 9, Manuscript 1123.

Pobeheim, H., Munk, B., Lindorfer, H. and Guebitz, G.M. (2011) 'Impact of nickel and cobalt on biogas production and process stability during semi-continuous anaerobic fermentation of a model substrate for maize silage', *Water Research*, vol 45, pp 781–787.

Pöschl, M., Ward, S. and Owende, P. (2010) 'Evaluation of energy efficiency of various biogas production and utilization pathways', *Applied Energy*, vol 87, pp 3305–3321.

Rehl, T. and Müller, J. (2011) 'Life cycle assessment of biogas digestate processing technologies', *Resources, Conservation and Recycling*, vol 56, pp 92–104.

Reinhold, F. and Noak, W. (1956) 'Laboratoriumsversuche über die Gasgewinnung aus landwirtschaftlichen Stoffen', in H. Liebmann (ed.) *Gewinnung und Verwertung von Methan aus Klärschlamm und Mist*, München: R. Oldenburg.

Resch, C., dos Santos, J.T.P., Kirchmayr, R., Braun, R. and Neureiter, M. (2007) 'Quantifying mass and energy losses of two different silo coverages for energy crops: Full scale investigations of silage quality at an anaerobic digestion plant' Poster presentation, 15th European Biomass Conference & Exhibition, May 7–11, 2007, Berlin, Germany.

Resch, C., Braun, R. and Kirchmayr, R. (2008) 'The influence of energy crop substrates on the mass-flow analysis and the residual methane potential at a rural anaerobic digestion plant', *Water Science and Technology*, vol 67, no 1, pp 73–81.

Rösch, C. and Skarka, J. (2008) 'Maisfelder statt Wiesen? Grünland im Spannungsfeld verschiedener Nutzungskonkurrenzen', *Technikfolgenabschätzung – Theorie und Praxis*, vol 17, no 2, pp 31–40, www.itas.fzk.de/tatup/082/rosk08a.htm, accessed 10 October 2011.

Salter, A. and Banks, C.J. (2009) 'Establishing an energy balance for crop-based digestion', *Water Science and Technology*, vol 59, no 6, pp 1053–1060.

Schattauer, A., Abdoun, E., Weiland, P., Plöchl, M. and Heiermann, M. (2011) 'Abundance of trace elements in demonstration biogas plants', *Biosystems Engineering*, vol 108, pp 57–65.

Schittenhelm, S. (2008) 'Chemical composition and methane yield of maize hybrids with contrasting maturity', *European Journal of Agronomy*, vol 29, pp 72–79.

Scholwin, F., Weidele, T. and Gattermann, H. (2005) 'Gasaufbereitung und Verwertungsmöglichkeiten', in Fachagentur Nachwachsende Rohstoffe e.V. (ed.) *Handreichung Biogasgewinnung und -nutzung*, pp 117–136., Gülzow: Fachagentur Nachwachsende Rohstoffe e.V.

Schumacher, B. Oechsner, H. and Senn, T. (2006) 'Vorbehandlung von Biomasse für die Biogasproduktion', in Ostbayerisches Technologie-Transfer-Institut e.V. (OTTI) (ed.) *Proceedings: 15. Symposium Bioenergie – Festbrennstoffe, Flüssigkraftstoffe, Biogas*, pp 352–357, 23/24 November 2006, Kloster Banz, Bad Staffelstein. Regensburg: OTTI e.V.

Schumacher, B. Oechsner, H. Senn, T. and Jungbluth, T. (2010) 'Life cycle assessment of the conversion of *Zea mays* and × *Triticosecale* into biogas and bioethanol', *Engineering in Life Sciences*, vol 10, no 6, pp 577–584.

Seppälä, M., Paavola, T., Lehtomäki, A., Pakarinen, O. and Rintala, J. (2008) 'Biogas from energy crops – optimal pre-treatments and storage, co-digestion and energy balance in boreal conditions', *Water Science and Technology*, vol 58, no 9, pp 1857–1863.

Shiferaw, B., Prasanna, B.M., Hellin, J. and Bänziger, M. (2011) 'Crops that feed the world 6. Past successes and future challenges to the role played by maize in global food security', *Food Security*, vol 3, pp 307–327.

Somitsch, W. (2007) 'Prozesstechnische und biochemische Wirkungsweise von Betriebhilfsmitteln in der Methangärung', in Ostbayerisches Technologie-Transfer-Institut e.V. (OTTI) (ed.) *Proceedings: 16. Symposium Bioenergie – Festbrennstoffe, Flüssigkraftstoffe, Biogas*, pp 164–169, 22/23 November 2006, Kloster Banz, Bad Staffelstein. Regensburg: OTTI e.V..

Stürmer, B., Schmid, E. and Eder, M.W. (2011) 'Impacts of biogas plant performance factors on total substrate costs', *Biomass and Bioenergy*, vol 35, pp 1552–1560.

Toews, T. (2009) 'Ökonomie', in A. Vetter, M. Heiermann, and T. Toews (eds) *Anbausysteme für Energiepflanzen. Optimierte Fruchtfolge + effiziente Lösungen*, pp 227–286, Frankfurt am Main: DLG-Verlags-GmbH.

van Groenigen, J.W., Kasper, G.J, Velthof, G.L., van den Pol-van Dasselaar, A. and Kuikman, P.J. (2004) 'Nitrous oxide emissions from silage maize fields under different mineral nitrogen fertilizer and slurry applications', *Plant and Soil*, vol 263, pp 101–111.

Vervaeren, H., Hostyn, K., Ghekiere, G. and Willems, B. (2010) 'Biological ensilage additives as pretreatment for maize to increase the biogas production', *Renewable Energy*, vol 35, pp 2089–2093.

Walla, C. and Schneeberger, W. (2008) 'The optimal size for biogas plants', *Biomass and Bioenergy*, vol 32, pp 551–557.

Weiland, P. (2000) 'Anaerobic waste digestion in Germany: Status and recent developments', *Biodegradation*, vol 11, pp 415–421.

Weiland, P. (2006) 'Biomass digestion in agriculture: A successful pathway for the energy production and waste treatment in Germany', *Engineering in Life Sciences*, vol 6, no 3, pp 302–309.

Weiland, P. (2008) 'Gärrestaufbereitung' in Ostbayerisches Technologie-Transfer-Institut e.V. (OTTI) (ed.) *Proceedings: 17. Symposium Bioenergie – Festbrennstoffe, Flüssigkraftstoffe, Biogas*, pp 356–361, 20/21 November 2008, Kloster Banz, Bad Staffelstein. Regensburg: OTTI e.V.

Weiland, P. (2010) 'Biogas production: current state and perspectives', *Applied Microbiology and Biotechnology*, vol 85, pp 849–860.

Wright, J.D. (1988) 'Ethanol from biomass', *Chemical Engineering Progress*, vol 84, no 8, pp 62–74.

Zhong, W., Zhang, Z., Qiao, W., Fu, P. and Liu, M. (2011) 'Comparison of chemical and biological pretreatment of corn straw for biogas production by anaerobic digestion', *Renewable Energy*, vol 36, pp 1875–1879.

Chapter 6

Suitability of microalgae and seaweeds for biomethane production

John A.H. Benzie[1,2*] *and Stephen Hynes*[1]

[1] Environmental Research Institute, University College Cork,
Lee Road, Cork, Ireland
[2] School of Biological Earth and Environmental Sciences,
University College Cork, North Mall Campus, Cork, Ireland
*Corresponding author email: j.benzie@ucc.ie

Introduction

Algae have been recommended as a fuel source for more than a century, most notably during world wars (McHugh, 2003). They were also the subject of increased attention after the oil crisis in the 1970s with the last major research programmes from that era ending in the 1980s (Sheehan et al., 1998). Algae are of interest again today as a renewable, carbon-neutral source of energy – or more correctly a source with potentially better greenhouse gas performance than other sources through their ability to capture CO_2 (Bruton et al., 2009; FAO, 2010a). As with other non-food biomass (known as second- or third-generation biofuel feedstocks), marine and freshwater biomass is attractive because it is thought it will not impact on food prices in the way that first-generation feedstocks (based on agricultural crops) are projected to do. Algae are also considered to promise higher productivity, lower greenhouse gas profiles and better sustainability than 1st generation feedstocks (de Schamphelaire and Verstraete, 2009; Clarens et al., 2010). This review will show some of these perceptions are ill-founded and that major impediments to the use of algal resources as commercial sources of biogas remain to be overcome.

Today, biogas production from algae largely uses seaweed harvested from the wild and unwanted algal production resulting from the eutrophication of estuaries and coastal zones by agricultural and human waste (Bruton et al., 2009). Most activity is best described as being research or pilot-scale investigations with negligible commercial applications. Several reviews are available which collate the available information and assess the potential of and technical constraints to using this resource (Chynoweth and Isaacson, 1987; Chynoweth, 2002; Bruton et al., 2009; FAO, 2009, 2010a; Mata et al., 2010; Brennan and Owende, 2010). This chapter summarizes the key information on present knowledge and potential future developments on the use of algae as a source of biomethane.

Major characteristics of algal resources

Algal types

Two main types of algal resources are identified on the basis of their size and mode of life, microalgae (phytoplankton) and macroalgae (seaweeds). The different properties of these two groups affect their accessibility, cultivation and harvest (Table 6.1).

Table 6.1 Major characteristics distinguishing the nature, cultivation and harvest of microalgae and macroalgae

	Microalgae (Phytoplankton)	Macroalgae (Seaweed)
Diversity	> 20,000 species	Approx. 10,000 species
Size	0.001 mm – 2 mm	> 2 mm – 70 m
Growth form	Planktonic, free floating	Benthic, attached
Wild habitat	Marine, brackish, saline and freshwaters	Mainly marine, shallow water
Cultivation	Land based open ponds and raceways Closed photobioreactors	Coastal marine aquaculture Grown mainly on ropes (longlines) Some land-based tank culture
Harvest method	Sedimentation Floatation Filtration Centrifugation	Hand cutting and collection Mechanical cutting and collection
Number of species principally used for food and energy production	< 50	< 20
Number of species in large-scale culture production	< 10	< 10

Units are mm – millimetres, m – metres. Data are derived from McHugh (2003), Guiry and Guiry (2010) and Mata et al. (2010).

Microalgae are a highly diverse group with more than 20,000 species known to date comprising a heterogeneous mix of photosynthetic prokaryotes and eukaryotes (Guiry and Guiry 2010). Small and mostly single celled, they mainly live in the water column in marine, brackish and fresh waters and in highly saline lakes. The large number of species, strains, and populations with different characteristics means that a key focus of early work, and a continuing issue, is identifying species or types that have the optimal characteristics for fuel production (FAO, 2010a). Preliminary work has been carried out on perhaps a few hundred species but much effort has focused on a relatively few species, including genetic selection and the production of transgenic strains (FAO, 2010a).

In contrast, macroalgae range from a few millimetres to several tens of metres in size, with most of those accessed for fuel being tens of centimetres to several metres in size (Braune and Guiry, 2011). They are usually attached to the sea-bed and live mainly, although not exclusively, in coastal marine habitats. Also diverse, with about 10,000 named species only a relatively small number are subject to wild collection, and an even smaller number subject to large-scale cultivation in aquaculture (McHugh 2003). Of the three main groups of seaweeds, brown, green and red, only the first two have been used in any volume for energy production (Guiry and Guiry, 2010).

A major difficulty in providing a succinct review in this field is that data on both growth rates and production of algae in aquaculture is fragmented, and both these and gas yields have been tested and/or reported in a wide variety of ways with little replication for a specific process or for comparison of different feedstocks. This makes it difficult or impossible to summarize data as simple means for a given species or method of production, and to develop more detailed assessments of technical conditions (see also previous reviews cited above). For that reason summary tables use figures that are reasonable examples and generally representative of those in the literature.

Properties related to bioenergy production

The two algal groups also differ from terrestrial crops and from each other in properties that affect their processing for energy uses and, in the context of this review, the extent to which they are potential targets for biomethane production (Table 6.2 and Figure 6.1). The large maximal light conversion efficiencies in algae are one of the reasons for their attraction as productive biological systems, theoretical maximums in productivity levels showing better performance than terrestrial examples. However, these levels of production are unlikely to be met in routine commercial scale production for a variety of reasons as demonstrated in a range of studies for microalgae and macroalgae (Tredici 2010).

Lignin does not break down in anaerobic digestion and its absence in algae is an advantage for their use in anaerobic digestion. However, microalgae store excess energy as lipids and oils rather than sugars (Hu et al., 2008). The high lipid levels that suit microalgae to the production of biofuels through oil and biodiesel production (Wijffels and Barbosa, 2010; Wijffels et al., 2010) are detrimental to anaerobic digestion as they, and the high protein levels some microalgae also produce, can inhibit the digestion process (Sialve et al., 2009; Mata et al., 2010). In contrast, the absence of lignin, low lipid content and high level of fermentable carbohydrates (equivalent to terrestrial sources) make macroalgae more suitable as a feedstock for anaerobic digestion and biogas production (Figure 6.1).

Microalgae

Work on microalgae for energy production has focused on biofuels such as ethanol or biodiesel and relatively little has been published on anaerobic digestion of microalgae for biogas production (Sialve et al., 2009). Therefore, discussion of this group will be relatively brief.

Access to resource and cultivation systems

The small size, diverse composition and highly dispersed nature of wild microalgae populations mean their collection from the wild is largely uneconomic. Algae blooms composed largely of single species do occur in the wild and those found in salt or freshwater lakes are geographically restricted and those waters are sufficiently calm to allow harvest by sieving using boats or barges (e.g. as in Klamath Lake in Oregon, http://www.algae4oil.com/algae_for_health.htm). These places are also closer to processing plants and markets and blooms are easier to locate than blooms in the open sea. Harvesting from the marine environment is rare because blooms and wave conditions are unpredictable and harvest is therefore unpredictable, expensive and difficult. Methods of harvest have changed little since they were described by Golueke and Oswald (1965) almost fifty years ago.

Most microalgae biomass is produced by aquaculture using several systems such as ponds and raceways that are open to the environment and a range of photobioreactors where the algae are enclosed in vats, tubes and closed systems to produce biomass on commercial scales (Richmond, 2004; Ripplinger, 2008). The sustained growth of single strains is difficult to

Table 6.2 Representative data illustrating the major differences in composition and productivity between microalgae, macroalgae and two examples of terrestrial crops, relevant to their anaerobic digestion to produce biogas.

	Microalgae	Macroalgae	Grass	Corn
Lipid	10–65[a]	0–2[a]	3–10[c]	4[g]
Protein	20–60[a]	12–19[a]	7–33[c]	8[g]
Total fermentable carbohydrates	11–47[b]	20–60[a]	42–78[c]	80[d]
Lignin	0[a]	0[a]	3–7[c]	15[d]
Dry matter	20[a]	15[a]	25[c]	86[h]
Ash content	10[a]	25[a]	15[c]	5[h]
Productivity (t ha^{-1} y^{-1})	20–75[a]*	11–45[e]	12[c]	18[f]
Photo efficiency (%) (practical maximum)	6[a]	6[a]	<1[a,i]	<1[a,i]

Data are percentage dry weight for all components except productivity which is given as dry tonnes per hectare per year (t ha^{-1} y^{-1}) and photoefficiency is given as a percentage (%). Data from [a] Bruton et al., 2009; [b] Dismukes et al., 2008; [c] Murphy et al., 2011; [d] Howard et al., 2003; [e] Chynoweth, 2002; [f] Hirning et al., 1987; [g] Nebraska Corn Board, 2007; [h] Preston, 2007; [i] Araya et al., 2011. Notes: dry matter for microalgae is after first dewatering or filtration steps; ash content for algae varies strongly between fresh and saltwater species. *The productivity range for algae, derived from Bruton et al. (2009), is conservative in that it focuses on potential year-round production and omits results from short-term runs that can be far greater for some species and experimental results that can be highly variable (see Figure 6.2 and related text).

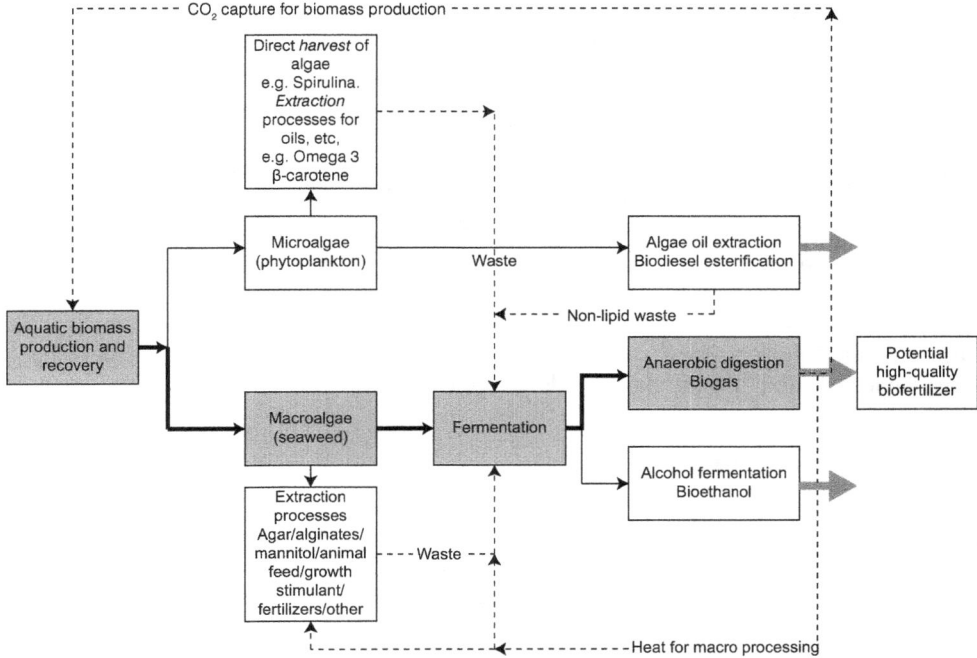

Figure 6.1 Major processing pathways for energy production from algae. Oil extraction or biodiesel production generally has to access the resource directly. Biogas and ethanol production require a fermentation step but can use the resource directly or as waste after extraction of some other products. The likely major pathway for biogas production is highlighted in grey. Concepts for the use of algae for CO_2 sequestration by injection of CO_2 derived from sources such as power plants are restricted largely to microalgae given the need usually for closed systems into which to inject the gas.

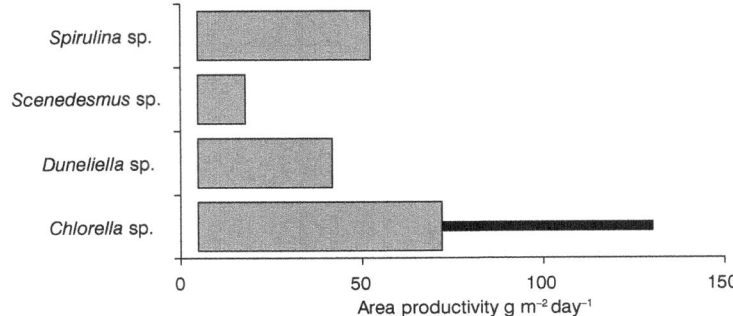

Figure 6.2 Representative results for the range in microalgae productivity observed in a number of types of production systems. Despite theoretical predictions of differences in production rate in different systems, the figures obtained from the literature showed a broad overlap of values from all production systems (ponds, raceways, closed systems and photobioreactors). Therefore values are not given separately for different production systems, except for outliers in production value for *Chlorella* (indicated by the black bar) obtained from bioreactors. Data are from Brennan and Owende (2010) and Mata et al. (2010). To convert g m^{-2} day^{-1} to dry tonnes per hectare per year (tDM ha^{-1} y^{-1}) in order to compare data in Table 6.2, multiply the g m^{-2} day^{-1} values by 3.65.

achieve in open systems because of a risk of contamination by other strains and by pathogens or predators. Closed systems greatly reduce the risk of contamination, but increase the capital and operational costs. The aim of the production systems is to efficiently convert solar power into usable biomass. In many cases the production of lipids and oils for fuels can be managed by specific actions such as shock loading of nutrients that increase lipid production (Rodolfi et al., 2009). In addition to solar radiation, adequate supplies of nutrients and carbon dioxide are required. Artificial addition of carbon dioxide can increase growth rates by up to five times and several species are well suited to utilize CO_2 from power plants (Edwards, 2008). A photosynthetic efficiency of 3–6 per cent is considered achievable using photobioreactors giving a dry yield of up to 75 tDS ha^{-1} y^{-1} by Subitec in Stuttgart (Bruton et al., 2009). A wide range in productivity is observed for all species and within different culture systems (Figure 6.2). While there are some examples of high productivity, both large- and small-scale culture systems often produce low values of about 1–2 g m^{-2} day^{-1} and rarely achieve mid-range values (around 20 g m^{-2} day^{-1}) in large-scale systems.

Harvesting methods

The fact that microalgae are suspended in water means that a variety of techniques including sedimentation, flotation, filtration and centrifugation are used to concentrate and separate the algae for further processing. Once separated, the algae can be further dewatered before drying. This last step, where it is used, is one of the most energy demanding parts of processing.

Microalgae based bioenergy products

Biodiesel, bioethanol

Significant investment in the sector has resulted in practical techniques for biofuel production, and trial use of aviation fuels (Hendricks et al., 2011). Biofuel research concentrates on species which produce high lipid content with experimental lipid contents of up to 70 per cent, although commercial scale pilots have achieved maximum yields of only 30 per cent (Bruton et al., 2009). A key part of the process for biodiesel production is the release of lipids from the algal cell (Sialve et al., 2009). Strain selection, and genetic manipulation is being used to increase lipid production. See reviews by Mata et al. (2010) and Brennan and Owende (2010) for details of biodiesel and bioethanol production.

Biogas production and yield

Unprocessed microalgae do not provide the best substrate for biogas production by fermentation as the high lipid content can lead to a build up of volatile fatty acids and inhibit the process (Mata et al., 2010). They are better used as a co-feedstock in small amounts similar to grease trap waste. However, the basic composition of microalgae after removal of the lipid fraction is suitable for biogas production through anaerobic digestion (Chynoweth, 2002). Breakdown of the cell wall to release oils also improves digestion, and species and strains with lower lipid levels give reasonable biogas yields. Given the high value of the lipid and other products for which microalgae are grown, the most likely development is the use of micro-algal by-product in anaerobic digestion alone, or with other wastes, after processing for high-value lipid extraction. Trials of a variety of species have provided a range of yields in the range 0.2–0.6 m^3 CH_4 kg^{-1} VS, and around 60 per cent methane content comparable to that from terrestrial crops (Table 6.3). The addition of paper as a source of carbon to balance carbon to nitrogen ratios has been found to improve yields from fishpond microalgal sludge (Yen and Brune, 2007).

Case studies and industrial applications

There is no commercial or industrial scale anaerobic digestion of microalgae or microalgae waste. Commercial, industrial-scale production of microalgae has been achieved, but is focused on food supplement production such as, for example β-carotene by Seambiotic in Israel, although some large-scale sites, such as Jerez in Spain, produce feedstock for biodiesel, and some facilities such as SunChem in Switzerland may utilize spent biomass for local heating of the production facility using biogas (Bruton et al., 2009; FAO 2010a). It is important to note that an additional advantage of some microalgae is their use to remove CO_2 from sources such as power station effluent gases. This leads to the view that algae (but importantly these

Table 6.3 Representative figures for yield and CH_4 content of biogas from anaerobic digestion of microalgae

Species	Yield (m^3 CH_4 kg^{-1} VS)	CH_4 content (% of total biogas produced)
Chlorella kessleri [a]	0.335	65
Dunaliella salina [a]	0.505	64
Scendesmus obliquus [a]	0.287	62
Chlamydomonas reinharrtii [a]	0.587	66
Spirulina maxima [b]	0.25–0.5	–

Sources: [a] Mussgnung et al., 2010; [b] Salerno and Nurdogan, 2009.

microalgae – not all algae) improve the carbon footprint of power generation but through sequestration of CO_2 rather than a better CO_2 profile of algae themselves. There is still a dearth of detailed lifetime analyses of algae for meaningful assessments of their lifetime carbon profiles to be made.

Macroalgae

Access to resource and cultivation systems

Seaweeds can form dense populations that have been successfully harvested by collecting algae exposed at low tide or collecting fragments floating in the water (McHugh 2003). Key limiting factors include access (where tides and shoreline allow material to be collected), sufficient quantity and growth of the algal population (to allow economic harvest) and environmental concerns (that sustainable harvest can be achieved and that ecosystem services are not adversely affected by collection). The major interest in seaweed collection traditionally has been as a source of fertilizer and of unique high-value food additives. Their use in energy production has largely focused on unwanted growth of algae resulting from coastal pollution from nutrient runoff, e.g. green tides of *Ulva* in Japan (Matsui et al., 2006).

The unpredictable and seasonal availability of wild material provides a challenge in accessing large volumes of material for particular species, and has led to the development of aquaculture to meet demands of the latter markets in particular. The difficulties of effective management of wild sourced harvests and the growth of aquaculture operations, has resulted in a worldwide decline in wild harvest over the last couple of decades from 48 per cent of total seaweed supply (3.6 million T wet weight (Tww)) to 6 per cent (1 million Tww). This reflects the greater efficiency of operation, lower costs and improved mechanization available using aquaculture, and the benefits of culturing the species desired and strains that are particularly productive. Aquaculture operations involve the growth of seaweeds on stakes in tidal zones or on ropes (longlines) and rafts suspended in the water. An excellent summary of seaweed culture is given by McHugh (2003). Data on the growth rates and productivity of seaweeds are scattered and interpretation of these is made difficult given the variation between studies in season, site, species and strain, and differences in the details of the growth measurements made. Nevertheless, the sample of representative information in Figure 6.3 shows yields have a wide range, and that these are comparable to, or greater than, those of terrestrial crops.

Harvesting methods

Harvesting of large volumes of seaweed requires the construction of specific machinery and the use of a boat. Mechanical harvesters have been developed to collect the large individuals of wild *Laminaria* spp. by twisting the long stalk around a rotating hook and pulling free from the holdfast (the Scoubidou system), or by trawling dredges to cut the canopy (see Bruton et al., 2009 for photographs of these). Seaweeds drifting in the water are also collected by mechanical harvesters but most wild seaweeds are harvested manually. This limits the volume and increases the expense of harvest. However, it permits choice of the individual alga and the portion to be cut, allowing for more precise management for sustainable harvests and cleaner product. Aquaculture operations permit mechanical collection from longlines but hand harvest is still prevalent in most countries with large-scale aquaculture operations (e.g. China and Indonesia).

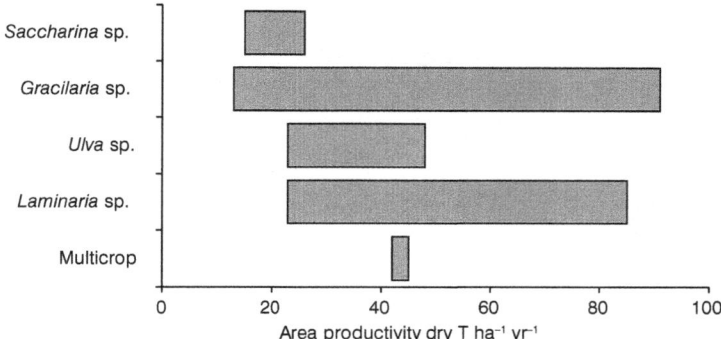

Figure 6.3 Representative results for the range in macroalgae productivity observed over a range of types of production systems. Data are from Kelly and Dworjanyn (2008), Habib et al. (2008) and Bruton et al. (2009).

Macroalgae based bioenergy products

Mechanical harvest in particular results in feedstock with rubbish such as plastic, driftwood, stones, sand and shells, while drift seaweed can have large amounts of sand (30 per cent by harvest wet weight) (Bruton et al., 2009). Some harvesters are designed to work in shallow water to remove foreign material through partial suspension in water, and others use salad washing equipment. Even so, additional screening is required, particularly prior to chopping or milling the feedstock. Fragmenting the seaweed provides a larger surface area to increase the rates of reaction in the digester. Although anaerobic digestion and fermentation has a high tolerance to foreign material, the advantages of chopping and the costs of screening have to be balanced for optimal production. The seaweed may also need to be washed to remove surface salt and may be dewatered or dried prior to some processing, e.g. bioethanol production.

Biodiesel, bioethanol

The relatively low lipid content and relatively high fermentable sugar content of macroalgae mean they are more suited to bioethanol than biodiesel production. The relatively high water content of macroalgae can interfere with the esterification of lipids and dewatering represents an additional cost of biodiesel production. In contrast, pilot-scale research carried out in Norway (Horn, 2000) showed that bioconversion to ethanol of seaweeds is feasible due to the high level of sugars (mannitol and laminaran) they store. These are also a by-product of alginate extraction suggesting waste from that industry could be used for ethanol production. A review of ethanol production from seaweeds can be found in Horn et al. (2000).

Biogas production and yield

The gross composition of macroalgae is sufficiently similar to other terrestrial biomass (Table 6.2) that the general processes and designs of anaerobic digesters used for terrestrial material would be expected to work on macroalgae. Experimental work has focused largely on the use of continuously stirred tank reactors (CSTRs) with overall retention times of

Table 6.4 Representative figures for yield and CH_4 content of biogas from anaerobic digestion of macroalgae and some data on H_2S production characteristic of some species of macroalgae

Species	Biological yield (m^3 CH_4 kg^{-1} VS)	CH_4 content (% of total biogas produced)	H_2S (% of total biogas produced)
Laminara saccharina	0.23	51	–
Laminara digitata	0.31	63	0.7
Ascophyllum nodosum	0.11	50	–
Ulva lactuca	0.15–0.2	66	1–5.1
Ulva sp.	0.15–0.2	50–60	2–6
Ulva sp. 60% + sludge 40%	0.31	62	–

Source: Guiry and Blunden (1991).

1–60 days (mostly 20–60 days). The compilation of the results of several experiments on more than 15 species of red, green and brown seaweed by Morand et al. (1991) confirmed reasonable rates of production (0.11–0.31 m^3 CH_4 kg^{-1} VS with a 50–62 per cent methane content) using temperatures in the range of 15–37°C (mostly 25–35°C). It is not possible given the limited range of testing for any one species to determine more accurate optima for individual species or processes. However, practical operations would require a process that was robust to a range of variations in feedstock composition and production parameters. The summation of experiments in the early 1980s has shown robustness in trials over the range of species and conditions noted above with biogas yields and quality similar to that of terrestrial crops (Table 6.4) and later experiments provide similar results (e.g. Vergara-Fernández et al., 2008). Chynoweth et al. (1987) had considered that the early CSTRs, with their vigorous mixing, were unsuitable for larger scale energy production as the high loadings they require would result in reduced biomass conversion and system instability. Tests with vertical flow reactors (VFRs) had a methane yield greater than 75 per cent (Chynoweth et al., 1987). However, seaweeds have not yet been trialled in state-of-the-art commercial CSTR systems (Kelly and Dworjanyn, 2008) that address some of the problems in earlier CSTR designs.

The anaerobic digestion process for seaweed

Seaweed feedstock can vary considerably in biomass composition depending on species, site, time of harvest, and, in the case of cultured crops, the nature of the rearing system or strain used for cultivation (Critchley and Ohno, 1998). The differences in composition between lots of the same species can dramatically affect the performance and stability of the digestion process. Levels of light and addition of fertilizers to seaweeds may also affect their biodegradability and methane yields (Chynoweth and Isaacson, 1987). The digestion of seaweed has a number of parameters that differ from traditional digestion feedstocks, which need to be considered. Inhibition of methanization can result from high concentration of substances such as phenols, heavy metals, sulphides, salts and volatile acids. Sulphur, although an element needed for methane fermentation, can also act as an inhibitor. It is not a problem reported in the digestion of brown algae but the green algae, *Ulva*, can contain large quantities of sulphur under certain conditions, up to ten times those found in brown algae (Table 6.4) and this can inhibit fermentation (Briand and Morand, 1997).

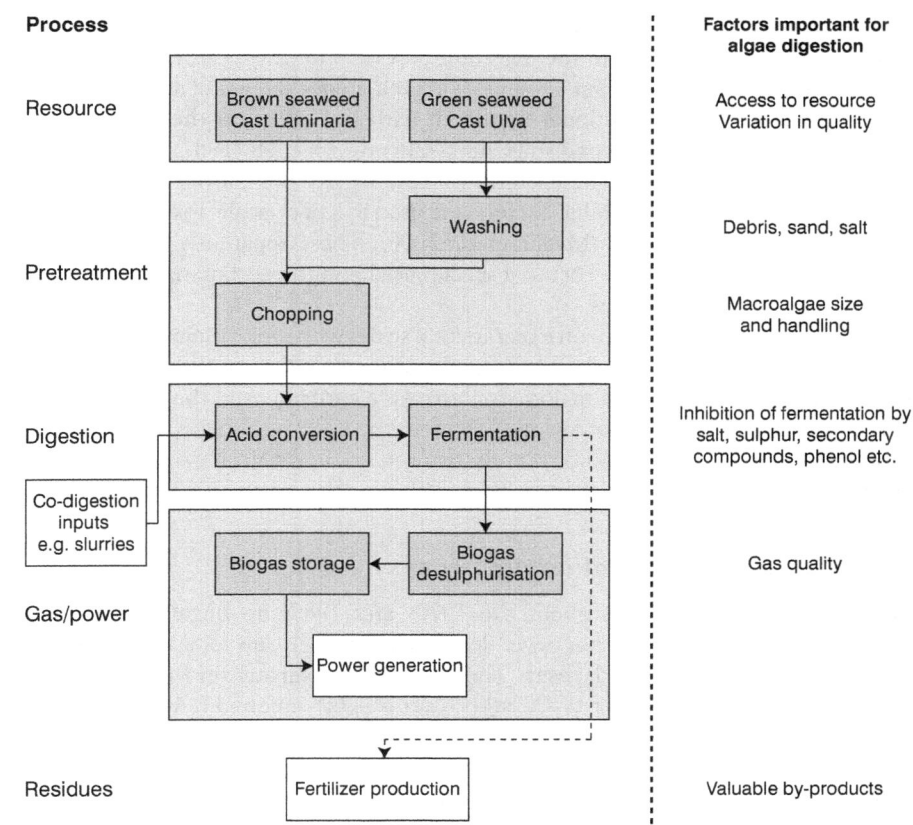

Figure 6.4 The classic two-stage anaerobic digestion process used for macroalgae (centre diagram) with specific factors that need to be considered for algae processing listed to the right, for each stage of the process, given in the list to the left of the larger box it identifies.

Pre-processing can assist with some of these issues. Surface salt and sand can be removed by washing but, in some trials, desalting has led to a decrease in methane production, possibly due to the loss of fermentable products along with the salts (Kelly and Dworjanyn, 2008). A general strategy is to use dilution of the marine biomass through mixing it with other feedstocks such as slurries to reduce the amounts of salt, sulphur or other inhibitory compounds to acceptable levels. Another approach is to introduce bacteria that may be more tolerant of the inhibitory materials such as salt and sulphur. This can be achieved by adding marine sediments from an area of decaying seaweeds to give a fast start to the digestion process. The use of marine bacteria can accelerate and increase biogas production when added to the digester (Kelly and Dworjanyn, 2008), but Morand et al. (1991) report that a marine inoculum has no greater final effect than co-digestion with slurries or traditional inoculums. These results no doubt reflect the differences in performance of the wild bacterial communities collected from the marine sediments and their performance relative to the bacterial community acclimated to conditions in the digester.

Acclimatization of the fermenting biota can be achieved by increased retention time of the microorganisms. This allows the bacterial community to function successfully at relatively high concentrations of inhibitory compounds even at levels, which if introduced suddenly, would inhibit or perturb the digestion process. However, this extends processing time, particularly if done in the main fermentation tank, hence the use of a two-stage process where the first stage allows additional treatment with less effect on the main fermenting community acclimating in the second stage tank. Chynoweth et al. (1987) reported that salt appears to inhibit the thermophillic bacteria which adapted only partially to the digestion environment. However, thermophillic bacteria are used in a pilot-scale Tokyo plant discussed in the case studies section below (Matsui et al., 2006). These apparently conflicting results may also reflect differences in the bacterial species used and the acclimatization achieved in particular operating circumstances.

Most work has been undertaken on experimental scales with tank volumes from <1 litre to several tens of cubic metres, with the majority of larger scale pilot studies lasting only several weeks to a few months. By far the majority of tests were carried out in the previous research peak investigating energy from macroalgae (Chynoweth, 2002) although some larger volume digestion tests have been done more recently (Bruton et al., 2009). Nevertheless, the state of the art has not progressed significantly since Chynoweth's review (2002) a decade ago.

Case studies and industrial applications

Despite significant research throughout the 1970s and 1980s an ongoing industry using macroalgae for biogas production has never developed, but there are some examples of large-scale digestion of waste seaweed deposits. The Tokyo Gas Company (in Matsui et al., 2006) reported gas yields for *Laminaria* of 17–22 m^3 t^{-1} (wet weight – inferred from Yokoyama et al., 2007) using a two-stage digestion process with a 55°C second stage with 15–22 day retention and a 2–3 day retention time in a 25–35°C first stage pre-fermentation. The biogas required refinement (sulphur removal) and was mixed with city gas for use in a combined heat and power unit, which provided heat to run the digestion process and also to dry the residue for use as a fertilizer.

The use of the residues from higher value extraction of brown seaweeds, such as agar extraction and alginate extraction, provides significant biomass with a high polysaccharide content, which is suitable for biomethane production. This has been trialled (cited in Morand et al., 1991) and developed further by Sopex of Belgium who completed an 800 m^3 reactor to treat 12 t a day of waste from agar production in Morocco. However no data is available on the operational efficiencies of this plant. These studies have demonstrated that large upscaling of brown and green seaweed digestion is possible. However long-term operational data on a large scale has not been documented reliably and further research is required on this topic.

Other uses and biorefinery concepts in microalgae and macroalgae

At present, the major barrier to microalgae bioenergy is the high cost of production, which is much greater than other biogas feedstocks (Figure 6.5). Macroalgae from the sea have much lower costs but these are still six to ten times higher than agricultural and forestry sources (Carlson et al., 2007). The high cost of production of microalgae is manageable given the high values of the materials supplied to the neutraceutical and other industries for which it is grown (see inset in Figure 6.7). In order to overcome the economic barriers for the use of algae for energy supply, integration with other high-value enterprises will be required.

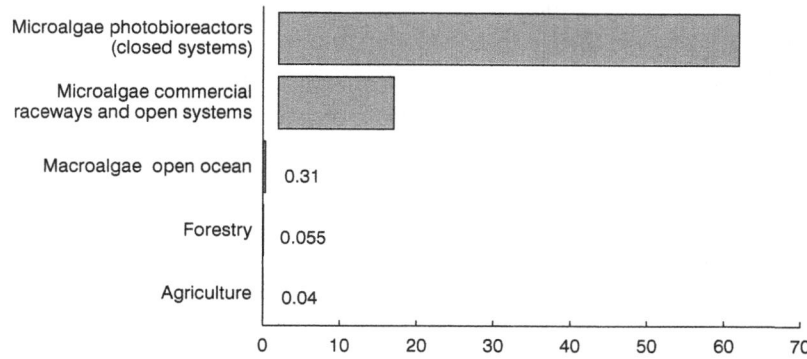

Figure 6.5 Present costs of producing biomass for energy from microalgae using photobioreactors and commercial raceways are much higher than terrestrial sources. Macroalgae from marine sources appear to be competitive, but this may reflect the use of unwanted growth arising from pollution. Data are from the EPOBIO Project (Carlson et al., 2007). There is little reliable data on the cost of gas (as $ per m^3) from different algal biomass types because of the lack of commercial production. However, pilot schemes such as the Tokyo Gas Company have indicated that no major technological developments are required to obtain yields from algae similar in range to agricultural biomass (Matsui et al., 2006). This suggests that the cost of processing the biomass delivered to an anaerobic digestion facility would be similar in all categories of biomass (with some exceptions for sulphur removal for algae) and therefore would not change the major pattern of costs of production of gas from source given in the figure. Note: this graph is used principally to indicate the relative costs of production of the source biomass, and to illustrate that algae are not a "free" resource. The production of algae for biogas would also compete with other uses of algae that provide greater value to producers (see discussion in the text and Figure 6.7).

In these biorefinery models (Figure 6.6), algae are processed for higher value materials prior to energy production and the waste from anaerobic digestion is used to make products such as fertilizers (Kerner et al., 1991; Bruton et al., 2009; Mata et al., 2010). It is too difficult to separate out market values for individual product groups, but values for total industry production can give gross values for the two main algal sectors (see Figure 6.7). Order of magnitude calculation gives the gross value of macroalgae at US$470 t^{-1} (total value of US$7.4 billion divided by total world production of 15.8 million tonnes) and that for microalgae at US$80,000–90,000 t^{-1} (total value of US$5–6 billion divided by total world production of 68,400 tonnes) for present industry uses. A requirement to supply algae for energy alone at prices less than US$50 t^{-1} would not be attractive. For the near future, therefore, economic scenarios most likely to match the scale of available resources and infrastructure investment will be restricted to multipurpose biorefinery models producing energy as a by-product.

In the case of microalgae, any biogas production is likely to be limited, since extraction for nutraceuticals uses much if not all of the algae, extraction of lipids for food additives will leave little for biodiesel production and manufacture of energy by-products is likely to focus on ethanol. In contrast, macroalgae biogas production is likely to be more frequent in circumstances where the gas is burned on site to provide thermal energy for use in the processing plant, or to produce electricity for export. In the longer term, it is possible that ethanol or biogas itself may be exported from the site for use as transport fuels but this use of biogas awaits the development of infrastructure suitable for gas-fuelled transport.

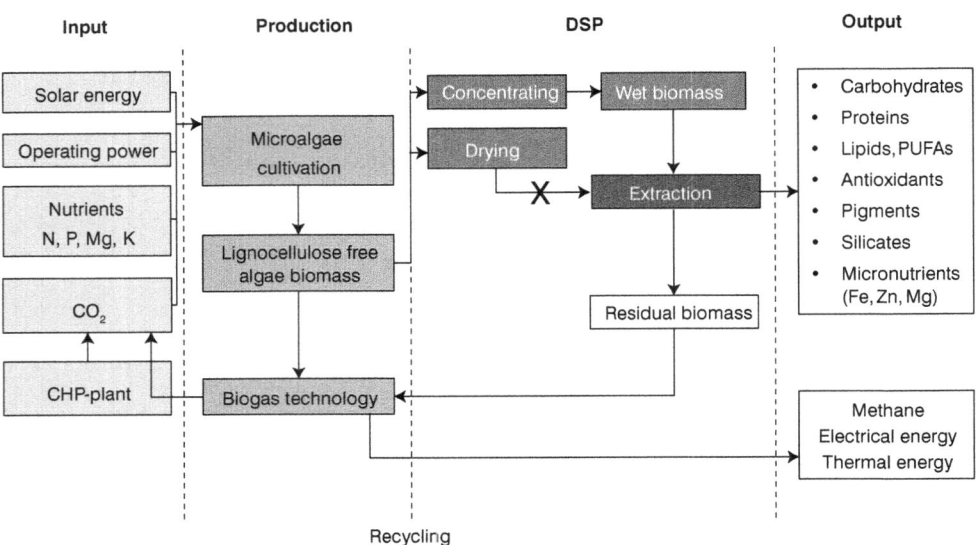

Figure 6.6 Given present costs and alternative uses of algae, developments in the near future will be restricted to multipurpose biorefinery models in which energy is one of several outputs derived from processing algae, with a focus on producing higher value materials. This example is from Schmid-Staiger (2009).

Future trends

Currently microalgae are used to produce high-value food additives, such as omega oils, antioxidants and specific proteins, materials for the pharmaceutical, cosmetics and biotechnology industries, and feeds in the aquaculture industry. Macroalgae are the largest aquaculture product by volume worldwide (FAO, 2010b) supplying unique additives such as carageenan and agars to the food industry, and others used in human, animal and aquaculture nutrition, and as fertilizers. Order of magnitude calculations indicate far greater amounts of algae than are harvested today would need to be used to completely replace fuel supplies for significant energy sectors.

For example, replacing Europe's transport energy needs in 2009 translates to an area far greater than that available for aquaculture in currently used shorelines worldwide (Figure 6.7). For this reason, concepts have been proposed for open-ocean aquaculture to provide the vast amounts of biomass required for large-scale biomethane production (Chynoweth, 2002; Reith et al., 2005). The multiple challenges of large-scale farming including novel engineering approaches, infrastructure costs and in particular husbandry issues such as propagule or seedling production, nutrient supply to the seaweed and disease control, and relevant governance issues have yet to be addressed in any detail (Buck, 2007; FAO, 2009; UNESCO 2009).

The major attraction of algae for energy production is the view that they will ultimately provide a cheaper and more sustainable supply of gas than other sources. This view reflects perceptions that algae have higher light conversion efficiencies and potentially greater productivity than land plants, lower greenhouse gas profiles, and that they will not impact on food prices (Bruton et al., 2009). As pressure increases on land use with rising population and inflating energy costs, the advantages of algae over land based biofuel production, including

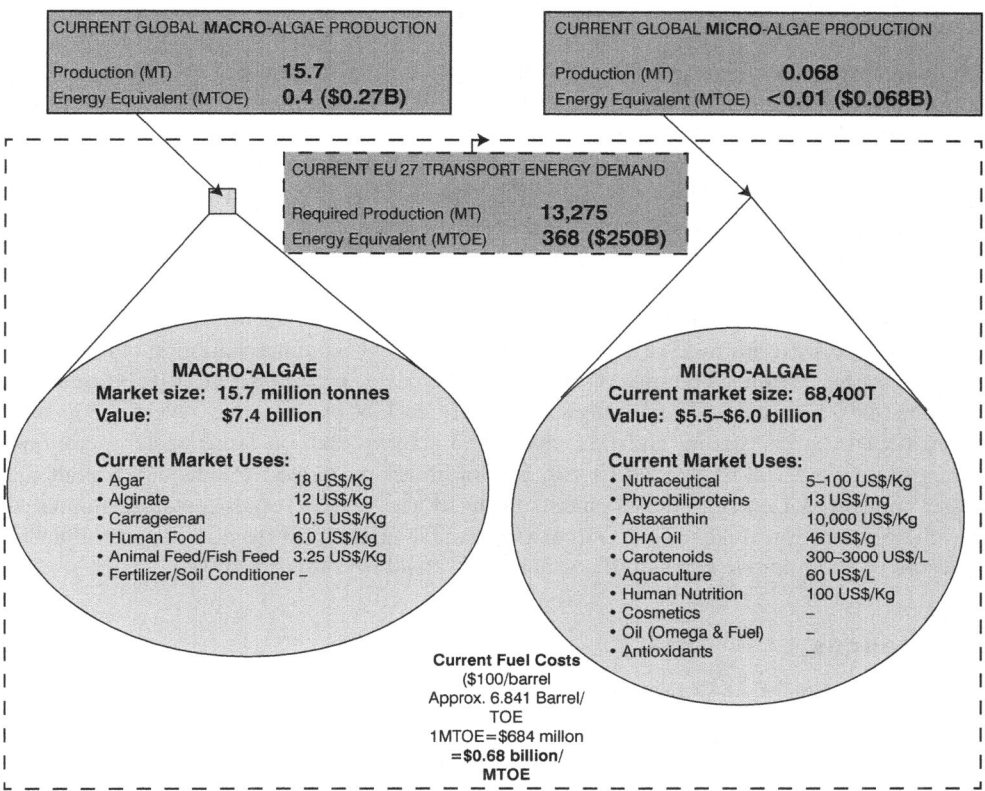

Figure 6.7 The total world production of algae in 2009, currently used for other valuable products, would provide only about 0.1 per cent of the total transport energy used in Europe in 2009. To meet this demand 900 times the current area under culture would be needed. Data on European energy from Eurostat (2011), algae production and market sizes from FAO (2010b) and cost ranges for a variety of products from FAO (2010a) and Bixler and Porse (2011).

their use as nutrient and carbon sinks, are considered to act as an incentive to large-scale commercial algae biomass production. However, as in terrestrial crops other important uses of algae will influence their availability for energy production, and there are limitations in the areas of wild growth and aquaculture technology. Tredici (2010) has summarized arguments as to why their light conversion maxima, and high production limits are unlikely to be met. In addition, the previous section has demonstrated that algal resources are already used for other important processes including food and health products, and that increasing production areas in the open sea will require significant investment and the development of novel technologies – from both ocean engineering and biological perspectives. Considerable research will be needed to understand the extent to which these issues can be overcome and the approaches to algal utilization are realistic and sustainable. It is clear though that the use of algae is feasible on their own or as mixed feedstock with other wastes. Using seaweed growing from eutrophic pollution in coastal areas (Bruton et al., 2009) or capture systems for nutrients (nitrogen and phosphate) from human, animal and aquaculture waste (Reith et al., 2004; Mulbry et al., 2008) in rural locations is certainly practical on small, local scales, although economically challenging.

Conclusions

The present review has identified that algae are a suitable resource in terms of basic composition and general productivity to allow for their inclusion as a feedstock for biogas production. Basic parameters of biogas production for a range of species and conditions have been reported for experimental scales. However, data on both growth rates and production of algae in aquaculture and gas yields are fragmented. They have been tested and reported in such a wide variety of ways that it is difficult to develop more detailed assessments of technical conditions. A few primary studies are continually recycled in a growing number of reviews. Data from larger pilot studies are few and of limited duration except the Tokyo plant. Basic information including yields and optimization of operating conditions over medium to longer terms need to be determined. Large-scale commercialization of biogas has not yet become a reality. The state of the art for macroalgae has progressed little beyond that described by Chynoweth (2002) a decade ago. The lack of progress may reflect the greater investment in infrastructure and time needed to achieve data on larger scale operations. Much more work will be needed to gain the information required to adequately plan the larger concepts for offshore developments outlined above. The recent research stimulated by renewed interest in algae as a source of renewable energy will need to attract sufficient investment and improve focus in the coming years to address these points.

References

Araya, A., Stroosnijder, L., Girmay, G. and Keesstra, S.D. (2011) 'Crop coefficient, yield response to water stress and water productivity of teff (*Eragrostis tef* (Zucc.)),' *Agriculture and Water Management*, vol 98, pp. 775–783.

Bixler, H.J. and Porse, H. (2011) 'A decade of change in the seaweed hydrocolloids industry', *Journal of Applied Phycology*, vol 23, pp. 321–335.

Braune, W. and Guiry, M.D. (2011) *Seaweeds: A colour guide to common benthic green, brown and red algae of the world's oceans*, Koenigstein: Koeltz Books.

Brennan, L. and Owende, P. (2010) 'Biofuels from microalgae – A review of technologies for production, processing, and extractions of biofuels and coproducts' *Renewable and Sustainable Energy Reviews* vol 14, pp. 557–577.

Briand, X. and Morand, P. (1997) 'Anaerobic digestion of *ulva* sp. 1 Relationship between *ulva* composition and methanisation', *Journal of Applied Phycology*, vol 9, pp. 511–524.

Bruton, T., Lyons, H., Lerat, Y., Stanley, M. and Rasmussen, M.B. (2009) 'A review of the potential of marine algae as a source of biofuel in Ireland', *Sustainable Energy Ireland (SEI) Report*.

Buck, B. (2007) *Farming in a high energy environment: potential and constraints of sustainable offshore aquaculture in the German Bight (North Sea)*, Bremerhaven: University of Bremen; Alfred Wegener Institute for Polar and Marine Research.

Carlson, A.S., Van Beilen, J.B., Moller, R. and Clayton, D. (2007) *Outputs from the EPOBIO Project September 2007 micro and macro algae: Utility for industrial applications*. Newbury: CPL Press.

Chynoweth, D. (2002) *Review of biomethane from marine biomass*, Department of Agricultural and Biological Engineering, University of Florida. Gainesville, FL.

Chynoweth, D.P. and Isaacson, R. (1987) *Anaerobic digestion of biomass*, London: Elsevier.

Chynoweth, D.P., Fannin, K.F. and Srivastava, V.J. (1987) 'Biological gasification of marine algae', in Bird, K.T. and Benson, P.H. (eds) Seaweed cultivation for renewable resources, *Developments in aquaculture and fisheries science*, vol 16, Amsterdam: Elsevier.

Clarens, A.F., Resurreccion, E.P., White, M.A., and Colosi, L.M. (2010) 'Environmental life cycle comparison of algae to other bioenergy feedstocks', *Environmental Science and Technology*, vol 44, pp. 1813–1819.

Critchley, A.T. and Ohno, M. (1998) *Seaweed resources of the world*, Yokosuka: JICA.

De Schamphelaire, L., and Verstraete, W. (2009) 'Revival of the biological sunlight-to-biogas energy conversion system', *Biotechnology and Bioengineering* vol 103, pp. 296–304.

Dismukes, G.C., Carrieri, D., Bennette, N., Anayev, G.M. and Posewitz, M.C. (2008) 'Aquatic phototrophs: efficient alternatives to land-based crops for biofuels', *Current Opinion in Biotechnology*, vol 19, pp. 235–240.

Edwards, M. (2008) *Green algae strategy – end biowar and engineer sustainable food and biofuels*, Tempe, AZ: LuLu Press.

Eurostat (2011) *Energy, transport and environment indicators*, Eurostat Pocket Book 2011 edition, Bruseels: Publications Office of the European Union.

FAO (2009) 'Algae-based biofuels – a review of challenges and opportunities for developing countries', Environment and natural resources management working paper no 33, Rome: FAO.

FAO (2010a) 'Algae-based biofuels – applications and co-products', Environment and natural resources management working paper no 44, Rome: FAO..

FAO (2010b) *The state of world fisheries and aquaculture 2009*. FAO Rome, Italy.

Golueke, C.G. and Oswald, W.J. (1965) 'Harvesting and processing sewage grown planktonic algae', *Journal of Water Pollution Control Federation*, vol 37, pp. 471–498.

Guiry, M.D. and Blunden, G. (eds) (1991) *Seaweed resources of europe. uses and potential*, Chichester: J. Wiley and Sons.

Guiry, M.D. and Guiry, G.M. (2010) *AlgaeBase*. World-wide electronic publication, National University of Ireland, Galway. http://www.algaebase.org; searched on 12 January 2011.

Habib, M.A.B., Parvin, M., Huntington, T.C. and Hasan, M.R. (2008) 'A review on culture, production and use of Spirulina as food for humans and feeds for domestic animals and fish', FAO Fisheries and Aquaculture Circular 1034. Rome: FAO Fisheries and Aquaculture Department.

Hendricks, R.C., Bushnell, D.M. and Shouse, D.T. (2011) 'Aviation fueling: a cleaner, greener approach', *International Journal of Rotating Machinery*, vol 2011, pp. 1–13 doi:10.1155/2011/782969.

Hirning, H., Hellevang, K. and Helm, J. (1987) 'Equivalent weights of grain and oilseeds, AE-945'. North Dakota Cooperative Extension Service, North Dakota State University, Fargo, Available at: http://www.ext.nodak.edu/extpubs/ageng/machine/ae945w.htm (accessed 18 Aug. 2004, verified 3 Jan. 2009).

Horn, S.J. (2000) 'Bioenergy from brown seaweeds' Department of Biothechnology, Norwegian University of Science and Technology.

Horn, S.J., Aasen, I.M. and Ostgaard, K. (2000) 'Ethanol production from seaweed extract', *Journal of Industrial Microbiology*, vol 25, pp. 249–254.

Howard, R.L., Abotsi, E., Jansen van Rensburg, E.L. and Howard, S. (2003) 'Lignocellulose biotechnology: issues of bioconversion and enzyme production'. *African Journal of Biotechnology*, vol 2, pp. 602–619.

Hu, Q., Sommerfeld, M., Jarvis, E., Ghirardi, M., Posewitz, M., Seibert, M. and Darzins A. (2008) 'Microalgal triacylglycerols as feedstocks for biofuel production: perspectives and advances', *Plant Journal*, vol 54, pp. 621–639.

Kelly, M.S. and Dworjanyn, S. (2008) *The potential of marine biomass for anaerobic biogas production: a feasibility study with recommendations for further research*, Edinburgh: Scottish Association for Marine Science, The Crown Estate.

Kerner, K.N., Hanssen, J.F. and Pedersen, T.A. (1991) 'Anaerobic-digestion of waste sludges from the alginate extraction process', *Bioresource Technology*, vol 37, pp. 17–24.

Mata, T.M., Martins, A.A. and Caetano, N.S. (2010) 'Microalgae for biodiesel production and other applications', *Renewable and Sustainable Energy Reviews*, vol 14, pp. 217–232.

Matsui, T., Toshiji, A., Yoji, K., Atsushi, S. and Saito, H. (2006) 'Methane fermentation of seaweed biomass', American Institute of Chemical Engineers Conference, Session 412, Sustainable Nonfuel products / Production Systems from Biomass Resources, San Francisco, Tokyo Gas Company.

McHugh, D. (2003) *A guide to the seaweed industry*, Fisheries Technical Paper No. 441. Rome: FAO.

Morand, P., Carpentier, B., Charlier, R., Maze, J., Orlandini, M., Plunkett, B. and de Waart, J. (1991) 'Bioconversion of seaweeds', in Guiry, M.D. and Blunden, G.B. (eds) *Seaweed resources of europe. uses and potential*, Chichester: J. Wiley and Sons.

Mulbry, W., Kondrad, S. and Buyer, J. (2008) 'Treatment of dairy and swine manure effluents using freshwater algae: fatty acid content and composition of algal biomass at different manure loading rates', *Journal of Applied Phycology*, vol 20, pp. 1079–1085.

Murphy, J., Braun, R., Weiland, P. and Wellinger, A. (2011) 'Biogas from crop digestion', Task 37 Energy from Biogas brochure, IEA Bioenergy, International Energy Agency.

Mussgnung, J.H., Klassen, V., Schluter, A. and Kruse, O. (2010) 'Microalgae as substrates for fermentative biogas production in a combined biorefinery concept,' *Journal of Biotechnology*, vol 150, pp. 51–56.

Nebraska Corn Board (2007) *2007–2008: Nebraska corn board quality report*' Lincoln, NE: Nebraska Corn Board.

Preston, R.L. (2007) 'Feed composition', *Beef magazine*. Available at http://beefmagazine.com/nutrition/feed-composition-tables/beef_feed_composition/. Accessed April 22, 2012.

Reith, J.H., van Zessen, E., van der Drift, A., den Uil, H., Snelder, E., Balke, J., Matthijs, H.C.P., Mur, L.R. and van Kilsdonk, K. (2004) 'Microalgal mass cultures for co-production of fine chemicals and biofuels and water purification'. CODON Symposium 'Marine Biotechnology; An ocean full of prospects ?', Wageningen: ECN Biomass.

Reith, J.H., Deurwaarder, E.P., Hemmes, K., Kamermans, P. and Brandenburg, W. (2005) *Bio-Offshore: Grootschalige teelt van zeewieren in combinatie met offshore windparken in de Nordzee*. Petten: ECN Energy Research Centre of the Netherlands.

Richmond, A. (2004) *Handbook of microalgal culture*, Oxford: Blackwell Publishing.

Ripplinger, P. (2008) 'Large scale production of microalgae'. Personal communication, Subitec GmbH. www.subitec.com.

Rodolfi, L., Zittelli, G.C., Bassi, N., Padovani, G., Biondi, N., Bonini, G. and Tredici, M.R. (2009) 'Microalgae for oil: strain selection, induction of lipid synthesis and outdoor mass cultivation in a low-cost photobioreactor', *Biotechnology and Bioengineering*, vol 102, pp. 101–112.

Salerno, M. and Nurdogan, Y. (2009), 'Biogas production from algae biomass harvested at wastewater treatment ponds', 2009 Bioenergy Engineering Conference, Washington, October 11–14.

Schmid-Staiger, U. (2009) 'Algae biorefinery concepts', National Workshop on Biorefineries 15 September 2009, Worms, Germany.

Sheehan, J., Dunahay, T., Benemann, J. and Roesler, P. (1998) *A look back at the U.S. Department of Energy's aquatic species program: biodiesel from algae*, Golden, CO: Department of Energy.

Sialve, B., Bernet, N. and Bernard, O. (2009) 'Anaerobic digestion of microalgae as a necessary step to make microalgal biodiesel sustainable,' *Biotechnology Advances*, vol. 27, pp. 409–416.

Tredici, M.R. (2010) 'Photobiology of microalgae mass cultures: understanding the tools for the next green revolution', *Biofuels*, vol 1, pp. 143–162.

UNESCO (2009) 'The ethics of adoption and development of algae-based biofuels'. Bangkok: L. McGraw.

Vergara-Fernández, A., Vargas, G., Alarcón, N., and Velasco, A. (2008) 'Evaluation of marine algae as a source of biogas in a two-stage anaerobic reactor system', *Biomass and Bioenergy*, vol 32, pp. 338–344.

Wijffels, R.H. and Barbosa, M.J. (2010) 'An outlook on microalgal biofuels', *Science*, vol 329, pp. 796–799.

Wijffels, R.H., Barbosa, M.J. and Eppink, M.H.M. (2010) 'Microalgae for the production of bulk chemicals and biofuels', *Biofuels, Bioproducts and Biorefining*, vol 4, no 3, pp. 287–295.

Yen, H.-W., and Brune, D.E. (2007) 'Anaerobic co-digestion of algal sludge and waste paper to produce methane', *Bioresource Technology*, vol 98, pp. 130–134.

Yokoyama, S., Jonouchi, K., and Imou, K. (2007) 'Energy production from marine biomass: Fuel cell power generation driven by methane produced from seaweed', *International Journal of Applied Science, Engineering and Technology*, vol 4, no 3, pp. 168–171.

Chapter 7

Organic wastes for biomethane production

Anoop Singh

Department of Scientific and Industrial Research (DSIR),
Ministry of Science and Technology, Technology Bhawan,
New Mehrauli Road, New Delhi 110016, India
Email: apsinghenv@gmail.com

Introduction

The production of organic waste is associated with people and is increasing with increasing population and living standards. The main form of organic waste includes household waste, garden waste, forest waste, human and animal excreta, municipal solid waste (organic fraction), agricultural refuse, slaughterhouse waste, etc. These all require proper disposal to reduce greenhouse gas (GHG) emissions. In landfills, these wastes are broken down by microbial activity into a liquid leachate containing microbes, chemicals and rotting material. When this leachate reaches a water body or enters ground water, it can be hazardous for human beings.

The economy of developing countries requires materials and resources to be used to their full potential. Hence, the idea of 3R (Reclaim, Recycle and Reuse) is disseminated to promote efficient and sustainable use of materials and resources. The use of organic residues as raw materials for the production of biofuels is an effective option for adopting the 3R strategy to address environmental concerns related to fossil fuel resources, to mitigate climate change, reduce dependence on imported fossil fuels and enhance cleaner production chains based on local and renewable sources (Cherubini and Ulgiati, 2010; Nigam and Singh, 2011; Prasad et al., 2007a, b).

Anaerobic digestion (AD) of organic waste produces not only methane for energy but also slurry for fertilizing agricultural crops. By replacing chemical fertilizers, the slurry generated during AD of organic waste, reduces emissions incurred in agriculture due to the manufacture and use of chemical fertilizers (Singh et al., 2010, 2011). This chapter is focused on technologies involved to enhance the biomethane production potential of different organic wastes.

Types of organic waste

The types of organic wastes that can be utilized for AD are discussed below. These are named mainly on the basis of the source of organic residue.

Household or domestic waste

Household or domestic waste comprises food scraps, paper, kitchen and garden waste such as grass cuttings or trimmings from bushes and hedges. Domestic waste is often mixed with non-organic materials such as glass and plastic packaging, which cannot be digested. The screening and separation of such waste at source is beneficial for AD.

Table 7.1 Average collectable slurry production during housing period from different animals.

Animal	Collectable slurry (t/a/head)
Cattle	508
Pigs	1.45
Sheep	0.05
Poultry	0.14

Source: Singh et al., 2010

Animal and human waste

Animal waste contains litter, urine and faeces, and is best suited for AD without any loss of nutrients. Human faecal residue is produced in large quantities in urban and rural areas. The handling of sewage has a high health risk, as raw sewage contains pathogens that can cause serious illness. The utilization of sewage for AD must therefore follow health and safety procedures. Singh et al. (2010) calculated the collectable slurry production during housing of different animals (Table 7.1). This quantity of slurry may be less than actual production because it is based on an average housing period of only 20 weeks for cattle and 6 weeks for sheep, while from pig and poultry all slurry can be collected.

Biodegradable municipal waste (BMW)

BMW is the fraction of municipal waste that can undergo biological decomposition and is typically composed of food and garden waste, wood, paper, cardboard and textiles. The rate of municipal waste generation is closely linked to gross domestic product (GDP) and personal consumption, while it is less closely linked to population growth (Le Bolloch et al., 2007). The average quantity of BMW generation in different countries is summarized in Table 7.2.

Agricultural waste

Agricultural residue is the waste which remains after the harvesting and processing of crops, e.g. stalks, husks, foliage, etc. These residues are rich in hemicelluloses and lignin and can be digested very well with a pre-treatment to remove/reduce the lignin content of the residue. The production of agricultural residue is increasing, as agricultural productivity increases and reached 3758 million tonnes in 2001 from 3448 million tonnes in 1991 (Table 7.3), i.e. a 9 per cent increase in agricultural waste production in one decade.

Slaughterhouse waste

The slaughtering of animals produces large quantities of organic waste that have the potential for methane production by anaerobic digestion. Slaughtering of cattle generates about 33 per cent waste (rumen, stomach, intestinal content, animal low-risk waste and blood) along with about 5 per cent fat and tallow, which can be utilized safely for AD with some handling precautions (Singh et al., 2010). Sheep and goat slaughtering generates about 12 per cent waste of their body weight (Liu, 2006; Clottery, 1985). Slaughtering of poultry generates about 0.5 kg per head waste of their live body weight (Singh et al., 2010).

Table 7.2 Biodegradable municipal waste generation in different countries.

Country	BMW (kg/cap/day)
Austria	1.59
Bangladesh	0.5
Bolivia	0.44
Bulgaria	1.42
Canada	1.34
China	0.8
Estonia	1.21
Honduras	0.4
India	0.3–0.6
Indonesia	0.8–1.0
Japan	1.3
Nepal	0.2–0.5
Netherlands	1.7
Nigeria	0.8
Pakistan	0.6–0.8
Philippines	0.3–0.7
South Africa	2.74
Sri Lanka	0.2–0.9
Sudan	0.8
UK	1.56
USA	3.12

Source: Ronteltap et al., 2009

Techniques to improve the biomethane potential

A number of techniques can be used for enhancing the biomethane potential of organic residue, e.g. digester design, operating conditions, pre-treatments, co-digestion, etc. Some important techniques are discussed below.

Digester type

Nizami and Murphy (2010) assessed numerous available technologies and combinations of AD; from one-stage batch dry systems to two-stage wet continuous systems; from one-stage continuous wet systems to two-stage systems incorporating a batch dry reactor coupled with a second-stage high-rate reactor. They concluded that optimal configurations can only be established by operating the different configurations in parallel, in real time, digesting the same grass silage feedstock. The design of digester has a great impact on the biomethane production potential of organic waste with one type of digester most efficient for one type of organic residue but not the most efficient for another residue

Table 7.3 The world crop residue production in 1991 and 2001.

Crop	Residue production (10^6 Mt)	
	1991	*2001*
Barley	254	212
Corn	479	609
Millet	44	44
Oats	34	27
Rice	780	890
Rye	41	35
Sorghum	87	87
Wheat	826	875
Other cereals	18	23
Beans	18	17
Broad beans	5	4
Chick peas	11	6
Groundnut	23	35
Lentils	2	3
Peas	16	11
Pulses	60	52
Soybeans	238	305
Linseed	3	2
Rapeseed	41	54
Safflower	1	1
Seed cotton	90	24
Sesame	4	6
Sunflower	23	21
Sugarbeet	76	59
Sugarcane	264	314
Tubers	47	59
Potato	67	77
Sweet potato	31	34
Total	3448	3758

Source: Lal, 2005

(Singh et al., 2011). Nizami and Murphy (2010) examined three digestion configurations, viz. dry continuous, wet continuous and a two-phase system (sequential leach bed reactor complete with upflow anaerobic sludge blanket) for biomethane production from grass and found that biomethane production varied by 12 per cent.

Digester operating conditions

The performance of AD can be optimized by monitoring the variations in parameters such as temperature, pH, C:N ratio, agitation, organic loading rate (OLR), hydraulic retention time (HRT), etc. Anaerobes are most active in the mesophilic (30–40°C) and thermophilic (50–60°C) temperature range (Nizami et al., 2010, 2011; Singh et al., 2010, 2011). The length of fermentation period is dependent on temperature (Yadvika et al., 2004). The biomethane production potential is greatly affected by OLR, which is varied according to the type and size of digester and feedstock (Lehtomäki, 2006; Nizami and Murphy, 2010). HRT is the average time spent by the input slurry inside the digester before it comes out. In tropical countries like India, HRT varies from 30–50 days, while in colder climate countries HRT may exceed 100 days. Short HRT might risk washout of an active bacterial population (incomplete digestion) while longer HRT requires a larger digester volume (Yadvika et al., 2004). It is important to maintain the optimal C:N ratio inside the digester. Generally microorganisms utilize carbon 25–30 times faster than nitrogen, thus a 25–30:1 C:N ratio would be optimal for AD (Yadvika et al., 2004). The optimal C:N ration can be maintained by mixing different feedstocks. Shimizu (1992) revealed that high bacterial concentrations could be retained in the AD by the addition of metal cations. Wong and Cheung (1995) found that plants with a high content of metals (Cr, Cu, Ni and Zn) produced a higher CH_4 yield than the control.

Pre-treatment of waste

Organic residue contains both soluble and insoluble substrates. Soluble substrates include monomers (monosaccharides, amino acids, and long-chain fatty acids) that are readily solubilized, absorbed into microorganisms and metabolized. Insoluble substrates include macromolecules (disaccharides, oligosaccharides, proteins, and lipids) that require enzymatic hydrolysis to break them down to their constituent monomers (Lai et al., 2009; Nizami et al., 2010). The methane production from AD of plant residues can be increased by pre-treatment of the substrate in order to break the polymer chains to more easily accessible soluble compounds (Lehtomäki, 2006). Pre-treatment by a physical, chemical or biological process, or a combination of these, increases surface area and reduces lignin content and polymerization of cellulose (Fan et al., 1981). Different pre-treatment methods, their impact on feedstock and pros and cons are summarized in Table 7.4.

The most important physical pre-treatment of crop biomass is particle size reduction, which increases the available surface area and release of intracellular components (Palmowski and Müller 1999). Chemical pre-treatments include treatments with acids, alkalis, solvents or oxidants. In biological pre-treatments, either microbes and/or microbial enzymes are used for partial degradation of complex molecules such as lignocellulose.

Composting is a bio-oxidative process involving the mineralization and partial humification of organic matter (Delgenès et al., 2003). Lignin degradation has been reported in the thermophilic stage of composting (Tuomela et al., 2000). When designing appropriate pre-treatment methods for anaerobic digestion of crop biomass, the costs, practicability and environmental impacts of pre-treatments, as well as the losses of organic matter and energy content of substrates during pre-treatments, need to be weighed against the overall benefits of pre-treating the biomass (Lehtomäki et al., 2004; Sun and Cheng 2002). Lehtomäki et al. (2004) conducted experiments on different pre-treatment of grass for methane production

Table 7.4 Different pre-treatments, their effects and pros and cons.

Treatment	Possible changes in biomass	Pros and cons
Physical treatments • Milling: ball milling; two-roll milling; hammer milling; colloid milling; vibro energy milling • Irradiation: gamma-ray irradiation; electron-beam irradiation; microwave irradiation • Others: hydrothermal; high pressure steaming; expansion; extrusion; pyrolysis	• Increase in surface area • Increase in accessible surface area and pore size • Decrease in cellulose crystallinity • Decrease in degrees of polymerization	• Most of the methods are highly energy demanding • Most of them cannot remove the lignin • It is preferable not to use these methods for industrial applications • No chemicals are generally required for these methods
Chemical/physico-chemical treatments • Explosion: Steam explosion; ammonia fibre explosion (AFEX); CO_2 explosion; SO_2 explosion • Alkali: sodium hydroxide; ammonia; ammonium sulfite • Acid: sulfuric acid; hydrochloric acid; phosphoric acid gas:; chlorine dioxide; nitrogen dioxide; sulfur dioxide • Oxidizing agents: hydrogen peroxide; wet oxidation; ozone • Solvent extraction of lignin: ethanol-water extraction; benzene-water extraction; ethylene glycol extraction; butanol-water extraction; swelling agents	• Increase in accessible surface area • Partial or nearly complete delignification • Decrease in cellulose crystallinity • Decrease in degrees of polymerization • Partial or complete hydrolysis of hemicelluloses	• These methods are among the most effective and include the most promising processes for industrial applications • Usually rapid treatment rate • Typically need harsh conditions • There are chemical requirements
Biological treatment Fungi and actinomycetes	• Delignification • Reduction in degree of polymerization of cellulose • Partial hydrolysis of hemicellulose	• Low energy requirement • No chemical requirement • Mild environmental conditions • Very low treatment rate • Did not consider for commercial application

Source: Taherzadeh and Karimi, 2008

and concluded that methane production can be increased by up to 17 per cent (Table 7.5). Biological pre-treatment offers advantages such as low energy or chemical requirement. However, currently few facilities use biological pre-treatment processes.

Co-digestion

Fang et al. (2011) used three reactor experiments to investigate biogas process efficiency and stability: by co-digestion of sugar beet pulp (SBP), desugared molasses (DMo) and manure in different ratios. They found that DMo was potentially inhibiting the biogas process and the co-digestion of SBP and DMo was only successful when highly diluted with manure or

Table 7.5 Effect of pre-treatments on the methane production, values in parenthesis are the percentage change in methane production from untreated grass.

Pre-treatment	Methane potential after pre-treatment ($m^3CH_4\ kg^{-1}VS_{added}$)
Before pre-treatment	0.23
NaOH 2% 24 h 20°C	0.25 (9)
NaOH 2% 72 h 20°C	0.27 (17)
3% Ca(OH)$_2$ + 4% Na$_2$CO$_3$ 24 h 20°C	0.24 (4)
3% Ca(OH)$_2$ + 4% Na$_2$CO$_3$ 72 h 20°C	0.27 (17)
Autoclaving	0.26 (13)
Pre-incubation in water 24 h 35°C	0.26 (13)
Enzyme 24 h 35°C	0.27 (17)
White rot fungi 21 h 20°C	0.24 (4)
Composting 7 days	0.19 (–17)

Source: Lehtomäki et al., 2004, Lehtomäki, 2006

water. In contrast, SBP was shown to be a good substrate for biogas production, producing a methane yield of 280 mL-CH$_4$/gVS in a thermophilic continuously operated reactor, co-digesting 50 per cent of SBP with cow manure. Astals et al. (2011) conducted an experiment to co-digest pig manure and glycerine at different ratios and found that 80:20 per cent pig manure:glycerine produced the greatest amount of methane.

Biomethane potential and bottlenecks of different organic waste

Biomethane potential of different organic wastes

Biomethane production potential varies from one type of organic waste to another, depending upon the release of volatile solids during AD, digester type, and digester operating conditions (pH, temperature, loading rate, etc.). The average biomethane potential of different organic waste is presented in Table 7.6 and varies significantly from 14 to 600 m^3 biogas per tonne of feedstock due to differences in their constitution (cellulose, hemicellulose, lignin, etc.). Methane production of a specific crop residue is affected by the chemical composition of the residue, which changes as the plant matures as affected by the timing and conditions of harvest (Lehtomäki et al., 2008; Gunaseelan, 1997).

Biomethane potential of municipal wastes

The biomethane potential of biological municipal waste is calculated on the basis of quantity of BMW generation reported by Ronteltap et al. (2009) per capita per day and is presented by country in Table 7.7.

Nepal produces the least BMW per capita per day (0.35 kg/capita/day) and the USA produces the most (3.12 kg/capita/day), about 10 times more than Nepal. China has the highest potential to produce biomethane from BMW because of its population size, with

Table 7.6 Biomethane production potential of organic waste.

Waste material	Biogas production potential ($m_n^3 t^{-1}$)
Maize	107.25
Barley	233.1
Rye	266.625
Triticale	246.5
Sugar beet	167.2
Hemp	83.2
Alfalfa	132.3
Rice straw	156.6
Human excreta	80
Cattle slurry	22
Pig manure	22
Sheep manure	54.4
Poultry litter	44.5
Slaughterhouse waste (cattle/pig/sheep)	156
Slaughterhouse waste (poultry)	110–140
Grass (hay)	100–118
Grass with chicken manure	350
Paper with chicken manure	400–500
Sewage sludge	600
Wheat straw	125
Water hyacinth	14.7
Leaf litter	600

Source: Vogler, 1981; Brikmose 2000; Salminen and Rintala, 2002; Murphy and Power, 2006; Tricase and Lombardi, 2009; Singh et al., 2010, 2011; Bond and Templeton, 2011

the USA second due to a relatively high rate of BMW generation. The total potential of biomethane production from BMW generated from the listed countries is 112 thousand million m³ per year, i.e. equivalent to about 101 MT oil equivalents.

Biomethane potential of agricultural residues

Lal (2005) reported the global quantity of agricultural crop residue production for 2001 and the biomethane potential of different crop residues is reported by various researchers (Vogler, 1981; Brikmose 2000; Salminen and Rintala, 2002; Murphy and Power, 2006; Tricase and Lombardi, 2009; Singh et al., 2010, 2011; Bond and Templeton, 2011; Rabelo et al., 2011) (Table 7.8). The total crop residue production of crops listed in Table 7.8 is about 3000 MT for the year 2001. If all of this were used for the production of biomethane, it could generate about 192 thousand million m³ biomethane, i.e. equivalent to about 520 Mt oil equivalent, although in practice much would need to be left in field to recycle nutrients and condition soil.

Table 7.7 Country-wise biomethane potential of biological municipal waste (BMW).

Country	BMW (kg/cap/day)	*Population (M)	BMW (Mt/y)	**Potential (Mm³ biomethane/y)
China	0.8	1343.24	392.23	31868.37
USA	3.12	313.85	357.41	29039.76
India	0.45	1205.07	197.93	16082.04
Indonesia	0.9	248.22	81.54	6625.15
Japan	1.3	127.37	60.44	4910.51
Nigeria	0.8	170.12	49.68	4036.1
South Africa	2.74	48.81	48.81	3966.21
Pakistan	0.7	190.29	48.62	3950.3
UK	1.56	63.05	35.9	2916.93
Bangladesh	0.5	161.08	29.4	2388.51
Philippines	0.5	103.77	18.94	1538.71
Canada	1.34	34.3	16.78	1363.06
Netherlands	1.7	16.73	10.38	843.45
Sudan	0.8	34.21	9.99	811.63
Austria	1.59	8.22	4.77	387.6
Sri Lanka	0.55	21.48	4.31	350.36
Nepal	0.35	29.89	3.82	310.25
Bulgaria	1.42	7.04	3.65	296.47
Bolivia	0.44	10.29	1.65	134.27
Honduras	0.4	8.3	1.21	98.46
Estonia	1.21	1.27	0.56	45.57
Total		4146.6	1378.01	111963.71

Sources: * The World Fact Book (2012); ** Singh et al. (2010); Brown (2004); Murphy and Power (2006)

Table 7.8 World-wide biomethane potential of some agricultural residues.

Crop	Residue production (Mt)	Potential (m³ bio-methane/t)	Total potential (Mm³)
Barley	212	233.1	49417.2
Corn	609	107.3	65315.3
Rice	890	156.6	139374.0
Rye	35	266.6	9331.9
Wheat	875	125.0	109375.0
Sugarbeet	59	167.2	9864.8
Sugarcane	314	613.0	192482.0
Total	2994		575160.1

The available quantity of organic residue has great potential to play a major role in energy supply. The conversion of biomethane from complex organic biomass (e.g., agricultural and animal waste) has relatively low conversion efficiency and presently costs more per unit than gas from natural-gas deposits but has the advantage of turning a pollution problem into an energy resource (Rittmann, 2008).

Conclusions

Organic wastes are generated in huge quantities in every part of the world. Digestion of organic waste can be done by different types of AD plant. It is necessary to determine the appropriate digester and optimal operating conditions for each type of organic waste. Pretreatment of mainly plant residues enhances biomethane production significantly and co-digestion can also increase biomethane yield.

References

Astals, S., Ariso, M., Galí, A. and Mata-Alvarez, J. (2011) 'Co-digestion of pig manure and glycerine: Experimental and modelling study', *Journal of Environmental Management*, vol. 92, pp. 1091–1096.

Bond, T. and Templeton, M.R. (2011) 'History and future of domestic biogas plants in the developing world', *Energy for Sustainable Development*, vol. 15, pp. 347–354.

Brikmose T. (2000) *Centralised biogas plants in Denmark*. Aarhus: The Danish Agricultural Advisory Centre, The National Department of Crop Production.

Brown, K. (2004) *Bioenergy in Ireland*, Dublin: Sustainable Energy Ireland.

Cherubini, F. and Ulgiati, S. (2010) 'Crop residues as raw materials for biorefinery systems – A LCA case study', *Applied Energy*, vol. 87, pp. 47–57.

Clottery, J.A. (1985) 'Manual for the slaughter of small ruminants in developing countries', Rome: Food and Agriculture Organization of the United Nations, http://www.fao.org/docrep/003/X6552E/X6552E00.htm, accessed 10 March 2012.

Delgenès, J.P., Penaud, V. and Moletta, R. (2003) 'Pretreatments for the enhancement of anaerobic digestion of solid wastes', in: J. Mata-Alvarez (ed.), *Biomethanization of the organic fraction of municipal solid wastes*, IWA Publishing, London, pp. 201–228.

Fan, L.T., Gharpuray, M.M. and Lee, Y.-H. (1981) 'Evaluation of pretreatments for enzymatic conversion of agricultural residues', *Biotechnology and Bioengineering Symposium*, vol. 11, pp. 29–45.

Fang, C., Boe, K. and Angelidaki, I. (2011) 'Anaerobic co-digestion of by-products from sugar production with cow manure', *Water Research*, vol 45, pp. 3473–3480.

Gunaseelan, V.N. (1997) 'Anaerobic digestion of biomass for methane production. A review', *Biomass and Bioenergy*, vol 13, pp. 83–114.

Lai, T.E., Koppar, A.K., Pullammanappallil, P.C. and Clarke, W.P. (2009) 'Mathematical modeling of batch, single stage, leach bed anaerobic digestion of organic fraction of municipal solid waste', in J. Kallrath, P.M. Pardalos, S. Rebennack, M. Scheidt (eds), *Optimization in the energy industry*, Berlin: Springer.

Lal, R. (2005) 'World crop residues production and implications of its use as a biofuel', *Environment International*, vol 31, pp. 575–584.

Le Bolloch, O., Cope, J., Meaney, B. and Kurz, I. (2007) *National waste report 2006*, Johnstown Castle: Environmental Protection Agency.

Lehtomäki, A. (2006) *Biogas production from energy crops and crop residues*, Jyväskylä: University of Jyväskylä.

Lehtomäki, A., Viinikainen, T.A., Ronkainen, O.M., Alen, R. and Rintala, J.A. (2004) 'Effect of pretreatments on methane production potential of energy crops and crop residues', in S.G. Guiot, S.G.

Pavlostathis, and J.B. van Lier (eds), *Proceedings the 10th World IWA Congress on Anaerobic Digestion*, pp 1016–1021, London: IWA Publishing.

Lehtomäki, A., Viinikainen, T.A. and Rintala, J.A. (2008) 'Screening boreal energy crops and crop residues for methane biofuel production', *Biomass and Bioenergy*, vol 32, pp. 541–550.

Liu, D. (2006) 'Waste management in meat plant' in Y.H. Hui (ed.). *Handbook of food science and engineering*, vol. 3, pp. 147.1–47.10. Boca Raton: Taylor & Francis.

Murphy, J.D. and Power, N.M. (2006) 'A technical, economic and environmental comparison of composting and anaerobic digestion of biodegradable municipal waste', *Journal of Environmental Science and Health Part A Toxic/Hazardous Substances & Environmental Engineering*, vol 41, no 5, pp. 865–879.

Nigam, P.S. and Singh, A. (2011) 'Production of liquid biofuels from renewable resources', *Progress in Energy and Combustion Science*, vol 37, pp. 52–68.

Nizami, A.S. and Murphy, J. (2010) 'What type of digester configurations should be employed to produce biomethane from grass silage?', *Renewable and Sustainable Energy Reviews*, vol 14, pp. 1558–1568.

Nizami, A.S., Thamsiriroj, T., Singh, A. and Murphy, J.D. (2010) 'The role of leaching and hydrolysis in a two phase grass digestion system', *Energy and Fuels*, vol 24, pp. 4549–4559.

Nizami, A.S., Singh, A. and Murphy, J. (2011) 'Design, commissioning and start-up of a sequentially fed leach bed reactor complete with upflow anaerobic sludge blanket digesting grass silage', *Energy and Fuels*, vol 25, no 2, pp. 823–834.

Palmowski, L. and Müller, J. (1999) 'Influence of the size reduction of organic waste on their anaerobic digestion', in J. Mata-Alvarez, F. Cecchi, and A. Tilche (eds), *Proceedings 2nd International Symposium on Anaerobic Digestion of Solid Waste*, pp. 137–144. London: IWA Publishing.

Prasad, S., Singh, A. and Joshi, H.C. (2007a) 'Ethanol as an alternative fuel from agricultural, industrial and urban residues', *Resources, Conservation and Recycling*, vol 50, pp. 1–39.

Prasad, S., Singh, A., Jain, N. and Joshi, H.C. (2007b) 'Ethanol production from sweet sorghum syrup for utilization as automotive fuel in India', *Energy and Fuels*, vol 21, no 4, pp. 2415–2420.

Rabelo, S.C., Carrere, H., Maciel Filho, R., and Costa, A.C. (2011) 'Production of bioethanol, methane and heat from sugarcane bagasse in a biorefinery concept', *Bioresource Technology*, vol. 102, no. 17, pp. 7887–7895.

Rittmann, B.E. (2008) 'Opportunities for renewable bioenergy using micro organisms', *Biotechnology Bioengineering*, vol. 100, pp. 203–212.

Ronteltap, M., Khadka, R., Sinnathurai, A.R. and Maessen, S. (2009) 'Integration of human excreta management and solid waste management in practice', *Desalination*, vol 248, pp. 369–376.

Salminen, E. and Rintala, J. (2002) 'Anaerobic digestion of organic solid poultry slaughterhouse waste—a review', *Bioresource Technology*, vol 83, pp. 13–26.

Shimizu, C. (1992) 'Holding anaerobic bacteria in digestion tank', JP Patent 4341398.

Singh, A., Smyth, B.M. and Murphy, J.D. (2010) 'A biofuel strategy for Ireland with an emphasis on production of biomethane and minimization of land-take', *Renewable and Sustainable Energy Reviews*, vol 14, no 1, pp. 277–288.

Singh, A., Nizami, A.S., Korres, N.E. and Murphy, J.D. (2011) 'The effect of reactor design on the sustainability of grass biomethane', *Renewable and Sustainable Energy Reviews*, vol 15, no 3, pp. 1567–1574.

Sun, Y. and Cheng, J. (2002) 'Hydrolysis of lignocellulosic materials for ethanol production. A review', *Bioresource Technology*, vol 83, pp. 1–11.

Taherzadeh, M.J. and Karimi, K. (2008) 'Pretreatment of lignocellulosic wastes to improve ethanol and biogas production: A review', *International Journal of Molecular Sciences*, vol 9, pp. 1621–1651.

The World Fact Book (2012) Central Intelligence Agency, US. https://www.cia.gov/library/publications/the-world-factbook/rankorder/2119rank.html Accessed on 10 August 2012.

Tricase, C. and Lombardi, M. (2009) 'State of the art and prospects of Italian biogas production from animal sewage: technical-economic considerations', *Renewable Energy*, vol 34, pp. 477–485.

Tuomela, M., Vikman, M., Hatakka, A. and Itävaara, M. (2000) 'Biodegradation of lignin in a compost environment: a review', *Bioresource Technology*, vol 72, pp. 169–183.

Vogler, J. (1981) *Work from waste: recycling wastes to create employment*, London: Intermediate Technology Publications.

Wong, M.H. and Cheung, Y.H. (1995) 'Gas production and digestion efficiency of sewage sludge containing elevated toxic metals', *Bioresource Technology*, vol 54, no 3, pp. 261–268.

Yadvika, S., Sreekrishnan, T.R., Kohli, S. and Rana, V. (2004) 'Enhancement of biogas production from solid substrates using different techniques-a review', *Bioresource Technology*, vol 95, pp. 1–10.

Chapter 8

Industrial residues for biomethane production

Markus Ortner,[1*] *Bernhard Drosg,*[1] *Elitza Stoyanova*[1] *and Günther Bochmann*[1]

[1]Institute for Environmental Biotechnology, BOKU – University of Natural Resources and Life Sciences, Vienna
Konrad Lorenz Str. 20, A-3430 Tulln an der Donau, Austria
*Corresponding author email: markus.ortner@boku.ac.at

Introduction

As a result of increasing energy costs and costs for the disposal and treatment of industrial residues, the interest of industrial companies in using renewable energy sources is increasing. An efficient and proper residue management system is necessary in the food processing and beverage industries such as dairies, breweries or abattoirs (slaughterhouses), due to several environmental concerns such as global climate change and diminishing fossil fuel resources. These industries produce various organic residues with high energy content, and biogas fermentation technology is an attractive option to cope with these residues. Food-processing factories, particularly abattoirs or breweries, use a large number of energy-intensive processes at different temperatures. The basis for designing and dimensioning such factories is always the peak energy load, which should guarantee a constant energy supply and a secure production. In most cases, only a part of the supplied energy is used, and as a consequence, such factories are neither cost nor energy efficient.

The main running costs in industrial processes are – apart from manpower – the energy supply (natural oil/gas) and the disposal and treatment costs of the residues. Certain industrial plants such as breweries, bioethanol plants or dairies are in the fortunate situation of being able to gain revenue by selling parts of their residues as animal feed. In general, the revenues are not high but at least it helps keep the disposal costs down.

Rendering of animal proteins has been an accepted pathway for treatment of slaughterhouse wastes for a long time. After the appearance of BSE (bovine spongiform encephalopathy) in Europe, the European Commission banned rendered animal protein from being fed to farmed animals in 2000 (European Commission, 2000, Decision 2000/766/EC). The result was a tremendous increase in disposal fees of slaughterhouse wastes and an additional financial burden on the abattoir industry.

Anaerobic digestion (AD) is an adequate and well-known technology to treat industrial organic residues almost regardless of their consistency. The utilisation of industrial organic residues by AD is an appropriate way to improve both the process and the economic efficiency of an industrial factory. Anaerobic digestion produces renewable energy in the form of biogas. Furthermore, it enables a controlled stabilisation of the organic material, reduces greenhouse gas emissions and contributes to the closing of nutrient cycles.

The fermentation of process-specific waste materials to biogas yields a highly combustible gas consisting of 55–70 per cent methane. It can be used in a combined heat and power plant (CHP) to generate heat and electricity, to substitute fossil fuels for heat and steam generation or as vehicle fuel if upgraded to biomethane.

The implementation of an AD unit onsite has some advantages. The biggest are the reduction of disposal costs, the current national subsidies by the green energy law when generating electricity by CHP and the possibility of waste heat integration (CHP) onsite into the production process. In addition, the effluent of the anaerobic digester, the digestate, represents a high-quality agricultural fertiliser, which can be used directly or as a processed concentrate. The reduction of greenhouse gas emissions is an additional advantage. On the other hand, there are limitations and technical challenges regarding the composition of the residues. The main issues that have to be considered are: the lack of essential microelements; process instabilities (foaming, low degradation rates and gas yields) caused by the digestion of protein-rich material such as slaughterhouse waste; and foaming during the pasteurisation step when using lipid- and protein-rich fractions. Another issue is that certain materials such as brewers' spent grains are rich in components that are difficult to degrade due to their cellulose and/or hemicellulose content. In this case, a pre-treatment step is necessary to make these components available to anaerobic degradation. Another very important issue in industrial biogas processes is the utilisation of the digestate, the effluent from a biogas plant. Especially if the organic by-products are accumulated in very large amounts (e.g. bioethanol plants) an optimised strategy for the utilisation of the digestate can become decisive. The integration of the waste heat of the CHP into the production process represents a further challenge.

In the following paragraphs, concrete examples of industries and their potential for anaerobic treatment as well as their main critical issues (bottlenecks) are described in detail. These examples stand for the main challenges in the AD of industrial residues. Within the selected industrial residues, aspects of high nitrogen impact, need of pre- and post-treatment techniques as well as other inhibiting or limiting factors will be discussed.

Anaerobic digestion of slaughterhouse waste

Slaughterhouse wastes (SW) are parts of slaughtered animals which are not intended for direct human consumption or animal feed. It is estimated that humans directly consume only 68 per cent (%w/w) of a chicken, 62 per cent of a pig and 54 per cent of a cow (Freudenreich & Bach, 1993). Large amounts of animal by-products accumulate and have to be disposed of. Certain parts can be recycled and used in human food, cosmetics or pharmaceuticals, such as gelatine derived from bones. An overview of the annual amounts of animal by-products derived from pigs, cattle and poultry in several selected countries and worldwide (total) can be found in Table 8.1.

The usual treatment is a rendering process in carcass plants. Although the rendering process is very energy intensive and expensive, the sale of meat and bone meal as an animal feed additive was a valuable source of income for slaughterhouses until new legislation was introduced in 2000 in response to the BSE outbreak. The European Union immediately banned rendered animal proteins from human and animal food-chains and enacted a law in 2002 for safe and proper disposal of slaughterhouse wastes (animal by-product (ABP) regulation EC 1774/2002 replaced in 2009 by EC 1069/2009). As a result, a serious protein gap emerged in Europe as meat and bone meal was no longer available as a protein source; and meat and bone meal turned from being a valuable product to a problematic waste. The disposal of this waste put an economic burden on agriculture as well as on all other

Table 8.1 Overview of estimated amounts of animal by-products (ABP) deriving from slaughter (t/a) (FAO 2012).

	Pigs	Cattle	Poultry
Austria (2010)	126,500	214,200	28,920
China (2010)	14,835,000	n.a	3,332,000
Egypt (2009)	n.a.	275,400	n.a.
Germany (2010)	1,265,000	1,071,000	273,200
India (2006)	n.a.	746,640	156,400
Italy (2009)	303,600	826,200	n.a.
Poland (2009)	437,000	n.a.	n.a.
Turkey (2009)	n.a.	459,000	288,000
World (2010)	26,162,500	56,640,688	36,763,636

sectors linked to meat production. This act does, however, allow alternative pathways for the treatment of this waste material, such as the utilisation in anaerobic digestion systems if approved pre-treatment steps are applied, depending on the by-product category (according to the potential risk to animals, the public or to the environment). The categories described in the following paragraphs are mainly related to slaughterhouse waste and wastewater. It should be mentioned that the animal by-product act regulates the safe disposal of the entire spectrum of animal by-products including dairy by-products, kitchen and canteen waste and organic waste, which will not be detailed here.

Risk categories of animal by-products

Category I: Materials (i.e. spinal cord, brain, eyes of cattle) that present the highest risk such as TSE (transmissible spongiform encephalopathy) or scrapie and have to be completely disposed of by incineration.

Category II: Includes all materials that do not fit into category I or III and present a risk of contamination with other animal diseases. These may not be used in feed, but can be recycled for other uses (e.g. biogas or composting) after appropriate treatment (sterilisation at 133°C and 3 bar for a minimum 20 min, particle size < 50 mm). Exceptions include intestinal contents, manure or milk that can be used in a biogas plant without any sanitation steps.

Category III: Materials (i.e. by-products derived from healthy animals slaughtered for human consumption, blood) may be used in the production of animal feeds following appropriate treatment in approved processing plants. The treatment comprises pasteurisation at 70°C for 60 min minimum with a required particle size ≤ 12 mm).

Pathways of treatment of slaughterhouse waste

There are different pathways for treating slaughterhouse wastes: either the transformation to electrical and thermal energy or the supply of valuable compounds for biotechnological and chemical transformation to chemical precursors. The last option is still not fully developed and further research is necessary for industrial scale. The most common way at the moment is the utilisation of rendered meat and bone meal as a secondary fuel in cement plants or waste incineration plants. The meal has similar heat value to lignite and shows very good burning characteristics. Rendered grease is mainly used in the rendering plant as a substitute for heavy fuel oil for heat generation. Alternatively it can be transformed into biodiesel. Slaughterhouse waste is considered to be an excellent substrate for fermentation processes. For instance, the biotechnological production of certain chemical precursors such as polyhydroxyalkanoates (PHA) is an alternative option, as is the formation of bio composites (Braunegg, 2006).

Composting of category III material is also a feasible way to process slaughterhouse waste.

Anaerobic digestion of slaughterhouse waste

Last but not least, the anaerobic treatment of slaughterhouse waste in biogas plants is considered to be a challenging and promising alternative. Due to the high protein and lipid content, SW is considered to be a very good substrate for biogas production with expected high yields of methane. In theory, proteins are able to deliver biogas containing 60 per cent methane, while lipids produce biogas containing 72 per cent methane. However, in practice a lot of limitations restrict the applicability of SW.

The most significant limitations are the slow hydrolysis rate of certain particulates which are difficult to degrade, as well as foaming and floatation caused by lipid degradation resulting in a biomass washout and different inhibitory effects caused by several intermediates (i.e. long chain fatty acids (LCFA), hydrogen sulphide (H_2S) or ammonia (NH_3)) formed during the degradation process (Chen et al., 2008; Salminen et al., 1995; Angelidaki & Ahring, 1993).

For these reasons, it is difficult to digest this material as single substrate. Therefore, slaughterhouse wastes are commonly used as co-substrates in the agro-industrial sector together with canteen waste, manure and/or energy crops.

Integration of AD technology into an abattoir

Few AD plants use SW as a single substrate, but one of these biogas plants is located in St Martin (Austria). The biogas plant in St Martin was erected in 2003 and is operated only with SW derived from the nearby pig abattoir with an annual capacity of 500,000 heads. By the time of construction, this plant was the first abattoir worldwide utilising wastes in mono-fermentation. The main idea was to reduce its running costs in terms of energy supply and waste disposal by implementing an AD plant onsite (using also some additional rumen content from a cattle abattoir nearby). The overall treatment capacity of the AD plant is 12,000 t organic residues per annum, which covers the waste fractions accumulated during the slaughter process. The substrate consists of blood, rumen and rumen content derived from the nearby cattle abattoir, grease, stomach content, colon and wastewater from the slaughtering facilities (see Table 8.2).

The plant employs conventional two-stage fermentation with CSTRs (continuously stirred tank reactors). As shown in Figure 8.1, there are three main fermentation tanks, which are loaded in parallel, followed by a third one and a final storage tank for the digestate. According to the European Directive (1069/2009 EC), the material is minced to a maximum particle size of 12 mm and collected in a separate buffer tank followed by pasteurisation at 70°C for 60 min. After passing the recuperator, substrate is pumped into the two main fermenters. The operation temperature is 38°C; higher temperatures are not recommended because the high concentrations of lipids can cause foaming and associated operational problems. The biogas produced in the AD plant is directed to an external biological desulphurisation

Table 8.2 Chemical characterisation of slaughterhouse waste processed in the biogas plant.

Substrate	TS (%)	VS (%)	COD (g/kg)	TKN (g/kg)	Relative amount
Blood (pig)	18.2	16.7	265	28.0	+++
Blood (cattle)	18.5	17.0	260	27.5	+++
Colon (pig)	24.4	22.0	575	10.0	+
Stomach content (pig)	24.4	23.5	408	5.3	++
Rumen content	13.6	13.0	187	3.7	++
Omasum	19.3	18.4	698	15.1	+
Fat scrubber material	6.1	5.4	157	1.3	+++

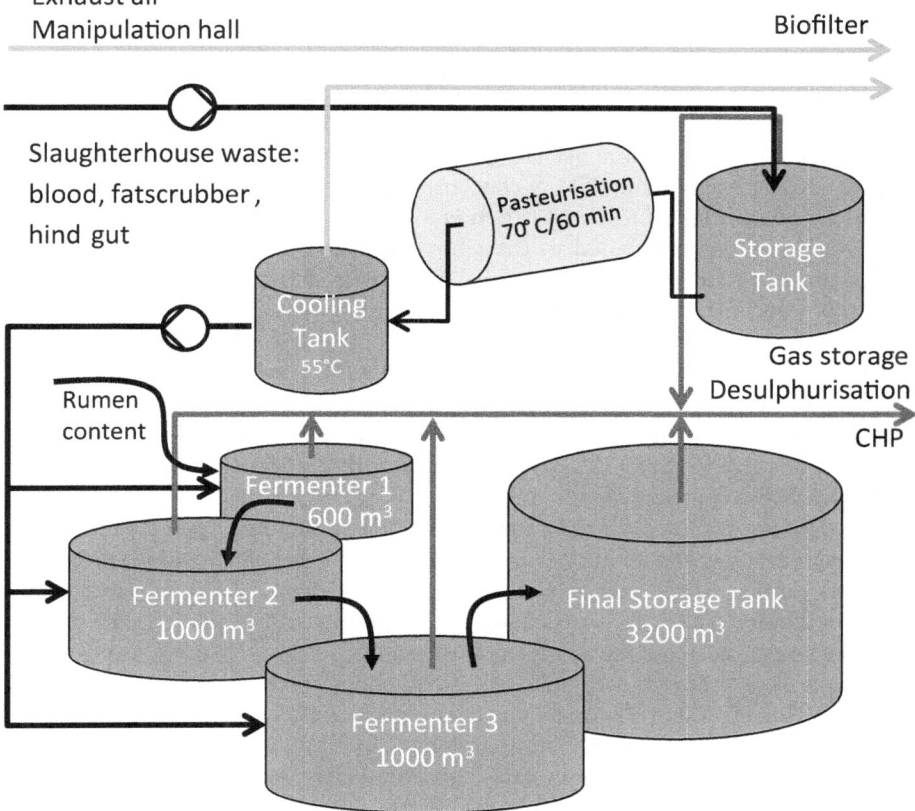

Figure 8.1 Process scheme of the AD-plant

unit and combusted afterwards in a combined heat and power plant (CHP) with an average monthly output of approximately 300 MWh electricity and 300 MWh heat. The generated electricity covers about 43 per cent of the abattoir's electricity demand. About 25 per cent of the waste heat generated from the CHP is used for the biogas plant including the sanitation unit and the desulphurisation unit. Most heat (about 75 per cent) is used in the abattoir, covering about 90 per cent of the abattoir's heat demand. The overall degree of energetic self-sufficiency of the abattoir is about 55 per cent.

The biogas plant at St Martin is a pioneer project in terms of mono-fermentation of SW and is the result of intensive research activities in this field over the last seven years and the willingness of the facility owner to embrace alternative treatment technologies.

Limitations and bottlenecks

Although there are many advantages, there are certain critical issues that have to be considered when using SW in a mono-fermentation. The high protein content of SW (up to 33 g TKN/kg, mainly from the blood) leads to a high ammonia concentration during the degradation process. Free ammonia is well known and described as a potential inhibitor of anaerobic digestion. Among the different groups of microorganisms involved in the degradation, the methanogens are the most sensitive to ammonia inhibition, which causes them to slow their growth (Kayhanian, 1994).

Methanogen strains isolated from AD sludge, such as *Methanospirillum hungatei*, show very high sensitivity to ammonia inhibition. Other strains such as *Methanobacterium thermoautotrophicum*, *Methanobacterium formicicum* or *Methanosarcina barkeri* were found to be less sensitive to higher ammonia (10 g/l NH_4) concentrations (Jarrell et al., 1987; Goberna et al., 2010). Ammonium (NH_4^+) concentrations have been reported to start being inhibitory from between 1.7 g/l to 14 g/l (Chen et al., 2008). The differences in inhibitory concentrations are mainly attributed to the various physiochemical conditions, such as pH, temperature and different biomass acclimation periods. In terms of free ammonia, it is believed that concentrations above 100 mg/l cause inhibition. High concentrations of ammonium (> 5 g/l NH_4^+) lead to a shift from the aceticlastic to the syntrophic acetate degradation pathway. In the syntrophic pathway, acetate is converted to hydrogen and carbon dioxide by syntrophic acetate oxidising bacteria (SAO), followed by a subsequent utilisation by hydrogenotrophic bacteria. This slows down the degradation process due to the higher doubling time of SAO (28 days) compared with aceticlastic bacteria (about 12 days) (Schnürer & Nordberg, 2008).

There are different techniques to counteract ammonia inhibition. Applied and feasible methods include stripping, precipitation, biomass retention, biomass immobilising on zeolite, addition of an ion-exchanger (i.e. zeolite), antagonistic cations such as magnesium or calcium ions or dilution with other liquid wastewater. Dilution is not really recommended as it increases the process cost and the volume of wastewater.

Fermentation of protein-rich material such as SW should take place under mesophilic conditions (between 35–40°C). Thermophilic digestion is not recommended as higher levels of free ammonia at this temperature cause inhibition. In addition, at high temperatures, foaming and intensified biomass washout have been observed (Ortner, 2010).

To guarantee full heat integration within the facility, heat generated at a constant rate in the CHP should be transferred from periods of low energy demand to periods of high energy demand. That can be achieved by using a hot water storage tank in combination with intelligent recovery networks.

Another critical issue during mono-fermentation is the inefficient supply of essential micro elements to the bacterial biomass. This can cause severe constraints in terms of substrate degradation rate, which results in reduced biogas production. Due to the high concentrations of hydrogen sulphide, the bioavailability of trace elements such as nickel, cobalt or molybdenum is reduced significantly due to the formation of poorly soluble metal-sulphide precipitates.

The supplementation of trace elements may help to counteract this insufficiency. It is important that addition happens in a well-balanced way; otherwise it can have the opposite effect as overdosing may poison the microbial community.

Economics

The costs of industrial AD plants are higher than the costs of a conventional agricultural biogas plant, attributable to the installation of the sanitation unit and comprehensive exhaust air treatment units. In principle, digestate shows good fertilising potential, but if digestate cannot be used as an organic fertiliser in a direct way, post-treatment units (evaporation, filtration, separation) are required. This may further increase the overall costs.

Generally speaking, the economic efficiency of such an AD plant is strongly dependent on the national legal situation. That means a lot of factors have to be considered, whereby the green energy law and its applicable feed-in tariffs, national subsidies and the national disposal costs play the most important role in the decision.

Integration of AD technology into bioethanol production

In bioethanol production processes, very large amounts of organic by-products are accumulated, which are almost all suitable for anaerobic digestion. In grain bioethanol plants, typically all stillage fractions are anaerobically degradable (Drosg et al., 2013; Rosentrater et al., 2006; Cassidy et al., 2008).

In sugar cane bioethanol plants, either sugar cane molasses (after the recovery of sugar) or the cane juice directly, can be used for ethanol fermentation. Either way, the liquid effluents (vinasse, stillage) are suitable substrates for anaerobic digestion (Nguyen et al., 2009; Harada et al., 1996; Yeoh 1997; Cail and Barford 1985), whereas the solid bagasse is mainly incinerated for energy recovery.

The rest of this chapter will focus on the production of bioethanol from grains, which is the prevailing process in Europe and the USA. In the USA, the current world leader in bioethanol production, bioethanol production increased almost tenfold in the last decade (Renewable Fuels Association, 2011). The current increase in bioethanol production capacity of the USA, the EU and the world is shown in Table 8.3. It lies between 46 per cent and 60 per cent in a period of only two years.

Due to such a high quantity of bioethanol produced, large amounts of by-products are also accumulated. The dry-grind bioethanol process from grains produces up to 5.6 t of stillage

Table 8.3 Examples of increase in bioethanol production capacities from 2008 to 2010 (Renewable Fuels Association, 2011).

Country	2008 (million m^3)	2010 (million m^3)	Increase (%)
USA	34	50	47
EU	2.8	4.5	61
World	65	95	46

per cubic metre of ethanol (Drosg et al., 2008). The state-of-the-art stillage treatment process is drying to animal feed. This consumes a considerable amount of energy, since grain stillage has a water content of about 85–90 per cent. As the bioethanol industry becomes more prominent, there will be a greater need for implementing industrial anaerobic digestion processes. Anaerobic digestion can be a valuable option, depending on the price of animal feed and energy. Since dry-grind grain ethanol production is the prevailing process in the USA, it can be estimated that roughly 280 million t/a of stillage is accumulated in the USA's domestic ethanol production. Using anaerobic digestion on the annual stillage produced in the USA, roughly 16.3 billion Nm^3/a of methane could be recovered. Translated to the European Union, about 25 million t/a of stillage are accumulated with a methane potential of approximately 1.5 billion Nm^3/a.

State-of-the-art stillage treatment

In the dry-grind bioethanol production process, as the prevailing process for grain ethanol production, ethanol, carbon dioxide and animal feed are produced. This process is described in detail by Senn and Pieper (2001) and Bothast and Schlicher (2005). The stillage accumulates as liquid by-product after distillation of the fermentation broth (the so-called 'beer'). In the-state-of-the-art process (see Figure 8.2) it is separated by centrifuges to thin stillage (liquid phase) and wet cake (solid phase). The liquid phase is concentrated via vacuum evaporation to syrup and mixed with the wet cake. This mixture is finally dried to animal feed called DDGS (distillers' dried grains with solubles).

Characterisation of stillage fractions

The composition of the stillage fractions in a dry-grind bioethanol plant can vary depending on the input mixture of grains. In Table 8.4, the mean values of the stillage fractions of a large-scale bioethanol plant are shown. In this plant, mainly wheat and corn were used as substrates in varying ratios. Due to this fact, the values shown in Table 8.4 are highly variable.

Limitations and bottlenecks

Digestate accumulation

The integration of a biogas process into a biofuel production process multiplies the problem of by-product (digestate) accumulation compared with food production processes. In small and medium-sized biogas plants, especially if they are strongly linked to agriculture, the state-of-the-art utilisation of digestate is as fertiliser applied to land. The European Nitrate Directive 91/676/CEE limits the application of nitrogen per ha and year, and there is considerable potential of nitrate leaching when digestate is not applied during the time of plant demand (Goberna et al., 2011). As a consequence, the land area that is needed for digestate application increases steadily by increasing biogas plant size and transport costs will increase drastically. In addition, if the utilised raw materials (crops) are not purchased regionally, which is often the case in bioethanol facilities, the willingness of local farmers to utilise the digestate as fertiliser is uncertain, even though digestate is a valuable fertiliser. As a consequence, when integrating biogas technology in such a large-scale process as a bioethanol plant, it is clear that the management of the digestate will

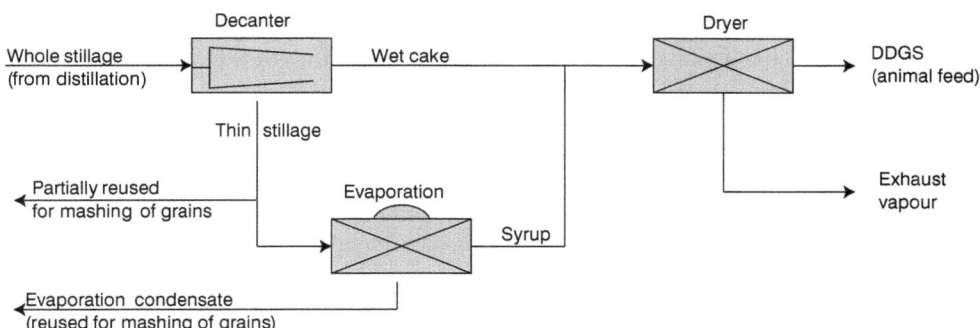

Figure 8.2 State-of-the-art stillage treatment process in dry-grind bioethanol plants (Drosg et al. 2013)

Table 8.4 Mean values of standard parameters of the stillage fractions in a dry-grind bioethanol plant. (Drosg et al. 2013)

Fraction	pH (–)	TS (%)	VS (% of TS)	COD (g/kg)	TKN (g/kg)	BMP (Nm³ CH₄/t COD)	BMP (Nm³ CH₄/t VS)
Whole stillage	4.5	13	91	175	7	290	469
Thin stillage	4.2	8	89	100	4	303	500
Wet cake	4.4	31	97	458	19	267	425
Syrup	4.3	27	89	348	11	298	470
Condensate	3.1	< 0.01	–	10	< 0.01	292	–

Note: TS = total solids; VS = volatile solids; COD = chemical oxygen demand; TKN = total Kjeldahl nitrogen; BMP = biomethane potential

be a key issue. In some cases, it will be possible to directly apply the digestate as fertiliser in the region, especially if the stillage management at a bioethanol plant has already been land application. At many other bioethanol facilities, a detailed digestate treatment strategy will have to be implemented. A variety of technologies are available for digestate treatment (Fuchs and Drosg 2010, 2013), such as solid–liquid separation by presses and centrifuges, evaporation, membrane purification, etc. The aim of these technologies is to produce process water and a nutrient concentrate. This nutrient concentrate can either be further processed to a marketable bio-fertiliser or the decrease of the water content can allow larger distances for economic land application. At the moment, although some industrial-scale digestate treatment facilities already exist, digestate treatment can still not be considered state-of-the-art technology in industrial biogas processes. Apart from that, additional investment and energy will be needed for digestate treatment. However, to what extent depends strongly on the type of technology applied. The selection of a suitable technology is highly dependent on boundary conditions like:

- availability of waste heat for evaporation;
- flux rates and life time of membranes for membrane purification;
- regionally available agricultural areas for land application;
- market value of the produced fertiliser products;
- quality of process water in the case of parallel animal feed production.

In the case of digestate from bioethanol fractions, one interesting option for digestate treatment can be solid–liquid separation by centrifuges, where the efficiency of the suspended solids removal can be increased by precipitating agents. Another alternative is the evaporation of digestate, if waste heat at the bioethanol plant is available. The produced condensate can be reused, together with evaporated ammonia, as process water and nutrients in yeast fermentation. By adjusting the pH in the digestate prior to evaporation, the concentration of ammonia in the condensate can be regulated to the amounts needed in yeast fermentation. In the literature (Tiejun and Xiaomei 2010) it is mentioned that the reuse of digestate in ethanol fermentation is possible and even beneficial due to its nutrient content. For the recirculation of digestate fractions, it is important to know that highly volatile fatty acid concentrations can have a negative effect on yeast fermentation. In addition, possible legal restrictions will have to be checked.

The suitable digestate treatment technology depends strongly on whether parts of the stillage fractions are still processed to animal feed. In this case, digestate can influence the smell as well as the colour of the animal feed. In addition, endospores from sporulating bacteria might become a problem. If no animal feed is produced in the process, the influence of recirculating the digestate (or digestate fractions) on animal feed quality need not be considered. However, a nitrogen sink (e.g. denitrification process, ammonia stripping) will be necessary in the process to reduce the recirculation of nitrogen loads.

Nitrogen impact

High nitrogen concentrations in biogas feedstocks can have a negative effect on process stability in anaerobic digestion (for details on nitrogen inhibition in industrial residues see the example of abattoirs above). By-products from bioethanol plants show increased nitrogen concentrations since a part of the carbon in the biomass has already been transformed by the yeasts to produce ethanol and CO_2 and removed from the process. Among the stillage fractions in a bioethanol plant, the condensate has the lowest nitrogen concentration. In this fraction, practically no nitrogen is present (< 0.01 g/kg) so it will have to be added for anaerobic digestion. Thin stillage has a TKN of about 4 g/kg. Pilot-scale trials (500 l) for almost two years showed that a stable digestion process is possible for thin stillage, if the process is optimised. It can be assumed that also whole stillage can be degraded in a stable process although the nitrogen concentration is already quite high (7 g/kg). In syrup (11 g/kg) and especially wet cake (19 g/kg) the nitrogen concentration is too high for direct anaerobic digestion. Therefore, these stillage fractions would have to be diluted or a nitrogen sink (denitrification process, membrane extractor process) would have to be integrated.

Potential for energy recovery by biogas process integration

The potential for energy recovery by anaerobic digestion from each stillage fraction is given in Table 8.5 along with annual methane production and energy supply of biogas plants. These results are compared with the energy demand of the bioethanol plant which was estimated according to Murphy and Power (2008) and Lurgi GmbH (2006).

Clearly, anaerobic digestion has a very high potential for energy recovery in bioethanol production. Table 8.5 shows that using all the available stillage, more than 100 per cent of the energy demand of the bioethanol plant can be provided by biogas. For thin stillage the

Table 8.5 Potential for energy recovery per stillage fraction in a large-scale bioethanol plant (Drosg et al., 2013)

Stillage fraction	Biogas plant – energy recovery		Bioethanol plant	
	Methane production (10^6 Nm^3/yr)	Energy supply (GWh$_{therm}$/yr)	Energy demand (GWh$_{therm}$/yr)	Coverage by biogas (%)
Whole stillage	60	522	395	128
Thin stillage	25	215	507	41
Wet cake	34	294	495	57
Syrup	24	209	507	40
Condensate	1.7	15	608	2.5

potential was 41 per cent of energy coverage. In addition, it is a very promising substrate, especially because it is rapidly and easily degradable and contains no bulky material.

Wet cake can provide 57 per cent energy recovery, and syrup 40 per cent. However, both stillage fractions show very high nitrogen concentrations. The reason for the small difference between thin stillage and syrup, is the volatile substances that evaporate in the concentration step from thin stillage to syrup. Condensate shows by far the lowest potential (2.5 per cent coverage). Nevertheless, anaerobic condensate treatment can be integrated quite easily into the process, since it contains no particles. In addition, the concentration of potentially inhibitory metabolites for yeasts (e.g. acetic acid) can be reduced by anaerobic digestion of the condensate.

In real-scale processes, energy recovery will be lower than suggested in Table 8.5. The reason is that energy demand for digestate management has not been considered. If land application of digestate is not possible, considerable amounts of energy will be needed for digestate treatment. Different technologies are available, but no state-of-the-art treatment has evolved up to now. For this reason the energy demand for digestate treatment has not been integrated into this estimation of the potential. It will have to be estimated separately for every case study.

Organic residues from breweries

Beer production is one of the first biotechnological processes, producing ethanol by fermentation with a final concentration of 4.5 to 6.0 per cent. Malt is used as the typical feedstock. Beside that, other raw materials such as corn, rice or barley are applied as well. During beer production, various organic residues accumulate, mainly non-degradable components of malt and yeasts. In 2009, worldwide 1809 million hectolitres (hl) beer were produced. The biggest producers are China, the USA, Brazil, Russia and Germany (Barth Report, 2010).

Characterisation of organic residues

During the brewing process, approximately 25 kg (FM)/hl of paste-like and solid residues accumulate. As presented in Table 8.6, the organic wastes consist basically of brewers' spent

Table 8.6 Amounts of brewery residues and biogas yield (Pesta et al., 2006).

Residue	Amount (kg/hl)	Biogas yield (Nm³/t FM)
Malt dust	0.05–0.25	600
Brewers' spent grains (BSG)	18.0–20.0	120
Cold break	0.1–0.3	400
Hot break	0.4–2.0	400
Yeast	2.0–2.6	60
Wastewater	350–400	0.32

grains (BSG), break, yeast and wastewater. Due to the consistency of the residues (solid, paste or liquid) anaerobic digestion seems to be a feasible technology for treatment. The biogas yield of these organic fractions ranges from 60 to 600 Nm³/t FM.

There are two more residues which have to be considered, diatomaceous earth and labels. In most breweries diatomaceous earth is used for filtration. Both materials are not suitable for anaerobic digestion and have to be removed. Currently, most of the residues from breweries are used as an animal feed. Beside that, there is an increasing amount of BSG, cold and hot break and yeast, which is used as a co-substrate for biogas production. In 2010, about one-third of the total amount of BSG in Austria were used in biogas plants.

Pre-treatment overview

The four steps of anaerobic digestion are hydrolysis, acidogenesis, acetogenesis and methanogenesis. The time needed for the degradation of biomass to biogas, or macromolecules to mainly methane and carbon dioxide, varies depending on the nature of the chemical bonding of the carbohydrate in the biomass (Noike et al., 1985). The microorganisms in anaerobic digestion convert simple molecules into biogas. Starch is used by the plants as energy storage and is therefore easily degradable by bacteria. In contrast, cellulose is used to maintain the structure of the plant and is for that reason difficult to break down. The breakdown of cellulose is further complicated by the bonds between different cellulose chains, and between cellulose, hemicelluloses and lignin. Lignin cannot be degraded by anaerobic bacteria.

In recent years, different pre-treatment technologies have been developed to increase the degradability of carbon, particularly in ligno-cellulosic material. There are a huge number of pre-treatment technologies, and it is often difficult to decide which of these are suitable.

In general, pre-treatment technologies can be divided into physical, chemical and biological processes. Physical pre-treatment comprises milling, extruding or thermal techniques. An example for a combined physical treatment process is steam-explosion. Chemical pre-treatment technologies include the addition of alkali, acid or organic solvents. Among the biological pre-treatment technologies addition of enzymes as well as a multi-stage digestion including a pre-acidification step have to be mentioned. Generally, a biological pre-treatment process increases the digestion rate, while chemical and physical treatment leads to higher gas yields and higher digestion rate. During both treatments, inhibitory substances can be formed. The energy demand of pre-treatment technologies varies greatly (Wellinger et al., 2012).

Pre-treatment technologies of BSG

Each residue from brewing needs different retention times for complete degradation by anaerobic consortia. While yeast, cold and hot break is degraded rapidly, BSG needs a higher retention time for a complete degradation. The reason for that is the chemical composition of BSG; it contains cellulose, hemicellulose and lignin (see Table 8.7). BSG consists of high amounts of holocellulose (Table 8.7) and needs 40 to 60 days for total degradation. Due to this problem, different pre-treatment technologies have been analysed.

The influence of milling of BSG was analysed and evaluated in several research projects. Voigt et al. (2009) reduced the retention time of BSG during anaerobic digestion to 24–27 days (OLR 4.9 kg VS/(m^3·d)) in a three-stage system. During one-stage digestion with a retention time of 45 days, a higher OLR (3.4 kg VS/(m^3·d)) and a higher gas yield (+16 per cent) was realised. The three-stage process includes two acidification steps. Voigt presented energy recovery of 25 per cent of the total energy demand of a brewery. This low amount of recovery can be explained due to the high energy input for the milling process (Voigt et al., 2009).

In 2003, the von Nordenskjöld company presented a milling process in combination with a two-step anaerobic digestion system (including an acidification and methanogenesis step). An organic loading rate of approximately 4 kg/m^3·d was realised. Following the anaerobic digestion process by aerobic treatment, COD concentration of wastewater could be reduced below 100 mg/l (von Nordenskjöld 2008; von Nordenskjöld and Stippler 2003). The ATRES company patented a multistage anaerobic digestion process and demonstrated, together with the Enbasys company, the digestion of BSG at the Weihenstephaner Brewery in Bavaria, Germany. The patented hydrolysis process allows a reduction of the hydraulic retention time down to 7 days with the disadvantage of incomplete exploitation of the biogas potential (Pesta 2009).

Bochmann et al. (2007) showed the effect of enzymes on the anaerobic digestion of BSG. An increase of volatile fatty acids by about 50 per cent due to a high hydrolysis rate and thus a high degradation rate of BSG could be observed at 40°C. During a continuous digestion process higher gas quantity and quality was measured during a hydraulic retention time of 40 days (Bochmann et al., 2007).

A combined process, pressing by a belt press and subsequent combustion of BSG, was evaluated by the Montan University in Leoben, Austria and the Austrian brewery Gösser. The result showed an increase of total solids (TS) from 20 to 42 per cent by the pressing process. Before the subsequent combustion was carried out, the TS was increased again to 55 per cent by a drying process. The liquid fraction of the pressing process was digested in the UASB of the brewery. Combustion of the solid fraction partially supplied the thermal energy demand of the brewery (Herfellner et al., 2006).

Table 8.7 Composition of brewers' spent grains (Narziß 1995; Kanaucho et al. 2001; Mussatto & Roberto 2005; Böchzelt et al. 2002).

Cellulose	16.2–25.4%
Hemicellulose	21.8–28.4%
Lignin	11.9–27.8%
Proteins	15.2–28.0%
Fats	5.5–10.6%
Ash	2.4–6.2%

Through thermo-chemical pre-treatment the biogas yield of BSG could be increased by 28 per cent. A total biogas yield of 155 Nm^3/t FM could be observed (Bochmann et al., 2010). Chemical and mechanical pre-treatment of BSG was analysed by Sezun et al. (2010). Acid and alkali pre-treated BSG produced more gas than mechanically pre-treated substrate. In both studies, inhibition occurred during the anaerobic degradation process of the pre-treated BSG. This resulted in a lower degradation rate and lower gas yield caused by the formation of bacteriostatic compounds, such as furfurals.

Energy supply of breweries by AD

Many breweries use anaerobic digestion technology for the treatment of wastewater, but not for solid or paste-like wastes. In some European countries brewery residues are used as a co-substrate in biogas plants. Currently, in Austria about one-third of the total BSG are used in biogas plants. Through anaerobic digestion of the total amount of BSG accumulated in Austria, approximately 21 million Nm^3 CH$_4$ per year (equal to 210 GWh) can be generated.

The production of beer requires thermal and electrical energy of about 26.8 or 9.9 kWh/hl beer, respectively. Using all residues in a brewery for anaerobic digestion, up to 17.9 kWh/hl can be generated. As a consequence, approximately 50 per cent of the energy demand can be covered.

Economics

An onsite anaerobic digestion unit offers the opportunity of energy recovery by biogas. A large-scale brewery producing 1,000,000 hl annually has an accumulation of 25,000 tons/year of solid or paste-like waste (25 kg waste/hl). In this case, a total digester volume of 3000 to 5000 m^3 is needed; investment costs of €2.5 to 3.0 million are required (Walla & Schneeberger, 2008).

Currently, BSG are mainly sold as animal feed. In Austria revenues range between 5 to 15 €/t FM. If BSG are used for the production of biogas, higher revenues can be expected.

An important point in terms of economics is the accumulation of digestate. After the digestion process, 15,000 to 18,000 t of digestate (1 million hl brewery) accumulate with a total solid content of 4–5 per cent. According to different national laws, and due to the composition of the digestate, it can be used as fertiliser. Another option is digestate treatment, where additional costs of approximately 5–8 €/t have to be considered. Anaerobic digestion of BSG has still to be optimised in economic terms.

Anaerobic digestion of residues from olive oil production

Another important source for renewable energy generation from industrial residues is waste from olive oil production: olive mill wastewater (OMW) and olive mill solid waste (OMSW). World olive oil production has increased from 2.51 million tons in 2000 to 3.27 million tons in 2010 (FAO 2012). The main olive oil producers are concentrated in the Mediterranean area: Spain 36 per cent, Italy 27 per cent, Greece 15 per cent, Tunisia and Syria 6 per cent, Turkey 4 per cent (Buckland & Gonzales, 2010).

Argyropoulos, D., Psallida, C. and Spyropoulos, C.G. (2006) 'Generic normalization method for real-time PCR application for the analysis of the mannanase gene expressed in germinating tomato seed', *FEBS Journal*, vol 273, pp. 770–777.

Basen, M., Sun, J. and Adams, M.W.W. (2012) 'Engineering a hyperthermophilic archaeon for temperature-dependent product formation', *mBio*, vol 3, no. 2. e00053-12.

Bayer, E.A., Belaich, J.P., Shoham, Y., and Lamed, R. (2004) 'The cellulosomes: multienzyme machines for degradation of plant cell wall polysaccharides', *Annual Reviews of Microbiology*, vol 58, pp. 521–554.

Bayer, E.A., Lamed, R. and Himmel, M.E. (2007) 'The potential of cellulases and cellulosomes for cellulosic waste management', *Current Opinion in Biotechnology*, vol 18, pp. 237–245.

Béguin, P. and Lemaire, M. (1996) 'The cellulosome: an exocellular, multiprotein complex specialized in cellulose degradation', *Critical Reviews in Biochemistry and Molecular Biology*, vol 31, pp. 201–236.

Benedict, M.N., Gonnerman, M.C., Metcalf, W.W. and Price, N.D. (2012) 'Genome-scale metabolic reconstruction and hypothesis testing in the methanogenic archaeon *Methanosarcina acetivorans* C2A', *Journal of Bacteriology*, vol 194, pp. 855–865.

Berg, I.A., Kockelkorn, D., Ramos-Vera, W.H., Say, R.F., Zarzycki, J., Hügler, M., Alber, B.E. and Fuchs, G. (2010) 'Autotrophic carbon fixation in archaea', *Nature Reviews in Microbiology*, vol 8, pp. 447–460.

Bertani, G. (1999) 'Transduction-like gene transfer in the methanogen *Methanococcus voltae*', *Journal of Bacteriology*, vol 181, pp. 2992–3002.

Bertani, G. and Baresi, L. (1987) 'Genetic transformation in the methanogen *Methanococcus voltae* PS', *Journal of Bacteriology*, vol 169, pp. 2730–2738.

Bertin, L., Bettini, C., Zanaroli, G., Frascari, D. and Fava, F. (2012) 'A continuous-flow approach for the development of an anaerobic consortium capable of an effective biomethanization of a mechanically sorted organic fraction of municipal solid waste as the sole substrate', *Water Research*, vol 46, pp. 413–424.

Briones, A. and Raskin, L. (2003) 'Diversity and dynamics of microbial communities in engineered environments and their implications for process stability', *Current Opinion in Biotechnology*, vol 14, pp. 270–276.

Brochier-Armanet, C., Forterre, P. and Gribaldo, S. (2011) 'Phylogeny and evolution of the *Archaea*: one hundred genomes later', *Current Opinion in Microbiology*, vol 14, no 3, pp. 274–281.

Brummell, D.A. (2006) 'Cell wall disassembly in ripening fruit', *Functional Plant Biology*, vol 33, pp. 103–119.

Bugg, T.D.H., Ahmad, M., Hardiman, E.M. and Singh, R. (2011) 'The emerging role for bacteria in lignin degradation and bio-product formation', *Current Opinion in Biotechnology*, vol 22, pp. 394–400.

Bustin, S.A. (2002) 'Quantification of mRNA using real time reverse transcription PCR (RT-PCR): trends and problems', *Journal of Molecular Endocrinology*, vol 29, pp. 23–39.

Clavero, T. and Razz, R. (2002) 'Effects of biological additives on silage composition of mott dwarf elephant grass and animal performance', *Revista Científica*, vol 12, pp. 313–316.

Dalby, P.A. (2011) 'Strategy and success for the directed evolution of enzymes', *Current Opinion in Structural Biology*, vol 21, pp. 1–8.

DeSantis, T.Z., Hugenholtz, P., Larsen, N., Rojas, M., Brodie, E.L., Keller, K., Huber, T., Dalevi, D., Hu, P. and Andersen, G.L. (2006) 'Greengenes, a chimera-checked 16S rRNA gene database and workbench compatible with ARB', *Applied Environmental Microbiology*, vol 72, pp. 5069–5072.

Drake, H.L., Gößner, A.S. and Daniel, S.L. (2008) 'Old acetogens, new light', *Annals of the New York Academy of Sciences*, vol 1125, pp. 100–128.

Dugat-Bony, E., Peyretaillade, E., Parisot, N., Biderre-Petit, C., Jaziri, F., Hill, D., Rimour, S., and Peyret, P. (2012) 'Detecting unknown sequences with DNA microarrays: explorative probe design strategies,' *Environmental Microbiology*, vol 14, pp. 356–371.

Durand, H., Clanet, M. and Tiraby, G. (1988) 'Genetic improvement of *Trichoderma reesei* for large scale cellulase production', *Enzyme and Microbial Technology*, vol 10, pp. 341–346.

(Lessner et al., 2010). The engineered M. *acetivorans* had 80 times the esterase activity of wild type M. *acetivorans* and grew well, producing methane, when supplied only methyl acetate and methyl propionate esters as carbon and energy sources, using up more than 97 per cent of these substrates.

All these examples provide a clear glimpse of the potential these tools offer to improve the AD process, but the effort to overcome the challenges to date also demonstrate the considerable amount of work that will be required to achieve practical outcomes. Careful consideration will also need to be given to the use of any genetically modified organisms, before their application to large-scale biomethane production. The establishment of monitoring and destruction protocols and/or the use of pure enzymes or of cells in bioreactors immobilized on specific nanosurfaces preventing diffusion to the environment, could be solutions to this issue.

Conclusions

Molecular techniques have established that far greater diversity of micro-organisms exist and participate in AD processes then had been thought prior to their application. They have allowed greater capability to search for new enzymes and new strains that could be better adapted to operate in particular environmental conditions. Some significant new discoveries have been made. Molecular genetics has provided accurate tools to detect and monitor species, even in complex mixtures, and to trace the biological processes occurring in AD. This has demonstrated considerable differences between microbial communities in different fractions of the digestate (e.g. liquid and solid fractions and biofilms) and between digesters fed different substrates or having different environmental circumstances. Considerable temporal changes have been observed and these dynamics have led to questions as to how best to manage microbial populations (see Chapter 14). These advances have been great but have served to demonstrate how much more work is needed to understand these complex systems.

The use of a variety of approaches to isolate and characterize genes, alter them through various mutagenic processes, to select for these and to transfer genes into different species, has facilitated the description and understanding of metabolic pathways, even of uncultured species. Work has proceeded in developing improved and genetically engineered strains of *Fungi*, *Bacteria* and *Archaea*, with some genetically engineered *Fungi* and *Bacteria* already used to produce enzymes safely on an industrial scale. The cost of development is likely to limit engineering of processes specifically tailored for biogas production to industrial scale in the near future. However, advances in technology for altering methanogens, key to biogas production, suggest more rapid advances in this area are more likely.

References

Al Seadi, T., Rutz, D., Prassl, H., Kotther, M., Finsterwalder, T., Volk, S. and Janssen R. (2008) *Biogashandbook*. BiGEast Project. Biogas for Eastern Europe. University of Southern Denmark.

Anderson, I., Ulrich, E.L., Lupa, B., Susanti, D., Porat, I., Hooper, S.D., Lykidis, A., Sieprawska-Lupa, M., Dharmarajan, L., Goltsman, E., Lapidus, A., Saunders, E., Han, C., Land, M., Lucas, S., Mukhopadhyay, B., Whitman, W.B., Woese, C., Bristow, J. and Kyrpides, N.. (2009) 'Genomic characterization of Methanomicrobiales reveals three classes of methanogens', *PLoS One*, vol 4: e5797.

providing increased oligosaccharide product and peroxidizes with increased thermostability (e.g. cited in Howard et al., 2003), and enhanced production of a variety of alcohols and esters (e.g. cited in Kumar et al., 2008). Further examples are given in a number of reviews (e.g. Sharma et al., 2009; Tracy et al., 2012; Menon and Rao, 2012).

Concepts to improve the digestion of lignocellulosic material have included genetic engineering of the substrate so that it is more easily broken down using alterations to the lignin biosynthesis in maize (e.g. Grabber et al., 2010), the inhibition of lignin production on transgenic poplar trees (Hu et al., 1999) or the production of processing enzyme by the feedstock plant (Shen et al., 2012). In the latter case it is important to protect the growing plant from the processing enzyme until it is needed, and this was done by Shen et al. (2012) by engineering a self-splicing peptide from a thermotolerant bacterium, meaning the inhibitor in the hybrid enzyme only comes off at processing temperatures of over 60°C.

It is clear that anaerobic digestion could benefit from the research focused essentially on biofuels such as biodiesel, where there may be the economic and practical justification of engineering feedstock to increase the utilization of substrates such as cellulose, hemicellulose or organic acids. However, since much biogas production is focused on the use of wastes from a variety of processes, including mixed wastes, this may not be a suitable strategy to focus on biogas production *per se*. Similarly, the large amounts of enzymes (e.g. cellulases, lipases and proteases) produced by engineered fungi and bacteria for other industrial uses (a sample list can be seen in Menon and Rao, 2012) have been found effective when added as pre-treatments to AD, but the costs are considered prohibitive for most AD applications (Parawira, 2012).

While a range of tools has been available for sophisticated and nuanced transformation of *Bacteria* for many years, there are sufficient differences between this group and the *Archaea* that meant many of those methods could not be used in the *Archaea* (Mevarech and Werczberger, 1985; Bertani and Baresi, 1987; Gernhardt et al., 1990; Eggen, 1994; Bertani, 1999). The development of reliable methods to transfer genes among the *Archaea* required considerable work to understand basic aspects of their molecular apparatus and modify techniques accordingly (Lange and Ahring, 2001; Rother and Metcalf, 2005). In the last decade efficient genetic manipulation systems have been developed (Rother and Metcalf, 2005; Porat and Whitman, 2009) and the basic biochemistry of methane production has been elucidated, including some sophisticated models (Benedict et al., 2012). A detailed review for gene transfer for hydrogenotrophic methanogens including protocols and plating methods is provided in recent reviews (Sarmiento et al., 2011; Kohler and Metcalf, 2012).

Now work is beginning to appear that, while still at the experimental stage, has relevance to biogas production. The hyper-thermophilic archaeon, *Pyrococcus furiosus*, optimum growth at 100°C, has been engineered recently to express a lactate dehydrogenase (*ldh*) gene from the moderately thermophilic bacterium, *Caldicellulosiruptor bescii*, optimal growth at 78°C, controlled by a cold shock promoter (Basen et al., 2012). At 98°C, the engineered strain ferments sugar to acetate and hydrogen as end products but at 72°C the *ldh* promoter is activated and lactate is produced instead, providing proof of principle for the use of such a control switch in applications for biofuel production.

A potential route to overcoming interruptions in digester function at high loading rates of biomass is engineering methanogens to allow them to utilize a wider range of substrates. A broad-specificity esterase from the bacterium *Pseudomonas veronii*, which hydrolyses a variety of esters, including methyl acetate and methyl propionate, was successfully expressed in the archaeon *Methanosarcina acetivorans* C2A, under control of an *M. acetivorans* gene promoter

Rational protein design (RD) consists of a set of molecular biology techniques, such as site-directed mutagenesis (SDM), where changes are made in the DNA to alter specific amino-acids responsible for specific substrate–enzyme interactions, activity, specificity, docking and stability to improve performance. Clearly, considerable information on the target enzyme is essential for the effective application of RD, as is the availability of a number of gene transfer techniques and the absence of the latter for many years has delayed the use of this technique in the *Archaea* (Kohler and Metcalf, 2012).

Directed evolution has been used to improve the capability of an archaeal species to survive processes required for replication in the laboratory and so provide a means to develop improved gene transfer methods and therefore a better understanding of methanogen metabolism (Ehlers et al., 2005 – see also Zhang et al., 2002). The low plating efficiency of *Methanosarcina mazei* strain Gö1 on solid medium was improved by selecting for a spontaneous mutant of *M. mazei* that showed significantly higher resistance to mechanical stress during spreading on agar plates and went on to develop improved transformation processes (Kohler and Metcalf, 2012).

Genetic engineering and gene transfer

Genetic engineering is used effectively in producing biofuels by inserting genes from one species into another and the production of their heterologous proteins to improve the production of desired enzymes or other end products (Sharma et al., 2009). Approaches include the suppression of pathways not leading to the production of the desired product (e.g. alcohols), the introduction of one or more genes to allow more efficient production of the desired product, or the addition of genes allowing an organism with a useful metabolism to survive in the fermenter environment (Menon and Rao, 2012). Technical details of the wide variety of approaches needed to effectively insert and express genes are beyond the scope of the present chapter but are given in recent reviews (Sarmiento et al., 2011; Kohler and Metcalf, 2012).

Much of this activity has focused on model organisms such as yeast *Saccharomyces cerevisiae* and bacteria *Escherichia coli* given the amount of genetic and metabolic knowledge of these species and the availability of a wide range of molecular genetics tools (cassettes of genes of known function linked together with promoters and the use of a variety of transfection pathways to introduce the DNA effectively). However, these strategies have been used to develop genetically engineered strains of fungi (e.g. *Trichoderma, Aspergillus*) or bacteria (e.g. *Clostridium, Bacillus* and others) to secrete large quantities of enzyme either as part of the fermentation community in the digester, or for the industrial production of enzymes that are added to other fermentation vats (Vitikainen et al., 2010; Menon and Rao, 2012; Zou et al., 2012). The main model species used for genetic modifications in research and commercial applications, *Escherichia coli* and *Saccharomyces* species, have limitations in substrate utilization and therefore need a lot of genetic modification for use in biofuel fermenters (Tracy et al., 2012). Although not investigated to the same extent, *Clostridia* offer advantages in terms of substrate utilization, diversity, biosynthetic capabilities and good tolerance to toxic substrates and metabolites. *Clostridium acetobutylicum* has been used as a model for metabolic engineering and applications in butanol production and is also a good example for fermentation improvements for biofuels (Papoutsakis, 2008) and biogas production. In addition, further examples from this extensive field include the development of thermostable hybrid glucanases between *Bacillus amyloliquefaciens* and *B. macerans*, engineered cellulases

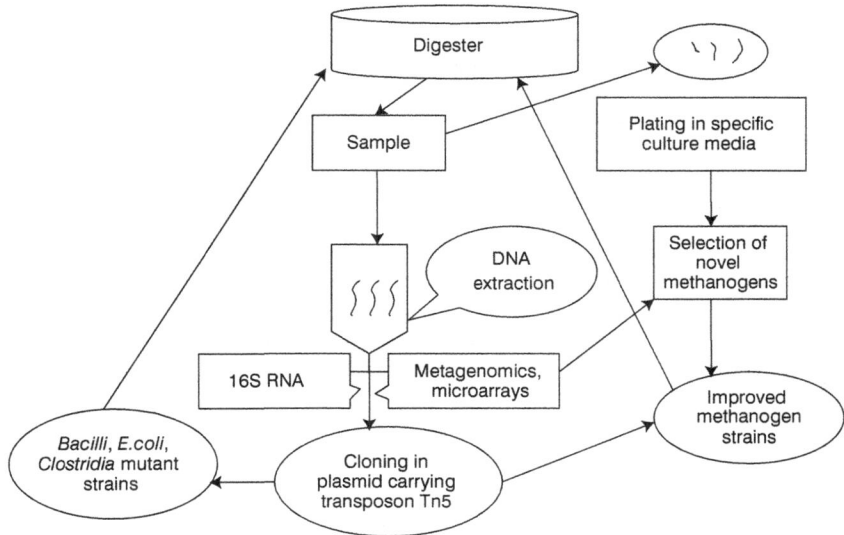

Figure 15.5 Illustrative example of possible paths in directed evolution of microbial communities for biomethane production combined with molecular methods. The presented steps can be repeated for many evolutionary cycles until the best strains are obtained.

mutagenesis produced twice the cellulose as QM 6a, with further mutagenesis providing strain QM 9414 that produced even more, and later strains such as CL 847 exhibited a further four-fold increase in cellulase productivity in cellulose media and increased β-d-glucosidase specific activity following several generations of selection. Many of these strains have been sequenced and compared in detail to better understand the nature of the genetic changes that resulted in improved production and to help guide further work to improve gene expression and enzyme function (Vitikainen et al., 2010).

Clostridium acetobutylicum is important in butanol production and has been the subject of much research (Papoutsakis, 2008). Several rounds of mutagenesis using NTG (*N*-methyl-*N'*-nitro-*N*-nitrosoguanidine) as a selective agent produced strain *C. acetobutylicum* EA 2018 that does not produce spores and has greater capability of solvent production. Genomic comparison with strain ATCC 824 identified many variations including a total of 46 deletion sites and 26 insertion sites which may contribute to the hyper-butanol producing capability of the EA 2018 strain (Hu et al., 2011). In addition, transcriptomic profiling of strain EA 2018 revealed changes of expression-level in several key genes related to solvent formation compared with strain ATCC 824. Additional examples are given by Kumar et al. (2008).

Recently, methodologies that provide a more structured and rapid approach to the formation and production of improved biocatalysts such as directed evolution and rational designs have been developed (Illanes et al., 2012). Directed evolution focuses on producing a large number of mutations in the gene of interest (instead of randomly within the organism as in older techniques) by using techniques such as error-prone PCR and DNA shuffling. This diversification step stage is followed by a selection step that automatically eliminates all non-functional mutants, and/or a screening step where high performing mutants are identified and isolated. The process can be iterated until the desired change is reached or until no further change is elicited, and the evolved protein is characterized (Dalby, 2011).

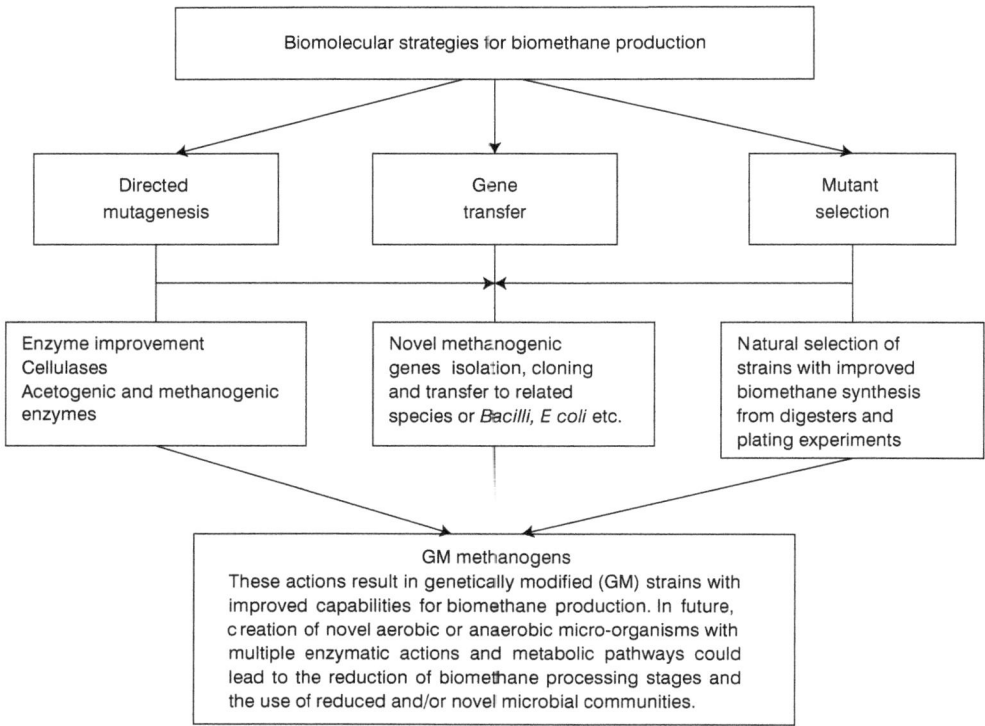

Figure 15.4 Bio-molecular strategies for improved biomethane production based on genetic technologies for the creation of novel microbial strains. These approaches are not mutually exclusive and strategies for improvement can utilize a number of these (see Figure 15.5).

radioactive methods which induced random changes in the genome, not necessarily in the gene or genes important for the process being improved. These processes also damaged many other key metabolic pathways and so random mutagenesis was therefore rather inefficient, although it did allow progress to be made. Once molecular methods permitted the isolation and characterization of given genes and their protein products, it became possible to target mutagenesis to particular genes, as discussed later.

Strain and mutant selection has been applied to some of the main fungal (Sharma et al., 2009), bacterial (Tracy et al., 2012) and archaeal (Kohler and Metcalf, 2012) groups that participate in AD, although not necessarily with AD as the primary target, since the production of industrial enzymes and other biofuels have greater economic value. An excellent example illustrating this approach is the fungus *Trichoderma reesei* which plays a major role in cellulose degradation and, because of its ability to secrete large amounts of enzyme, has become the major means of industrial cellulase production (Vitikainen et al. 2010). All industrial strains have been derived from a single natural isolate, QM 6a, in the 1940s, but samples of this have been subjected many times in various laboratories to random mutagenesis and the subsequent mutants selected to detect those mutants with superior performance, either in terms of their greater productivity, enhanced enzyme stability and so on (Durand et al., 1988). Strain QM 9123 developed in the 1970s from irradiation

2011). So this work has already demonstrated that a great diversity of species are involved in AD, but that there are some subsets of species that recur, or dominate, particularly in the acetogenic and methanogenic stages of the cycle (Sanz and Köchling, 2007; Talbot et al., 2008; Nelson et al., 2011). Their composition is strongly influenced by environmental factors such as temperature, pH, feedstock, the concentration of volatile fatty acids and operating conditions that affect any of these (e.g. Krakat et al., 2011; Nelson et al., 2011, 2012; Lee et al., 2012b). The work has also revealed considerable spatial and temporal dynamics in the microbial communities, not always related to major changes in digester function, and high degrees of stochasticity in the identity of members of microbial communities developed in a particular place at a particular time (e.g. Krakat et al., 2011; Lee et al., 2012b). This raises important questions as to the need to manage diversity to maintain functional redundancy in the community to achieve better robustness and stability of digester function (Briones and Raskin, 2003). The topic of microbial diversity in anaerobic digestion is dealt with in much greater detail in Chapter 14.

Molecular genetics strategies for improving biomethane production

This section deals with applications that can modify the AD process to improve performance, and this is at an early stage of development. Improvements can be made through using strains with better performance found in nature and/or improving strains further by selecting them in appropriate environments, with or without the use of additional genetic processes such as mutagenesis (e.g. Howard et al., 2003). Finally, engineering improved strains by altering the genome and/or the enzymes active in a particular process is well established for a range of fermentative processes and offers the potential to improve biofuel and biogas production (Rittmann et al., 2008; Menon and Rao, 2012). Much of this work is directed at the use of feedstocks for production of other types of energy, such as biofuels rather than biogas, that are more likely to repay the investment required. Nevertheless, some of the applications developed may assist biogas production, and the knowledge base developed during these endeavours will assist any studies ultimately focused on biogas.

Methods for improvement such as directed mutagenesis, gene transfer and the selection of improved mutants are not mutually exclusive and strategies for improvement can involve combinations of any or all of these techniques (Figure 15.4). This will become clear when particular examples are discussed in the sections below.

Strain and mutant selection, directed mutagenesis (directed evolution) and rational design

At its simplest, strain selection involves testing new species or strains in a digestion environment and monitoring the performance of the digester and particular aspects of the performance of the strain such as testing the amount of relevant enzyme(s) they produce and the efficiency of those enzymes on the substrates of interest (Howard et al., 2003; Menon and Rao, 2012). The best performing strains are maintained and amplified for further use and possible crossing with other high performing strains to obtain even better performance.

The next step in complexity was the use of mutagenesis to provide a wide range of novel genotypes from a chosen strain in the hope that some of the new mutants would have better performance (Vitikainen et al., 2010). Mutation was induced originally by chemical or alkaline pH, or H_2-rich environments) that existing industrial enzymes cannot (Simon and Daniel, 2011).

Some examples relevant to AD include the discovery of novel methanogenic Archaea from the genus Methanobrevibacter associated with the flagellated protist species Dinenympha and Microjoenia in the gut of the termites, Reticulitermes speratus and Hodotermopsis sjoestedti (Tokura et al., 2000). Genes extracted from libraries made from combined extracts from shipworm, Tereso navalis, elephant faeces and a biogas plant, obtained new halo- and thermo-tolerant cellulases with excellent activity (Ilmberger et al., 2012). A metagenomic study of an anaerobic digester community discovered a novel aminotransferase that acted only on coenzyme A (CoA) esters and provided a new pathway for lysine fermentation (Perret et al., 2011). New genome sequences provide the basis for developing information on the metabolic capability and potential pathways the sequenced organism is capable of undertaking and therefore of potentially new microbial or gene variants that could be useful in AD as in, for example, the sequence of the mesophilic methanogen Methanocella paludicola (Sakai et al., 2011). Comparative genomics of the Methanomicrobiales revealed similarities and differences in substrate processing and gas-producing genes between taxa and the identification of 2–4 membrane bound hydrogenases whose potential role could be discussed given their spatial relationship to other genes (Anderson et al., 2009). In a final example, Fonknechten et al. (2009) report the identification of the last four genes involved in the anaerobic degradation of L-ornithine in Clostridium and its detection in a number of taxa.

Monitoring of the AD process

In addition to the fundamental difficulty of isolating species in the first place, there have been three major difficulties in the traditional identification of isolates, particularly methanogens: (1) uncertainty that the characters used to define the phenotype of the species were representative given that most species descriptions were based on examination of only a few strains, (2) the strong likelihood that a new isolate may represent an undescribed species or subspecies since only a fraction of the methanogens in nature has been described, and (3) the inherent difficulty of identifying methanogens with limited catabolic pathways and extreme genetic diversity using only a few phenotypic characters (Whitman et al., 2006).

As gene sequencing provides unique species identifiers, molecular approaches have revolutionized the capacity to identify microbes found in reactors. Quantitative PCR and FISH can be used to assess their abundance and the latter can be used determine their location (Rittmann et al., 2008; Wirth et al. 2012). Narihiro and Sekiguchi (2011) summarize the design of DNA primers for use in PCR to detect particular species, or defined groups of species (genera, families or Archaea versus Bacteria). These tools can then be used to determine the presence of a species, or group of species (genus or family etc.) in any wild habitat or digester, and an increasing number of studies are being carried out in anaerobic digesters (e.g. Schlüter et al., 2008; Nelson et al., 2011; Wirth et al., 2012).

The speed with which the field is advancing is demonstrated by the fact that analysis of 10 restriction endonuclease digest patterns of the 16S gene were used to discriminate 26 methanogenic species almost ten years ago (Wright and Pimm, 2003). Only five years later, 369 different 16S rRNA gene probes were spotted on a glass chip and were used to study bacterial communities during the degradation process of organic waste composting with the aid of pyrosequencing and metagenomics technology (Schlüter et al., 2008). Now MPS methods have detected several thousand species in anaerobic digesters (Nelson et al.,

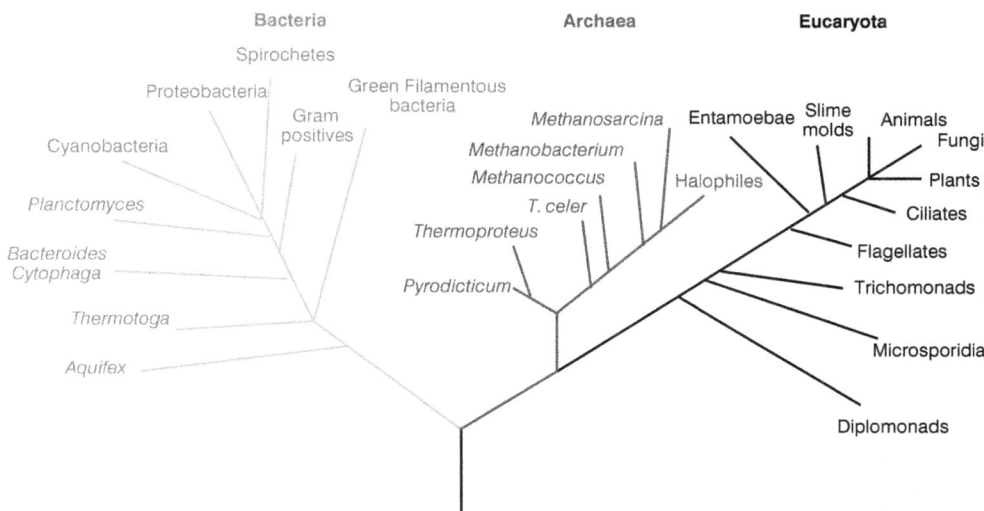

Figure 15.3 Universal tree of life based on small-subunit ribosomal (ssu) rRNA sequences. All known methane-producing organisms are members of the domain *Archaea*. The *Methanosarcineae* are the most metabolically and environmentally diverse methanogens. They are also members of the lineage giving rise to the anaerobic sulfate-reducing *Archaeoglobus fulgidus* and the aerobic halophilic *Halobacterium* species.

et al., 2011). Such analyses of multiple enzymes have shown that all methanogens have the same set of homologous enzymes and cofactors required for methanogenesis which indicates their common origin (Whitman et al., 2006; Berg et al. 2010). Their arrangement in the genome also indicates they are unlikely to be passed as a functional gene set between bacteria (horizontal transfer), a finding that is important in developing effective strategies for genetic improvement of the AD process (Reeve et al., 1997). The identification of methanogens as strictly anaerobic, methane-producing *Archaea* provides improved targeting of searches for strains and genes related to this pathway.

Another major result of extensive surveys of microbes was the discovery of far greater numbers of species (*Bacteria* and *Archaea*) in many habitats than had previously been thought – millions instead of thousands of species (Simon and Daniel, 2009, 2011). More than 32 genera of methanogens have been described with four phyla of *Archaea* now described (Brochier-Armanet et al., 2011).

Exploration for novel capability

The high diversity of microbial species provides the opportunity to discover novel metabolic functions in newly discovered strains and to search for enzymes of known function by screening for genes coding for such enzymes (Krause et al., 2006; Simon and Daniel, 2009; Uchiyama and Miyazaki, 2009; Dugat-Bony et al., 2012). It is also clear that significant numbers of novel proteins and pathways are likely to be discovered by extending the phylogenetic scope of genomic surveys (Wu et al., 2009). There have been a number of searches in, for example, deep sea hydrothermal vents, hot springs, acidic mines, marine plankton and animal guts for enzymes that can degrade a particular substrate but which can operate in specific environmental conditions (e.g. high temperature, high pressure, acidic or of sequences of the same gene from many different species) or which genes from various metabolic pathways are switched on (if the array is made of sequences from many different genes). More detailed descriptions of the design strategies for microarrays and on overcoming some of the potential problems of such screens can be found in Franke-Whittle et al. (2009a, b), Uchiyama and Miyazaki (2009), Simon and Daniel (2009, 2011) and Dugat-Bony et al. (2012). A description of a microarray specifically designed for anaerobic microbes is given by Franke-Whittle et al. (2009a).

Gene expression

The extent to which genes are active (gene expression) is studied by measuring the levels of messenger RNA (mRNA) transcripts they produce under different culture conditions or after genetic modifications. The mRNA sequence is complementary to that of the DNA from which it is transcribed in the first step in producing a protein from a gene. The techniques used to quantify mRNAs usually involve real-time PCR assays (Bustin, 2002; Argyropoulos et al., 2006), microarrays (Franke-Whittle et al., 2009a) or direct sequencing of the whole set of mRNA transcripts (transcriptomics) (Zakrzewski et al., 2012). These techniques can be used to identify unknown genes that may be expressed in a given environment, and which may indicate novel processes or to search for genes of known function in novel environments or from new species and better understand metabolic interactions within a system (for an example applied to AD see Zakrzewski et al., 2012).

Understanding and monitoring microbial diversity in AD for biogas production

This section deals with some of the fundamental findings from the use of molecular genetics approaches that have illuminated the process of AD and those applications that help identify, locate and monitor particular species, and those which can identify genes related to particular metabolic functions and so detect locate and monitor those functions.

Basic classification, identification of species and metabolic capability

By choosing genes that are found in every organism and which are involved in basic aspects of metabolism, so that they change only slowly [e.g. small subunit ribosomal RNA (SSU rRNA) or 16S ribosomal (16S RNA) sequences], it is possible to obtain unique identifiers for given taxonomic groups. By comparing the extent of differences in sequences between species their degree of relationship can be ascertained and their evolutionary relationships (phylogeny) determined. Molecular genetics demonstrated that two major groupings of microbes exist that differ markedly from each other and in their metabolic capabilities – the *Bacteria* and the *Archaea*. The latter is more closely related to the *Eukaryotes* which include the protozoa, fungi, and all animals and plants (Figure 15.3). The *Archaea* includes all methane-producing microbes, whereas the *Bacteria* and *Eucarya* (e.g. fungi) provide many species contributing to the earlier stages of AD.

Phylogenetic reconstruction is more accurate the greater number of genes that can be included, and if aspects of the organization of the genome can be included. Now that the complete genome sequences have been obtained from a diverse number of *Archaea* (more than 100), it is possible to probe their evolutionary history in greater detail (Brochier-Armanet

Laboratory (EMBL) and GenBank]. These data have been used to establish information on the sequences of genes of known function, different sequences of regulatory DNA that are known to have different functions (such as the stop and start points for a gene, sites for the attachment of various molecules regulating gene function). With the assistance of this knowledge, software has been developed to automatically annotate novel gene sequences (all these types of analysis being collectively called bioinformatics) (e.g. DeSantis et al., 2006). Bacterial genomes are small (approximately 3 million base pairs) and can be sequenced in a few hours, and annotated in a few days. To give an example of the explosion in this information the Genomes OnLine Database (GOLD) had information on more than 11,000 genome sequencing projects, 8473 of which are bacterial and 329 archeal in 2011, compared with < 500 in total a decade ago (Pagani et al., 2012).

MPS can sequence a mixture of different genomes. The associated software stitches together the matching overlaps into longer DNA sections which will fall into groups of sequence that are identical (or very similar) to each other but different from other groups of sequence so providing several, automatically separated consensus sequences for different genomes. The identity of these can be checked against databases of known species to identify those present in the mixture. Those with no match indicate the presence of a novel genome(s). When this process is applied to samples of a community of micro-organisms it is called metagenomics (see Simon and Daniel, 2009, 2011).

PCR

The second major advance was the development of polymerase chain reaction (PCR), which allowed specific sections of DNA to be amplified from a sample using specifically engineered primer DNAs. The presence or absence of amplified sequence could be used to determine the presence or absence of a species or given gene which a primer had been designed to match, or could provide specific DNA segments for further sequencing to determine the identity of the species or genotype. A range of methods such as RFLP, RAPD, AFLP and real-time PCR make use of PCR technology to detect and quantify genetic variation and each differs in their ease of use, accuracy and quantitative nature (see Table 15.2 for definition of the acronyms and more detail). A number of techniques have been used to separate and visualize the DNA produced by PCR with denaturing gradient gel electrophoresis (DGGE) being commonly used in microbial studies. See Sanz and Köchling (2007) and Talbot et al. (2008) for deeper discussion of all these approaches.

Probes and microarrays

Another set of technologies are based on adding fluorescent molecules to isolated sections or engineered stretches of DNA. These bind uniquely to their complementary sequences in samples and so can identify the location of that DNA in sample sub-fractions, tissues or cells [e.g. fluorescent in-situ hybridization (FISH)] (for details see Sanz and Köchling, 2007; Talbot et al., 2008). FISH can be used to count the cells and to identify where they exist – in a biofilm, sludge or whether they are suspended in fluid (for specific discussion in relation to methanogens, see Kumar et al., 2011). A similar process of binding fluorescently labelled gene products to different engineered complementary sequences, each sequence in a different spot in a structured array, is used to assess which sequences are present in a sample (microarrays). This can be used to determine which species are present (if the array is made

Technique	Description	Application
FISH Fluorescent In-Situ Hybridization	A method for the detection of DNA or mRNA using probes (each of 40–50 base pairs), each probe tagged with fluorophores or targets for antibodies. Sample tissues or cells are treated to allow entry of the probes and to allow their attachment to target sequences, excess probes are removed and the attached probes visualized.	Allows the location of the target DNA or RNA sequences to be determined. This can identify gene expression sites within cells, identify cells of a given species or taxonomic rank (e.g. *Bacteria* versus *Archaea*). Allows counting identified cells either as species (if the probes are species specific) or those cells with a particular gene active (if the probe targets a particular protein production).
DGGE Denaturing Gradient Gel Electrophoresis	DNA subjected to increasingly denaturing conditions fragments into single strands with discrete portions of the fragment becoming single-stranded in a narrow range of denaturing conditions, making it possible to detect small differences in DNA sequences. Acrylamide gels with a gradient of a denaturing agent separate fragments of DNA that differ in sequence. Temperature can also be used in separate process called Thermal Gradient Gel Electrophoresis (TGGE).	A technique often used in microbial studies to determine the number and dominance of different types of DNA (usually 16S RNA) present in a mixed sample. Bands from the gels could be cut out and the eluted DNA sequenced to identify the species concerned.
Metagenomics	Study of microbial communities in ecological environments or digesters based on genome analysis using technologies like pyrosequencing.	Metagenomics provide data from massive parallel sequencing of DNA fragments resulting in the identification and classification of micro-organisms from samples obtained from natural environments, digesters etc.
Transcriptomics (= expression profiling)	Study of the expression level of messenger RNAs in a sample of cells. One method compares sequence reads to a genome map of the species studied (or a closely related one). The other uses massive parallel sequencing approaches to develop *de novo* transcriptome assembly (e.g. using RNAseq).	Identifies the genes being actively expressed in the sample analysed, so identifying new gene transcripts or transcripts of known genes from new species, and the level of their expression in a given environment. Specific experiments can allow the likely role of genes to be identified depending on their responses.
Gene cloning	Gene cloning is the production of multiple copies of a gene. Genetic engineering is the process of cloning genes into new organisms with the help of DNA restriction enzymes capable of cutting plasmid or genomic DNA at distinct sites where the gene is inserted.	Cloning technology is the basis to establish transformation systems for cells in order to alter gene expression and create organisms with new properties.

Table 15.2 Summary of the main techniques for molecular genetic analysis. Rittmann et al. (2008) provide a listing of the techniques associated with key findings or milestones related to microbial energy. Further details can be obtained in reviews by Sanz and Köchling (2007), Talbot et al. (2008), Uchiyama and Miyazaki (2009), Simon and Daniel (2009, 2011), Dugat-Bony et al. (2012), Lee et al. (2012a), and Pinto and Raskin (2012).

Analytical method	Technology	Applications
RFLP Restriction Fragment Length Polymorphism	Organisms can differ based on the length of the fragments produced when their DNA is digested with a restriction enzyme. The restriction fragments are separated according to length by agarose gel electrophoresis. The resulting gel may be further analysed by Southern blotting using specific probes.	The similarity of the patterns generated can be used to differentiate even strains of the same organism.
AFLP Amplified Fragment Length Polymorphism	The technique is based on the selective PCR amplification of restriction fragments from a total digest of genomic DNA. Restriction fragments are analysed by automated capillary electrophoresis.	Specific co-amplification of large numbers of restriction fragments can be done without previous knowledge of restriction sites. More organisms to be differentiated compared with RFLP.
RAPD Random Amplification of DNA	A of PCR with DNA sequences amplified randomly. RAPD is applied by using several arbitrary, short primers (8–12 nucleotides) to DNA, hoping that some parts of DNA will be amplified. Resulting patterns are analysed by agarose gel electrophoresis	Less laborious than AFLP, it is possible to differentiate strains of the same organism, but not always successful as it is based on random primer sequences, and results are not always reproducible.
RT-PCR Real-Time PCR	A of PCR based on recording fluorescence intensities of DNA probes in real time during DNA amplification with specific primers.	Fast and accurate analysis for identification, quantification of gene transcripts and gene expression studies.
16S rRNA	16S ribosomal RNA is part of the 30S small subunit of prokaryotic ribosomes. Universal primer pairs are used to amplify the 16Sr RNA genes providing the phylogenetic information.	The 16Sr RNA genes coding are used in phylogenetic analysis as they are highly conserved between different species of *Bacteria* and *Archaea*.
Microarrays	A selection of short sections of DNA elements are attached to a solid surface in the form of microscopic spots that hybridize to target sequences and are detected by fluorometry.	Expression of a large number of genes, metabolic pathways and identification of key responsive genes can be analysed simultaneously in a semi-quantitative manner. For a limited number of species there are oligo-arrays where all expressed gene sequences are represented.

This has allowed major advances in the study of the genetic and molecular basis of the physiology and gene function of micro-organisms (e.g. Kotsyurbenko et al., 2001; Rittmann et al., 2008).

The capacity to sequence DNA (deoxyribose nucleic acid) reliably and cheaply, and to do this for particular genes, has provided the means to identify micro-organisms and to determine the functional genes underlying biochemical processes without the need to culture individual species. The progress in molecular biology and information science has allowed:

- routine, reliable identification of the species involved in AD;
- identification of the genes and enzymes responsible for given biochemical or physiological processes;
- targeted exploration for new candidate genes and species (better performing species/strains/enzymes, or those that can maintain their function in particular environments);
- monitoring the dynamics of microbial populations during the AD process;
- optimizing the AD process by a) choice of known strains with high performance for given functions, b) engineering improved strains through processes of genetic selection or genetic engineering.

The following sections will review each of these applications after summarizing the principal methods of molecular genetics relevant to AD.

Methods in genetic analysis

A detailed description of the many molecular biological or molecular genetics methods that now exist is beyond the scope of this review. The aim is to outline the main methods that are pertinent to the analysis of AD (Table 15.2). Readers interested in more detail should refer to the papers or reviews indicated in the relevant sections.

Sequencing

The principal advance has been the ability to sequence DNA rapidly and reliably and there are now a range of techniques that can do this. Earlier gel-based techniques that sequenced relatively long sections of DNA have been replaced by methods that sequence short strands of many different sections of DNA in parallel (massively parallel sequencing – MPS). The latter techniques take advantage of the huge increase in computing capability developed over the last two decades to use automated software to match the small overlaps in the many short sequences produced and stitch together longer sequences. Sequencing the total DNA which characterizes an organism (its genome) can now be done in a few hours, although putting together the jigsaw of many small sequences into one complete sequence takes much longer. The ability to achieve genome sequencing based on material from single cells has provided a way to obtain information on unculturable microbes (Pagani et al., 2012). Details of these techniques are reviewed extensively (e.g. Lee et al., 2012a; Pinto and Raskin, 2012).

Data banks

The information available for many sequences in many organisms has been accumulated in large databases [e.g. the DNA Data Bank of Japan (DDBJ), the European Molecular Biology

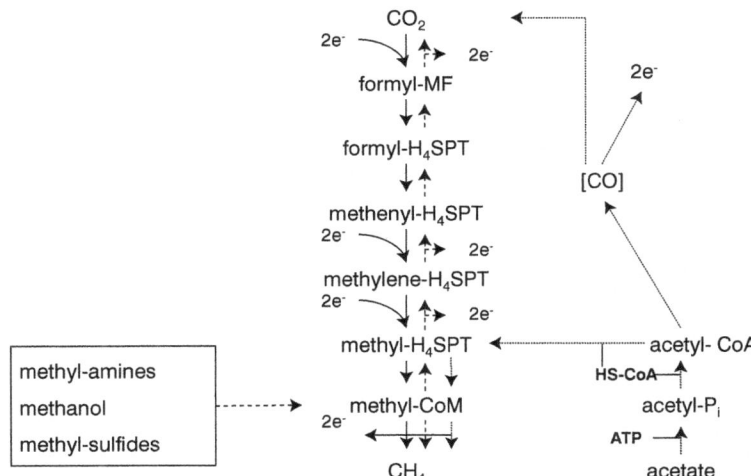

Figure 15.2 A summary of three major pathways for methanogenesis illustrating the multiple steps required in each process, every one of which involves one or more enzymes and a variable number of co-enzymes. Many methanogens can reduce CO_2 to methane using electrons derived by oxidizing H_2 (the hydrogenotrophic pathway – solid arrows in the central vertical column from CO_2 to CH_4). Others can utilize C-1 compounds (listed in box on the left) with one molecule of the C-1 compound being oxidized to provide electrons for reducing three additional molecules to methane (the methylotrophic pathway – dashed arrows). Other methanogens split acetate into a methyl group and an enzyme-bound CO, with the CO subsequently oxidized to provide electrons for the reduction of the methyl group to methane (the acetotrophic, also called acetoclastic pathway – on the right of the diagram and light solid arrows). Most methanogens possess only one of the three methanogenic pathways but *Methanosarcina* species possess all three. All three pathways converge on the reduction of methyl-CoM to methane (CH4) (bottom of central column). CoM, coenzyme M; H4SPT, tetrahydropterin; MF, methanofuran.

chemistry, biology and materials science have assisted this advance (e.g. isotope analytical techniques, computing and molecular imaging) but molecular genetics techniques developed in the last twenty years have revolutionized the field.

Molecular genetics applications in AD for biomethane production

It is clear that major efficiencies could be gained if those species or strains that produced the most effective enzymes for a given substrate or digester environment could 1) be identified and preferentially maintained (Bertin et al., 2012) and/or 2) be specifically engineered for greater efficiency (Vitikainen et al., 2010; Menon and Rao, 2012). To achieve the first goal, however, requires the ability to accurately describe the community of micro-organisms which exist in reactors, to understand their metabolic capability and therefore their role in the process. The second implies, in addition, the means to alter this capability reliably and safely.

The ability to isolate and characterize proteins and genes, to mutate these specifically, to knock out particular genes and to engineer them into different genetic backgrounds and into other species has helped elucidate metabolic pathways (Reeve et al., 1997) and the function of particular genes and enzymes of many organisms (Reeve et al., 1997; Perret et al., 2011).

to produce biogas. Acetogenic bacteria utilize the reductive acetyl-CoA pathway, perhaps better known as the Wood–Ljungdahl pathway as their main mechanism for transformation of CO_2 to cellular carbon and energy conservation and for synthesis of acetyl-CoA (Ragsdale and Pierce, 2008). This pathway enables the use of hydrogen as an electron donor and carbon dioxide as an electron acceptor in addition to being a building block for biosynthesis. In fact, bacteria are identified as acetogenic on the basis of their possessing this anaerobic metabolic pathway in addition to their capability to produce acetate as an end product of CO_2 reduction (Drake et al., 2008).

There are many reactions that lead to the reduction of CO_2 to carbon monoxide (CO) and the attachment of the CO to a methyl group catalysed by CO dehydrogenase. The coupling of the methyl group and the CO is catalysed by acetyl CoA synthetase. Obligate hydrogen (H_2)-producing acetogenic bacteria are capable of producing acetate and H_2 from higher fatty acids. Few pure strains of these types of acetogens have been isolated as H_2 production inhibits their growth. Examples include *Syntrophobacter wolinii*, which degrades propionate only when it is cultured with a H_2-using organism (so reducing toxic H_2 levels) in the absence of light or exogenous electron acceptors (such as O_2, sulfate, or nitrate) and *Sytrophomonos wolfei*, a butyrate decomposer (McInerney et al., 1981). It is the demanding requirements for metals, cofactors, anaerobiosis and substrates with low reducing potential such as H_2 or CO that restricts the Wood–Ljungdahl pathway to a few anoxic niches (Berg et al., 2010).

The methanogenesis pathways

The importance of the transformations carried out in earlier stages of the AD process is that methanogens can use only a limited number of substrates for methane production. These include CO_2 together with H_2, formate, material with methyl groups and acetic acid (acetate). Methanogens are classified on the basis of the type of substrates they use as, either, hydrogenotrophs, methylotrophs or acetotrophs and each type uses a different biochemical pathway to produce methane (Figure 15.2). Organic substances like carbohydrates, proteins, long-chain fatty acids and alcohols, are not utilized directly by methanogens, with the exception of some rare hydrogenotrophs that can also use 2-propanol, 2-butanol and cyclopentanol as electron donors. Methanogens also use the Wood–Ljungdahl pathway to conserve energy while growing in H_2 and CO_2 (Lapado and Whitman, 1990; Ferry, 1999). Hydrogenotrophic methanogens use the acetyl-CoA pathway with CO_2 but do not produce a mixture of methane and acetate as end products.

The pathways of methanogenesis have been extensively studied and comprise several steps each of which requires the action of at least one enzyme and more detailed descriptions of these can be obtained in a number of reviews (Reeve et al., 1997; Graham and White, 2002; Whitman et al., 2006). These studies demonstrate the scale and depth of biochemistry, biology and molecular science that has been required to elucidate these systems.

A major impediment to elucidating the main microbial species underpinning the AD process has been the need to isolate and culture species of microbes in order to characterize their physiological and biochemical processes and so determine their specific status. The sensitivity of anaerobic microbes to the oxygen in the atmosphere proved particularly challenging and is one of the main reasons why it took 60 years to understand the fundamentally important Wood–Ljungdahl pathway. Drake et al. (2008) provide a readable history of this discovery which provides a good illustration of the difficulties that had to be overcome and the range of technologies required to understand their function. Many technological advances in physics,

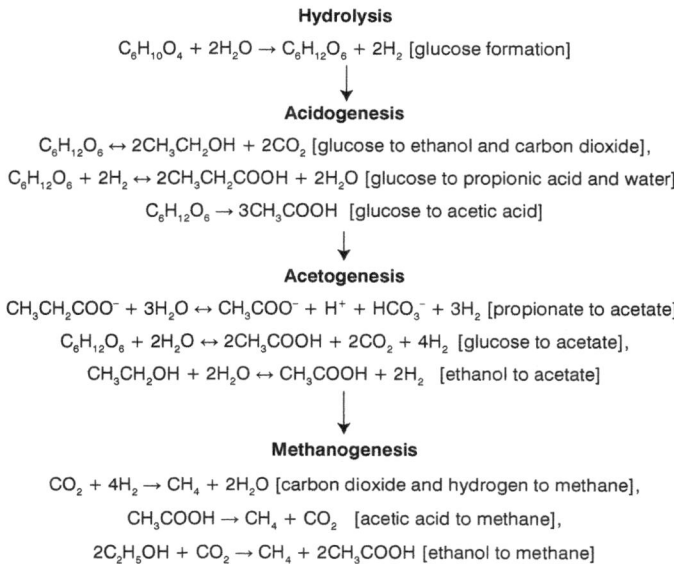

Figure 15.1 The main stages and major representative chemical reactions during methanogenesis (in this diagram largely based on starting with a carbohydrate).

Biochemical pathways in biomethane production

Biomethane formation is generally a multiple stage process involving various biochemical reactions in each stage (McCarty, 1982; Khalid et al., 2011) (Figure 15.1). The first step is mainly hydrolysis of polysaccharides (e.g. starch, cellulose, hemicellulose etc.), into soluble sugars and degradation of proteins to amino acids and fats to glycerol and triglycerides. The second step is acidogenesis where fermentation products from the previous stage are converted to volatile fatty acids (such as acetic, propionic and butyric acid), carbon dioxide, hydrogen, alcohols and other minor compounds. Products from the acidogenesis stage, such as volatile fatty acids and alcohols enter the third, acetogenesis, stage where they are converted by acetogenic micro-organisms into hydrogen, carbon dioxide and acetic acid. Due to the long generation time of these bacteria this is a limiting step in the AD production process. Finally, products from previous stages, such as hydrogen, acetic acid (acetate) and alcohols enter the fourth stage of methanogenesis, a strictly anaerobic process. In this stage, they are converted by slow-growing *Archaea* to methane (CH_4) (70 per cent v/v), carbon dioxide (CO_2) (30 per cent v/v) plus some other products such as ammonia (NH_3) or hydrogen sulphide (H_2S).

There are, then, multiple pathways using a range of enzymes to transform the input feedstock to simple sugars and fatty acids that are transformed in the acidogenesis stage to volatile fatty acids and alcohols. The later stages of acetogenesis and methanogenesis are worth considering in a little more detail given their importance for the formation of biogas and the involvement of some particular biochemical pathways in that process.

The acetogenesis process

This process is crucial for AD in that acetogenic bacteria transform a wide range of precursors into acetate and this molecule is used as a key substrate for methanogenic bacteria

Proteins	Proteases hydrolyse these to amino acids	Bacteria e.g. *Bacteroides, Butyrivibrio, Clostridium, Fusobacterium, Selenomonas* and *Streptococcus*.
	Amino acids then transformed to fatty acids and ammonia	Bacteria e.g. *Clostridium, Peptococcus, Selenomonas, Campylobacter* and *Bacteroides*
Lipids	Lipases and β-oxidation to acyl-CoA dehydrogenase	Bacteria e.g. *Anaerovibrio, Clostridia, Micrococcus, Syntrophomonas*

Sources: Van Soest et al. (1991), Saha (2000), Clavero and Razz (2002), Brummell (2006), Kumar et al. (2008). See also more detailed listings of enzymes, including those for lignin, in Howard et al. (2003) and Bugg et al. (2011).

Table 15.1 Major substrates used in anaerobic digestion and examples of the principal types of enzymes or enzyme groups used in their breakdown to fermentable compounds (usually sugars).

Substrate	Principal biochemical structure	Processed by these major enzymes or enzyme groups	Micro-organism producing enzymes
Lignin	Aromatic heteropolymer consisting of phenylpropanoid aryl-C3 units joined variously by carbon–carbon or ether links	Laccases, lignin peroxidases and manganese peroxidises	Ascomycete and basidiomycete fungi, (white and brown rot) e.g. Botrytis, Phanerochaete Phlebia / Bacteria e.g. Actinomycetes: Nocardia, Rhodococcus, Microbacterium, Streptomyces / α-Protobacteria: Brucella, Ochrobactrum, Paracoccus, Sphingobium / Pseudomonas, γ-Protobacteria: Acinitobacter, Burkhloderia, Citrobacter, Escherichia, Pseudomonas
Celluloses	Homopolysaccharides composed of β-D-glucopyranose units, linked by β-(1→4)-glycosidic bonds	Cellulases (3 main groups, endoglucanases, cellobiohydrolases (exoglucanases), β-glucosidases)	Filamentous fungi e.g. Trichoderma (especially T. reesei) / Fusarium, Penicillium, Phanerochaete ,Aspergillus / Bacteria e.g. Bacillus, Clostridium, Pyrococcus, Streptomyces
Hemicelluloses	Heterogeneous polymers consisting of pentoses (D-xylose, D-arabinose), hexoses (D-mannose, D-glucose, D-galactose) and sugar acids	Hemicellulases (e.g. in xylan degradation, endo-1,4-β-xylanase, β-xylosidase, α-glucuronidase, α-L-arabinofuranosidase, acetylxylan esterase)	Fungi e.g. Trichoderma, Aspergillus, Fusarium, Penicillium / Bacteria e.g. Bacillus, Butyrivibrio, Clostridium, Fusobacterium
Pectins	Polysaccharides: backbone of homo-galacturonic acid regions with neutral sugar side chains made from L-rhamnose, arabinose, galactose and xylose. L-rhamnose residues in the backbone carry side-chains containing arabinose and galactose	Pectin-degrading enzymes, (polymethylgalacturonase, (endo-) polygalacturonase, pectin depolymerase, pectinase, exopolygalacturonase) all hydrolyse the polygalacturonic acid chain of the pectin polymer by the addition of a water molecule	Fungi e.g. Aspergillus, Fusarium, Penicillium, / Bacteria e.g. Bacillus, Clostridium, Fusobacterium

Substrates used in AD

A prime source used for AD which is a major component of many of the plant derived feedstocks described above is lignocellulosic material. Lignocellulose is mainly composed of varying amounts of lignin, cellulose, and hemicellulose – for example in wood their proportions range from 18–35 per cent, 45–55 per cent, and 24–40 per cent respectively (Malherbe and Cloete, 2003). It has evolved to be highly resistant to degradation by herbovore, fungi, or microbes and its binding with the other components often prevents or retards their degradation (Himmel et al., 2007), but these organisms have developed specialist enzymes and biochemical processes to achieve breakdown of cellulosic biomass. The effectiveness of the degradation process is dependent on the concerted action of many enzymes in the same environment. The exoglucanases or cellobiohydrolases, endoglucanases and β-glucosidase enzymes secreted by the filamentous fungi, *Trichoderma reesei*, have a synergistic action to catalyse the hydrolysis of lignocellulosic materials. Their levels of activity are dependant on physical parameters such as pH, temperature, adsorption, the presence of enabling elements like nitrogen and phosphorus and other micronutrients or inhibiters such as phenolic compounds, salt and heavy metals (Goyal et al., 1991, Bayer et al., 2007). Bacterial cellulases exist as discrete multi-enzyme complexes, called cellulosomes, which consist of multiple subunits to provide an integrated attack on a variety of enzyme targets (Béguin and Lemaire, 1996; Bayer et al., 2004). Mechanical, thermal or chemical pre-treatments of feedstock are aimed at breaking down bonds with lignin molecules but there is also a concerted search for enzymes, and genetically engineered biomass and fermenting organisms that can assist this process (Menon and Rao, 2012; Zhang et al., 2012).

Bioconversion of the other main components of lignocelluloses (mainly cellulose and hemicellulose) into fermentable sugars is a necessary first step in the production of industrially important products (McCarty, 1982). The main enzymes or enzyme groups involved in their breakdown are listed in Table 15.1. Cellulose is the most abundant renewable biomass. Cellobiose is the smallest repetitive unit of cellulose and can be converted into glucose residues through the action of cellulose-hydrolysing enzymes (i.e. cellulases) (Saha, 2000; Kumar et al., 2008). Hemicellulose is the second most abundant renewable biomass and accounts for 25–35 per cent of lignocellulosic materials and is converted into sugars by a range of hemicellulose-hydrolysing enzymes (i.e. hemicellulases) (Saha, 2000; Clavero and Razz, 2002; Kumar et al., 2008). Hemicelluloses in hardwood contain mainly xylans, while in softwood glucomannans are most common (Kumar et al., 2008). Pectins, the third main component (Van Soest et al., 1991), can form up to half of the polymeric content of the cell walls in pulps and fruits such as citrus and apples and are hydrolysed by pectin-degrading enzymes (Brummell, 2006; Kumar et al., 2008).

Proteins are generally utilized as a substrate after being hydrolysed to amino acids by proteases, which are then degraded to fatty acids such as acetate, propionate, and butyrate, and to ammonia, reactions mediated by additional enzymes. Lipids are converted by lipases to long-chain fatty acids which are further degraded by p-oxidation to produce acetyl CoA (Miyamoto, 1997).

This area has been the subject of much research over the last decade or more and several reviews provide more detail concerning the chemical structure of lignocellulose and other feedstocks, and more detailed discussion of the range of micro-organisms and the various enzymes they use to degrade their substrates (e.g. Kumar et al., 2008; Sánchez, 2009; Bugg et al., 2011; Menon and Rao, 2012; Parawira, 2012; Zhang et al., 2012).

Chapter 15

The role of molecular biology in optimizing anaerobic digestion and biomethane production

Dimitris N. Argyropoulos,[1*] *Charoula C. Psallida*[1] *and John A.H. Benzie*[2,3*]

[1] Genetic Identification Laboratory, Hellenic Agricultural Organization "DEMETER", Sofokli Venizelou 1, 14123, Lykovrissi, Greece, [2] Environmental Research Institute, University College Cork, Lee Road, Cork, Ireland; [3] School of Biological Earth and Environmental Sciences, University College Cork, North Mall Campus, Cork, Ireland
*Corresponding authors email: dimargy@yahoo.com; j.benzie@ucc.ie

Introduction

Biomethane production is the result of a fermentative process known as anaerobic digestion (AD), in which micro-organisms decompose complex biological materials with the parallel release of methane and carbon dioxide (McCarty, 1982; Al Seadi et al., 2008; Khalid et al., 2011). The biochemical and genetic characteristics of the micro-organisms and the substrates used in AD are key factors that determine the efficiency of this process. Understanding and manipulating these components and their interactions will be important for the development of future technologies aimed at optimizing large-scale production of methane biogas for commercial use (Rittmann et al., 2008).

Modern biotechnology techniques have provided powerful tools with which to speed analysis of microbial communities and improve understanding of the biochemical pathways by which the raw materials are transformed (Talbot et al., 2008). Molecular genetics also offers the potential to improve efficiency of production by manipulating the species, strains, pathways and genes used in particular processes (Kumar et al., 2008; Rittmann et al., 2008). The present chapter will give an overview of these advances and their future potential but begins with a summary of the main biochemical processes involved in AD, in order to set the framework for their application.

Biochemistry in AD

Sources used for biomethane production are solid or liquid substrates, usually sewage, animal slurry, silages (corn, grass, etc.), or organic wastes that provide the basic ingredients for fermentation or digestion (McCarty, 1982). Polymers such as carbohydrates, proteins, and lipids provide the carbon source for formation of biomethane through the collaborative action of a variety of micro-organisms that include fermentative microbes (hydrolytic, and acidogens), hydrogen-producing, acetate-forming microbes (acetogens) and methane-producing microbes (methanogens), the latter producing the reduced end-products.

Weiss, A., Jérôme, V., Freitag, R. and Mayer, H.K. (2008) 'Diversity of the resident microbiota in a thermophilic municipal biogas plant', *Applied Microbiology and Biotechnology*, vol 81, pp. 63–73.

Westerholm, M., Muller, B., Arthurson, V. and Schnurer, A. (2011) 'Changes in the acetogenic population in a mesophilic anaerobic digester in response to increasing ammonia concentration', *Microbes and Environments*, vol 26, pp. 347–353.

Wirth, R., Kovács, E., Maróti, G., Bagi, Z., Rákhely, G. and Kovács, K.L. (2012) 'Characterization of a biogas-producing microbial community by short-read next generation DNA sequencing', *Biotechnology for Biofuels*, vol 5, p. 41.

Zakrzewski, M., Goesmann, A., Jaenicke, S., Jünemann, S., Eikmeyer, F., Szczepanowski, R., Al-Soud, W.A., Sørensen, S., Pühler, A. and Schlüter, A. (2012) 'Profiling of the metabolically active community from a production-scale biogas plant by means of high-throughput metatranscriptome sequencing', *Journal of Biotechnology*, vol 158, pp. 248–258.

Zhang, D., Li, J., Guo, P., Li, P., Suo, Y., Wang, X. and Cui, Z. (2011a) 'Dynamic transition of microbial communities in response to acidification in fixed-bed anaerobic baffled reactors (FABR) of two different flow directions', *Bioresource Technology*, vol 102, pp. 4703–4711.

Zhang, H., Banaszak, J.E., Parameswaran, P., Alder, J., Krajmalnik-Brown, R. and Rittmann, B.E. (2009) 'Focused-pulsed sludge pre-treatment increases the bacterial diversity and relative abundance of acetoclastic methanogens in a full-scale anaerobic digester', *Water Research*, vol 43, pp. 4517–4526.

Zhang, J., Wei, Y., Xiao, W., Zhou, Z., and Yan, X. (2011b) 'Performance and spatial community succession of an anaerobic baffled reactor treating acetone-butanol-ethanol fermentation wastewater', *Bioresource Technology*, vol 102, pp. 7407–7414.

Zhang, Y., Cañas, E.M.Z., Zhu, Z., Linville, J.L, Chen, S. and He, Q. (2011c) 'Robustness of archaeal populations in anaerobic co-digestion of dairy and poultry wastes', *Bioresource Technology*, vol 102, pp. 779–785.

Zumstein, E., Moletta, R. and Godon, J.J. (2000) 'Examination of two years of community dynamics in an anaerobic bioreactor using fluorescent-PCR single strand conformation polymorphism analysis', *Environmental Microbiology*, vol 2, pp. 69–78.

Zwietering, M.H., Jongenburger, I., Rombouts, F.M. and Van T Riet, K. (1990). 'Modeling of the bacterial growth curve', *Applied and Environmental Microbiology*, vol 56, no 6, pp. 1875–1881.

Shin, S., Lee, S., Lee, C., Hwang, K. and Hwang, S. (2010) 'Qualitative and quantitative assessment of microbial community in batch anaerobic digestion of secondary sludge', *Bioresource Technology*, vol 101, pp. 9461–9470.

Shin, S.G., Yoo, S., Hwang, K., Song, M., Kim, W., Han, G. and Hwang S. (2011a) 'Dynamics of transitional acidogenic community along with methanogenic population during anaerobic digestion of swine wastewater', *Process Biochemistry*, vol 46, no 8, pp 1607–1613.

Shin, S.G., Zhou, B.W., Lee, S., Kim, W. and Hwang, S. (2011b) 'Variations in methanogenic population structure under overloading of pre-acidified high-strength organic wastewaters', *Process Biochemistry*, vol 46, pp. 1035–1038.

Singhania, R.R. (2009), 'Cellulolytic enzymes', in P. Singh neé Nigam and A. Pandey (eds), *Biotechnology for Agro-Industrial Residues Utilisation*, pp. 371–381, Amsterdam: Springer Science+Business Media.

Song, M., Shin, S.G. and Hwang, S. (2010) 'Methanogenic population dynamics assessed by real-time quantitative PCR in sludge granule in upflow anaerobic sludge blanket treating swine wastewater', *Bioresource Technology*, vol. 101, no 1 supplement, pp. S23–S28.

Sötemann, S.W., van Rensburg, P., Ristow, N.E., Wentzel, M.C., Loewenthal, R.E. and Ekama, G.A. (2005) 'Integrated chemical/physical and biological processes modelling Part 2 – Anaerobic digestion of sewage sludges', *Water SA*, vol 31 pp. 545–568.

Sousa, D.Z., Pereira, M.A., Stams, A.J.M. and Alves, M.M. (2007) 'Microbial communities involved in anaerobic degradation of unsaturated or saturated long chain fatty acids (LCFA)', *Applied Environmental Microbiology*, vol 73, pp. 1054–1064.

Supaphol, S., Jenkins, S.N., Intomo, P., Waite, I.S. and O'Donnell, A.G. (2011) 'Microbial community dynamics in mesophilic anaerobic co-digestion of mixed waste', *Bioresource Technology*, vol 102, pp. 4021–4027.

Talbot, G., Roy, C.S., Topp, E., Kalmokoff, M.L., Brooks, S.P.J., Beaulieu, C., Palin, M.-F. and Masse, D.I. (2010) 'Spatial distribution of some microbial trophic groups in a plug-flow-type anaerobic bioreactor treating swine manure', *Water Science and Technology*, vol 61, no 5, pp. 1147–1155.

Tang, Y.Q., Shigematsu, T., Ikbal, Morimura, S. and Kida, K. (2004) 'The effects of microaeration on the phylogenetic diversity of microorganisms in the thermophilic anaerobic municipal solid-waste digester', *Water Research*, vol 38, pp. 2537–2550.

Tang, Y.Q., Shigematsu, T., Morimura, S. and Kida, K. (2007) 'Effect of dilution rate on the microbial structure of a mesophilic butyrate-degrading methanogenic community during continuous cultivation', *Applied Microbiology and Biotechnology*, vol 75, pp. 451–465.

Tang, Y.Q., Ji, P., Hayashi, J., Koike, Y., Wu, X.L. and Kida, K. (2011) 'Characteristic microbial community of a dry thermophilic methanogenic digester: its longterm stability and change with feeding', *Applied Microbiology and Biotechnology*, vol 91, pp. 1447–1461.

Tomei, M.C., Braguglia, C.M., Cento, G. and Mininni, G. (2009) 'Modeling of anaerobic digestion of sludge', *Critical Reviews in Environmental Science and Technology*, vol 39, pp. 1003–1051.

Trosvik, P., Rudi, K., Næs, T., Kohler, A., Chan, K-S., Jakobsen, K.S. and Stenseth, N.C. (2008) 'Characterizing mixed microbial population dynamics using time-series analysis', *ISME Journal*, vol 2, pp. 707–715.

Uyanik, S., Sallis, P.J. and Anderson, G.K. (2002) 'The effect of polymer addition on granulation in an anaerobic baffled reactor (ABR). Part II: Compartmentalization of bacterial populations', *Water Research*, vol 36, pp. 944–955.

Wagner, A.O., Malin, C., Lins, P. and Illmer, P. (2011) 'Effects of various fatty acid amendments on a microbial digester community in batch culture', *Waste Management*, vol 31, pp. 431–437.

Walter, A., Knapp, B.A., Farbmacher, T., Ebner, C., Insam, H. and Franke-Whittle, I.H. (2012) 'Searching for links in the biotic characteristics and abiotic parameters of nine different biogas plants', *Microbial Biotechnology*, vol 5, pp. 717–730.

Wang, H., Vuorela, M., Keränen, A.-L., Lehtinen, T.M., Lensu, A., Lehtomäki, A., and Rintala, J. (2010) 'Development of microbial populations in the anaerobic hydrolysis of grass silage for methane production', *FEMS Microbiology Ecology*, vol 72, pp. 496–506.

Rademacher, A., Zakrzewski, M., Schlüter, A., Schönberg, M., Szczepanowski, R., Goesmann, A., Pühler, A. and Klocke, M. (2012b) 'Characterization of microbial biofilms in a thermophilic biogas system by high-throughput metagenome sequencing', *FEMS Microbiology Ecology*, vol 79, pp. 785–799.

Ramirez, I., Volcke, E.I.P. and Steyer, J.P. (2008) 'Modeling and monitoring of microbial diversity in ecosystems: Application to biological wastewater treatment processes', *Proceedings of the 17th World Congress of the International Federation of Automatic Control*, Seoul, Korea, July 6–11, pp. 13605–13610.

Ramirez, I., Volcke, E.I.P., Rajinikanth, R. and Steyer, J.-P. (2009) 'Modeling microbial diversity in anaerobic digestion through an extended ADM1 model', *Water Research*, vol 43, pp 2787–2800.

Rapaport, A., Dochain, D., and Harmand, J. (2009). 'Long run coexistence in the chemostat with multiple species', *Journal of Theoretical Biology*, vol 257, pp. 252–259.

Rastogi, G., Ranade, D.R., Yeole, T.Y., Patole, M.S., and Shouche, Y.S. (2008) 'Investigation of methanogen population structure in biogas reactor by molecular characterization of methyl-coenzyme M reductase A (mcrA) genes', *Bioresource Technology*, vol 99, pp. 5317–5326.

Regueiro, L., Veigaa, P., Figueroaa, M., Alonso-Gutierreza, J., Stamsb, A.J.M., Lemaa, J.M. and Carballaa, M. (2012) 'Relationship between microbial activity and microbial community structure in six full-scale anaerobic digesters', *Microbiological Research*, vol 167, pp. 581–589.

Ritari, J., Koskinen, K., Hultman, J., Kurola, J.M., Kymäläinen, M., Romantschuk, M., Paulin, L. and Auvinen, P. (2012) 'Molecular analysis of meso- and thermophilic microbiota associated with anaerobic biowaste degradation', *BMC Microbiology*, vol 12, pp. 121 http://www.biomedcentral.com/1471-2180/12/121.

Rittmann, B.E., Krajmalnik-Brown, R. and Halden, R.U. (2008) 'Pre-genomic, genomic and postgenomic study of microbial communities involved in bioenergy', *Nature Reviews in Microbiology*, vol 6, pp. 604–612.

Rivière, D., Desvignes, V., Pelletier, E., Chaussonnerie, S., Guermazi, S., Weissenbach, J., Li, T., Camacho, P. and Sghir, A. (2009) 'Towards the definition of a core of microorganisms involved in anaerobic digestion of sludge', *ISME Journal* vol 3, pp. 700–714.

Sang, N.N., Soda, S., Inoue, D., Sei, K. and Ike, M. (2009) 'Effects of intermittent and continuous aeration on accelerative stabilization and microbial population dynamics in landfill bioreactors', *Journal of Bioscience and Bioengineering*, vol. 108, pp. 336–343.

Sbarciog, M. and Vande Wouwer, A. (2012) 'Some considerations about control of multispecies anaerobic digestion systems', *Proceedings of the 7th Vienna International Conference on Mathematical Modelling (MATHMOD)*, 15–17 February. Vienna, Austria.

Scherer, P., Lehmann, K., Schmidt, O. and Demirel, B. (2009) 'Application of a fuzzy logic control system for continuous anaerobic digestion of low buffered, acidic energy crops as mono-substrate', *Biotechnology & Bioengineering*, vol 102, pp. 736–748.

Schlüter, A., Bekel, T., Diaz, N.N., Dondrup, M., Eichenlaub, R., Gartemann, K.H., Krahn, I., Krause, L., Kromeke, H., Kruse, O., Mussgnug, J.H., Neuweger, H., Niehaus, K., Puhler, A., Runte, K.J., Szczepanowski, R., Tauch, A., Tilker, A., Viehover, P. and Goesmann, A. (2008) 'The metagenome of a biogas-producing microbial community of a production-scale biogas plant fermenter analysed by the 454-pyrosequencing technology', *Journal of Biotechnology*, vol 136, pp. 77–90.

Schoen, M.A., Sperl, D., Gadermaier, M., Goberna, M., Franke-Whittle, I., Insam, H., Ablinger, J. and Wett, B. (2009) 'Population dynamics at digester overload conditions', *Bioresource Technology*, vol 100, no 23, pp. 5648–5655.

Shigematsu, T., Tang, Y., Kobayashi, T., Kawaguchi, H., Morimura, S. and Kida, K. (2004) 'Effect of dilution rate on metabolic pathway shift between acetoclastic and non-acetoclastic methanogens in chemostat cultivation', *Applied Environmental Microbiology*, vol 70, pp. 4048–4052.

Shigematsu, T., Tang, Y., Kawaguchi, H., Ninomiya, K., Kijima, J., Kobayashi, T., Morimura, S. and Kida, K. (2003) 'Effect of dilution rate on structure of a mesophilic acetate-degrading methanogenic community during continuous cultivation', *Journal of Biosciences and Bioengineering*, vol 96, pp. 547–558.

Masci, P., Bernard, O., Grognard, F., Latrille, E., Sorba, J.B. and Steyer, J.P. (2009) 'Driving competition in a complex ecosystem: application to anaerobic digestion', *Proceedings of the ECC Conference, Budapest, Hungary*.

McCarty, P.L. (1982) 'One hundred years of anaerobic treatment', In D.E. Hughes, D.A. Stafford, B.I. Wheatley, W. Baader, G. Lettinga, E.J. Nyns, W. Verstraete, and R.L. Wentworth (eds.), *Anaerobic Digestion, 1981*, pp. 3–21, Amsterdam: Elsevier Biomedical Press.

McGarvey, J.A., Miller, W.G., Zhang, R., Ma, Y. and Mitloehner, F. (2007) 'Bacterial population dynamics in dairy waste during aerobic and anaerobic treatment and subsequent storage', *Applied Environmental Microbiology*, vol 73, 193–202.

McHugh, S., Carton, M., Mahony, T. and O'Flaherty, V. (2003) 'Methanogenic population structure in a variety of anaerobic bioreactors', *FEMS Microbiology Letters*, vol 219, pp. 297–304.

McMahon, K.D., Stroot, R.G., Mackie, R.I. and Raskin, L. (2001) 'Anaerobic codigestion of municipal solid waste and biosolids under various mixing conditions – II: Microbial population dynamics', *Water Research*, vol 35, pp. 1817–1827.

McMahon, K.D., Zheng, D.D., Stams, A.J.M., Mackie, R.I. and Raskin, L. (2004) 'Microbial population dynamics during start-up and overload conditions of anaerobic digesters treating municipal solid waste and sewage sludge', *Biotechnology and Bioengineering*, vol 87, pp. 823–834.

Miyamoto, K. (ed.) (1997) 'Renewable biological systems for alternative sustainable energy production', *FAO Agricultural Services Bulletin* 128, http://www.fao.org/docrep/W7241E/W7241E00.htm, accessed April 2012.

Munk, B., Bauer, C., Gronauer, A. and Lebuhn, M. (2010) 'Population dynamics of methanogens during acidification of biogas fermenters fed with maize silage', *Engineering in Life Sciences*, vol 10, pp. 496–508.

Nelson, M.C., Morrison, M. and Yu, Z. (2011) 'A meta-analysis of the microbial diversity observed in anaerobic digesters', *Bioresource Technology*, vol 102, pp. 3730–3739.

Nelson, M.C., Morrison, M., Schanbacher, F., and Yu, Z. (2012) 'Shifts in microbial community structure of granular and liquid biomass in response to changes to infeed and digester design in anaerobic digesters receiving food-processing wastes', *Bioresource Technology*, vol 107, pp. 135–143.

Nettmann, E., Bergmann, I., Pramschufer, S., Mundt, K., Plogsties, V., Herrmann, C. and Klocke, M. (2010) 'Polyphasic analyses of methanogenic archaeal communities in agricultural biogas plants', *Applied Environmental Microbiology*, vol 76, pp. 2540–2548.

O'Reilly, J., Lee, C., Chinalia, F., Collins, G., Mahony, T. and O'Flaherty, V. (2010) 'Microbial community dynamics associated with biomass granulation in low-temperature (15°C) anaerobic wastewater treatment bioreactors', *Bioresource Technology*, vol 101, pp. 6336–6344.

Oz, N.A., Ince, O., Turker, G. and Ince, B.K. (2012) 'Effect of seed sludge microbial community and activity on the performance of anaerobic reactors during the start-up period', *World Journal of Microbiology and Biotechnology*, vol 28, pp. 637–647.

Padmasiri, S.I., Zhang, J., Fitch, M., Norddahl, B., Morgenroth, E. and Raskin, L. (2007) 'Methanogenic population dynamics and performance of an anaerobic membrane bioreactor (AnMBR) treating swine manure under high shear conditions', *Water Research*, vol 41, no 1, pp. 134–144.

Palatsi, J., Illa, J., Prenafeta-Boldú, F.X., Laureni, M., Fernandez, B., Angelidaki, K. and Flotats, X. (2010) 'Long-chain fatty acids inhibition and adaptation process in anaerobic thermophilic digestion: batch tests, microbial community structure and mathematical modelling', *Bioresource Technology*, vol 101, pp. 2243–2251.

Peleg, M. and Corradini, M.G. (2011) ' Microbial growth curves: What the models tell us and what they cannot', *Critical Reviews in Food Science and Nutrition*, vol 51, pp. 917–945.

Pobeheim, H., Munk, B., Móller, H., Berg, G., and Guebitz, G.M. (2010) 'Characterization of an anaerobic population digesting a model substrate for maize in the presence of trace metals', *Chemosphere* vol 80, pp. 829–836.

Rademacher, A., Nolte, C., Schönberg, M. and Klocke, M. (2012a) 'Temperature increases from 55 to 75°C in a two-phase biogas reactor result in fundamental alterations within the bacterial and archaeal community structure', *Applied Microbiology and Biotechnology*, vol 96, pp. 565–576.

Krakat, N., Westphal, A., Schmidt, S. and Scherer, P. (2010c) 'Anaerobic digestion of renewable biomass: thermophilic temperature governs methanogen population dynamics', *Applied and Environmental Microbiology*, vol 76, pp. 1842–1850.

Krakat, N., Schmidt, S. and Scherer, P. (2011) 'Potential impact of process parameters upon the bacterial diversity in the mesophilic anaerobic digestion of beet silage', *Bioresource Technology*, vol 102, pp. 5692–5701.

Krause, L., Diaz, N.N., Edwards, R.A., Gartemann, K.-H., Krömeke, H., Neuweger, H., Pühler, A. and Goesmann, A. (2008) 'Taxonomic composition and gene content of a methane-producing microbial community isolated from a biogas reactor', *Journal of Biotechnology*, vol 136, pp. 91–101.

Kröber, M., Bekel, T., Diaz, N.N., Goesmann, A., Jaenicke, S., Krause, L., Miller, D., and Schlüter, A. (2009) 'Phylogenetic characterization of a biogas plant microbial community integrating clone library 16S-rDNA sequences and metagenome sequence data obtained by 454-pyrosequencing', *Journal of Biotechnology*, vol 142, pp. 38–49.

Lee, C., Kim, J., Hwang, K., O'Flaherty, V. and Hwang, S. (2009) 'Quantitative analysis of methanogenic community dynamics in three anaerobic batch digesters treating different wastewaters', *Water Research*, vol 43, pp. 157–165.

Lee, C., Kim, J., Shin, S.G., O'Flaherty, V. and Hwang, S. (2010) 'Quantitative and qualitative transitions of methanogen community structure during the batch anaerobic digestion of cheese-processing wastewater', *Applied Microbiology and Biotechnology*, vol 87, pp. 1963–1973.

Lee, S.-H., Kang, H-J., Lee, Y.H., Lee, T.J., Han, K., Choi, Y. and Park, H.-D. (2012) 'Monitoring bacterial community structure and variability in time scale in full-scale anaerobic digesters', *Journal of Environmental Monitoring*, vol 14, pp. 1893–1905.

Lerm, S., Kleyböcker, A., Miethling-Graff, R., Alawi, M., Kasina, M., Liebrich, M. and Würdemann, H. (2012) 'Archaeal community composition affects the function of anaerobic co-digesters in response to organic overload', *Waste Management*, vol 32, pp. 389–399.

Leven, L., Eriksson, A.R. and Schnürer, A. (2007) 'Effect of process temperature on bacterial and archaeal communities in two methanogenic bioreactors treating organic household waste', *FEMS Microbiology Ecology*, vol 59, pp. 683–693.

Lin, J., Zuo, J., Ji, R., Chen, X., Liu, F., Wang, K. and Yang, Y. (2012) 'Methanogenic community dynamics in anaerobic co-digestion of fruit and vegetable waste and food waste', *Journal of Environmental Sciences*, 24, pp. 1288–1294.

Lin, P.-Y., Whang, L.-M., Wua, Y.-R., Rena, W.-J., Hsiaoa, C.-J., Lia, S.-L. and Chang, J.-S. (2007) 'Biological hydrogen production of the genus *Clostridium*: Metabolic study and mathematical model simulation', *International Journal of Hydrogen Energy* vol 32, pp. 1728–1735.

Liu, F.H., Wang, S.B., Zhang, J.S., Zhang, J., Yan, X., Zhou, H.K., Zhao, G.P. and Zhou, Z.H. (2009) 'The structure of the bacterial and archaeal community in a biogas digester as revealed by denaturing gradient gel electrophoresis and 16S rDNA sequencing analysis', *Journal of Applied Microbiology*, vol 106, pp. 952–966.

Liu, W.T., Chan, O.C. and Fang, H. (2002) 'Microbial community dynamics during start-up of acidogenic anaerobic reactors', *Water Research*, vol 36, no 13, pp 3203–3210.

Lübken, M., Gehring, T. and Wichern, M. (2010) 'Microbiological fermentation of lignocellulosic biomass: current state and prospects of mathematical modeling', *Microbiological Biotechnology*, vol 85, pp. 1643–1652.

Madden, P., Chinalia, F.A., Enright, A., Collins, G. and O'Flaherty, V. (2010) 'Perturbation-independent community development in low-temperature anaerobic biological wastewater treatment bioreactors', *Biotechnology and Bioengineering*, vol 105, no. 1, pp. 79–87.

Martin-Gonzalez, L., Castro, R., Pereira, M.A., Alves, M.M. and Font, X. (2011) 'Thermophilic co-digestion of organic fraction of municipal solid wastes with FOG wastes from a sewage treatment plant: reactor performance and microbial community monitoring', *Bioresource Technology*, vol 102, pp. 4734–4741.

Hori, T., Haruta, S., Ueno, Y., Ishii, M. and Igarashi, Y. (2006) 'Dynamic transition of a methanogenic population in response to the concentration of volatile fatty acids in a thermophilic anaerobic digester', *Applied and Environmental Microbiology*, vol 72, 1623–1630.

Hwang, K., Song, M., Kim, W., Kim, N. and Hwang, S. (2010) 'Effects of prolonged starvation on methanogenic population dynamics in anaerobic digestion of swine wastewater', *Bioresource Technology*, vol 101, no 1 supplement, pp. S2–S6.

Ike, M., Inoue, D., Miyano, T., Liu, T.T., Sei, K., Soda, S. and Kadoshin, S. (2010) 'Microbial population dynamics during startup of a full-scale anaerobic digester treating industrial food waste in Kyoto eco-energy project', *Bioresource Technology*, vol 101, pp. 3952–3957.

Jaenicke, S., Ander, C., Bekel, T., Bisdorf, R., Dröge, M., Gartemann, K.-H., Jünemann, S., Kaiser, O., Krause, L., Tille, F., Zakrzewski, M., Pühler, A., Schlüterand, A. and Goesmann, A. (2011) 'Comparative and joint analysis of two metagenomic datasets from a biogas fermenter obtained by 454-pyrosequencing', *PLoS ONE*, vol 6 no 1, art. no. e14519.

Kampmann, K., Ratering, S., Baumann, R., Schmidt, M., Zerr, W. and Schnell, S. (2012a) 'Hydrogenotrophic methanogens dominate in biogas reactors fed with defined substrates', *Systematic and Environmental Microbiology*, vol 35, pp. 404–413.

Kampmann, K., Ratering, S., Kramer, Schmidt, M., Zerr, W. and Schnell, S. (2012b) 'Unexpected stability of Bacteroidetes and Firmicutes communities in laboratory biogas reactors fed with different defined substrates', *Applied and Environmental Microbiology*, vol 78, pp. 2106–2119.

Karadagli, F. and Rittmann, B.E. (2007) 'A mathematical model for the kinetics of *Methanobacterium bryantii* M.o.H. considering hydrogen thresholds', *Biodegradation*, vol 18, pp. 453–464.

Karakashev, D., Batstone, D.J., Trably, E., and Angelidaki, I. (2006) 'Acetate oxidation is the dominant methanogenic pathway from acetate in the absence of Methanosaetaceae', *Applied and Environmental Microbiology*, vol 72, pp. 5138–5141.

Khalid, A., Arshad, M., Anjum, M., Mahmood, T. and Dawson, L. (2011) 'The anaerobic digestion of solid organic waste', *Waste Management*, vol 31, no 8, pp. 1737–1744.

Kim, J., Shin, S.G., Han, G., O'Flaherty, V., Lee, C. and Hwang, S. (2011) 'Common key acidogen populations in anaerobic reactors treating different wastewaters: Molecular identification and quantitative monitoring', *Water Research*, vol 45, no 8, pp. 2539–2549.

Kim, W., Hwang, K., Shin, S.G., Lee, S. and Hwang, S. (2010a) 'Effect of high temperature on bacterial community dynamics in anaerobic acidogenesis using mesophilic sludge inoculum', *Bioresource Technology*, vol 101, supplement, ppS17–S22.

Kim, W., Lee, S., Shin, S.G., Lee, C., Hwang, K. and Hwang, S. (2010b) Methanogenic community shift in anaerobic batch digesters treating swine wastewater', *Water Research*, vol 44, pp. 4900–4907.

Klocke, M., Mähnert, P., Mundt, K., Souidi, K., and Linke, B. (2007) 'Microbial community analysis of a biogas-producing completely stirred tank reactor fed continuously with fodder beet silage as mono-substrate', *Systematic and Applied Microbiology*, vol 30, pp. 139–151.

Klocke, M., Nettmann, E., Bergmann, I., Mundt, K., Souidi, K., Mumme, J., and Linke, B. (2008) 'Characterization of the methanogenic Archaea within two-phase biogas reactor systems operated with plant biomass', *Systematic and Applied Microbiology*, vol 31, pp. 190–205.

Kobayashi, T., Yasuda, D., Li, Y-Y., Kubota, K., Harada, H. and Yu, H-Q. (2009) 'Characterization of start-up performance and archaeal community shifts during anaerobic self-degradation of waste-activated sludge', *Bioresource Technology*, vol 100, pp. 4981–4988.

Krakat, N., Schmidt, S. and Scherer, P. (2010a) 'The mesophilic fermentation of renewable biomass – does hydraulic retention time regulate diversity of methanogens?' *Applied and Environmental Microbiology*, vol 76, pp. 6322–6326.

Krakat, N., Westphal, A., Satke, K., Schmidt, S. and Scherer, P. (2010b) 'The microcosm of a biogas fermenter: comparison of moderate hyperthermophilic (60°C) with thermophilic (55°C) conditions', *Engineering and Life Science*, vol 10, pp. 520–527.

Chelliapan, S., Wilby, T., Yuzir, A. and Sallis, P.J. (2011) 'Influence of organic loading on the performance and microbial community structure of an anaerobic stage reactor treating pharmaceutical wastewater', *Desalination*, vol 271. pp. 257–264.

Chen, Y., Cheng, J.J. and Creamer, K.S. (2008) 'Inhibition of anaerobic digestion process: A review', *Bioresource Technology*, vol 99, pp. 4044–4064.

Cheon, J., Hidaka, T., Mori, S., Koshikawa, H. and Tsuno, H. (2008) 'Applicability of random cloning method to analyze microbial community in full-scale anaerobic digesters'. *Journal of Bioscience and Bioengineering*, vol 106, pp. 134–140.

Chouari, R., Le Paslier, D., Dauga, C., Daegelen, P., Weissenbach, J. and Sghir, A. (2005) 'Novel major bacterial candidate division within a municipal anaerobic sludge digester', *Applied Environmental Microbiology*, vol 71, pp. 2145–2153.

Cirne, D.G., Lehtomäki, A., Björnsson, L. and Blackall, L.L. (2007) 'Hydrolysis and microbial community analyses in two-stage anaerobic digestion of energy crops', *Journal of Applied Microbiology*, vol 103, no 3, pp. 516–527.

da Silva, T.L., Roseiro, J.C. and Reis, A. (2012) 'Applications and perspectives of multi-parameter flow cytometry to microbial biofuels production processes', *Trends in Biotechnology*, vol 30, no 4, pp. 225–232.

Demirel, B. and Scherer, P. (2008) 'The roles of acetotrophic and hydrogenotrophic methanogens during anaerobic conversion of biomass to methane : a review', *Reviews in Environmental Science and Biotechnology*, vol 7, pp. 173–190.

Dewil, R., Lauwers, J., Appels, L., Gins, G., Degrève, J. and Van Impe, J. (2011) 'Anaerobic digestion of biomass and waste: current trends in mathematical modeling', *Preprints of the 18th IFAC World Congress* Milano (Italy) August 28–September 2, pp. 5024–5033.

Donoso-Bravo, A., Mailier, J., Martin, C., Rodríguez, J., Aceves-Lara, C.A. and Vande Wouwer, A. (2011) 'Model selection, identification and validation in anaerobic digestion: A review', *Water Research*, vol 45, pp. 5347–5364.

Ellis, J.T., Tramp, C., Sims, R.C. and Miller, C.D. (2012) 'Characterization of a methanogenic community within an algal fed anaerobic digester', *ISRN Microbiology* vol 2012.

Feng, X.M., Karlsson, A., Svensson, B.H. and Bertilsson, S. (2010) 'Impact of trace element addition on biogas production from food industrial waste – linking process to microbial communities', *FEMS Microbiology and Ecology*, vol 74, pp. 226–240.

Franke-Whittle, I.H., Goberna, M., Pfister, V. and Insam, H. (2009) 'Design and development of the ANAEROCHIP microarray for investigation of methanogenic communities', *Journal of Microbioogical Methods*, vol 79, pp. 279–288.

Garcia, S.L., Jangid, K., Whitman, W.B. and Das, K.C. (2011) 'Transition of microbial communities during the adaption to anaerobic digestion of carrot waste', *Bioresource Technology*, vol 102, pp. 7249–7256.

Goberna, M., Insam, H., and Franke-Whittle, I.H. (2009) 'Effect of biowaste sludge maturation on the diversity of thermophilic bacteria and archaea in an anaerobic reactor', *Applied and Environmental Microbiology*, vol 75, pp. 2566–2572.

Goberna, M., Gadermaier, M., Garcia, C., Wett, B. and Insam, H. (2010) 'Adaptation of methanogenic communities to the cofermentation of cattle excreta and olive mill wastes at 37°C and 55°C', *Applied Environmental Microbiology*, vol 19, pp. 6564–6571.

Griffin, M.E., McMahon, K.D., Mackie, R.I. and Raskin, L. (1998) 'Methanogenic population dynamics during start-up of anaerobic digesters treating municipal solid waste and biosolids', *Biotechnology and Bioengineering*, vol 57, pp. 342–355.

Henson, M.A. (2003) 'Dynamic modeling of microbial cell populations', *Current Opinion in Biotechnology*, vol 14, pp. 460–467.

Hoffmann, R.A., Garcia, M.L., Veskivar, M., Karim, K., Al-Dahhan, M.H. and Angenent, L.T. (2008) 'Effect of shear on performance and microbial ecology of continuously stirred anaerobic digesters treating animal manure', *Biotechnology and Bioengineering*, vol 100, pp. 38–48.

biochemical pathways (Rittmann et al., 2008). Ultimately biogas production may move in the direction of reducing the number of microorganisms involved in a process to a few highly engineered taxa, but a far greater understanding of the key biochemical processes and the organisms that currently provide the catalytic mechanisms to process the appropriate substrates will be required to do so (Rittmann et al., 2008). The finding that a number of taxa with redundant metabolic performance may be crucial to robust digester function (Briones and Raskin, 2003) may not fundamentally contradict this approach, but does suggest greater thought may be required as to how to achieve such a goal. This is particularly so given the number and proportion of taxa which are found that are new to science and whose biology will need to be characterised.

The methods are available to document this diversity and measure population change in response to environment, and interest in developing suitable models to address these issues is now established. The challenge will be to undertake this work efficiently in order to identify and extract the salient information concerning process given the number of parameters, and considerable stochasticity that characterises these systems.

References

Ahring, B.K., Ibrahim, A.A. and Mladenovska, Z. (2001) 'Effect of temperature increase from 55 to 65°C on performance and microbial population dynamics of an anaerobic reactor treating cattle manure', *Water Research*, vol 35, pp. 2446–2452.

Antoni, D., Zverlov, V.V. and Wolfgang, H.S. (2007) 'Biofuels from microbes', *Applied Microbiology and Biotechnology*, vol 77, no 1, pp. 23–35.

Batstone, D.J. (2006) 'Mathematical modelling of anaerobic reactors treating domestic wastewater: Rational criteria for model use', *Reviews in Environmental Science and BioTechnology*, vol 5, pp 57–71.

Batstone, D.J., Keller, J., Angelidaki, I., Kalyuzhnyi, S.V., Pavlostathis, S.G., Rozzi, A., Sanders, W.T.M., Siegrist, H. and Vavilin, V.A. (2002) 'Anaerobic Digestion Model No. 1 (ADM1)', *IWA Task Group for Mathematical Modelling of Anaerobic Digestion Processes*. London: IWA Publishing.

Bauer, C., Korthals, M., Gronauer, A. and Lebuhn, M. (2008) 'Methanogens in biogas production from renewable resources: A novel molecular population analysis approach', *Water Science and Technology*, vol 58, pp. 1433–1439.

Blesgen, A. and Hass, V.C. (2010) 'Efficient biogas production through process simulation', *Energy Fuels*, vol 24, pp. 4721–4727.

Blume, F., Bergmann, I., Nettmann, E., Schele, H., Rehde, G., Mundt, K. and Klocke, M. (2010) 'Methanogenic population dynamics during semi-continuous biogas fermentation and acidification by overloading', *Journal of Applied Microbiology*, vol 109, pp. 441–450.

Briones, A. and Raskin, L. (2003) 'Diversity and dynamics of microbial communities in engineered environments and their implications for process stability', *Current Opinion in Biotechnology*, vol 14, pp. 270–276.

Briones, A.M., Daugherty, B.J., Angenent, L.T., Rausch, K.D., Tumbleson, M.E. and Raskin, L. (2007) 'Microbial diversity and dynamics in multi- and single-compartment anaerobic bioreactors processing sulfate-rich waste streams', *Environmental Microbiology*, vol 9, pp. 93–106.

Cardinali-Rezende, J., Debarry, R.B., Colturato, L.F.D.B., Carneiro, E.V., Chartone-Souza, E. and Nascimento, A.M.A. (2009) 'Molecular identification and dynamics of microbial communities in reactor treating organic household waste', *Applied Microbiology and Biotechnology*, vol 84, pp. 777–789.

Cardinali-Rezende, J., Colturato, L.F.D.B., Colturato, T.D.B., Chartone-Souza, E., Nascimento, A.M.A. and Sanz, J.L. (2012) 'Prokaryotic diversity and dynamics in a full-scale municipal solid waste anaerobic reactor from start-up to steady-state conditions', *Bioresource Technology*, vol 119, pp. 373–383.

anaerobic digestion processes, and which is the main metabolic pathway used in popular models such as ADM1 (Krakat et al., 2010c).

Modelling has led to greater insights into the anerobic digestion process and has been used to develop a number of advances in process control. For example, a model based on classic Monod kinetics, but including also substrate thresholds and survival r-mode data for the H_2-oxidising methanogen *Methanobacterium bryantii* M.o.H., accurately described key process parameters and in addition a starvation response allowed the minimum substrate concentration needed to maintain steady-state biomass to be determined (Karadagli and Rittmann, 2007). The ADM1 model was used to investigate two scenarios of population dynamics at digester overload conditions in which the methanogenic community composition in the digester was identified using the ANAEROBECHIP microarray (Schoen et al., 2009). The study found that conventional parameters such as the organic loading rate appeared unsuitable for process description under dynamic conditions.

With respect to the issue of microbial communities and their population dynamics, while some studies have found they can usefully collapse information from a number of microbial subpopulations into one (e.g. Sötemann et al., 2005), others are beginning to examine the effect of multiple species and to model how digester process could be used to select more effective communities (Masci et al., 2009; Ramirez et al., 2008, 2009). Models changing operational conditions frequently seems to provide a more diverse community and lead to process stability (Ramirez et al., 2008). Manipulating VFA levels can force a more rapid selection process to achieve a more efficient and robust community, and more stable process output (Masci et al., 2009).

The anaerobic models are complex even when focused on process dynamics, let alone attempting to take account of a richer population dynamics, so the process of including all this complexity will be challenging. Nevertheless, now that the richness of the population diversity is being realised and tools to track it are available, it is clear that modelling will play a key role in understanding the key parameters and processes involved in new management approaches.

Future trends

The preceding sections have demonstrated how surveys over the last two decades, but with far greater resolving power more recently, have demonstrated that microbial communities are highly diverse and their composition is strongly influenced by temperature, pH, feedstock, the concentration of VFAs and operating conditions that affect these (e.g. Krakat et al., 2011; Nelson et al., 2011, 2012; Lee et al., 2012). Some patterns have emerged concerning the identification of two types of community dominated by acetogens depending on acetate concentration (e.g. Walter et al., 2012), but recent research has also emphasised the role of hydrogenotrophic methanogens working with syntrophic bacteria as dominant members of effective gas-producing communities based on many feedstocks (e.g. Cardinali-Rezende et al., 2012; Regueiro et al., 2012). Although a relatively small number of groups have been shown to have a dominant role in AD the total biomass and their relative proportions and that of different species within these change dynamically over time (Nelson et al., 2011; Briones and Raskin, 2003). That this dynamic change may underpin reliable digester function through the redundancy of function in many species (Briones and Raskin, 2003) is a crucial finding which raises an important issue for future strategies of biogas development.

Genetic engineering of microbes to provide them with key biochemical capabilities provides a pathway for more controlled and efficient energy production focused on particular

A final and critical part in mass balance modelling is the rate equation, as the other terms are fixed by hydraulics, and this is where biological processes are included. For biological activity it is usually described by the Monod kinetic (Equation 14.3) where the specific growth rate μ is related to the substrate concentration S.

$$\mu = \mu_{max} (S/(K_s + S)) \tag{Eq. 14.3}$$

where μ is the specific growth rate of the microorganisms,
μ_{max} is the maximum specific growth rate of the microorganisms,
S is the concentration of the limiting substrate,
K_s is a constant equal to the concentration of the limiting substrate at $\mu/\mu_{max} = 0.5$.

The values of μ_{max} and K_s are empirical coefficients of the Monod equation and they are different among species. The kinetics of individual species has been determined for a number of culturable species (e.g. Lin et al., 2007) and can be obtained from the literature (e.g. see summaries in Tomei et al., 2009). However, most models do not include the dynamics of individual microbial populations that participate in the process but assume only one microbial population involved with the process, or each main step of the process [i.e. hydrolysis + acidogenesis + acetogenesis producing the substrates for methanogens, which then produce the final gas product] (e.g. Sötemann et al., 2005; Sbarciog and Vande Wouwer, 2012).

One of the most successful models of anaerobic digestion is the ADM1 originally proposed by the IWA Task Group for Mathematical Modelling of Anaerobic Digestion Processes to unify a number of main modelling approaches then available (Batstone et al., 2002). It is a structured and highly complex model with 32 dynamic state concentration variables which describe bacterial groups and archaea catalysing 19 biochemical kinetic processes that are coupled to 105 kinetic and stoichiometric parameters, which has been developed further since as described by Batstone (2006). Nevertheless, parameters for anaerobic digestion remain poorly standardised, the assessment of parameter variability is limited, most popular anaerobic models are very complex, and the information contained in parameter values and model structure is not always easily amenable to interpretation when taking decisions about process management (Batstone, 2006). Ways in which key processes not included in the earlier versions (e.g. VFA inhibition) could be incorporated are being tackled (Palatsi et al. 2010; Dewil et al., 2011) and variations of this model have been used to include larger numbers of microbial taxa in assessments (Ramirez et al., 2008, 2009).

Progress in software engineering and control systems has provided a fuzzy logic control (FLC) system for operation of biogas reactors using energy crops developed at the Hamburg University of Applied Sciences (HAW Hamburg) (Scherer et al., 2009). The FLC system uses measured parameters such as pH, the methane (CH_4) content, and the specific gas production rate (spec. GPR = m^3/kg VS/day) to control inputs with the objective of avoiding stabilisation of pH using buffering supplements, like lime or manure. The developed FLC system can cover most applications, such as a careful start-up process and a recovery strategy after a severe reactor failure. This FLC system was used in a long-term experiment where beet silage and beet juice were digested in a thermophilic biogas digester operated continuously for more than seven years, and some key aspects of microbial dynamics were investigated (Krakat et al., 2010a,b,c, 2011) as discussed in previous sections (see Figure 14.3). While not a result of the modelling their observations on the role of acetoclastic methanogens contradicts the view that acetotrophic *Euryarchaeota* are the predominant organisms in the

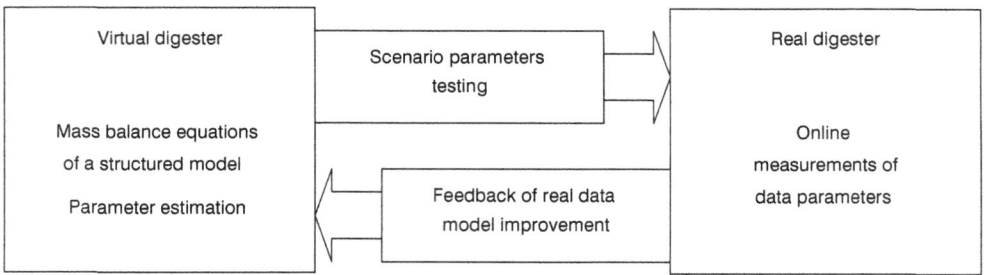

Figure 14.7 Generic scheme for development of models of the dynamics in a digester. Examples detailing the parameters that can be involved in complex systems can be found in Tomei et al. (2009) and Lübken et al. (2010).

Models of anaerobic digestion

Most models of biomethane digesters estimate biomethane production based on mass balance by substrate utilisation, and the large literature on this topic has been reviewed recently (Dewil et al., 2011; Donoso-Bravo et al., 2011). The types of biochemical models used in anaerobic systems range from simple steady state models, through one- and two-step dynamic models to complex fully structured dynamic models. The general approach most commonly used is to set a number of differential equations according to the objectives of the project and to describe the dynamics in relation to the desirable products of the process. In biomethane production this could be the population and type of various microorganisms participating in the process, initial carbon source levels, environmental conditions (pH, temperature) etc. and how these factors are related to the yield of biomethane or other intermediate products (examples are found in Tomei et al., 2009; Lübken et al., 2010). Differential equations that describe the process can be set and values for the parameters can be determined numerically and from raw data evaluation. The general process of model development is illustrated in Figure 14.7, but is discussed in greater detail by Tomei et al. (2009) and Donoso-Bravo et al. (2011) who compare and examine the selection of appropriate models for anaerobic digestion, methods for measuring or estimating key variables and kinetic parameters and processes for model validation.

The generic mathematical model that can describe a process is the mass balance equation (Equation 14.2) for a specific state variable. This type of equation describes accumulation and reaction within a system in relation to flow across the system boundaries.

$$(dMsys/dt) = m_{in} - m_{out} + r \qquad \text{(Eq. 14.2)}$$

where $Msys$ is the mass in the system (unit mass),
m_{in} is mass flow rate in,
m_{out} is mass flow rate out,
r is the overall generation rate.

The overall generation rate r is the sum of all the different rates influencing the compound being modelled. Simple models may have only a handful of variables but the more sophisticated may have closer to twenty or more (Batstone, 2006; Tomei et al., 2009). Some simplifying assumptions can be made though for anaerobic systems. Most are dilute systems in which concentration changes have no impact on volume, and the concentration of a component in the stream out of the reactor is the same as that within a well-mixed reactor.

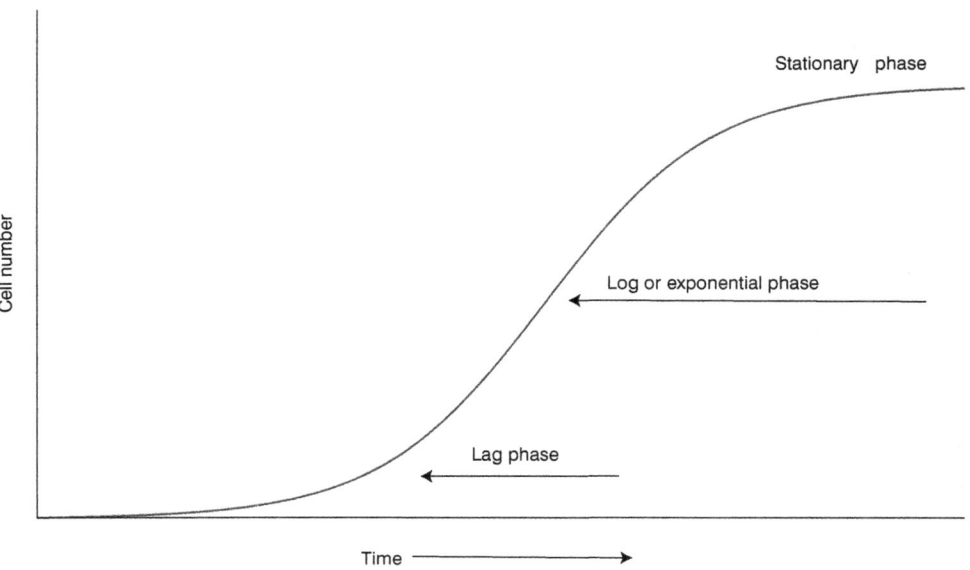

Figure 14.6 Sigmoidal curve of bacterial growth during batch culture

Population growth models

For a simple batch culture three phases of bacterial growth are indentified, the lag phase at the start of the culture where growth is slow, the log phase where the maximum growth rate is reached and the stationary phase where population reach a plateau followed by the death phase (Zwietering et al., 1990). The first three phases can be graphically described by a sigmoidal curve (Figure 14.6) described by the formula

$$\lg(N_t) = \lg(N_0) + t \ \lg 2 \tag{Eq. 14.1}$$

where N_t is cell population at time t,
N_0 is the starting population at time zero,
μ is the specific growth rate constant for a cultured organism and is equal to the reciprocal of the doubling period (generation time) of the cells.

Equation 14.1 is the most simplified form used to describe the dynamics of microbial population growth but cannot take account of cell mortality and provides little insight with respect to the mechanisms driving population change (see Peleg and Corradini (2011) for a discussion of some of these issues). The approach is used successfully to describe the growth of individual organisms and to calculate their maximum growth rate, but in the case of complex communities there is a need to take account of mortality and competition between species to model them effectively (Briones and Raskin, 2003; Trosvik et al., 2008). Approaches to these issues will need to access models related to community population dynamics for which studies using chemostats have long played a role, and are continuing to receive attention although not necessarily in the specific context of AD (e.g. Rapaport et al., 2009). More detailed discussion of these issues is beyond the scope of this chapter but can be found in reviews by Peleg and Corradini (2011) and Lübken et al. (2010).

al., 2004). There are no focused meta-analyses of a broader range of environments or variables available at present. Other reviews have identified some trends such as the sensitivity of *Methanosaeta* not only to high acetate levels but also to other VFAs, ammonia and sulphur, and for higher temperatures to favour some *Methasarcina* and *Methanobacteria* (Demirel and Scherer, 2008); that overloading can tip the balance to bacteria or fungi involved in hydrolysis and acidogenesis (e.g. Ritari et al., 2012) leading to the accumulation of VFAs and communities adapted to high levels of these (Tang et al., 2007; Chelliapan et al., 2011). Many recent studies have emphasised the role of syntrophic bacteria interacting with hydrogenotrophic methanogens, demonstrating the importance of this metabolic pathway for methane production (e.g. Karakashev et al., 2006; Krakat et al., 2010c; Kampmann et al. 2012a; Cardinali-Rezende et al., 2012; Regueiro et al., 2012). However it is still impossible to derive more generalisations from the literature other than these very broad ones given the huge variation between experiments.

In addition, the majority of studies have used denaturing gradient gel electrophoresis (DGGE) to identify major groupings and subsets of primers to look at more specific groups, but discuss changes mainly at genus or higher taxonomic level. The dynamic changes at these levels have been shown to be large, but it is clear that even more changes at lower taxonomic levels are likely masked by these approaches (e.g. Nelson et al., 2012: > 6000 bacterial spp. in AD, but 10s to 100s in individual digesters). The large differences at lower taxonomic levels between reactors, and over time within reactors even when environmental circumstances appear similar, demonstrate at least a high level of stochasticity in these systems, or far more complex controls that are far from understood. In fact the variety may provide a redundancy of function that permits continued reactor function in the face of perturbations and stochastic change (Briones and Raskin, 2003).

Kinetics and modelling in methanogenesis

Models are formulated in order to describe an existing situation mathematically and to test different scenarios using these, so allowing a faster, more cost-effective assessment of circumstances that may be more demanding to test in reality (Blesgen and Hass, 2010; Donoso-Bravo et al., 2011; Peleg and Corradini, 2011). Modelling the dynamics of populations has a long history in biology dating back to the Malthusian model and the Lotka–Voltera models for predator–prey equations (Lübken et al., 2010). Simple early models of microbial population dynamics which described the growth and division of a microbial cell into two daughter cells have developed into sophisticated models, some additionally assessing a number of aspects of cellular process (Henson, 2003; Peleg and Corradini, 2011). In combination with methods to provide real-time estimates of cell numbers with direct flow cytometry and measurement of cell characteristics with fluorescent technology using direct flow cytometry, these offer the means to provide improved production control for biofuels (da Silva et al., 2012).

Mathematical modelling of AD systems is also well established and a number of software innovations have been applied to industrial scale AD systems for some time (Scherer et al., 2009; Krakat et al., 2010c). To date this modelling has focused on bulk chemical and physical processes with highly simplified treatment of microbial dynamics (Dewil et al., 2011; Donoso-Bravo et al., 2011). However, following the development of molecular capability to track microbial populations in more detail, and the discovery of the rich dynamics in AD systems as a result, there is increasing interest and capability in modelling the microbial community in more detail (e.g. Henson, 2003; Ramirez et al., 2009).

2011b). Different bacterial communities developed in different compartments of a five-stage digester, with different dynamics also for the hydrogenotrophic archaea after sulphur loading (Briones et al., 2007). In contrast no major differences were seen in the archaeal communities in a two-stage digester fed with triticale waste (Klocke et al., 2008).

Patterns in community structure and relationship to environmental factors

Many studies use multivariate analyses such as principal components analyses or related techniques to help interpret complex results by trying to reduce the parameter space (e.g. Madden et al., 2010; Supaphol et al., 2011; Ritari et al., 2012). In a study of a range of factors over three seasons in nine full-scale mesophilic reactors fed principally cow manure, mixed cow–swine–chicken manure, sewage sludge or pomace, Walter et al. (2012) found no relationship of the methanogenic community with any variable except an association of *Methanosarcina* with higher concentrations of acetate and *Methanosaeta* with lower concentrations of acetate (Figure 14.5). This relationship has been noted in a number of previous reports of communities dominated by acetogenic methanogens (e.g. McMahon et

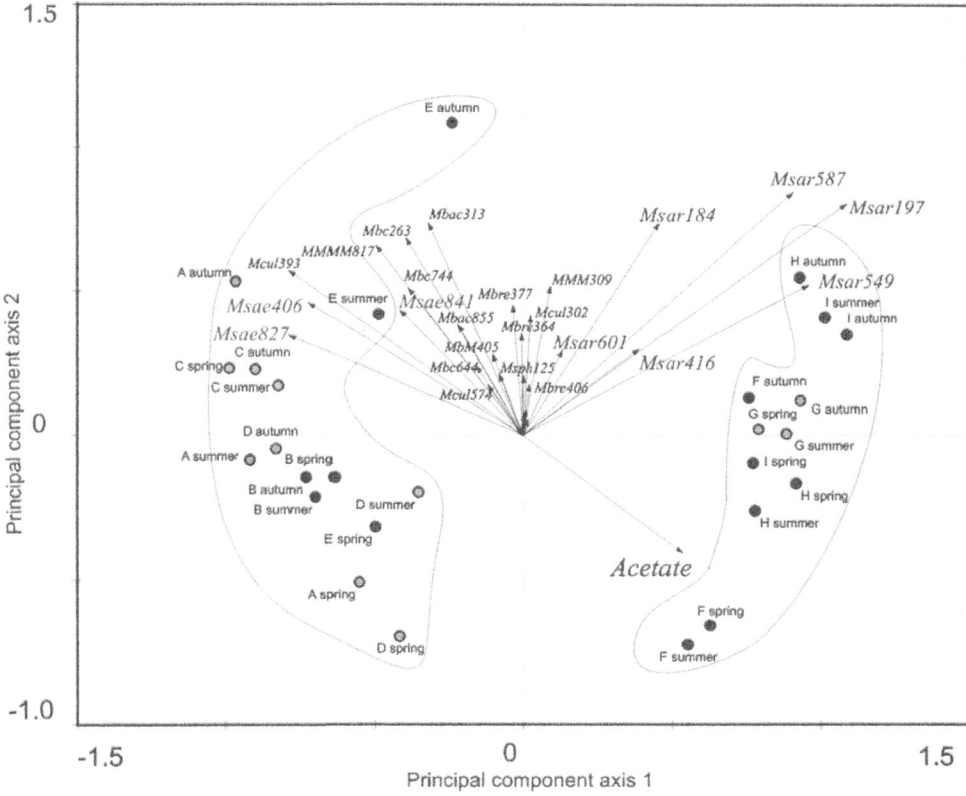

Figure 14.5 Principal component analysis showing the position in multivariate space of the sludge samples, and their separation into two groups, one with high acetate dominated by *Methanosarcina* species, and the other with low acetate dominated by *Methanosaeta* species. The vectors show acetate or the different species probes (Walter et al., 2012).

After increases in VFAs acetogenic *Methanosaetaceae* often decrease and *Methanosarcinaceae* (sometimes with hydrogenotrophic *Methanomicrobiales*) increase or *Methanobacteriales* become dominant (Padmasiri et al., 2007; Blume et al., 2010; Chelliapan et al., 2011). Many groups in the bacterial community changed and the archaeal aceticlastic *Methanosarcina thermophila* was replaced in one replicate reactor by hydrogenotrophic *Methanoculleus* sp. and in another, two *Methanospirillum* spp. became more common (Lerm et al., 2012). *Methanothermobacter wolfei* became dominant in formate rich feeds, while *Methanoculleus thermophilus* became dominant in feeds with acetate, propionate or butyrate (Wagner et al., 2011). Responses of particular families, genera (or species) to the VFA intervention vary depending on differences in the details of experimental treatment, including the particular composition of the existing community, illustrating the difficulty of detecting more general trends from the literature when studies differ in their taxonomic resolution, and other details of experimental condition.

Other factors (granulation, aeration, mixing, pre-treatments, digester design)

Microbial communities have been found to be quite different in biofilms or granular sludge compared with the liquid fraction with carbohydrate bacterial decomposers. The use of baffles, substrates and filters permits the development of attached biofilms, or settled granular sludges that provide additional diversity and metabolic capability in those systems (McHugh et al., 2003; Zhang et al., 2011a). *Methanocorpusculum* species are important in this process at low temperatures (Madden et al., 2010; O'Reilly et al., 2010).

Increased mixing can reduce the development of flocs and granules and high shear can break them up, so changing community structure. Increased mixing has been reported to shift dominance from acetoclastic *Methanosaeta concilii* to an acetoclastic *Methanisarcina* spp. (McMahon et al., 2001; Hoffmann et al., 2008). Micro-aeration successfully reduced sulphur levels and produced only relatively minor changes in the bacterial community but shifts in dominance from aceticlastic *Methanisarcina* to hydrogenotrophic *Methanoculleus* (Tang et al., 2004).

A range of other processes have been shown to change the bacterial and archaeal community such as addition of trace elements in maize waste (Munk et al., 2010) and in food industrial wastes (Feng et al., 2010). Focused-pulse electric treatment of municipal waste increased numbers of carbohydrate and cellulose degrading bacteria, including taxa (e.g. *Ruminococcus* sp.) only seen after use of this treatment, consistent with greater accessibility of those substrates (Zhang et al., 2009). In this regime, acetoclastic *Methanosaeta* sp. increased while hydrogenotrophic *Methanoculleus* sp. and those bacteria involved in syntrophic propionate fermentation with them decreased.

The nature of the digester design appears to have no major general effect on community structure (Cheon et al., 2008; Lee et al., 2012), i.e. there is no community associated particularly with single or multiple stage reactors, baffled or continuously stirred systems although multiple compartments can develop different communities (e.g. Uyanik et al., 2002; Klocke et al., 2008; Zhang et al., 2011b). The extent to which different communities developed in different compartments depended on the operation of the digester. For example, a four-compartment digester showed the first two compartments had similar communities under heavy loading rates while the second compartment developed communities more similar to the two later ones under low loading rates showing major shifts in several bacterial groups (Zhang et al., some groups tend to occur mainly in thermophilic conditions (e.g. *Methanothermobacter*) the response by others can be very different, implying a major influence of other factors or suggesting that species level differences may be important if the tracking has only been at a higher taxonomic level (as is often the case).

Principal shocks resulting in changes to digester performance

Another obvious area of interest is what happens to microbial communities when shocks are imposed on the system such as major shifts in the levels of organic loading, changes in feedstock, addition of fatty acids, trace elements (see references in Table 14.4) or the many materials that inhibit the AD process (Chen et al., 2008). Although minor ones may have relatively little effect, major changes in organic loading lead to marked shifts in the microbial community, and often major impact on the methanogens. Prolonged starvation reduced methanogen population size, diversity and the relative amounts of dominant taxa (Hwang et al., 2010). One of the few studies to include fungi in an analysis of sewage and biowaste digestion showed the number and diversity of fungi increased after loading while that of bacteria decreased and archaea remained similar, although there were some differences between mesophilic and thermophilic reactors (Ritari et al., 2012).

Changes in hydraulic retention times (HRTs) and dilution rates effectively vary loading effects. Short HRTs imply higher loads and require efficient processing given the shorter times substrates are made available, while dilution can reduce the level of organic load (Chelliapan et al., 2011). Short HRTs and high organic loading rates (OLRs) were associated with the presence of *Acidobacteria* populations and long HRTs and low ORTs with the presence of *Planctomycetes*, *Actinobacteria* and *Alcaligenaceae* in beet digestate (Krakat et al., 2011) as shown in Figure 14.3. A shift from low to high dilution rates resulted in a shift in the relative abundance of the four main bacterial phyla present including the loss of *Chloroflexi*, and a shift from hydrogenotrophic methanogens (*Methanoculleus*) to aceticlastic ones (*Methanosarcina*) in one case (Shigematsu et al., 2003, 2004) and a shift in dominance of bacterial phyla from *Proteobacteria* to *Firmicutes* and the dominance of *Methanoculleus* and aceticlastic *Methanosaeta* to a community with hydrogenotrophic *Methanospirillum* and aceticlastic *Methanisarcina*, in another (Tang et al., 2007). However, increased loading often increases VFA levels [which can also result from trace element limitation (Munk et al., 2010)], and induces effects related to these such as acidification and the interruption of gas production as they are toxic to methanogens. The influence of VFAs is seen clearly in the marked shifts in the bacterial community, with an increase in the *Chloroflexi*, when both acetate and propionate levels were high in a beet fodder system (see Figure 14.3 – day 1017).

Another excellent illustration of the clear influence VFA loading has on methanogens is shown in Hori et al. (2006) where hydrogenotrophic *Methanoculleus* sp. reduced and hydrogenotrophic *Methanothermobacter* sp. decreased while high VFA (acetate and propionate) levels persisted and the acetitrophic *Methanosarcina* sp. increased to persistently higher levels. It this case it was stated that pH, independently controlled from VFA level, impacted bacteria more than VFA levels which had their main impact on the archaea.

Community effects have been investigated either by direct addition of VFA's or by adding particularly fatty or lipid rich wastes (see Table 14.4). In summary, many show changes in the established populations of bacterial and archaeal groups, in the relative abundance of established species, loss and/or addition of groups after the event (Hori et al., 2006; Sousa et al., 2007; Martin-Gonzalez et al., 2011), but others report little effect (Palatsi et al., 2010).

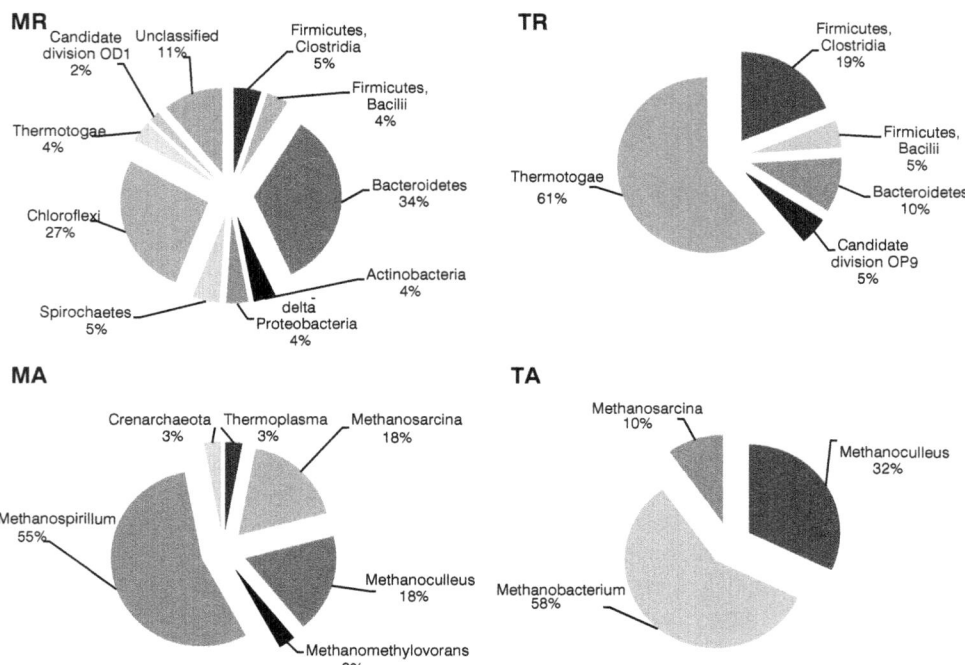

Figure 14.4 Different relative abundances of 16S rRNA gene sequences in (*top*) bacterial (R) and (*bottom*) archaeal (A) clones in a mesophilic (M) and a thermophilic (T) reactor. This example demonstrates the general finding that bacterial diversity is usually greater than archaeal diversity, and that an increase in temperature often results in reduced diversity in both groups. Based on Leven et al. (2007).

temperature change. This often results in a loss of diversity at higher temperatures (Leven et al., 2007), but not always (Ritari et al., 2012), and usually multiple changes in the relative abundance of taxonomic groups (Leven et al., 2007; Ritari et al., 2012), see Figure 14.4. For example, a mesophilic sludge undergoing thermal acidogenesis to 51°C resulted in mesophilic *Pseudomans mendocina* disappearing from the existing bacterial community at higher temperatures but *Clostridium hastiforme* and *Gracilobacter thermotolerans* originating from the mesophilic anaerobic sludge survived thermal acidogenesis (Kim et al., 2010a).

Among the archaea, *Methanosaeta* was replaced by *Methanocorpusculum* species at 15°C (Madden et al., 2010; O'Reilly et al., 2010). Cattle waste dominated by *Methanosarcina* showed different shifts in the archaeal community after the addition of olive oil mill wastes depending on temperature – operated at 37°C *Methanosarcina* numbers increased six-fold, and at 55°C *Methanobacterium*, *Methanoculleus* and *Methanothermobacter* dominated (Goberna et al., 2010). In contrast, in household waste at 37°C *Methanospirillum* dominated and *Methanosarcina* and *Methanoculleus* were common but at 55°C *Methanosarcina* dominated with *Methanobacterium* and *Methanoculleus* being common (Leven et al., 2007). A change in temperature of 55°C to 75°C resulted in *Methanobacterium*, *Methanoculleus* spp., and *Methanosaeta* declining, the maintenance of *Methanothermobacter* levels, and a rise in *Methanosarcina* sp. (Rademacher et al., 2012a). A shift in temperature from 60°C to 55°C and back was related to a clear shift in archaeal community structure and the morphology of the colonies in a beet digester (Krakat et al., 2010c). These examples also illustrate that while which are reduced to a relatively simplified set of fewer energy sources for methanogens (Regueiro et al., 2012). While some of the major phyla may be present throughout the period of reactor function (over years) there is often dynamic change in other phyla and in lower taxonomic units, even in the absence of identifiable major changes in digester performance (Zumstein et al., 2000).

Temporal shifts in community structure occur in the absence of major perturbations (Lee et al., 2012) and not always related to major changes in operation [as in beet waste fermented at 55–60°C (Krakat et al., 2010b)] or in response to particular experimental perturbations that may have been made in those studies (Madden et al., 2010). See Figure 14.3, comparing days 1600 and 1755 of a beet digester, where there are no obvious differences in treatment of a stably operating reactor for another example.

Influence of seed inocula

The goal of seeding new reactors with an inoculum from an established AD system is to provide a viable community for efficient digester function (McMahon et al., 2004). Most evidence suggests that the inoculum is ultimately replaced by a new community dependent on the feedstock and operating conditions of the reactor except where there is no pre-existing community, as illustrated in the following three studies.

In the first, two reactors using an identical laboratory feedstock were each inoculated with sludge of different microbial diversity and operated identically afterwards (Oz et al., 2012). They developed different final communities each related to their initial inoculum, implying a major influence of seed inoculum. However, both lost diversity (187 to 131 and 200 to 84 OTUs respectively, one losing all its methanogens), suggesting an interaction of operating environment with inoculum or strong stochastic effects. In the second, the communities developed on three different feedstocks differed markedly from those present in their seed sludge, both in the relative abundances at the phylum level and at lower taxonomic levels implying a dominant effect of feedstock and operating environment (Nelson et al., 2012). In this case, only 1–4 unclassified taxa in the phylum *Chloroflexi* were present in the inoculum from corn and potato waste, but 50–101 were present in the community developed on cheese waste; and 49–55 taxa from the phylum *Firmicutes* in the corn and potato inoculum reduced to 5–11 in the cheese waste community. Similarly, inocula composed of either a diverse assemblage of microbes from a variety of environments or from rumen samples had large numbers of bacterial taxa (5828–6020 respectively) and archaeal taxa (33–23 respectively) which reduced to < 200 and < 10 respectively in the digestate of carrot waste (Garcia et al., 2011).

Temperature

Temperature has attracted considerable attention as an independent variable with a major effect on AD microbial communities (Ahring et al., 2001; Cheon et al., 2008). In papers examining several reactors or several variables, temperature is often the variable which has the strongest identifiable impact on community composition (e.g. McHugh et al., 2003). By far the greatest number of studies have concerned mesophilic (25–45°C) and thermophilic (45–65°C) temperature ranges, but a few have dealt with cryophilic (< 20°C) and hyper-thermophilic ones (> 65°C) (Table 14.4).

Several papers comparing reactors operated at different temperatures report differences in their bacterial communities or changes in the communities in digesters subjected to

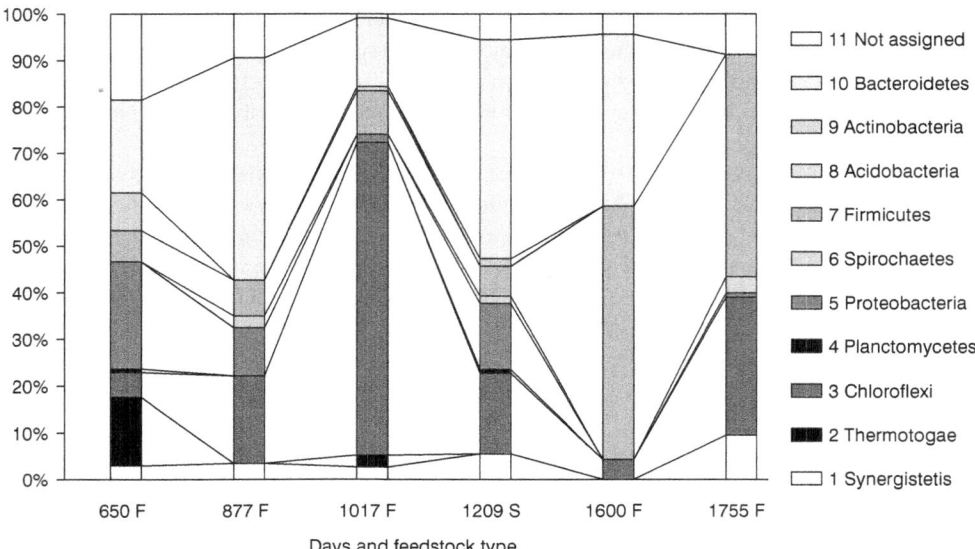

Figure 14.3 An example of the marked changes in bacterial diversity over time with a change in feedstock from fodder beet (F) to sugar beet silage (S) at day 1209 of operation. This was accompanied by a fall in pH to 6.5 which otherwise ranged from 7.1 to 7.3. The changes between sampling days can also be related to other operational differences. Organic loading rates were low, at one third that of other times, at days 650 and 1209. Acetate was less than one-third that at other times for days 650 and 877, propionate 8–10 times higher than other days on days 877 and 1017. Note that this means the changes between days 1600 and 1755 are not related to variation in any of these factors. Redrawn from data in Krakat et al. (2011).

difficulties in developing suitable microbial populations (e.g. McMahon et al., 2001, 2004). Marked changes are often observed as the bacterial community develops and provides the materials for growth of acetogens and methanogens (Shin et al., 2010; Supaphol et al., 2011; Cardinali-Rezende et al., 2012). A general pattern of slower development of archaeal populations and a decline in bacterial populations later is often observed. This may be only a few hours or days depending on reactor size, feedstock and the nature of the seed inoculum, but during the hydrolytic phase of a two-stage digester for biogas production from energy crops archaea started to appear in the hydrolytic stage between days 10 and 15, when the fraction of bacteria decreased (Cirne et al., 2007).

There is much variation in the detail of results of studies of reactor start up and associated changes in the microbial populations with much variation related to the acidity and particularly the volatile fatty acid (VFA) levels in the reactor (e.g. Liu et al., 2002; Kobayashi et al., 2009; Kim et al., 2011). Repeated shifts in the dominance of different aceticlastic and hydrogenotrophic methanogen groups were thought related to shifts in the amounts of different VFAs (Lee et al., 2010).

Variation has been observed in stably operating reactors over time in both archaeal and bacterial communities but variation in the bacterial community was far greater (Zumstein et al., 2000; Krakat et al., 2010b; Regueiro et al., 2012). There is no correspondence between the shifts in bacterial and archaeal communities (Regueiro et al., 2012). These findings should not be a surprise since the bacterial community deal with a diverse range of substrates

Feedstock type

About 50 different types of feedstock are listed in the studies in Table 14.4. All have the major phyla discussed in the previous section, but additional phyla are recorded in significant abundance in some reactors, e.g. *Thermotogae* in thermophilic reactors (Leven et al., 2007; Ritari et al., 2012), and *Crenarchaeota* in some mesophilic and thermophilic digesters (Leven et al., 2007; Zhang et al., 2011c). A number of generalist taxa are found in digesters fed by very different feedstocks (Kim et al., 2011) but many have sporadic occurrences (Cheon et al., 2008; Nelson et al., 2011). Almost all studies show greater diversity of bacterial than of archaeal communities, often by a factor of 2 to 10 (e.g. Klocke et al., 2007; Nelson et al., 2011; Garcia et al., 2011; Regueiro et al., 2012). Many record differences in the dominant phyla in different reactors, and certainly in the major groups that are found within these, depending on the feedstock (Lee et al., 2009, 2012; Cardinali-Rezende et al., 2012).

There are differences between feedstocks reflecting some of their specific characteristics. For example, animal wastes are relatively rich in ammonia compared with vegetable and grain feedstocks and support bacteria that process ammonia, or that are tolerant of it (Nettmann et al., 2010; Zhang et al., 2011c). Similarly, some animal and food wastes have more proteins and lipids which mean bacterial groups that process these materials are detected at greater levels (Leven et al., 2007; Garcia et al., 2011). Wastes rich in lipids and fats support bacteria able to process, or which are reliant upon volatile fatty acids, such as propionate (Wagner et al., 2011; Martin-Gonzalez et al., 2011).

These relationships are clearly demonstrated in studies where digesters acclimated to a particular feedstock are supplied with a different waste. For example, a digester using cellulose-rich feedstock (paper) at 53°C had a high bacterial diversity dominated by *Firmicutes* [mainly *Clostridium* (90 per cent) and *Halocella* cellulose degraders], *Bacteroidetes* and *Synergistetes* and an archaeal community dominated by hydrogenotrophic *Methanothermobacter*, *Methanobacterium*, *Methanoculleus* and acetoclastic *Methanosarcina* (Tang et al., 2011). When fed garbage waste the bacterial community changed to protein and fatty acid degraders and the *Archaea*, although represented by the same genera, changed in relative abundance. An example which clearly illustrates several effects of different factors on the microbial community, including a change in feedstock, is given by Krakat et al. (2011) where clear shifts in the bacterial community occurred when fodder beet was replaced by sugar beet silage (Figure 14.3 – compare days 650 and 1209 where conditions were the same except for feedstock).

There are also reports of relatively little change in communities in the face of change in feedstock as in the addition of poultry waste to that of dairy cattle (Zhang et al., 2011c). The stability of *Bacteriodetes* and *Firmicutes* communities fed different substrates (casein, starch and cream) were emphasised by Kampmann et al. (2012b) although changes in species composition within these groups were observed. However, it should be noted that in the first example both wastes are ammonia rich and have other relative similarities, although Kampmann et al. (2012b) compared casein, starch and cream that differ markedly in protein, carbohydrate and lipid fractions.

Temporal variation

The start up and development of a stable digester function has been the subject of many studies as this is a critical period, with failure or protracted start up being associated with

Mixing conditions / Aeration / Retention time / Dilution	Municipal solid waste, sewage sludge (4); Cattle manure (18) / Glucose and fatty acid sludge (32); Municipal solid waste (9) / Beet silage (28, 43) Glucose and fatty acid sludge (32) / Acetate, butyrate (7, 16)
Biomass granulation / Sulphate load / Trace elements / Electric pre-treatment / Ammonia loading / Polymer addition	Glucose and volatile fatty acid wastewater (35) / Corn syrup (11) / Food waste (23) / Municipal solid waste, sewage sludge (22) / Municipal waste (49) / Ice cream wastewater (5)

1. Griffin et al. (1998), 2. Zumstein et al. (2000), 3. Ahring et al. (2001), 4. McMahon et al. (2001), 5. Uyanik et al. (2002), 6. McHugh et al. (2003), 7. Shigematsu et al. (2003), 8. McMahon et al. (2004), 9. Tang et al. (2004), 10. Hori et al. (2006), 11. Briones et al. (2007), 12. Klocke et al. (2007), 13. Leven et al. (2007), 14. McGarvey et al. (2007), 15. Sousa et al. (2007), 16. Tang et al. (2007), 17. Cheon et al. (2008), 18. Hoffmann et al. (2008), 19. Klocke et al. (2008), 20. Goberna et al. (2009), 21. Lee et al. (2009), 22. Zhang et al. (2009), 23. Feng et al. (2010), 24. Goberna et al. (2010c), 25. Hwang et al. (2010), 26. Ike et al. (2010), 27. Kim et al. (2010b), 28. Krakat et al. (2010a), 29. Krakat et al. (2010b), 30. Krakat et al. (2010c), 31. Lee et al. (2010), 32. Madden et al. (2010), 33. Munk et al. (2010), 34. Nettmann et al. (2010), 35. O'Reilly et al. (2010), 36. Palatsi et al. (2010), 37. Shin et al. (2010), 38. Song et al. (2010), 39. Wang et al. (2010), 40. Chelliapan et al. (2011), 41. Garcia et al. (2011), 42. Kim et al. (2011), 43. Krakat et al. (2011), 44. Martin-Gonzalez et al. (2011), 45. Shin et al. (2011a, b), 46. Supaphol et al. (2011), 47. Tang et al. (2011), 48. Wagner et al. (2011), 49. Westerholm et al. (2011), 50. Zhang et al. (2011a), 51. Zhang et al. (2011b), 52. Zhang et al. (2011c), 53. Blume et al. (2010), 54. Cardinali-Rezende et al. (2012), 55. Kampmann et al. (2012a), 56. Kampmann et al. (2012b), 57. Lee et al. (2012), 58. Lerm et al. (2012), 59. Lin et al. (2012), 60. Nelson et al. (2012), 61. Oz et al. (2012), 62. Rademacher et al. (2012a), 63. Regueiro et al. (2012), 64. Ritari et al. (2012).

Table 14.4 Environmental variables whose effects on population dynamics of microbial communities in AD are being increasingly investigated as the cost and the technical difficulty of such studies decreases. The listings are not exhaustive but representative of an increasingly active research area. Sources are recent ones with some highly referenced older work. Nelson et al. (2011) provide a meta-analysis of data available at mid-2010.

Effect investigated	Feedstock (sources in parentheses)
Feedstock type	Municipal solid waste (1, 8, 37, 46, 54); Municipal waste water (57); Sewage, whey, glucose (21, 31, 42); Food waste (26); Fruit and vegetable waste (17, 59); Maize silage (53); Swine wastewater (25, 27, 38, 45); Wine distillation waste (2); Cattle manure (3, 14, 18); Grass silage (39); Pharmaceutical waste (40); Glucose and fatty acid sludge (32); Corn syrup (11); Biofuel waste (51); Wheat/Rye waste (19); Sewage, brewery waste, dairy sugar yeast, mixed fish and dairy (63); Kitchen, sewage, livestock waste (17); Household and garden waste (20); Triticale waste (19); Cattle, swine, maize, grass and grain waste (34); Cattle manure olive mill waste (24); Carrot waste (41); Cattle and swine manure with casein, starch and cream (55, 56); Food processing waste (60); Kitchen, sewage livestock waste (17); Molasses, milk and potato wastes (6); Cattle and poultry waste (52); Municipal lipid rich waste (15, 44); Beet silage (12); Paper, ethanol fermentation waste (47)
Temporal dynamics: start up to steady state / Seed sludge	Municipal solid waste (1, 8, 37, 54); Sewage, whey, glucose (21, 31, 42); Food waste (26); Fruit and vegetable waste (17, 59); Maize silage (53); Swine wastewater (27, 38, 45); Wine distillation waste (2); Cattle manure (14); Grass silage (39) / Food processing waste (60); Glucose, carbohydrate (61)
Loading levels (overloading, starvation) / fatty acid load	Municipal solid waste and sewage sludge (8, 58, 64); Maize silage (53); Pharmaceutical waste (40); Swine wastewater (25); Glucose and fatty acid sludge (32); Molasses (50); Maize silage (33); Manure (36); Glucose (10); Municipal solid waste (48)
Compartmentalization	Municipal solid waste (46); Corn syrup (11); Biofuel waste (51); Municipal waste water (57); Wheat/Rye waste (19); Pharmaceutical waste (40); Grass silage (39)
Large-scale systems	Municipal solid waste (46, 54); Sewage, brewery waste, dairy sugar yeast, mixed fish and dairy (63); Food waste (26); Municipal waste water (57); Kitchen, sewage, livestock waste (17); Household and garden waste (20); Triticale waste (19); Cattle, swine, maize, grass and grain waste (34)
Temperature	Municipal solid waste (1); Municipal waste water (57); Cattle manure olive mill waste (24); Biowaste and sewage sludge (64); Beet silage (29, 30, 43); Swine wastewater (27); Glucose and fatty acid sludge (32); Glucose and volatile fatty acid wastewater (35); Rye silage (62); Organic household waste (13); Cattle manure (3) / < 20°C (32, 35) / 25–45°C (1, 13, 24, 27, 57, 64) / 45–65°C (1, 3, 13, 24, 27, 29, 30, 43, 57, 62, 64) / 65°C (3, 62)

Table 14.3 Main patterns of bacterial and archaeal diversity in anaerobic digesters.

	%	Number of species (OTUs)	Current estimated coverage	Key notes / Main genera
Bacterial phyla		5926	61	
Bacteroidetes	14	705	66	70% unidentified / *Paludibacter*
Firmicutes	15	1352	59	86% class *Clostridiales*, 6% class *Bacilli*, 6% unidentified / *Acetivibrio, Clostridium,*
Proteobacteria	21	1590	60	30% unidentified, 169 known genera, α- β- δ- *protobacteria* total 86% / *Brachymonas, Rhodobacter, Thauera, Smithella, Syntrophobacter*
Chloroflexi	22	693	60	90% unclassified *Anaerolineaceae*, 8 known genera / e.g. *Sphaerobacter*
Synergistetes	6			*Cloacabacillus*
Plactomycetes	2			Mainly from two sites using *Planctomycetes* specific primers
Actinobacteria	2			*Iamia*
Unclassified	16	–	–	–
Archaeal phyla		296	90	
Euryarchaeota	95	–	–	27% class *Methanomicrobiales*, 6% class *Methanobacteriales* / *Methanosaeta, Methanospirillum, Methaobacterium, Methanoculleus, Methanolinea, Methanosarcina*
Crenarchaeota	3	–	–	–
Unclassified	2	–	–	–

Abstracted from Nelson et al. (2011)

Community structure in relation to digester environment and digester operation

As molecular tools become cheaper, and technically easier to apply, a greater range of substrates, digester designs, process conditions, and different subsets of the digestate (e.g. different particle sizes, flocs, granules, solute) are being investigated to assess how they affect microbial community compositions, and how this relates to digester performance (Table 14.4). Several tens of feedstocks have now been investigated, most using multiple primer sets designed to amplify bacterial and archaeal 16s rRNA genes, with more detailed sequencing and analysis of particular subgroups, usually the methanogens. Pyrosequencing studies are still relatively rare and few have examined non microbial organisms, an exception being Ritari et al. (2012) who reported that the main fungal phyla involved in anaerobic reactors are the *Ascomycota* and *Basidiomycota*.

Table 14.2 Two examples demonstrating how digestate communities can be very diverse, although many species found in a sample are relatively rare as indicated by the relatively small number of hits (No) for most species. Few species were found in common between the two data sets (those species are marked in bold).

Algal sludge 1-stage digester (batch)	No	Agricultural waste 1st of 2-stage digester	No
Methanosaeta concilii GP-6	946	Euryarchaeota	
No significant similarity found	262	**Methanoculleus marisnigri JR1**	4529
Methanosaeta concilii VeAc9	140	**Methanospirillum hungatei JF-1**	203
Methanobacterium sp. 169	84	Candidatus Methanoregula boonei 6A8	160
Methanobacterium formicicum DSM 1535	73	Methanosarcina acetivorans C2A	144
Methanospirillum hungatei JF-1	55	Methanosarcina barkeri str. fusaro	123
Methanobacterium formicicum NBRC 100475	51	Firmicutes (Clostridia)	
Methanosaeta harundinacea 8Ac	48	Clostridium thermocellum ATCC 27405	995
Methanolinea sp. TNR	44	Thermosinus carboxydivorans Nor1	665
Methanosarcina mazei strain MT	43	Pelotomaculum thermopropionicum SI	621
Methanobacterium subterraneum NBRC 105231	35	Alkaliphilus metalliredigens QYMF	583
		Desulfotomaculum reducens MI-1	497
Methanoregula formicicum SMSP	22	Clostridium cellulolyticum H10	483
Methanoculleus palmolei	12	Alkaliphilus oremlandii OhILAs	475
Methanosarcina sp. HB-1	12	Caldicellulosiruptor saccharolyticus DSM 8903	375
Methanobacterium ivanovii NBRC 104952	8	Clostridium phytofermentans ISDg	365
Methanobacterium formicicum S1 mrtA	6	Carboxydothermus hydrogenoformans Z-2901	359
Methanogenium organophilum	6	Desulfitobacterium hafniense Y51	332
Methanoculleus thermophilus	5	Clostridium kluyveri DSM 555	206
Methanosarcina thermophila	5	Desulfitobacterium hafniense DCB-2	193
Methanobacterium uliginosum DSM 2956 mrtA	5	Clostridium beijerinckii NCIMB 8052	187
		Clostridium tetani E88	185
Methanosarcina mazei	4	Clostridium novyi NT	155
Methanobacterium ferruginis	3	Clostridium acetobutylicum ATCC 824	151
Methanobacterium petrolearium	3	Ruminococcus obeum ATCC 29174	133
Methanoculleus marisnigri JR1	3	Ruminococcus gnavus ATCC 29149	130
Methanolinea tarda	3	Ruminococcus torques ATCC 27756	121
Methanosarcina mazei LYC	3	Eubacterium ventriosum ATCC 27560	120
Methanosarcina sp. T36	3	Halothermothrix orenii H 168	629
Methanobacterium sp. GH	2	Moorella thermoacetica ATCC 39073	542
Methanosaeta concilii DSM 3671	2	Thermoanaerobacter tengcongensis MB4	458
Methanosaeta thermophila PT	2	Thermoanaerobacter pseudoethanolicus ATCC 33223	341
Methanospirillum lacunae	2		
Methanobacterium palustre NBRC 105230	1	Thermoanaerobacter sp. X514	222
Methanoculleus chikugoensis	1	Firmicutes (Bacilli)	
Methanosaeta harundinacea	1	Symbiobacterium thermophilum IAM 14863	433
Methanothermobacter thermautotrophicus	1	Geobacillus thermodenitrificans NG80-2	165
		Bacillus halodurans C-125	161
		Bacillus sp. NRRL B-14911	152
		Geobacillus kaustophilus HTA426	151
		Bacillus coagulans 36D1	121
		Bacteroidetes	
		Parabacteroides distasonis ATCC 8503	600
		Parabacteroides merdae ATCC 43184	507
		Bacteroides vulgatus ATCC 8482	437
		Bacteroides capillosus ATCC 29799	266
		Bacteroides thetaiotaomicron VPI-5482	238
		Bacteroides fragilis YCH46	203
		Porphyromonas gingivalis W83	175
		Bacteroides caccae ATCC 43185	186
		Synergistetes	
		Syntrophomonas wolfei subsp. wolfei str. Goettingen	457
		Lentisphaerae	
		Victivallis vadensis ATCC BAA-548	299
		Thermotogae	
		Thermotoga lettingae TMO	130
		Petrotoga mobilis SJ95	123
		Spirochaetes	
		Treponema denticola ATCC 35405	126
Total species detected **35**		Total species detected **57**	

Data abstracted from Ellis et al. (2012) and Schlüter et al. (2008).

The few studies of digestates relevant to biogas production that are available have already confirmed suspicions from earlier analyses that these communities are highly diverse and differ markedly in digestates dominated by a different stages of chemical transformation. The two samples compared in Table 14.2 show the dominance of methanogenic bacteria in the sample from the algal digestate obviously in a mature stage of chemical transformation and likely dominated by methanogenesis (Ellis et al., 2012), and the wide range of other species in the early stage digestion (the first stage of a two-stage digester of agricultural feedstock) with a mix of early stage processes (Schlüter et al., 2008). The early stage data also show the wide variety and dominance of species from the genus *Clostridium* and related species present at this stage of the process, detail that may have been missed in earlier studies.

Both studies found large numbers of taxa (> 35–50), with only two of the more common species found in both data sets. Significantly, one of these studies revealed a species whose sequences were not found in established databases and is likely to be new to science (second listing for algal sludge 1). Novel bacteria, including new higher level groups, are being discovered in AD systems (e.g. Chouari et al., 2005) and dominant taxa within particular digesters can be taxa, like the key acetogenic bacteria in an ammonium loaded digester, that were unknown (Westerholm et al., 2011). In addition, comparison of these data sets shows that significant biomass of a methanogenic bacterium was found in the first stage digester.

Integrative analyses are still rare, but one meta-analysis of 16S rRNA gene data available up to May 2010 has confirmed some clear patterns in AD microbial diversity that had been emerging from individual studies through the last decade (Nelson et al., 2011). The principal findings are: 1) While sequences from 28 of the 30 bacterial phyla recognised were detected in the samples, only seven (*Chloroflexi, Proeobacteria, Firmicutes, Bacteroidetes, Synergistetes, Plactomycetes, Actinobacteria*) were represented by more than 1 per cent of the 19,388 sequences analysed, and only four (*Chloroflexi, Proteobacteria, Firmicute, Bacteroidetes*) were dominant, accounting collectively for 63 per cent of sequences (Table 14.3). Of the dominant four, *Proteobacteria, Firmicutes* and *Bacteroidetes* have been detected in almost all studies of AD. As the *Planctomycetes* were detected mainly from two sites in studies using *Planctomycetes* specific primers these were considered special cases and this group was not considered to be particularly important to AD. 2) Similarly, only two of the four current archaeal phyla were detected and one (*Euryarchaeota*) was dominant with 95 per cent of sequences. 3) Within each of the bacterial and archaeal phyla a small number of groups were abundant but a large number of taxa were unidentified in some groups. 4) The diversity of *Bacteria* was far greater than *Archaea*. 5) Rarefaction analysis estimated that that about 60 per cent of likely operational taxonomic units (OTUs) within the bacterial phyla and 90 per cent of archaeal phyla had been detected. This analysis demonstrates that much work remains to be done to finally detail the diversity in AD systems given the number of unidentified species, and the fact that many sequences cannot be assigned to higher level taxonomic units at present.

It also shows the considerable complexity of the AD community and the need for far more work to understand its function, particularly at species level. However, it does indicate that subgroups dominate the process and provides some encouragement that smaller communities could provide core groups that could provide the basis for efficient function (Rivière et al., 2009). A scan of literature published since this work (see Table 14.4) generally supports the findings, with some cases where additional groups play a larger role in individual digesters. The response of microbial communities to different environments and to changes in digester function is the focus of the next section.

Table 14.1 Examples of different microorganisms shown to be involved in different stages of AD.

Stage of AD	Major taxonomic entities identified
Hydrolysis and acidogensis	• Fungi *Trichoderma* (e.g. *T. reesei*), *Thermomonospora*, *Ralstonia* and *Shewanella*, *Penicillium*, *Aspergillus* and *Humicola* • Bacteria e.g. *Bacteroides*, *Butyrivibrio*, *Clostridium*, *Cellulomonas*, *Fusobacterium*, *Selenomonas*, *Streptococcus*, *Peptococcus* and *Campylobacter*. Actinomycetes such as *Streptomyces* • *Pseudomonas mendocina*, *Bacillus halodurans*, *Clostridium hastiforme*, *Gracilibacter thermotolerans*, and *Thermomonas haemolytica*. *Synergistete*.
Acetogenesis	• Most acetogens are in the phylum *Firmicutes* e.g. *Moorella thermoacetica*. • *Spirochaetes*. • δ-proteobacteria e.g. *Desulfotignum phosphitoxidans*. • *Acidobacteria* e.g. *Holophaga foetida* • Exclusively acetogenic bacteria e.g. *Acetobacterium* and *Sporomusa* • Genera with acetogenic and non-acetogenic species e.g. *Clostridium*, *Ruminococcus*, *Eubacterium*, *Thermoanaerobacter*, *Treponema*.
Methanogenesis	Exclusively anaerobic, methane-producing *Archaea* from the phylum *Euryarchaeota*, with • 6 orders: *Methanobacteriales*, *Methanococcales*, *Methanomicrobiales*, *Methanosarcinales*, *Methanopyrales*, *Methanocellales*, and • 31 genera e.g. *Methanosarcina*, *Methanobrevibacter/Methanobacterium Methanosaeta*

Sources: Miyamoto (1997), Cirne et al. (2007), Rastogi et al. (2008), Schlüter et al. (2008), Franke-Whittle et al. (2009), Sang et al. (2009), Singhania (2009), Kim et al. (2010a), Pobeheim et al. (2010), Song et al. (2010), Ellis et al. (2012)

et al., 2010; Zhang et al., 2009; O'Reilly et al., 2010; Talbot et al., 2010; Wang et al., 2010; Regueiro et al., 2012).

Classical work isolating and characterising microorganisms demonstrated that different groups of microorganisms have different metabolic capabilities that determine their role in AD (McCarty, 1982) (Table 14.1). Organic material is decomposed by heterotrophic microorganisms dominated by bacteria and fungi capable of hydrolysis and acidogenesis and a diverse bacterial consortium is responsible for acetogenesis. Methanogenesis is dominated by the *Archaea* (Khalid et al., 2011). Some bacteria involved have been isolated and characterised but thorough studies with pure plant biomass are missing, especially on hydrolytic and thermophilic processes (Cirne et al., 2007; Antoni et al., 2007).

However, traditional tools, reliant on culturing individual species, are not capable of detecting the majority of species present in AD let alone determining their changing proportions in the complex environmental conditions in the digestate. In contrast, the relatively new technique of massive parallel sequencing (MPS), allows sequencing of all the DNA present in a sample and, through comparison of those sequences with each other and sequences in established databases (bioinformatic analysis), the detection of how many different species are present, their identity, and a crude estimation of their relative abundance (see Chapter 15 for details). Since molecular tools became available there has been an increasing amount of work applying these to anaerobic digestion systems (Rittmann et al., 2008). Early PCR data often allowed identification only to higher order taxonomic levels such as genus, class or order, but it is clear new technologies such as MPS and metagenomics has the potential to identify the constituents of communities in finer detail.

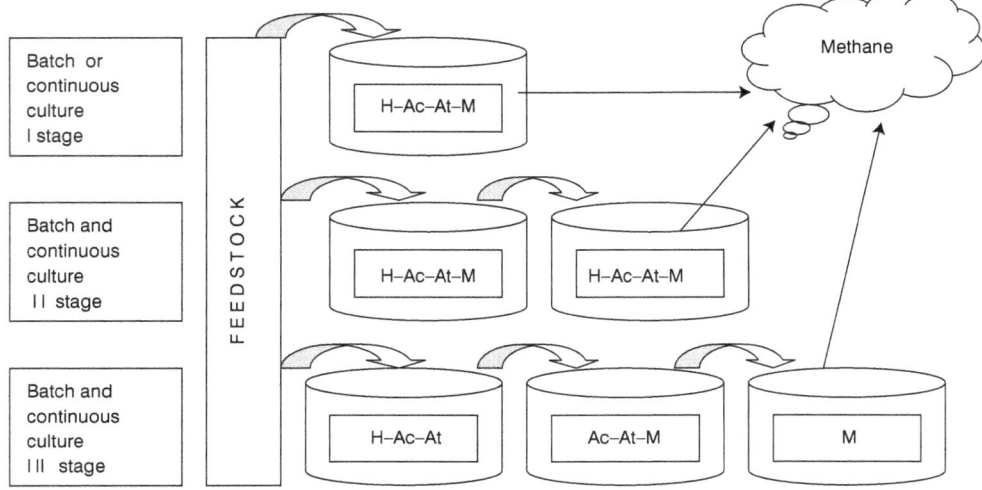

Figure 14.1 A simplified conceptual illustration of AD for I, II, and III stage digesters. Letters H, Ac, At, M indicate microbial activity related to the major stages of chemical transformation: hydrolysis (H), acidogenesis (Ac), acetogenesis (At) and methanogenesis (M) in each digester stage. Each chemical stage involves different microbial communities (see Tables 14.1 and 14.2), and separate physical containers facilitate particular reactions, therefore sustaining different microbiota. Most digester operations rely on the microorganisms that are introduced with the substrate and an inoculum of microorganisms from mature digestate from a previous run (arrows). The introduction of particular species of microorganisms with known function is rare.

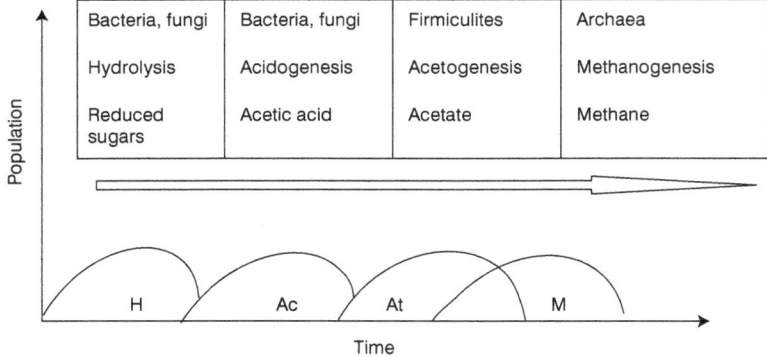

Figure 14.2 The four major stages of chemical transformation: hydrolysis (H), acidogenesis (Ac), acetogenesis (At) and methanogenesis (M). Schematic illustrations of the niche of hypothetical microbiota are used to illustrate how those with different ranges of capability (to use substrates and survive in the relevant environment), lead to different communities that will change with the balance of chemical transformation (utilising different substrates, and creating different conditions of, e.g., pH, and CO_2 concentration). Boundaries among different chemical stages are not defined and overlap is common, determined mainly from environmental conditions in digester, configuration of the digesting system (I, II or III step digestion) and the participating microbial species.

(Lee et al., 2010) and improved production by monitoring and controlling key populations (da Silva et al., 2012; Ramirez et al., 2008, 2009).

The present chapter provides an overview of the present state of the art concerning the identity of the microorganisms present in AD fermentation systems and their roles, and a brief consideration of the models of their dynamics that will lead ultimately to improved production control systems. It will not be possible to provide an exhaustive account but examples are given to illustrate key concepts and show the current and future trends.

AD fermentation systems and their microorganisms

AD systems usually range from one- to three-step digesters (Figure 14.1). In one-step digesters the whole process is carried out in one container. In the very commonly used two-step digesters, leachate (with some substrate) is taken from an initial reaction container as a feed into the next one, and in three-step digesters leachate from the second stage is used as feedstock for the third stage. Multiple stages provide discrete physical environments that may allow the development of more specialised microbial communities, with greater emphasis on particular stages in the chemical transformation of the substrate (Figures 14.1 and 14.2).

In the practical daily operation of most digesters, little attention is paid to the composition of the microbial community that is critical for the transformation of the substrate into biogas. The operation of this complex community is effectively treated as a black box. Present practice is to rely on microorganisms added with the substrate and/or the inoculation of complex communities from established digestates, and to monitor processes using key aspects of the physical and chemical environment, intermediate and final products (see Chapter 12 for more details). The lack of detailed information on the species that are most effective in undertaking particular roles means that few of known function are specifically added to digestates as a matter of course.

Microorganism communities involved in AD

The process of AD has four distinct phases of chemical transition (McCarty 1982; Demirel and Scherer, 2008): hydrolysis, acidogenesis, acetogenesis and methanogenesis (Figure 14.2). Hydrolysis breaks down complex organic substrates (e.g. carbohydrates, lipids or other insoluble organic material) into simple sugars, amino acids and fatty acids, which acidogenic bacteria convert into carbon dioxide, hydrogen, ammonia, and various organic acids during acidogenesis. Acetogenic bacteria then convert organic acids into acetic acid, ammonia, hydrogen, and carbon dioxide which methanogens convert to methane and carbon dioxide. Acetate and hydrogen can be produced in the first phase and are used directly by methanogens, while substances such as volatile fatty acids with long chain length need to be catabolised by acetogenic bacteria into compounds that can be utilised by methanogens.

It is important to stress that while different groups of microorganisms are involved in the four major stages of AD, more than one stage (or even all) can, in fact must, co-occur to permit their synergistic action. Micro-spatial structuring within the digestate also plays a critical role in the maintenance of overall function through providing niches for different communities. For example, the development of flocculent particles or biofilms on surfaces within digesters provides physical structure for different microbes (e.g. acetogens and methanogens) to occur in physical proximity and so exchange key metabolites, such as acetate, reducing or oxidising agents, with each other (McMahon et al., 2004; Madden

Chapter 14

Microbial communities and their dynamics in biomethane production

Dimitris N. Argyropoulos,[1*] *Theodoros H. Varzakas*[2] *and John A.H. Benzie*[3,4*]

[1]Genetic Identification Laboratory, Hellenic Agricultural Organization "DEMETER", Sofokli Venizelou 1, 14123, Lykovrissi, Greece, [2]Department of Food Technology, School of Agricultural Technology, Higher Technological Educational Institute of Kalamata, Antikalamos 24100, Kalamata, Hellas, [3]Environmental Research Institute, University College Cork, Lee Road, Cork, Ireland; [4]School of Biological Earth and Environmental Sciences, University College Cork, North Mall Campus, Cork, Ireland
*Corresponding authors email: dimargy@yahoo.com; j.benzie@ucc.ie

Introduction

Biomethane is the result of the anaerobic digestion of a wide range of substrates by the collaborative action of a diverse range of microorganisms whose number and identity at various stages of methanogenesis depends on the substrates used and the fermentation environment (Briones and Raskin, 2003; Demirel and Scherer, 2008). However, the digester has often been regarded as a black-box process converting substrate to biogas (Madden et al., 2010; Supaphol et al., 2011). This state of affairs is changing as it is realised that attention to the details of the fermentation process could lead to significant improvements in efficiency, control and industrial development of biogas production (Briones and Raskin, 2003; Rittmann et al., 2008).

Molecular tools have revolutionised the capacity to identify the microbes present in mixed populations found in reactors (e.g. through gene sequences providing unique species identifiers), assess their abundance (e.g. using quantitative PCR (polymerase chain reaction) or fluorescent in-situ hybridisation (FISH)) and determine their location (e.g. using FISH) (Rittmann et al., 2008; Wirth et al. 2012). Chapter 15 gives more information on these techniques and discusses genetic engineering approaches to more efficient anaerobic digestion (AD) production. However, a number of studies are already providing exciting new information on microbial communities and providing new visions for future work (e.g. Bauer et al., 2008; Krause et al., 2008; Weiss et al., 2008; Kröber et al., 2009; Liu et al., 2009; Cardinali-Rezende et al., 2009, 2012; Jaenicke et al., 2011; Rademacher et al., 2012a, 2012b; Zakrzewski et al., 2012).

Mathematical modelling of AD systems and their use in controlling industrial-scale AD systems is well established (Scherer et al., 2009; Dewil et al., 2011) with increasing interest and capability in modelling the microbial community in more detail (e.g. Henson, 2003; Ramirez et al., 2009). Advances in molecular biology, software development and computing power provide the possibility of monitoring performance through using probes for key species

Part IV
Molecular biology and population dynamics

Strik, D.P.B.T.B., Domnanovich, A.M., Zani, L., Braun, R. and Holubar, P. (2005) 'Prediction of trace compounds in biogas from anaerobic digestion using the MATLAB neural network toolbox', *Environmental Modelling & Software*, vol. 20, pp. 803–810.

Van Vooren, L. (2000) 'Buffer capacity based multipurpose hard- and software sensor for environmental applications', PhD Thesis, Faculty of Agricultural and Applied Biological Sciences, Ghent University.

Vasiliev, Y. (2011) 'Cube development for beginners', Oracle Articles, 2011 http://www.oracle.com/technetwork/articles/bi/cube-development-359587.html [Accessed September 2012].

Velickov, S. and Solomatine, D. (2000) 'Predictive data mining: practical examples', in *Proceedings of the 2nd Joint Workshop 'Artificial Intelligence in Civil Engineering'*, March, Cottbus, Germany.

Walpole, R.E. and Mayers, R.H. (1993) *Probability and statistics for engineers and scientists*, 5th edn, Englewood Cliffs, NJ: Prentice Hall.

Walpole, R.E., Myers, S.L. and Ye, K. (2002) *Probability and statistics for engineers and scientists*, 7th edition, Englewood Cliffs, NJ: Prentice Hall.

Ward, A.J., Hobbs, P.J., Hollimana, P.J., and Jones, D.L. (2008) 'Optimisation of the anaerobic digestion of agricultural resources', *Bioresource Technology*, vol. 99, no 17, pp. 7928–7940.

Werner, J.J., Koren, O., Hugenholtz, P., DeSantis, T.Z., Walters, W.A., Caporaso, J.G., Angenent L.T., Knight, R. and Ley, R.E. (2012) 'Impact of training sets on classification of high-throughput bacterial 16s rRNA gene surveys', *ISME Journal*, vol.6, no. 1, pp. 94–103.

Witten, I. and Eibe, F. (2005) *Data mining: practical machine learning tools and techniques*, 2nd edition, San Fransisco, CA: Elsevier & Morgan Kaufmann, .

Yao, X. (1999) 'Evolving artificial neural networks', In *Proceedings of the IEEE*, vol. 87, pp 1432–1447.

Ye, N. (2003) *The handbook of data mining*, London: Lawrence Erlbaum Associates.

Zaher, U. (2005) 'Modelling and monitoring the anaerobic digestion process in view of optimisation and smooth operation of WWTP's', PhD thesis, Ghent University, Ghent, Belgium.

Zhao, Y. (2012) 'R and data mining: examples and case studies', RDataMining.com: R and Data Mining, RDM, http://www.rdatamining.com [accessed July 2012].

Kimball, R. and Caserta, J. (2004) *The data warehouse etl toolkit: practical techniques for extracting, cleaning, conforming, and delivering data*, New York: John Wiley and Sons.

Knight, T.P. (2004) 'MARIA: A multilayered unsupervised machine learning algorithm based on the vertebrate immune system', PhD Thesis, University of Kent, Canterbury, UK.

Korres, N.E. (2005) *Encyclopaedic dictionary of weed science. Theory and digest*, Andover: Intercept and Paris: Lavoisier SAS.

Krakat, N., Westphal, A., Schmidt, S. and Scherer, P. (2010) 'Anaerobic digestion of renewable biomass: thermophilic temperature governs methanogen population dynamics', *Applied and Environmental Microbiology*, vol. 76, pp. 1842–1850.

Lee, S., Bellamy, D., Bettison, P., Chesshire, M. et al. (2009) 'Developing an implementation plan for anaerobic digestion', Report of the Anaerobic Digestion Task Group, DEFRA, July 2009, http://archive.defra.gov.uk/environment/waste/ad/documents/implementation-plan.pdf [Accessed August 2012]

Martinez, E., Marcos, A., Al-Kassir, A., Jaramillo, M.A. and Mohamad, A.A. (2012) 'Mathematical model of a laboratory-scale plant for slaughterhouse effluents biodigestion for biogas production', *Applied Energy*, vol. 95, pp. 210–219.

Nizami, A.S., Korres, N.E. and Murphy, J.D. (2009) 'A review of the integrated process for the production of grass biomethane', *Environmental Science and Technology*, vol. 43, no 22, pp 8496–8508.

Nsofor, G.C. (2006) 'A comparative analysis of predictive data-mining techniques', Thesis Presented for the Master of Science Degree, University of Tennessee, Knoxville.

Polit, M., Estaben, M., and Labat, P. (2002) 'A fuzzy model for an anaerobic digester, comparison with experimental results', *Engineering Applications of Artificial Intelligence*, vol. 15, pp 385–390.

Poncelet, P., Teisseire, M. and Masseglia, F. (2008) *Data mining patterns. new method and application*, Information Science Reference, Hershey, NY: IGI Global, .

Ramageri, B.M (2011) 'Data mining techniques and applications', *Indian Journal of Computer Science and Engineering*, vol. 1, no. 4, pp. 301–305.

Ravichandra Rao, I.K. (2003) 'Data mining and clustering techniques', DRTC Workshop on Semantic Web, 8–10 December 2003, DRTC, Bangalore, Paper K.

Rob, P., Coronely, C. and Crockett, K. (2008) *Data base systems: design, implementation and management*, London: Cengage Learning EMEA.

Schievano, A., D'Imporzano, G., Salati, S., and Adani, F. (2011) 'On-field study of anaerobic digestion full-scale plants (Part I): An on-field methodology to determine mass, carbon and nutrients balance', *Bioresource Technology*, vol. 102, no 17, pp. 7737–7744.

Sembiring, W.R., Zain, J.M., and Embong, A. (2010) 'Clustering high dimensional data using subspace and projected clustering algorithms', *International Journal of Computer Science & Information Technology (IJCSIT)*, vol. 2, no.4, pp. 162–170.

Shapiro-Piatetsky, G., Fayyad, U. and Smith, P. (1996) 'From data mining to knowledge discovery: An overview', in Fayyad, U.M., Piatetsky-Shapiro, G., Smyth, P. and Uthurusamy, R. (eds), *Advances in knowledge discovery and data mining*, pp. 1–35. Cambridge, MA: AAAI/MIT Press.

Sharvelle, S., Keske, C., Davis, J. and Lasker, J. (2011) 'Online decision tool for anaerobic digestion at Colorado cattle operations: user manual', Natural Resources Conservation Service, Colorado State University, December 2011.

Siegert, I. and Banks, C. (2005) 'The effect of volatile fatty acid additions on the anaerobic digestion of cellulose and glucose in batch reactors', *Process Biochemistry*, vol. 40, no 11, pp. 3412–3418.

Sjoberg, J. (2005) 'Mathematical neural networks. Train and analyse neural networks to fit your data', Wolfram Research, www.Wolfram.com [Accessed September 2012].

Sparks, A.H., Esker, P.D., Bates, M., Dall' Acqua, W., Guo, Z., Segovia, V., Silwal, S.D., Tolos, S. and Garrett, K.A. (2008) 'Ecology and epidemiology' in 'R: Disease Progress over Time', *The Plant Health Instructor*.

StatSoft (2002) 'Data mining techniques', Electronic textbook, 1984–2002, http://www.obgyn.cam.ac.uk/cam-only/statsbook/stdatmin.html#pdm, Accessed June 2012.

Cummings, T. (2011) 'Database normalization and denormalization', Presentation available online at http://www.slideserve.com/liam/database-normalization-and-denormalization [Accessed August 2012].

Douglas, C.M. and George, C.R. (2002) *Applied statistics and probability for engineers*, 3rd edn, New York: John Wiley.

Few, S. (2004) 'Dashboard confusion. perceptual edge', 20 March 2004. Available at http://www.perceptualedge.com/library.php#Whitepapers [Accessed November 2012].

Gibert, K., Sanchez-Marre, M. and Codina, V. (2010) 'Choosing the right data mining technique: Classification of methods and intelligent recommendation', in *Proceedings of 'Modelling for Environment's Sake', International Congress on Environmental Modelling and Software*, July 5–8, 2010, Ottawa, Ontario, Canada, S-23-03.

Giudici, P. (2003) *Applied data-mining: Statistical methods for business and industry*, Chichester: John Wiley.

Grieder, C., Dhillon, B.S., Schipprack, W. and Melchinger A.E. (2012) 'Breeding maize as biogas substrate in Central Europe: I. Quantitative-genetic parameters for testcross performance', *Theoretical and Applied Genetics*, vol. 124, pp. 971–980.

Grover, M., Kumar, R., Mondal, K.T. and Rajkumar, S. (2001) 'Computational models for sustainable development', *Indian Journal of Computer Science and Engineering*, vol. 2, no 1, pp. 55–60.

Gupta, S., Kumar, D. and Sharma, A. (2011) 'Data mining classification techniques applied for breast cancer diagnosis and prognosis', *Indian Journal of Computer Science and Engineering*, vol. 2, no 2, pp. 188–195.

Guwy, A.J., Hawkes, F.R., Wilcox, S.J. and Hawkes, D.L. (1997) 'Neural network and on-off control of bicarbonate alkalinity in a fluidised-bed anaerobic digester', *Water Research*, vol. 31, pp. 2019–2025.

Han, J. and Kamber, M. (2001) *Data mining: concepts and techniques*, San Francisco, CA: Morgan Kaufmann Academic Press.

Han, J. and Kamber, M. (2006) *Data mining: concepts and techniques*, 2nd edn, San Francisco, CA: Elsevier & Morgan Kaufmann Academic Press.

Hand, D.J. (1998) 'Data mining: statistics and more?', *The American Statistician*, vol. 52, no.2, pp. 112–118.

Hand, D., Mannila, H. and Smyth, P. (2001) *Principles of data mining*, Cambridge, MA: The MIT Press.

Hanke, J.E., Wichern, D.W. and Reitsch, A.G. (2001) *Business forecasting*, 7th edn, Hoboken, NJ: Prentice Hall.

Harwood, O., Lukas, P., Mezzullo, W., Scurlock, J., Digby, R., Collins, D., Thompson, P., Morton, C. et al. (2011) 'The case of crop feedstocks in anaerobic digestion', A join briefing paper by ADBA, CLA, NFU and REA, November 2011.

Holubar, P., Zani, L., Hager, M., Frochl, W., Radak, Z. and Braun, R. (2003) 'Start-up and recovery of a biogas-reactor using a hierarchical neural network-based control tool', *Journal of Chemical Technology and Biotechnology*, vol. 78, pp. 847–854.

Inmon, W.H. (1996a) 'The data warehouse and the data mining,' *Communications of the ACM*, vol. 39, no. 11, pp. 49–50.

Inmon, W.H. (1996b) *Building the data warehouse*, New York: John Wiley & Sons.

Inmon, W.H. (2005) *Building the data warehouse*, 4th edn, New York: Wiley Publishing.

Jackson, J. (2002) 'Data mining: A conceptual overview', *Communications of the Association for Information Systems*, vol. 8, pp. 267–296.

Jacobi, F.H., Ohl, S., Thiessen, E., and Hartung, E. (2012) 'NIRS-aided monitoring and prediction of biogas yields from maize silage at a full-scale biogas plant applying lumped kinetics', *Bioresource Technology*, vol. 103, no 1, pp. 162–172.

Kim, J., Park, C., Kim, T.H., Lee, M., Kim, S., Kim, S.W. and Lee, J. (2003) 'Effects of various pretreatments for enhanced anaerobic digestion with waste activated sludge', *Journal of Bioscience and Bioengineering*, vol. 95, no 3, pp 271–275.

References

Abellan, J., Cano, A., Masegosa, A.R. and Moral, S. (2007) 'A semi-naive Bayes classifier with grouping of cases', in *9th European Conference, ECSQARU*, Hammamet, Tunisia, pp. 477–488.

Abonyi, J. and Feil, B. (2007) *Cluster analysis for data mining and system identification*, Basel, Boston, and Berlin: Birkhauser.

Aha, D.W. (1995) 'Machine learning: An annotated bibliography for the 1995 AI and statistics tutorial on machine learning', in *Proceedings of the Fifth International Workshop on Artificial Intelligence and Statistics*, Fort Lauderdale, FL, January.

Ahmed, A., Otreba, M., Korres, N.E., Elhadi, H. and Menzel, K. (2010) 'Assessing building performance of naturally day-lit buildings using data mining', *Advanced Engineering Informatics* vol. 25, no. 2, pp 364–379.

Alijamaat, A., Khalilian, M., and Norwati, M. (2010) 'A novel approach for high dimensional data clustering', *Third International Conference on Knowledge Discovery and Data Mining*, pp. 264–267.

Amon, T., Amon, B., Kryvoruchko, V., Machmuller, A., Hopfner-Sixt, K., Bodiroza, V., Hrbek, R., Friedel, J., Potsch, E., Wagentristl, H., Schreiner, M. and Zollitsch, W. (2007) 'Methane production through anaerobic digestion of various energy crops grown in sustainable crop rotations', *Bioresource Technology*, vol 98, no 17, pp 3204–3212.

Anonymous (2005) *Introduction to data mining and knowledge discovery*, 3rd edn, Two Crows Corporation, www.twocrows.com [Accessed June 2012].

Anonymous (undated) 'Oracle data mining concepts',11g Release 1 (11.1), http://docs.oracle.com/cd/B28359_01/datamine.111/b28129/process.htm#CHDDIABC [Accessed August 2012].

Ayoub, N., Wang, K., Kagiyama, T., Seki, H. and Naka, Y. (2006) 'A planning support system for biomass-based power generation', in *Proceedings of 16th European Symposium on Computer Aided Process Engineering and 9th International Symposium on Process Systems Engineering*, pp. 1899–1904.

Batstone, D.J. (2006) 'Mathematical modelling of anaerobic reactors treating domestic wastewater: Rational criteria for model use', *Reviews in Environmental Science and BioTechnology*, vol. 5, pp 57–71.

Batstone, D.J., Keller, J., Angelidaki, I., Kalyuzhnyi, S.V., Pavlostathis, S.G., Rozzi, A., Sanders, W.T.M., Siegrist, H. and Vavilin, V.A. (2002) 'Anaerobic digestion model no. 1 (ADM1)', *IWA Task Group for Mathematical Modelling of Anaerobic Digestion Processes*. London: IWA Publishing.

Benefield, D.L., Judkins, J.F. and Weand, B.L. (1982), *Process chemistry for water and wastewater treatment*, Hoboken, NJ: Prentice-Hall.

Berry, M.J.A. and Linoff, G.S. (2000) *Mastering data mining. The art and science of customer relationship management*, New York: Wiley.

Berthold, M.R., Dill, F., Kotter, T. and Thiel, K. (2008) 'Supporting creativity: Towards associative discovery of new insights', in Washio, T., Suzuki, E., Ting, K.M. and Inokuchi, A. (eds) *Lecture notes in computer science*, 12th Pacific-Asia Conference, PAKDD 2008, Osaka, Japan, May 20–23, vol. 5012, pp. 14–25. Berlin and Heidelberg: Springer-Verlag.

Bowerman, B.L. and O'Connell, R.T. (1993) *Forecasting and time series. An applied approach*, 3rd edn, Duxbury Classic Series, Duxbury: Thompson Learning.

Bramer, M.A. (1999) *Knowledge discovery and data mining*. London: Institution of Electrical Engineers.

Burgess, J. and Witheford, S. (2012) 'Biogas in practice', KWS SAAT AG. http://www.kws.de/li/bk/yia/?countryfolder=aaaaaaaaaaathd&search=biogas&UNICODE=%26%2312398&3B%80%C4%D6%DC [Accessed September, 2012].

Chakrabarti, S., Ester, M., Fayyad, U., Gehrke, J., Han, J., Morishita, S., Piatetsky-Shapiro, G. and Wang, W. (2006) 'Data mining curriculum: A proposal (Version 1.0)', ACM SIGKDD, April 2009, http://www.sigkdd.org/curriculum.php [accessed August 2012].

Chaudhuri, S. and Dayal, U. (1997) 'An overview of data warehousing and OLAP technology', *Proceedings of the 1997 ACM/SIGMOD Conference*, pp. 65–75, New York: ACM Press.

Chen, L.Z., Nguang, S.K., Chen, X.D. and Li, X.M. (2004) 'Modeling and optimization of fed-batch fermentation processes using dynamic neural networks and genetic algorithms', *Biochemical Engineering Journal*, vol. 22, pp. 51–61.

intraclass similarities and minimisation of the interclass similarities based on some criteria defined on the characteristics of the objects under examination (Knight, 2004; Witten and Eibe, 2005). Good clusters show high similarity within a group and low similarity between patterns belonging to two different groups (Ye, 2003; Alijamaat et al., 2010). Unlike classification, the clusters are unknown as are the attributes on which the data will be clustered (Anonymous, 2005). Generally, it is recommended that the use of standardised data is most appropriate since problems caused by different scales of measurements can thereby be eliminated. There are four basic steps, namely i) data collection and selection of variables for analysis; ii) generation of a similarity matrix; iii) decision about the number of clusters and their interpretation; and iv) validation of cluster solution. The main outcome of a cluster analysis is known as a dendrogram or tree diagram. A dendrogram is displayed in the previous chapter (Figure 12.3) where a hierarchical cluster analysis was performed for the characterisation of the similarity of various physico-chemical characteristics of grass silages. In hierarchical clustering a concept of ordering is involved driven by the number of observations to be combined at a time or by the statistical difference (from 0) of distance between two clusters. The clusters could be formed either by omitting dissimilar observations (divisive method) or joining together similar observations (agglomerative method) as in the case of the example presented in Chapter 12. Various approaches have been proposed in the literature for developing classifiers by means of clustering, which can be summarised as i) single linkage (nearest neighbour approach), ii) complete linkage (furthest neighbour), iii) average linkage, iv) Ward's method, and v) centroid method. All these approaches differ in the definition of distance and what defines largest distance as statistically no-distance or zero-distance. Most of the time, the distance is based on Euclidean distance in the sample axes (Mahalanobis distance is for non-orthogonal samples) (Velickov and Solomatine, 2000; Ye, 2003; Chakrabarti et al., 2006; Ramageri 2011).

Conclusions

Data warehouse and data mining techniques, as significant parts of knowledge discovery in databases, have been shown to be important tools in different business domains. The authors of this chapter strongly believe that the integration of these techniques in the AD industry for optimisation procedures and maximisation of biogas/biomethane production merits further investigation. This is particularly true since these techniques are considered two of the most important domains in database and information systems. The various approaches (e.g. biological, chemical, etc.) for determining methane potentials lead to substantial uncertainty which results in occasional and repeatable sampling occasions. Even if the perfect type of time-space matrix of data could be achieved the analysis of multimodal, multiscale, spatiotemporal data would still be a challenge. In addition, the most popular anaerobic models are very complex, and the information contained in parameter values and model structure is not easily applicable when we need to take decisions about the process. Data warehouse and data mining techniques can be used to filter the huge amount of information produced by monitoring AD processes and in conjunction with the progress in software engineering and control systems can be proved invaluable tools towards the sustainability of biogas production.

regression since it can fit hyperbolic, logarithmic and exponential equations that define y as a function of x (or more than one x) by estimating parameters of the non-linear equation that minimise the residuals (Korres, 2005). Choosing the most appropriate model for the data under examination is a scientific decision, meaning that the chosen model must make sense in scientific terms. According to Korres (2005) assumptions and checks of non-linear regression consist of i) the scientific soundness of the model; ii) the variability of values around the curve must follow a normal distribution; iii) the variance has to be homogeneous independently of the values of y, while the errors must be independent; iv) the higher the R^2 value the better the fit of the curve. Some well-known non-linear growth models are i) the Malthus (exponential) model [$y(t) = y_0 e^{(rt)}$ where $y_0 = y(0)$ = initial response at time = 0, r = growth rate and t = time]; ii) the monomolecular model [$y(t) = k - (k - y_0)e^{(-rt)}$ where k = carrying size of the system, $y_0 = y(0)$ = initial response at time = 0, r = growth rate and t = time]; iii) the logistic model [$y(t) = k / [1 + (k|y_0 - 1)e^{(-rt)}]$]; iv) the Gompertz model [$y_{(t)} = k e^{\left[\log_e(y_0/k) e^{(-rt)}\right]}$] and v) the Richards model [$y(t) = k y_0 / [y_0 + (k^m | y_0^m)e^{(-rt)}]^{(1/m)}$]. The Malthus model assumes that the absolute rate of increase (dy/dt) is proportional to the response intensity (y). The monomolecular model assumes a carrying capacity of one (i.e. the rate of growth at any time is proportional to the resources yet to be achieved. A logistic regression model is used when the response variable is a binary, hence the absolute rate of response variable change depends on both values of the outcome (i.e. y and (1 − y)) at the time. The graph of y(t) versus t is an elongated S-shape and the curve is symmetrical about its point of inflexion. The Gompertz model assumes that the absolute rate of change depends on y and ln(1/y) and is very similar to the logistic model, although unlike the logistic model the curve is asymmetrical about the inflexion point (Jackson, 2002; Sparks et al., 2008; Ramageri, 2011; Zhao, 2012). The titration test, as mentioned in the previous chapter, is performed by adding small amounts of a strong base to a weak acid solution or by adding small amounts of a strong acid to a weak base solution and measuring the pH after each addition (Zaher, 2005). Consequently, a titration curve can be obtained by plotting the amount of base/acid added versus the change in pH. Examples of a titration curve are provided by Benefield et al. (1982) and Van Vooren (2000) in which the resulting curves can be described by the logistic and Gompertz equations respectively (i.e. S-shaped curves), which suggests that the pH does not change at a constant rate with the addition of strong base. Non-linear modelling for optimisation and control has also been applied using ordinary differential equations to model kinetic reactions in AD. This type of approach has been widely used and already mentioned in population kinetics modelling (see Chapter 15) from the thorough review of Batstone (2006) for anaerobic fermentation modelling to his proposition for the ADM1 model (Batstone et al., 2002). The centre of the methodology is mass balance differential equations and growth rate estimations. Limitations of this approach find unique support in data mining techniques as an alternative approach to predict parameter values and to provide real-time control (see Chapter 15).

Clustering

Clustering, an unsupervised learning data mining technique (Poncelet et al., 2008; Sembiring et al., 2010), is the identification of classes, also called clusters or groups, for a set of objects (or attributes or measurements or cases) without any prior knowledge of the relationships between them. The purpose of clustering is the maximisation of the

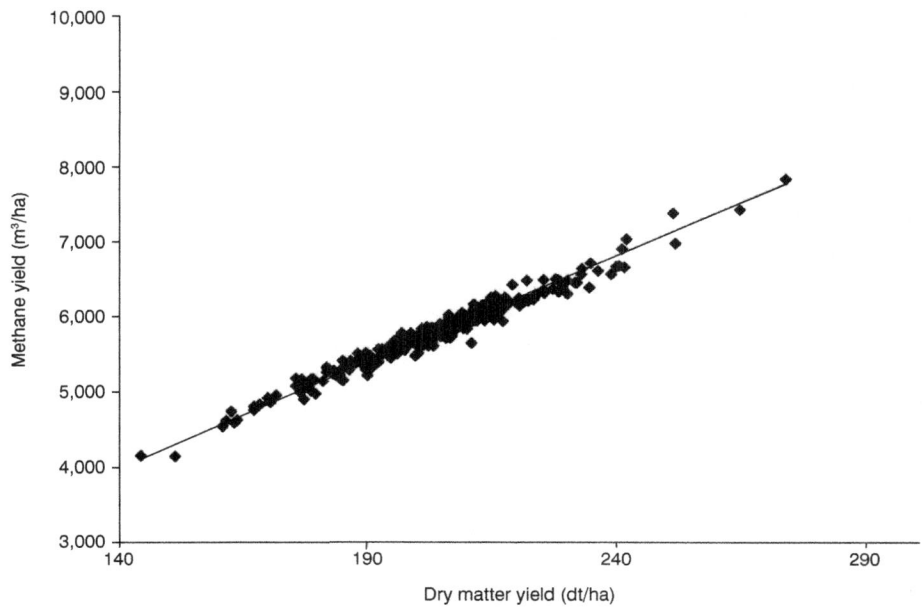

Figure 13.12 Relationship between maize dry matter (dry tonnes/ha) and methane yield

Regression analysis was applied for the control of AD based on data provided from near-infrared spectroscopy (NIRS). Jacobi et al. (2012) collected 520 days' NIR-spectra for feedstuff analysis. The biogas potential of the samples was calculated from the laboratory analysis results and NIRS-regression-models for various parameters were calibrated. Continuously gathered spectra, NIRS-models, actual plant-feeding data and degradation kinetics were used to calculate time series of theoretically expectable biogas reproduction. Results were validated against measured gas quantity.

It is worth mentioning that the term 'linear' implies that the coefficients of the independent variables are linear, independent of the nature of the model (i.e. it might be argued that polynomial models are not linear), because in statistics only the parameters (i.e. the coefficients of the independent variables) are considered in classifying the linearity or non-linearity of a model and not the independent variables (Walpole et al., 2002).

Non-linear regression analysis

Unfortunately, many real-life problems are not simply linear projections of previous values and hence are difficult to predict because they may depend on complex interactions of multiple predictor variables (e.g. as in the case of AD as shown in the previous chapter). Therefore, more complex techniques (e.g., non-linear regression models) may be necessary to predict future values (Ramageri, 2011). Non-linear regression is a technique for fitting a statistical model to values that cannot be described by a linear function. In other words, it is used when the effect of a stimulus x on the response y is not additive but multiplicative (Korres, 2005). Though linearisation by transformation of raw data is a common practice, the analysis of raw data is preferable since linear transformation distorts the experimental error (Korres, 2005). Non-linear regression is a more general technique than linear

Linear regression analysis in AD

In predictive data mining methods the aim is to build a model which will permit the value of one variable to be predicted from the known value(s) of other variable(s). The difference between the two predictive data mining techniques is that in classification the variable being predicted is categorical while in regression the variable is quantitative (Hand et al., 2001). The term 'prediction', in this section, is used in a general sense and no notion of a time continuum is implied, otherwise it is called time series prediction (i.e. a variable that varies with time) (Anonymous, 2005). In the case of the linear regression model, there is allowance for error (noise) in the relationship hence the relationship becomes $y=g(x)+e$, where $g(x)=\alpha x+b$, and e represents the error in the model. This defines the divergence between the predicted and the actual value, while α represents the weight that linearly combines with the input to predict the output (Nsofor, 2006). When the x variable is multiple, it is known as multiple linear regression. A simple linear regression predicts the value of a quantitative variable for a new instance as a linear equation of a single numerical variable; normality, linearity and homoscedasticity are prerequisites for its application. Multiple linear regression predicts the value of a quantitative variable for a new instance as a linear equation of several numerical variables and requires normality, linearity, homoscedasticity and independence for its application (Douglas and George, 2002; Korres, 2005; Gibert et al., 2010). The equation of multiple regression is in the form:

$$y = w_1x_1 + w_2x_2 + w_3x_3 + \cdots + w_ix_i + e \qquad (Eq.\ 13.3)$$

where w_i is the coefficient of the regression. Finally, a polynomial model with one predictor variable has the form

$$y=\alpha+bx+cx^2+e \text{ (second order or quadratic or curvilinear)} \qquad (Eq.\ 13.4).$$

In both cases (simple regression or multiple linear regression) several ways of estimating goodness of fit (i.e. how well a predictive model fits the observed data) exist. The most common statistics for this are regression coefficient (R^2), adjusted R^2, p-value, sum of squares, and r (the correlation coefficient) (Korres, 2005). Based on linear relationships, certain predictions called extrapolations (for predictions outside of the range of the observed data) or interpolations (for predictions within the range of the observed of data) can be made, cautiously though, particularly in the case of extrapolations, because outside of the range of the observed values possible relationships may change dramatically (Korres, 2005).

A representative example concerning the application of the simple linear regression model between maize dry matter and biomethane production (Burgess and Witheford, 2012; Grieder et al., 2012) was reproduced from the original data provided by Dr Grieder (personal communication) (Figure 13.12). More particularly, the means of 290 maize genotypes averaged over two years and three locations were used to prove the strong linear relationship between maize dry matter and biomethane yield. The regression coefficient (R^2) equals 0.98 indicating the small variance around the straight line plot, hence the strong relationship between these two parameters.

Schievano et al. (2011) established a simple methodology to study mass balances of full-scale anaerobic digestion plants. They verified a linear equation as a tool for calculating input–output flows. This tool allowed determining carbon and nutrients balances in three full-scale AD plants, necessary for digestate land applications as fertiliser. According to the authors, this approach is especially useful to evaluate the contribution of AD processes to elemental cycles, especially when digestate is applied to agricultural land.

Support vector machine

A support vector machine (SVM) algorithm is based on statistical 'learning' theory. SVM classification uses the concept of decision planes (i.e. it distinguishes between a set of objects having different class memberships) that define decision boundaries (Ahmed et al., 2010). SVM finds the vectors ('support vectors') that define the separators giving the widest separation of classes. Data records with N attributes can be thought of as points in N-dimensional space and SVM attempts to separate the points into subsets with homogeneous target values. Points are separated by hyperplanes in the linear case (linear kernel), and in the non-linear case (Gaussian) by non-linear separators. In the case of $N = 2$, the solution defines a straight line (linear) or a curve (non-linear) separating the differing classes of points in the plane (Ahmed et al., 2010). For both kernels, the tolerance (i.e. a measure that functions as a stopping mechanism when an algorithm should be satisfied with the result and consider the building process complete) usually is 0.001. A higher value will give a faster build but perhaps a less accurate model in a stopping mechanism. SVM is sensitive to missing values. A common treatment replaces a numerical missing value with the mean (average) for relevant attributes, whereas it replaces a categorical missing value with the mode (the most frequently occurring value) (Ahmed et al., 2010).

Regression

Regression, in its simplest form, involves building a predictive model to relate a predictor (i.e. an independent variable x) to a response (i.e. a dependent variable y) through a relationship of the form $y = \alpha x + b$ (Hand et al., 2001). The steepness (slope) of the curve is defined by the numerical value of α and represents the change in y by the corresponding change in x while the algebraic sign of α describes the relationship between x and y (i.e. positive, negative or no association when $\alpha > 0$, $\alpha < 0$ and $\alpha = 0$ respectively) (Korres, 2005). The other constant b is called the intercept and is the value of y when $x = 0$, in other words it defines the elevation of the curve as shown in Figure 13.11.

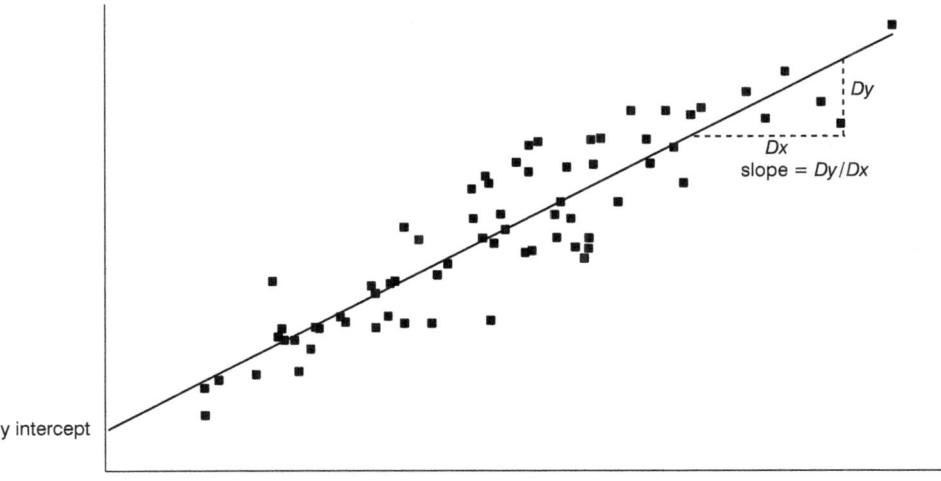

Figure 13.11 Regression curve showing data points and the prediction line

Fuzzy logic systems

According to Ward et al. (2008) fuzzy logic systems are empirical control systems for problem solving that can arrive at a definite conclusion from vague information using intermediate values rather than simple yes/no data. Applications of fuzzy logic have been reported by Polit et al. (2002) where fuzzy inference rules were used to predict specific bacterial growth rate based on temperature and pH data. More recent examples of fuzzy logic related with AD reported on the microbial kinetics where classical modelling could not anticipate microbial, morphological and physiological changes affecting methane production for long time periods (Krakat et al., 2010) (see Chapter 15). Fuzzy logic has also been combined with neural networks for on-line detection and analysis of a 120-litre fluidised bed reactor. The hybrid system was used for the treatment of wine distillery wastewater in order to handle foam forming and temperature regulation. Raw data from the process (pH, temperature, recirculation flow rate, input flow rate and gas flow rate) were pre-processed using fuzzy logic to build a vector of features (i.e. a pattern vector). This feature vector was classified into a pre-specified category which according to discrimination fuzzy rules was a state of the system. An artificial neural network was then used to classify the process states and to identify the faulty or dangerous ones.

Genetic algorithms

Genetic algorithms in data mining are not used to find patterns *per se*, but rather to guide the learning process of data mining algorithms such as neural nets (Anonymous, 2005) by formulating hypotheses about dependencies between variables in the form of association rules or some other internal formalism (Jackson, 2002). They act as a method for performing a guided search for good models in the solution space and they are called genetic algorithms because they loosely follow the pattern of biological evolution in which the members of one generation (of models) compete to pass on their characteristics to the next generation (of models) until the best (model) is found (Anonymous, 2005). With the progress in computational power this type of analysis is becoming more popular. Chen et al. (2004), for example, used genetic algorithms to generate an optimised feed rate profile in conjunction with neural-net techniques. The complex relationships between the manipulated feed rate and the biomass product was described by two recurrent neural sub-models, in which outputs of one sub-model were fed into another sub-model to provide meaningful information for the biomass prediction. Based on the neural network model, a modified generic algorithm was used to provide optimal feed rate and to generate a smooth feed rate profile. Similar work has been presented by Martinez et al. (2012) who proposed a mathematical model for slaughterhouse effluent biodigestion based on a continuous anaerobic digestion laboratory scale pilot-plant. The model reproduces the substrate degradation along with the bacterial population evolution using variables such as chemical oxygen demand (COD), acetic acid, propionic acid and methane. In the first step, a generic algorithm where the parameters were directly related to the measured variables was used. In the second step, a gradient descendent algorithm was used for a smooth adjustment of the whole set of parameters. The values of COD, acetic acid, propionic acid and methane which were obtained from the simulation of the model with the optimised parameters were similar to those directly obtained from the prototype.

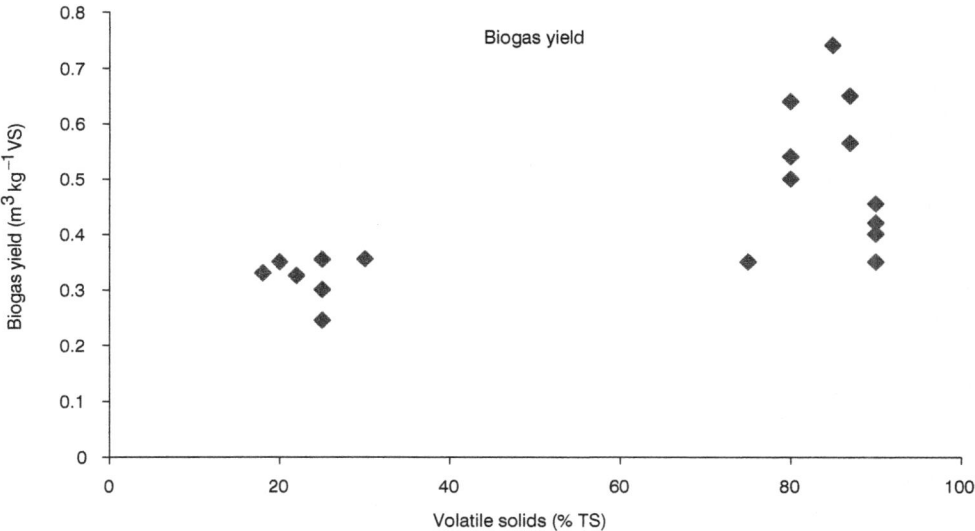

Figure 13.9 Classification of biogas yield data according to volatile solids concentration

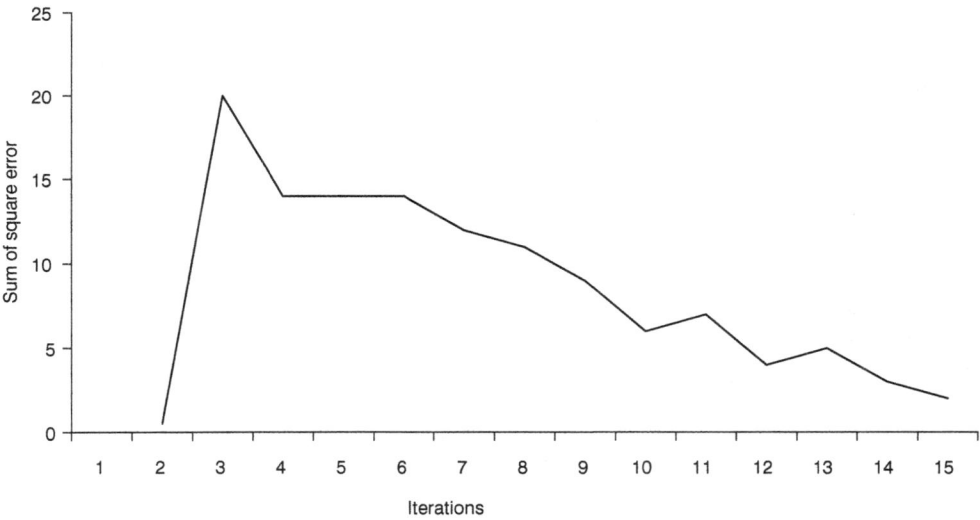

Figure 13.10 Summary of a training process

tank reactors which were operated in steady-state conditions and disturbed by pulse-like increase of the organic loading rate was used by Holubar et al. (2003). The trained network was used to control a lab-scale anaerobic continuous stirred tank reactor and it was possible to maintain a methane concentration of about 60 per cent at a rather high gas production rate of between 5 and 5.6 $m^3\ m^{-3}d^{-1}$.

at the root node and ends when after following the assertions down a terminal node (or leaf) is reached, at that point, a decision is made (Jackson 2002). A disadvantage of decision trees is that trees in the training process use up data very rapidly. They should never be used with small data sets although problems with very big data sets have been recorded (Gibert et al., 2010). They are also highly sensitive to noise in the data and they try to fit the data exactly, which is referred to as over-fitting (Jackson 2002). A series of decision trees could guide, for example, potential investors in anaerobic digestion through i) initial options; ii) levels and type of feedstock; iii) sources of information; iv) appropriate upgrade technologies; v) end uses of biogas etc. (Lee et al., 2009; Harwood et al., 2011; Sharvelle et al., 2011).

Neural networks

A neural network or neural net is typically a collection of neuron-like information processing units with weighted connections between these units (Anonymous, 2005). In most cases a neural network is an adaptive system hence in response to an obtained output it may be changed by a learning algorithm, which enables it to learn. Accordingly, neural networks can be categorised as supervised, unsupervised and reinforcement learning. In the first case, they often have a target output and rely on direct comparison between the actual output and the target output. This is calculated as an error function and is often formulated as the minimisation of the mean or sum squared error between the actual output and the desired output summed over all available data (Yao, 1999). Disadvantages of neural networks are the time-intensive effort during the development of the initial model since the input processing almost always requires the transformation of the raw data, and the requirement of technical knowledge by the user (Jackson, 2002).

An example, to demonstrate how a neural network, as a classification method, can be applied in AD, has been modified from Sjoberg (2005). Volatile solids and biogas yield of various feedstocks as presented in Chapter 12 have been used as input data whereas output data represent the class to which each feedstock belongs. Because there are two possible classes in this example, the output can be stored as 0 or 1 indicating the class to which a feedstock belongs. The plot, in which the measured values are plotted (Figure 13.9), clearly demonstrates that the data is divided into two classes. The successfully trained neural network classifier will return the correct group to which a feedstock belongs, given the feedstock's total solids and biogas yield.

A measure of fit, or performance index, to be minimised by the training algorithm, must be chosen for the training to proceed. For classification problems, the criterion is set to the number of incorrectly classified data samples, where the classifier has correctly classified all data when the criterion is zero. A perceptron (i.e. an algorithm for supervised classification of an input into one of two possible outputs) is the simplest linear neural network classifier (i.e. an algorithm that makes its predictions based on a linear predictor function combining a set of weights with the feature vector describing a given input) and was used to illustrate the progress of the training process here.

A summary of the training process is shown in the plot of sum squared error (SSE) against iteration number (Figure 13.10). It is obvious that the SSE tends toward 0 as the training goes through more iterations.

Strik et al. (2005) applied neural networks to describe trace gases in AD, whereas Guwy et al. (1997) applied them for the control of $NaHCO_3$ buffer addition. The application of hierarchical neural networks trained with data taken from four anaerobic continuous stirred

Table 13.3 Estimation of probability frequencies.

Attributes	Subject	Frequency	Attributes	Subject	Frequency
Temperature/High	Mesophilic/High	5/10	Temperature/Low	Mesophilic/Low	3/7
	Thermophilic/High	5/10		Thermophilic/Low	4/7
pH/High	Acidic/High	2/10	pH/Low	Acidic/Low	4/7
	Neutral/High	4/10		Neutral/Low	2/7
	Alkaline/High	4/10		Alkaline/Low	1/7
VFA/High	Low/High	1/10	VFA/Low	Low/Low	4/7
	Medium/High	4/10		Medium/low	2/7
	High/High	5/10		High/Low	1/7

Time series

A time series is a chronological sequence of observations on a particular variable recorded over a long period of time, usually at equal intervals (Bowerman and O'Connell, 1993; Korres, 2005). Time series forecasting predicts unknown future values based on a time-varying series of predictors where known results are being used to guide predictions made by time series (Anonymous, 2005). The mathematical expression of time series for the values $Y_1, Y_2, ..., Y_n$ of a variable Y (predictors) measured at time $t_1, t_2, ..., t_n$ is $Y = F(t)$. The analysis of time series consists of a description of its component movements, i.e. $Y = T \times C \times S \times I$, where T, C, S, I represent the components of a time series, namely trend, cyclic, seasonal and irregular fluctuations respectively (Korres, 2005). Trend is the long-term component of time series and refers to the upward or downward movement (i.e. growth or decline) that characterises a time series (Bowerman and O'Connell, 1993; Hanke et al., 2001). Long-run movements towards a wider application of AD technology might be determined by one, some or all of the factors such as technological changes in the industry, policy and legislation in renewable energy and environment, inflation or deflation, market growth etc. Cyclic fluctuations are those of which the pattern of increases and decreases (i.e. wavelike fluctuations around the trend) extends over a long period of time, possibly over many years. Seasonal variation is the shorter-term fluctuation that shows the changing pattern of natural seasons or other varied patterns such as the five- or seven-day work week, etc. Seasonality can be an important component in time series as in the case of feedstock production. Irregular or random fluctuations are the unexplained chance variations that remain when the other components have been identified and explained, as in the case of natural catastrophes that can affect the availability of feedstock.

Decision trees

A decision tree is a flowchart-like tree structure when each node denotes a test on an attribute value where each branch represents an outcome of the test, and tree leaves represent classes (Anonymous, 2005; Chakrabarti et al., 2006). Decision trees represent sets of decisions which generate rules for the classification of a dataset. A rule is a conditional statement in the form 'if-then-else' (Anonymous, 2005). The classification of a particular data item begins

Table 13.2 Training data of a hypothetical methane production in relation to process parameters temperature, pH and VFA.

Case	Temperature (°C)	pH	VFA (mg/l)	Methane production (m³/kgVS)
1	Mesophilic	Acidic	High	Low
2	Thermophilic	Acidic	High	High
3	Mesophilic	Alkaline	Medium	High
4	Thermophilic	Neutral	Low	High
5	Mesophilic	Alkaline	High	High
6	Mesophilic	Neutral	High	High
7	Thermophilic	Acidic	Medium	Low
8	Thermophilic	Alkaline	High	High
9	Mesophilic	Neutral	Low	Low
10	Thermophilic	Neutral	Medium	Low
11	Thermophilic	Alkaline	Medium	High
12	Thermophilic	Neutral	High	High
13	Mesophilic	Acidic	Low	Low
14	Mesophilic	Neutral	Medium	High
15	Thermophilic	Acidic	Low	Low
16	Mesophilic	Acidic	Medium	High
17	Thermophilic	Alkaline	Low	Low

$n=10$ and $n_c=5$. Note that the attribute Temperature shows two possible values whereas the attributes pH and VFA show three possible values. So $p=1/$(number of attribute values)$=0.5$ for Temperature and $p=1/$(number of attribute values)$=0.33$ for pH and VFA. The estimation of the $P(u_j)$ and $P(a_i|u_j)$ probabilities can be done, besides Eq. 13.2, based on their frequencies over the training data as the following: $P(u_j)=P(high)=10/(17)$ and $P(u_j)=P(low)=7/17$.

Accordingly, $P(a_i|u_j)=P(Mesophilic|High)=5/10$ whereas $P(a_i|u_j)=P(Mesophilic|Low)=3/7$ and so on (Table 13.3).

The classification of the 'methane production at mesophilic temperature in alkaline environment with low concentration of VFA' instance can be estimated as follows:

$$\max\{10/17 \times 5/10 \times 4/10 \times 1/10; 7/17 \times 3/7 \times 1/7 \times 4/7\} = \max\{0.0117, 0.0144\} = 0.0144$$

meaning that under these conditions (i.e. mesophilic temperature and alkaline environment and low VFA concentration) the production of biomethane, since $0.0144 > 0.0117$, will be classified as low. It is worth mentioning that poor estimates usually are due to the small number of observations. Naïve Bayes, besides the example demonstrated in this section, has been applied in taxonomic classification of 16S rRNA gene sequences generated in microbiome studies (Werner et al., 2012). Perhaps it is possible to extract useful information for anaerobic microorganisms and their population dynamics by the application of naïve Bayes.

Classification

Classification is a process of finding a set of models that describe and distinguish data classes or concepts of an object based on its attributes (Ravichandra Rao, 2003). DMN techniques create classification models via the examination of existing classified data (cases) through which they detect intelligently a predictive pattern. These existing cases could originate from a historical database or an experiment. Samples and/or observations with known attributes are examined to create a classifier (Anonymous, 2005) which is generalised to future cases (Velickov and Solomatine, 2000). For example, from a set of some representative AD plants, which serve as a training set, a classification model can be built, which concludes the performance of an AD plant based on its attributes which can be the whole set or part of its operational parameters. Some of the most important classification techniques are discussed in the following sections.

Naïve Bayes

Naïve Bayes is a probabilistic supervised classifier based on Bayesian theory (Abellan et al., 2007). In general, the Bayesian method utilises prior subjective knowledge of parameters in conjunction with the information provided by sample data (Walpole and Mayers, 1993). In the case of AD, for example, monitoring sensors are recording observations of specified process parameters (attributes) over long time periods and short intervals, e.g. on a daily basis. Naïve Bayes calculates, through these recording observations, conditional probabilities for target values by observing the frequency of attribute values and their combinations. The naïve Bayes (nb) classifier selects the most likely classification V_{nb} given the attribute values $a_1, a_2, \ldots a_n$. This results in:

$$V_{nb} = \mathrm{argmax}_{u_i \in V} P(u_j) \prod P(a_i | u_j) \quad \text{(Eq. 13.1)}$$

in which $P(a_i | u_j)$ is generally estimated using m-estimates as shown below

$$P(a_i | u_j) = \frac{n_c + mp}{n + m} \quad \text{(Eq. 13.2)}$$

where n = the number of training examples for which $u = u_j$; n_c = number of examples for which $u = u_j$ and $a = a_i$; p = a priori estimate for $P(a_i | u_j)$ and m = the equivalent sample size. The following example demonstrates the use of naïve Bayes in AD in which the operational parameters (attributes) taken into account are pH, temperature and VFA.

The biomethane production can be either low or high. In the previous chapter (Chapter 12) it was demonstrated that process parameters in AD vary highly, hence in this example attributes are categorised in the case of pH as acidic, neutral and alkaline when pH is below 7, equal to 7 or above 7 respectively; in the case of temperature as mesophilic and thermophilic and in the case of VFA as low (1000–2000 mg/l), medium (2000–3000 mg/l) and high (3000–4000 mg/l) according to Table 12.7.

Based on the data in Table 13.2, the classification (i.e. low or high) of a new instance, e.g. 'methane production under mesophilic temperature in alkaline environment with low VFA concentration', can be estimated.

In this example, looking at probability P(Mesophilic|High), there are ten cases where u_j = high and in five of those cases a_i = Mesophilic are recorded, so for P(Mesophilic|High),

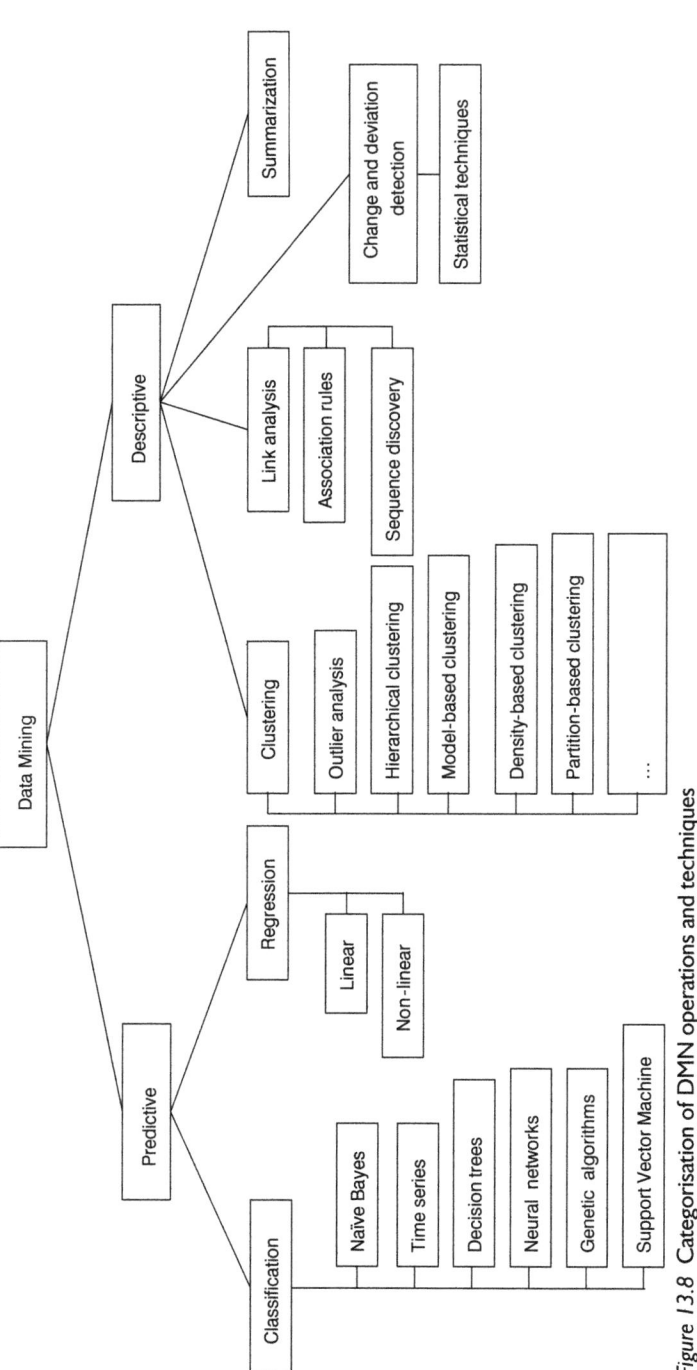

Figure 13.8 Categorisation of DMN operations and techniques

Categorisation of DMN techniques

Data mining systems can be categorised according to the type of databases mined, the type of knowledge mined and the techniques used for the accomplishment of the targeted purpose (Ravichandra Rao, 2003). DMN techniques are used for the extraction of patterns from DWH through data mining tasks. There are two types of data mining tasks, namely descriptive (i.e. they describe general properties of the existing data) and predictive (i.e. they attempt to perform predictions based on available data) (Gupta et al., 2011). DMN, as an interdisciplinary field, incorporate disciplines from statistics, database systems, machine learning, pattern recognition, neural networks, fuzzy systems and other soft computing techniques to achieve the desirable outcome (Velickov and Solomatine, 2000). While there are only few basic data mining operations (i.e. classification, regression, clustering, link analysis and summarisation) (Figure 13.8) there are many techniques that can be applied to make these operations possible (Velickov and Solomatine, 2000). This is the very reason why the categorisation of DMN techniques is always a sensitive subject. There are many categories of data mining techniques with inheritance with generalisation and specialisation of classes, going up and down respectively, whereas a particular class sometimes appears with many different names.

The rest of this chapter will focus on classification, regression and hierarchical clustering methods. The aim of the following section, without underestimating the importance of the techniques that have been intentionally omitted, is to provoke the reader to search for applications and develop his own skills in data analysis by providing some examples and conceptual case studies of the selected DMN techniques' application in AD.

Supervised vs. unsupervised (or predictive vs. descriptive) machine learning

Machine learning algorithms have been developed for tasks used to identify patterns in data or to optimise machine performance, production lines, computer programs or to model natural processes. DMN techniques are based on the utilisation of machine learning algorithms for solving various real-life mining problems (Bramer, 1999; Chakrabarti et al., 2006). These algorithms are divided into two types, namely supervised and unsupervised (Aha, 1995; Ahmed et al., 2010). Supervised learning generally results in predictive models, in contrast to unsupervised learning where the goal is pattern detection or discovering relationships (Ahmed et al., 2010). In addition, unsupervised learning is based solely on the correlations among input data and no information on correct output is available for learning (Yao, 1999). Reinforcement learning is a special case of supervised learning where the exact output is unknown and learning is based on whether the output is correct or not (Knight, 2004). Accordingly, the algorithms which support this learning process can be classified either as predictive or descriptive algorithms (Anonymous, 2005). Descriptive algorithms are used for exploratory data analysis to discover individual patterns, such as a priori for association rules and enhance K-means for clustering for an unsupervised learning process (Ahmed et al., 2010). Predictive algorithms focus on the creation of models that allow predicting observations from input data, such as decision trees for classifications and support vector machine (SVM) in regression models for supervised 'learning' processes (Ahmed et al., 2010).

format, but its applicability to the problem under question. Proper data cleansing and data preparation/integration are very important for data mining and a data warehouse can facilitate these activities (Jackson, 2002). However, a data warehouse will be of no use if it does not contain the data fitted to the particular data mining query. As mentioned earlier in this chapter, data mining is the process of finding correlations or patterns among numerous fields in large databases with reference to DWHs because of their great potential to extract the most important information from DWHs. The purpose that is served by the presence of integrated operational data in a convenient form is to facilitate the ability to perform detailed and advanced analyses. It would be very difficult and laborious to perform such analysis through the diverse operational systems that store large, often unstructured, amounts of data on a daily basis. Instead, well-structured and redundant data in a DWH can be very useful for performing DMN techniques, test scenarios and understanding business and operational circumstances. Especially in bioenergy production processes and in particular AD procedures, a well-structured DWH combined with DMN can prove very valuable. AD is characterised by a very high level of variability which affects performance and several types of instrumentation have been proposed to measure these variables, as mentioned in the previous chapter. Usually variables being measured vary; hence DMN techniques could have a significant role in the identification of key variables for AD optimum performance. DWH and DMN function as a loop where knowledge extracted by DMN is fed back to the data transformation in DWH. Eventually, performance is improved and stability is achieved. Correlations between variables and prediction of specific output values (e.g. VFA value) according to other variables (e.g. pH) are feasible with a well-structured DWH and DMN techniques for efficient data exploration. Well-known software tools, designed for advanced statistical analysis, use the data produced in the DWH to perform DMN. In addition, recent developments in OLAP tools whose initial purpose was for multi-dimensional reporting have included such algorithms. A critical question in the design of a DMN system is how to integrate or couple the DMN system with a DWH system. There are several approaches such as:

i   No coupling: by which the DMN system will not utilise any function of a DWH system other than obtaining data from a particular source, processing the data and then storing the mining results in another file.
ii Loose coupling: by which the DMN system will use some facilities provided by the DWH system, fetching data from a data repository managed by these systems, performing data mining and storing the mining results either in a file or in a designated place in a database or data warehouse.
iii Semi-tight coupling: which is linking a DMN system to the DWH system; efficient implementations of a few essential data mining primitives such as sorting, indexing, aggregation, histogram analysis, and pre-computation of some essential statistical measures can be provided in the DW system.
iv Tight coupling: by which the DMN system is smoothly integrated into the DWH system since it is treated as one functional component of an information system. Data mining queries and functions are optimised based on mining query analysis, data structures, indexing schemes, and query processing methods of the DWH system.

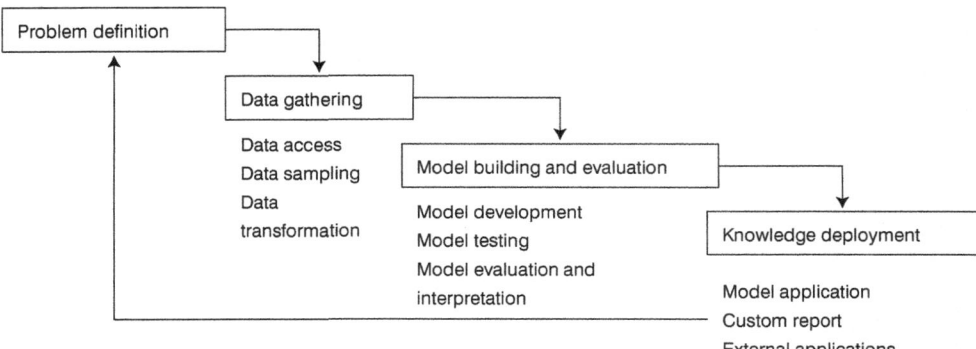

Figure 13.7 The data mining process. Based on Statsoft (2002) and Anonymous (undated)

Data mining and statistics

There is an overlap between data mining and statistics because most of the techniques used in data mining can be placed within a statistical framework (Anonymous, undated). Despite the overlapping between these analytical approaches, DMN techniques are not the same as traditional statistical techniques in which the user interaction is greatly required for the validation of the correctness of a model (Anonymous, undated). As a result, statistical methods can be difficult to automate. Moreover, statistical methods typically do not scale well to very large data sets. Statistical methods rely on testing hypotheses or finding correlations based on smaller, representative samples of a larger population. Data mining methods are suitable for large data sets and can be more readily automated. In fact, data mining algorithms often require large data sets for the creation of quality models.

Data mining and OLAP

The relationship between DMN and OLAP can be captured by the following statement 'the capability of OLAP to provide multiple and dynamic views of summarised data in a data warehouse sets a solid foundation for successful data mining' (Han and Kamber, 2001). Therefore, OLAP and data mining although different are complementary activities. The former supports activities such as data summarisation, cost allocation, time series analysis and what-if analysis. However, most OLAP systems do not have inductive inference (known as computational learning) capabilities beyond the support for time-series forecasting, whereas data mining systems do. OLAP systems provide a multi-dimensional view of the data, including full support for hierarchies. This view of the data is a natural way to analyse businesses and organisations. Data mining, on the other hand, usually does not have a concept of dimensions and hierarchies. Data mining and OLAP can be integrated in a number of ways. For example, data mining can be used to select the dimensions for a cube, create new values for a dimension, or create new measures for a cube. OLAP can be used to analyse data mining results at different levels of granularity.

Data mining and data warehousing

Data can be mined independently of their storage format (e.g. flat files, spreadsheets, database tables etc.) since the most important criterion for the data is not the storage

Measurements

Stage	COD (g/litre)	VFA (g/litre)	pH
Acetogenesis	8.820	224.60	24.85
Day 1	2.200	56.10	6.20
Day 2	2.220	56.00	6.25
Day 3	2.210	56.20	6.20
Day 4	2.190	56.30	6.20
Acidogenesis	16.482	240.30	19.60
Hydrolysis	24.158	343.18	16.39
Methanogenesis	7.640	179.00	28.00
Total	57.100	987.08	88.84

Figure 13.5 Pivot table where "Stage" and "Day" are the dimensions and COD, VFA, PH are attributes of facts

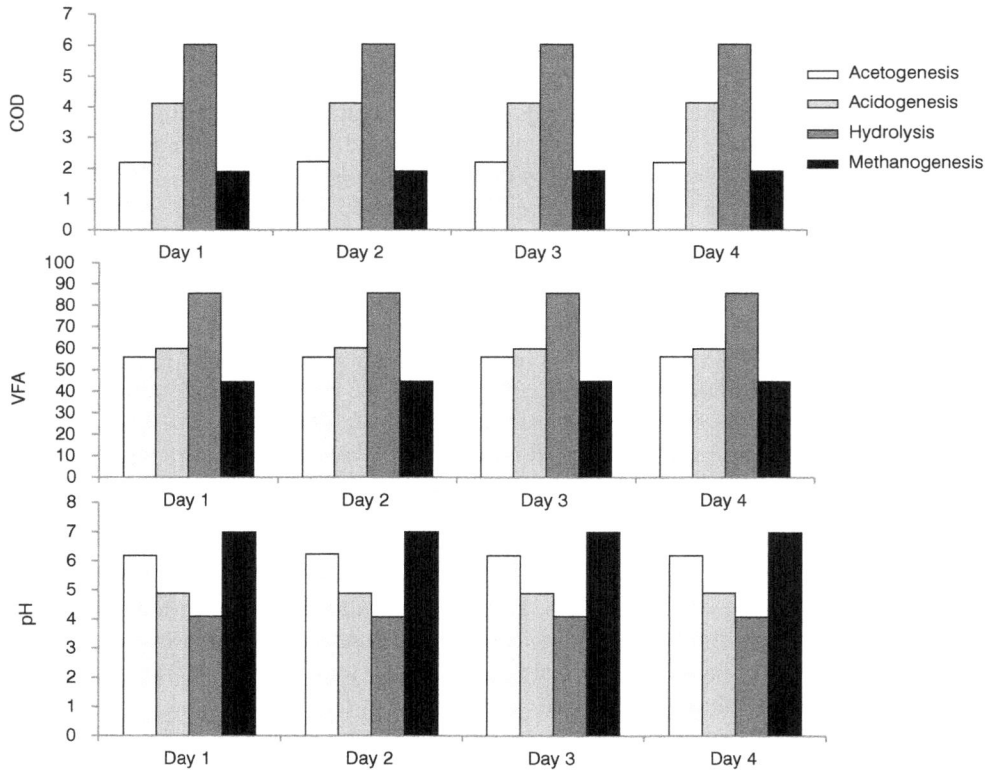

Figure 13.6 Performance of the selected attributes for each stage of AD based on the pivot table mentioned above

from the most summarised (up) to the most detailed (down). A pivot makes cross-tabulation, by rotating the cube to provide another representation of the data (Figure 13.5), while the slice-and-dice method makes range selection (slicing) on one or more dimension, like partitions. The dice operation is performed into two or more executive slices, i.e. slicing on two or more dimensions.

The graphical representation of the data presented in the pivot table (Figure 13.5) can assist operators to visualise operational conditions and/or determine problems during the AD phase (Figure 13.6).

Data mining

In many industries, it is commonly accepted that only a fraction of available information is incorporated into decision-making processes or when potentially crucial links between previously unconnected pieces of information are developed (Berthold et al., 2008). Additionally, quantitative analyses of strategies for utilising biomass energy sources have failed to propose a systematic approach to define the actual availability of energy from biomass (Ayoub et al., 2006). The DMN technique(s), in the KDD process, as mentioned earlier in this chapter, is an analytic process designed to explore large amounts of data for the discovery of patterns and systematic relationships between variables along with the validation of the findings through the application of the detected patterns to new subsets of data (Berry and Linoff, 2000; Statsoft, 2002; Giudici, 2003). DMN techniques extract information from large databases by the combination of various methods (e.g. statistical and artificial intelligence) with database management systems (Inmon, 1996a, 1996b; Witten and Eibe, 2005; Han and Kamber, 2006). According to Grover et al. (2001) DMN encompasses problems in diverse disciplines such as ecology, natural resources, atmospheric science, materials science and biological and environment engineering. The process of DMN techniques consists of the following stages: i) problem definition in terms of understanding project objectives and requirements, ii) data gathering, iii) model building and evaluation, and iv) knowledge deployment (Statsoft, 2002; Anonymous, undated) (Figure 13.7).

Data mining in the sphere of data analytical processes

There are many approaches for quantitative or qualitative data analysis based on the nature of data or variables (inputs) under investigation. Quantitative data analytics are generally divided into exploratory data analysis (i.e. discovery of new features in the data) and confirmatory data analysis (i.e. testing hypotheses) whereas qualitative data analysis is mainly used to draw conclusions from qualitative data through various approaches (e.g. soft system methodology, non-parametric tests etc.). The term data analytics, in information technology, has a special meaning in the context of IT audits, when the controls for an organisation's information systems, operations and processes are examined. Data analysis is used to determine whether the systems in place effectively protect data, operate efficiently and succeed in accomplishing an organisation's overall goals. In the following sections statistics, OLAP and data warehouse systems are discussed shortly in relation to data mining.

General principles of data warehouse and data mining in anaerobic digestion 237

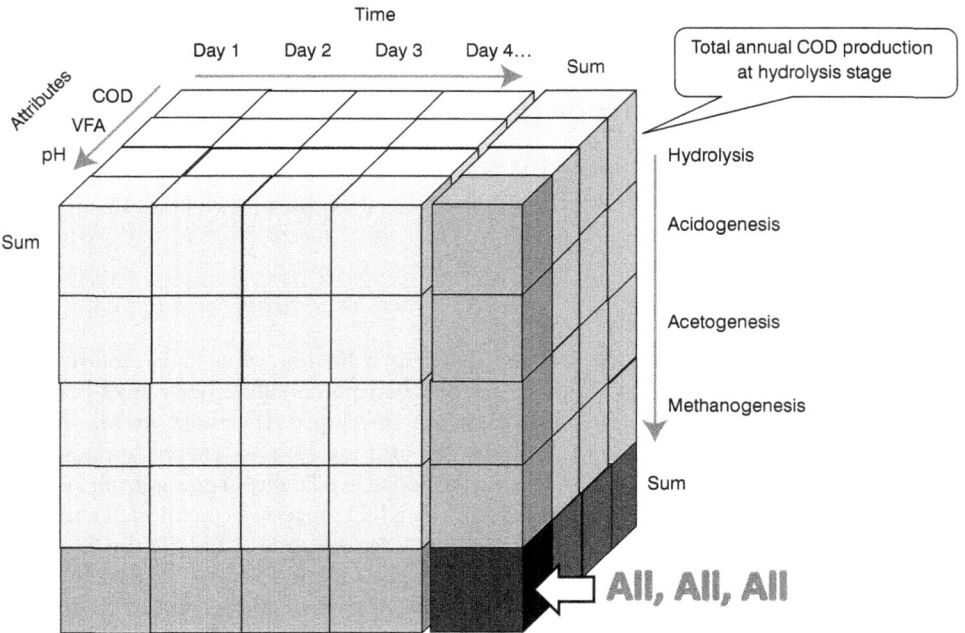

Figure 13.4 Representation of a data cube for AD

determine the measure. Thus, the multi-dimensional data defines a measure as a value in the multi-dimensional space of dimensions. Each dimension is described by a set of attributes. For example, a data cube can be defined for COD, volatile fatty acids (VFA) and pH concentration for each of the AD stages through the time dimension. The conceptual case study analysed in this chapter is based on pH, VFA and COD since these parameters indicate, as mentioned in the previous chapter, any disturbance in process performance and concentration of substrate organic compounds at each stage of digestion, i.e. hydrolysis, acidogenesis, acetogenesis and methanogenesis (Nizami et al., 2009). Increase in VFA and COD is accompanied by a reduction in pH as the substrate goes through hydrolysis and acidogenesis, the early phases of digestion (Nizami et al., 2009). In addition, high VFA concentration (> 1000–1500 mg/l) during late phases of AD, acetogenesis and methanogenesis, inhibits the process (Amon et al., 2007). According to Siegert and Banks (2005), pH changes when the VFA exceeds 4000 mg/l. The pH values vary with the digestion activities of each phase. For example, pH for stable hydrolysis and acidogenesis phases is between 4.0 and 6.5 and around 7.0 during the methanogenesis phase (Kim et al., 2003). The accumulation of COD (~ 10,000 mg/l) is significant in the first stage for hydrolysis and acidogenesis, as this will convert into methane in the second stage of methanogenesis (Nizami et al., 2009). Thus, COD measurement is significant for all of the digestion phases.

Cubes allow navigation through the data and viewing it from different perspectives. The particular representation supports roll-up, drill-down/up, slice-and-dice and pivot OLAP queries (Hand, 1998; Jackson, 2002) by allying the data content with conceptual models. The roll-up option can display data that increases in aggregation level, by computing all of the data relationships for one or more dimensions, mostly through defining a computational relationship. The drill-down/up displays details using queries for dimension table hierarchy, the surrounding dimensions contain context about the measurements. All hierarchies of a dimension are handled within a single table, which improves the performance of the DWH. Disadvantages associated with star data model include difficulty in maintaining the integrity of dimensions and fact table(s), as the ETL process is sometimes very complex for the star schema. If business requirements have changed then it is hard to change the underlining schemata.

In contrast to the star schema, the snowflake schema data is modelled following the degree of normalisation. The normalised structure divides data into tables. According to the degree of normalisation, tables are grouped together by subject areas that reflect general data categories and create several tables in a relational database (Rob et al., 2008).

The snowflake is more complex than the star schema because the tables which describe the dimensions are normalised. Normalisation of the dimension tables can impair the performance of a DWH. Whereas conventional databases can be tuned to match the regular pattern of usage, such patterns rarely exist in a DWH. Snowflaking will increase the time taken to perform a query, and the design goals of many DWH projects are to minimise these response times.

Mathematically, the object-oriented multi-dimensional model is denoted by $F\ (D_1, D_2, D_3, ..., D_n)$ and is made of a fact name and a list of dimensions. Each dimension $D_i\ (A_1, A_2, A_3, ..., A_n)$ is made up of dimension name D and a list of category attributes. The most common dimension that is used in most of the DWH cases is the time dimension where attributes can be months, days, hours, etc. A dimension example in AD could be the process dimension where attributes could be hydrolysis, acidogenesis, acetogenesis and methanogenesis. Dimensional data is descriptive data that allows structuring of fact data. Fact data is the plain bulk data that should be analysed by the DWH, e.g. the concentration of chemical oxygen demand (COD) in hydrolysis phase.

Each dimension can be organised into a hierarchy composed of numerous levels, each allowing to aggregate data at the desired level of abstraction. Hierarchies do not need to cascade and form strict patterns or trees. Each level in a dimension can have additional attributes that provide descriptive characteristics about the facts to narrow the search, filtering and classification of the facts data. These attributes also may be subject to change. These descriptive attributes and attributes describing the dimension hierarchy are called dimensional data.

Cubes

The use of a conceptual model that influences the front-end tools, database design and the query engines for OLAP is the multi-dimensional view of data in the warehouse known as data cubes, which are individually designed to solve a specific problem (Chaudhuri and Dayal, 1997; Vasiliev, 2011). The cells of a data cube contain data measures (i.e. facts), whereas the edges of a data cube represent the data dimensions (Figure 13.4). In a multi-dimensional data model, there is a set of numeric measures that are the objects of analysis (Chaudhuri and Dayal, 1997). Examples of such measures could be methane and/or CO_2 production or even return on investment. Each of the numeric measures depends on a set of dimensions, which provide the context for the measure (Chaudhuri and Dayal, 1997). The dimensions, for example, which are associated with methane production, can be the process parameters as described in the previous chapter (Chapter 12), along with the time dimension in which the measurements were taken. Altogether the dimensions are assumed to uniquely

a Conventional reporting: Two-dimensional reports like those in a simple Excel spreadsheet.
b Multi-dimensional reporting: Reports with more than two dimensions. Data can be viewed in multiple perspectives with pre-calculated and aggregated values.
c Dashboards: Visual displays of the most important information needed to achieve one or more objectives; consolidated and arranged on a single screen so the information can be monitored at a glance (Few, 2004).

The multi-dimensional data modelling concept

As stated above, DWHs are built upon common database systems and add support to import data from different data sources. They can structure and aggregate data, and pre-compute queries, to quickly meet different stakeholders' requests for analysing data. Data in the DWH is modelled multi-dimensionally with reference to star and snowflake schemata (Kimball and Caserta, 2004).

The star schema model is the simplest data warehouse schema, based on the de-normalisation concept (i.e. data is not necessarily spread in multiple tables on a relational mode). It is called a star schema because the data model resembles a star, with points radiating from a centre. The centre of the star consists of one or more fact tables and the points of the star are the dimension tables (Figure 13.3).

Some of the advantages of star schema data model are that the DWH is easier for an analyst to understand, use and maintain. The retrieval of data from a DWH that implements the star schema tends to operate quickly. Dimensional structures are easy to understand for business users, because the structure is divided into measurements/facts and context/dimensions. Facts are related to the organisation's business processes and operational system, whereas

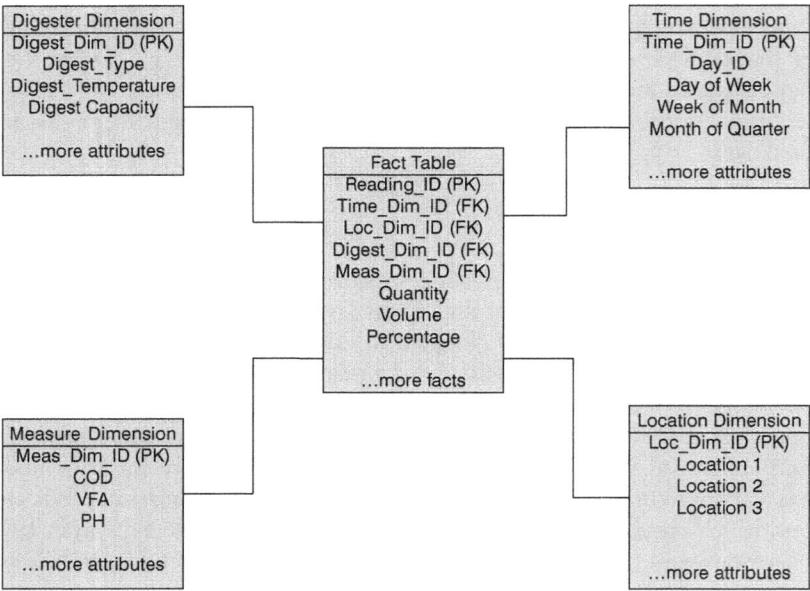

Figure 13.3 De-normalized multi-dimensional star schema for AD performance data management

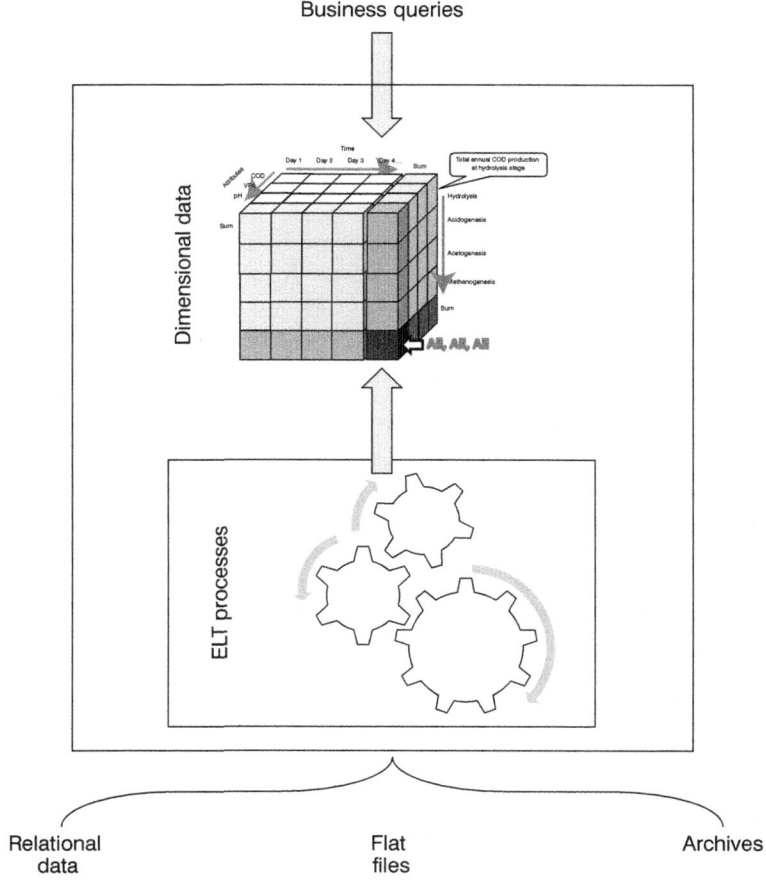

Figure 13.2 Gathering data from various sources and transforming them into useful information available for decision-making procedures

Different data sources like external tables from other databases, text files or programming interfaces (e.g. web interfaces) are usually supported by common DWH implementations to gather the required data from a broad range of data sources.

All source data is cleansed (i.e. unimportant parts are removed), transformed (i.e. data is restructured as needed) and consolidated (i.e. related data from different sources is combined), before loaded into the DWH.

Data cleansing is associated with the removal of irrelevant data or correction of corrupted data. For example, sensor data are initially loaded into the DWH from source files, regardless of data quality, as one of the main functions of this DWH is to maintain historical records. Data cleansing is performed inside the DWH for analysis purposes in compliance with business analysis requirements. Reporting through DWH is then available via multiple methods. Common methods of available reporting are:

Building an appropriate DWH model to support monitoring and analysis of AD processes is the basis for providing the decision makers integrated data to record history, monitor trends, measure variability and correlate all these figures through data mining techniques.

General principles and benefits of DWH
ETL and OLAP

A DWH is a user-free and transaction-free database system, in the sense that the stored data is not directly updated from daily transactions. It uses the operational systems, such as data-reading software equipped with sensors or an enterprise resource planning system, as sources for pulling data in a regular basis. The data is extracted from the source systems, transformed into the desired form and loaded into the DWH. This process is called an ETL (extract, transform and load) process (Kimball and Caserta, 2004). Then the data is available for analysis through online analytical process (OLAP) tools. OLAP tools enable users to view and analyse data in multi-dimensional perspectives. In contrast with OLTP systems, DWH and OLAP tools are mainly designed for queries and analyses and separate analysis workload from transactions workload (Figure 13.2).

DWH technology enables the multi-dimensional analysis of bioenergy production performance data to the finest level of detail, by using a separate database as a central repository, through which AD measurements and variability criteria can be methodologically structured, stored and analysed. DWH technology is able to provide a vast amount of historical, integrated, detailed and summarised data with emphasis on the context of information rather than the content. Timely reporting of aggregated and cleansed information can be automatically provided by online analytical processing materialised views, answering ad-hoc queries at operational levels, in support of strategic development decision making. Some of the benefits of the implementation of DWH are as follows: i) DWHs provide a common data model of interest regardless of the data source or its format. ii) DWHs keep historical records and are historically consistent to achieve better understanding of the business processes. iii) DWHs provide high query performance without interrupting the OLTP systems. DWHs often use de-normalised (Cummings, 2011) schemata (such as a star schema) to optimise query performance. iv) DWHs facilitate decision support applications such as trends and reports. DWHs are built to accommodate ad hoc queries. Typically, a DWH is optimised to perform well for a wide variety of possible query operations. v) A typical data warehouse query searches through thousands or millions of rows in a table or many tables. vi) DWHs are built separate from the OLTP systems. A DWH is updated on a regular basis through ETL operation using bulk data modification techniques (Cummings, 2011). The end users of the DWH do not directly update the DWH, so common data entry mistakes can be avoided. If such faulty entries exist in the OLTP systems, they can be tracked and corrected in the transform step of the ETL process. vii) Knowledge extraction through data mining on a single and integrated data source is greatly improved.

Data processing

The basic sequence of data processing in DWHs is shown in Figure 13.2. Existing data needs to be extracted from sources, transformed (Inmon, 2005) and loaded into the DWH.

of DWH through ETL (extract, transform and load) processes of the initial raw data along with data pre-processing where erroneous data are discarded and heterogeneous data may be integrated into a pre-processed database; iii) data transformation, where data are transformed or consolidated into forms appropriate for different mining algorithms; iv) data mining, where intelligent methods are applied for patterns recognition and discovery of useful features that represent the data; and finally v) knowledge acquisition through evaluation, interpretation and representation of the mined knowledge by the user.

However, as far as the authors are concerned, applications of DWH and DMN into renewable energy and particularly to biogas production and AD related issues are clearly lacking. If DWH and DMN techniques are used properly they can significantly assist in the optimisation of biogas production. This can be achieved by the acquisition of knowledge which can be easily incorporated into decision making processes throughout the whole biogas production chain. The scope of this chapter is to explain the general principles and concepts for both DWH and DMN techniques, aiming to ignite the interest of stakeholders to investigate the integration of these techniques into existing biogas production processes.

Data warehouse

The need for various monitoring and control systems in AD, together with the need of other operational applications that store data concerning the daily business transactions in the whole production process, form a very complex process in terms of information system requirements. Each system uses different techniques and multiple methods of storing data. For example, monitoring systems with sensors could archive measurements such as temperature and pH in a simple text file, while a transactional application would use a relational database to store transactions and inventory figures like quantities and cost of feedstock available. A variety of heterogeneous database systems with different structures, data types, etc., provides multiple sources of information, but the goal of exploiting this information in an integrated way in order to make intelligent decisions is hard to achieve. Thus, an integrated DWH system will be required in order to consolidate the appropriate data and support an easy-to-use decision support system (DSS). The design and the goals of a DWH are different from those on a transactional system which is often called an online transaction processing (OLTP) system. Some of the main differences are represented in Table 13.1.

Table 13.1 Differences between DWH and OLTP.

	OLTP system	Data warehouse
Source of data	Operational data	Consolidation data. Data comes from OLTP databases and other sources
Purpose of data	Designed for real-time business operations	Designed for planning, problem solving and decision support
Data	Snapshots of ongoing business processes	Multi-dimensional views of various kinds of business activities
Inserts and updates	Short and fast inserts and updates initiated by end users	Periodic long-running batch jobs refresh the data

Based on www.datawarehouse4u.info

Chapter 13

General principles of data warehouse and data mining in anaerobic digestion

Nicholas E. Korres,[1]* Anastasios Dekazos,* Dimitris N. Argyropoulos,[2] Ammar Ahmed[3] and Paul Stack[4]

[1]26 Grigoroviou str, Patisia, GR–11141, Athens, Greece; [2]Genetic Identification Laboratory, Hellenic Agricultural Organization "DEMETER", Sofokli Venizelou 1, 14123, Lykovrissi, Greece; [3]CCC Information Services Inc., 222 Merchandise Mart, Suite 900, Chicago, IL 60654, USA and DeVry University, 3300 N Campbell, Chicago, IL 60618, USA; [4]Department of Civil & Environmental Engineering, University College Cork, Cork, Ireland.
*Corresponding authors email: nkorres@yahoo.co.uk; tasos_dekazos@yahoo.com

Introduction

The importance of monitoring the parameters affecting the outcome of the biogas production was discussed in the previous chapter. Their contribution to the variability in the production process hence the uncertainty in biogas production was also explained. This uncertainty is mainly due to the large number of parameters affecting biogas production. Additionally the vast amount of historical/archival and/or real-time data collected throughout the whole anaerobic digestion (AD) production chain needs to be processed and applied in the decision making process for optimal biogas production efficiency. Advanced technologies have enabled the collection of large amounts of data which present the potential to discover useful information and knowledge that could not be accessed before (Ye, 2003). The data reflects the behaviour of the analysed system; therefore there is at least the theoretical potential to obtain useful information and knowledge from these data by the application of the knowledge discovery in databases (KDD) (Han and Kamber, 2001). It consists of various stages including data warehouse (DWH) and data mining (DMN) techniques and refers to the overall process of discovering knowledge from data (Shapiro-Piatetsky et al., 1996; Velickov and Solomatine, 2000; Abonyi and Feil, 2007) (Figure 13.1).

The basic stages of KDD as stated by Shapiro-Piatetsky et al. (1996) and Abonyi and Feil (2007) are: i) the collection of data and creation of various databases; ii) the development

Figure 13.1 Data warehouse and data mining in the knowledge discovery process. Based on Velickov and Solomatine (2000); Abonyi and Feil (2007)

Weilard, P. (2000) 'Anaerobic waste digestion in Germany—status and developments, *Biodegradation*, vol. 11, no 6, pp. 415–421.

Woolford, M.K. (1984) *The silage fermentation*, New York, Base:l Marcel Dekker.

Xie, S., Wu, G., Lawlor, P.G., Frost, J.P., and Zhan, X. (2012) 'Methane production from anaerobic co-digestion of the separated solid fraction of pig manure with dried grass silage', *Bioresource Technology*, vol 104, pp. 289–297.

Yadvika, Santosh, Sreekrishnan, T.R., Kohli, S. and Rana, V. (2004) 'Enhancement of biogas production from solid substrates using different techniques—a review', *Bioresource Technology*, vol. 95, no 1, pp. 1–10.

Yen, H-W. and Brune, D. (2007) 'Anaerobic co-digestion of algael sludge and waste paper to produce methane', *Bioresource Technology*, vol. 98, pp. 130–134.

Yu, H.W., Samani, Z., Hanson, A. and Smith, G. (2002) 'Energy recovery from grass using two-phase anaerobic digestion', *Waste Management*, vol. 22, no 1, pp. 1–5.

Zaher, U., Bouvier, J.C., Steyer, J.P. and Vanrolleghem, P.A. (undated) 'Titrimetric monitoring of anaerobic digestion: VFA, alkalinities and more'.

Zeng, M., Mosier, N.S., Huang, C.P., Sherman, D.M. and Ladisch, M.R. (2007) 'Microscopic examination of changes of plant cell structure in corn stover due to hot water pretreatment and enzymatic hydrolysis', *Biotechnology Bioengineering*, vol. 97, 265–278.

Zennaki, B.Z., Zadi, A., Lamini, H., Aubinear, M., and Boulif, M. (1996) 'Methane fermentation of cattle manure: effects of HRT, temperature & substrate concentration', *Tropicultural*, vol. 14, no 4, pp. 134–140.

Zhang, R., El-Mashad, H.M., Hartman, K., Wang, F., Liu, G, Choate, C. and Gamble, P. (2007) 'Characterization of food waste as feedstock for anaerobic digestion', *Bioresource Technology*, vol. 98, pp. 929–935.

Zhang, Y.H.P. (2008) 'Reviving the carbohydrate economy via multi-product lignocellulose biorefineries', *Journal of Industrial Microbiology and Biotechnology*, vol. 35, pp. 367–375.

Zimbardi, F., Viggiano, D., Nanna, F., Demichele, M., Cuna, D. and Cardinale, G. (1999) 'Steam explosion of straw in batch and continuous systems' *Applied Biochemistry and Biotechnology*, vol. 77–79, pp. 117–125.

Zupancic, G.D and Jemec, A. (2010) 'Anaerobic digestion of tannery waste: semi-continuous and anaerobic sequencing batch reactor processes', *Bioresource Technology*, vol. 101, no 1, pp. 26–33.

Zupancic, G.D. and Grilc, V. (2011) 'Anaerobic treatment and biogas production from organic waste', in S. Kumar and A. Bharti (eds) *Management of organic waste*, Intechweb.org, www.intechopen.com [accessed June 2012].

Tchobanoglous, G., Burton, F.L., and Stensel, H.D. (2003) *Wastewater engineering, treatment and reuse*, 4th edn, New York: Metcalf & Eddy and McGraw-Hill.

Tirumale, S. and Nand, K. (1994) 'Influence of anaerobic cellulolytic bacterial consortia in the anaerobic digesters on biogas production', *Biogas Forum III*, vol. 58, pp. 12–15.

Us, E. and Perendeci, N.A. (2012) 'Improvement of methane production from greenhouse residues: Optimization of thermal and H_2SO_4 pretreatment process by experimental design', *Chemical Engineering Journal*, vol. 181–182, pp. 120–131.

Uzodinma, E.O. and Ofoefule, A.U. (2009) 'Biogas production from blends of field grass (*Panicum maximum*) with some animal wastes', *International Journal of Physical Sciences*, vol. 4, no 2, pp. 91–95.

Van der Berg, L., and Kennedy, K.J. (1983) 'Comparison of advanced anaerobic reactors', in *Proceedings of III International Conference on Anaerobic digestion*, August, Boston.

Van Dorland, H.A., Wettstein, R., Leuenberger, H. and Kreuzer, M. (2006) 'Comparison of fresh and ensiled white and red clover added to ryegrass on energy and protein utilization of lactating cows', *Animal Sciences*, vol. 82, pp. 691–700.

van Lier, J.B., Tilche, A., Ahring, B.K., Macarie, H., Moletta, R., Dohanyos, M., Hulshoff Pol, L.W., Lens, P. and Verstraete, W. (2001) 'New perspectives in anaerobic digestion', *Water Science and Technology*, vol. 43, no 1, pp. 1–18.

Vandamme, E.J. (2009) 'Agro-industrial residue utilization for industrial biotechnology products', in P. Singh nee' Nigam and A. Pandey (eds) (2009) *Biotechnology for agro-industrial residues utilisation*, pp. 3–11. New York: Springer Science and Business Media.

Vandevivere, P. (1999) 'New and broad applications of anaerobic digestion', *Critical Review Environmental Science Technology*, vol. 29, pp. 151–173.

Vandevivere, P., De Baere, L. and Verstraete, W. (2003) 'Types of anaerobic digester for solid wastes' in J. Mata-Alvarez (ed.) *Biomethanization of the organic fraction of municipal solid wastes*, pp. 112–140. London: IWA Press

Varel, V.H., Hashimoto, A.G. and Chen, Y.R. (1980) 'Effect of temperature and retention time on methane production from cattle waste', *Applied Environmental Microbiology*, vol. 40, no 2, pp. 217–222.

Vartak, D.R., Angler, C.R., Ricke, S.C., and McFarland, M.J. (1997) 'Organic loading rate and bio-augmentation effects in psychrophilic anaerobic digestion of dairy manure', in *Proceedings of ASAE Annual International Meeting*, Minneapolis, Minnesota, USA, August 10–14, American Society of Agricultural Engineers.

Vasilian, A. and Trchunian, A. (2008) 'Effect of the medium redox potential on the growth and metabolism of anaerobic bacteria', *Biofizika*, vol. 53, no 2, pp. 281–293.

Vera, M.A., Nickel, K. and Neis, U. (2004) 'Disintegration of sewage sludge for better anaerobic digestion', In *Proceedings of Anaerobic Digestion 10th World Congress*, vol. 4, pp 2127–2128, Montreal, Canada, August.

Verma, S. (2002) 'Anaerobic digestion of biodegradable organics in municipal solid wastes', MSc Dissertation, Department of Earth & Environmental Engineering (Henry Krumb School of Mines), Fu Foundation School of Engineering & Applied Science, Columbia University, May 2002.

Verrier, D., Roy, F. and Albagnac, G. (1987) 'Two-phase methanization of solid vegetable waste', *Biological Waste*, vol. 22, pp. 163–177.

Wang, L., Zhou, Q. and Li, F.T. (2006) 'Avoiding propionic acid accumulation in the anaerobic process for biohydrogen production', *Biomass and Bioenergy*, vol. 30, pp. 177–182.

Ward, A.J., Hobbs, P.J., Holliman, P.J., and Jones, D.L. (2008) 'Optimization of the anaerobic digestion of agricultural resources', *Bioresource Technology*, vol. 99, no 17, pp. 7928–7940.

Wastesum (2006). 'Management and valorisation of solid domestic waste for the small urban communities in Morocco', LIFE-3rd Countries, 2007–2009, European Commission.

Webb, A.R. and Hawkest, F.R. (1985) 'The anaerobic digestion of poultry manure: variation of gas yield with influent concentration and ammonium-nitrogen levels', *Agricultural Wastes*, vol. 14, pp. 135–156.

Weiland, P. (2010) 'Biogas production: current state and perspectives', *Applied Microbiology and Biotechnology*, vol 85, pp. 849–860.

Sharma, V.K., Canditelli, M., Fortuna, F. and Cornacchia, G. (1997) 'Processing of urban and agro-industrial residues by aerobic composting: Review', *Energy Conversion and Management*, vol. 38, no 5, pp. 453–478.

Shuiwen, S., Premier, G.C., Guwy, A. and Dinsdale, R. (2007) 'Bifurcation and stability analysis of an anaerobic digestion model', *Nonlinear Dynamics*, vol. 48, no 4, pp 391–408.

Siegert, I. and Banks, C. (2005) 'The effect of volatile fatty acid additions on the anaerobic digestion of cellulose and glucose in batch reactors', *Process Biochemistry*, vol 40, no 11, pp 3412–3418.

Silverstein, R.A., Chen, Y., Sharma-Shivappa, R.R., Boyette, M.D. and Osborne, J.A. (2007) 'Comparison of chemical pretreatment methods for improving saccharification of cotton stalks', *Bioresource Technology*, vol. 98, pp. 3000–3011.

Smith, K.F., Culvenor, R.A., Humphreys, M.O. and Simpson, R.J. (2002) 'Growth and carbon partitioning in perennial ryegrass (*Lolium perenne*) cultivars selected for high water-soluble carbohydrate concentrations', *Journal Agricultural Science*, vol. 138, no 4, pp. 375–385.

Smith, P., Martino, D., Cai, Z., Gwary, D., Janzen, H. et al. (2008) 'Greenhouse gas mitigation in agriculture', *Philosophical Transactions of the Royal Society*, vol. B 363, pp. 789–813.

Smyth, B.M., Murphy, J.D. and O'Brien, C.M. (2009) 'What is the energy balance of grass biomethane in Ireland and other temperate northern European climates?' *Renewable and Sustainable Energy Reviews*, vol. 13, no 9, pp. 2349–2360.

Song, Y.C., Kwon, S.J. and Woo, J.H. (2004) 'Mesophilic and thermophilic temperature co-phase anaerobic digestion compared with single-stage mesophilic- and thermophilic digestion of sewage sludge', *Water Environmental Research*, vol. 38, no 7, pp. 1653–1662.

Spanjers, H. and van Lier, J.B. (2006) 'Instrumentation in anaerobic treatment-research and practice', *Water Science & Technology*, vol. 53, no 4–5, pp. 63–76.

Stamatelatou, K., Antonopoulou, G., Ntaikou, I. and Lyberatos, G. (2012) 'The effect of physical, chemical and biological pretreatments of biomass on its anaerobic digestibility and biogas production', in A. Mudhoo (ed.) *Biogas production: pretreatment methods in anaerobic digestion*, pp. 56–90. Cambridge, MA: John Wiley & Sons and Scrivener Publishing.

Steffen, R., Szolar, O. and Braun, R. (1998) 'Feedstocks for anaerobic digestion'. Institute for Agrobiotechnology, Tulln, & University of Agricultural Sciences, Vienna. Austria, www.adnett.org/dl_feedstocks.pdf, accessed August 2007.

Steyer, J.P. (2002) 'Evaluation of a four years experience with a fully instrumented anaerobic digestion process', *Water Science Technology*, vol. 45, no 4–5, pp 495–502.

Stinner, W., Möller, K., and Leithold, G. (2008) 'Effects of biogas digestion of clover/grass-leys, cover crops and crop residues on nitrogen cycle and crop yield in organic stockless farming systems', *European Journal of Agronomy*, vol. 29, no 2–3, pp. 125–134.

Sun, X.F., Xu, F., Sun, R.C., Fowler, P. and Bairdd, M.S. (2005) 'Characteristics of degraded cellulose obtained from steam-exploded wheat straw', *Carbohydrate Research*, vol. 340, pp. 97–106.

Sun, Y. and Cheng, J. (2002) 'Hydrolysis of lignocellulosic materials for ethanol production. A review', *Bioresource Technology*, vol. 83, pp. 1–11.

Sung, S. and T. Liu. (2003) 'Ammonia inhibition on thermophilic methanogens', *Chemosphere*, vol. 53, no 1, pp. 43–52.

Svoboda, I.F. (2003) 'Anaerobic digestion, storage, oligolysis, lime, heat and aerobic treatment of livestock manures', Provision of research and design of pilot schemes to minimise livestock pollution to the water environment in Scotland, QLC 9/2, FEC Services 2003.

Taherzadeh, M.J. and Karimi, K. (2008) 'Pretreatment of lignocellulosic wastes to improve ethanol and biogas production: A review', *International Journal of Molecular Sciences*, vol. 9, pp. 1621–1651.

Takashima, M., Shimada, K. and Speece, R.E. (2011) 'Minimum requirements for trace metals (iron, nickel, cobalt, and zinc) in thermophilic and mesophilic methane fermentation from glucose', *Water Environment and Research*, vol. 83, pp. 339–346.

Taniguchi, M., Suzuki, H., Watanabe, D., Sakai, K., Hoshino, K. and Tanaka, T. (2005) 'Evaluation of pretreatment with Pleurotus ostreatus for enzymatic hydrolysis of rice straw', *Journal of Bioscience and Bioengineering*, vol. 100, pp. 637–643.

Qi, B.C., Aldrich, C., Lorenzen, L. and Wolfaardt, G.W. (2005) 'Acidogenic fermentation of lignocellulosic substrate with activated sludge', *Chemical Engineering Communication*, vol. 192, no 7–9, pp. 1221–1242.

Rajeshwari, K.V., Balakrishnan, M., Kansal, A., Kusum Lata, and Kishore, V.V.N. (2000) 'State-of-the-art of anaerobic digestion technology for industrial wastewater treatment', *Renewable and Sustainable Energy Reviews*, vol. 4, pp. 135–156.

Rao, V.P., Baral, S.S., Dey, R. and Mutnuri, S. (2010) 'Biogas generation potential by anaerobic digestion for sustainable energy development in India' *Renewable and Sustainable Energy Reviews*, vol. 14, no 7, pp. 2086–2094.

Raven, R.P.J.M. and Gregersen, K.H. (2005) 'Biogas plants in Denmark: successes and setbacks', *Renewable and Sustainable Energy Reviews*, vol. 11, no 1, pp. 116–132.

Reith, J.H., Wijffels, R.H., and Barten, H. (2003) *Bio-methane & Bio-hydrogen*, Petten: Dutch Biological Hydrogen Foundation.

Riano, B., Molinuevo, B. and García-González, M.C. (2011) 'Potential for methane production from anaerobic co-digestion of swine manure with winery wastewater', *Bioresource Technology*, vol. 102, no 5, pp. 4131–4136.

Ridla, M., and Vehida, S. (1998) 'Effects of combined treatment of lactic acid bacteria and cell wall degrading enzymes on fermentation and composition of Rhodes grass (*Chloris gayana* Kunth)', *Asian-Australasian Journal of Animal Sciences*, vol. 11, pp. 522–529.

Ripley, L.E., Boyle, J.C. and Converse, J.C. (1986) 'Improved alkalimetric monitoring for anaerobic digestion of high strength wastes', *Journal of the Water Pollution Control Federation*, vol. 58, no. 5, pp. 406–411.

RISE-AT (Regional Information Service Centre for South East Asia on Appropriate Technology) (1998) *Review of Current Status of Anaerobic Digestion Technology for Treatment of Municipal Solid Waste*, Institute of Science and Technology Research and Development, Chiang Mai University, http://www.ist.cmu.ac.th/riseat/documents/adreview.pdf [accessed August June 2012].

Romano, R.T. and Zhang, R. (2008) 'Co-digestion of onion juice and wastewater sludge using an anaerobic mixed biofilm reactor', *Bioresource Technology*, vol. 99, no 3, pp. 631–637.

Sagagi, B S., Garba, B. and Usman, N.S. (2009) 'Studies on biogas production from fruits and vegetable waste', *Bayero Journal of Pure and Applied Sciences*, vol 2, no 1, pp. 115–118.

Sajko, H., Bedmarski, W., Prontos, J. and Mlynarezy, K. (1997) 'Quality and nutritive value of alfalfa and grass silage with biological addition' in *Proceedings of the XVIII International Grassland Congress*, Winnipeg. Canada, ID Number 848, Session 17, Forage Quality, pp. 17/49–17/50.

Salminen, E. and Rintala, J. (2002) 'Anaerobic digestion of organic solid poultry slaughterhouse waste – a review', *Bioresource Technology*, vol 83, pp. 13–26.

Sanchez, C. (2009) 'Lignocellulosic residues; biodegradation and bioconversion by fungi', *Biotechnology Advances*, vol. 27, no.2, pp. 185–194.

Sanchez, E.P., Weiland, P., and Travieso, L. (1992) 'Effect of hydraulic retention time on the anaerobic biofilm reactor efficiency applied to screened cattle waste treatment', *Biotechnology Letters*, vol. 14, no 7, pp. 635–638.

Santosh, Y., Sreekrishnan, T.R., Kohli, S. and Rana, V. (2004) 'Enhancement of biogas production from solid substrates using different techniques—a review', *Bioresource Technology*, vol. 95, no 1, pp. 1–10.

Sarath, G., Mitchel, R.B., Satler, S.E., Funnell, D., Pedersen, J.F., Graybosch, R.A. and Vogel KP (2008) 'Opportunities and roadblocks in utilizing oranges and small grains for liquid fuels', *Journal of Industrial Microbiology and Biotechnology*, vol.35, pp. 343–354.

Schnurer, A. and Jarvis, A. (2010) 'Microbiological handbook for biogas plants', Swedish Waste Management U2009:03, Swedish Gas Centre Report 207.

Shahriari, H., Warith, M., Hamoda, M. and Kennedy, K.J. (2012) 'Anaerobic digestion of organic fraction of municipal solid waste combining two pretreatment modalities, high temperature microwave and hydrogen peroxide', *Waste Management*, vol 32, pp. 41–52.

Sharma, S.K., Mishra, I.M., Sharma, M.P., and Saini, J.S. (1988) 'Effect of particle size on biogas generation from biomass residues', *Biomass*, vol. 17, pp. 251–263.

Nordberg, A. (1999) 'EU and National Legislation of Relevance to Anaerobic Digestion'.

Nordheim-Viken, H., and Volden, H. (2009) 'Effects of maturity stage, nitrogen fertilization and seasonal variation on ruminal degradation characteristics of neutral detergent fibre in timothy (*Phleum pratense*)', *Animal Feed Science & Technology*, vol. 149, pp. 30–59.

Nozhevnikova, A.N., Lotsyurbenko, O.R. and Parshina, S.N. (1999) 'Anaerobic manure treatment under extreme temperature conditions', *Water Science Technology*, vol. 40, no 1, pp. 215–221.

Nuri, A., Keskin, T. and Yuruyen, A. (2008) 'Enhancement of biogas production from olive mill effluent (OME) by co-digestion', *Biomass and Bioenergy*, vol. 32, no 12, pp. 1195–1201.

O'Kiely, P., Moloney, A. and O'Riordan, E.G. (2002) 'Reducing the cost of beef production by increasing silage intake', in *Beef production*, Series No.51, Project No. 4622, ARMIS No. 4622, Grange Research Centre, December 2002.

Ortenblad, H. (2000) 'Anaerobic digestion: Making energy and solving modern waste problems'. AD-Nett report.

Owen, W.E., Stuckey, D.C., Healy, J.B., Young, L.Y. and McCarty, P.L. (1979) 'Bioassay for monitoring biochemical methane potential and anaerobic toxicity', *Water Research*, vol. 13, pp. 485–492.

Pakarinen, O., Lehtomaki, A., Rissanen, S. and Rintala, J. (2008) 'Storing energy crops for methane production: Effects of solids content and biological additive', *Bioresource Technology*, 99, 7074–7082.

Palatsi, J., Viñas, M., Guivernau, M., Fernandez, B. and Flotats, X. (2011) 'Anaerobic digestion of slaughterhouse waste: Main process limitations and microbial community interactions', *Bioresource Technology*, vol. 102, pp. 2219–2227.

Palmowski, L. and Müller, J. (1999) 'Influence of the size reduction of organic waste on their anaerobic digestion', in J. Mata-Alvarez, F. Cecchi, and A. Tilche (eds), in *Proceedings 2nd International Symposium on Anaerobic Digestion of Solid Waste*, pp. 137–144. London: IWA Publishing.

Parawira, W. (2004) 'Anaerobic treatment of agricultural residues and wastewater. Application of high-rate reactors', PhD Thesis, Department of Biotechnology, Lund University, Sweden.

Parawira, W., Murto, M., Zvauya, R. and Mattiasson, B. (2006) 'Comparison of the performance of a UASB reactor and an anaerobic packed-bed reactor when treating potato waste leachate', *Renewable Energy*, vol. 31, pp. 893–903.

Parawira, W., Read, J.S., Mattiasson, B. and Bjornsson, L. (2008) 'Energy production from agricultural residues: High methane yields in pilot-scale two-stage anaerobic digestion', *Biomass and Bioenergy*, vol. 32, no 1, pp. 44–50.

Pavlostathis, S.G. and Gossett, J.M. (1985) 'Alkaline treatment of wheat straw for increasing anaerobic biodegradability', *Biotechnology and Bioengineering*, vol. 27, pp. 334–344.

Peltola, R.J., Laine, V.H., Koutola, H. and Kymalainen, M.A.L. (2004) 'Impact grinding as a pretreatment method for biowaste and sludge', In *Proceedings of Anaerobic Digestion 10th World Congress*, vol. 4, pp. 2129–2132, Montreal, Canada.

Petersson, A., Thomsen, M.H., Hauggaard-Nielsen, H. and Thomsen, A.B. (2007) 'Potential bioethanol and biogas production using lignocellulosic biomass from winter rye, oilseed rape and fababean', *Biomass Bioenergy*, vol. 31, no 11–12, pp. 812–819.

Peyraud, J.L., Astigarraga, L. and Faverdin, P. (1997) 'Digestion of fresh perennial ryegrass fertilized at two level of nitrogen by lactating cows', *Animals Feed Science Technology*, vol. 64, pp. 155–171.

Prasad, S., Singh, A. and Joshi, H.C. (2007) 'Ethanol as an alternative fuel from agricultural, industrial and urban residues', *Resources, Conservation and Recycling*, vol. 50, pp. 1–39.

Prochnow, A., Heiermann, M., Drenckhan, A. and Schelle, H. (2005) 'Seasonal pattern of biomethanisation of grass from landscape management', *Agricultural Engineering International The CIGR E Journal*, vol. 7, Manuscript EE 05 011.

Provenzano, M.R., Iannuzzi, G., Fabbri, C., and Senesi, N. (2011) 'Qualitative characterization and differentiation of digestates from different biowastes using FTIR and fluorescence spectroscopies', *Journal of Environmental Protection*, vol. 2, pp. 83–89.

Pullammanappallil, P.C., Svoronos, S.A., Chynoweth, D.P. and Lyberatos, G. (1998) 'Expert system for control of anaerobic digesters', *Biotechnology and Bioengineering*, vol. 58, no 1, pp 13–22.

Malik, R.K., Singh, R. and Tauro, P. (1987) 'Effect of inorganic supplementation on biogas production' *Biological Wastes*, vol. 21, no 2, pp. 139–142.

Marchaim, U. (1992). 'Biogas Processes for Sustainable Development'. *FAO Agricultural Services Bulletin 95*. Food and Agricultural Organization of the United Nations.

Martínez-Sibaja, A., Posada-Gomez, R., Alvarado-Lassman, A. and Sebastia-Cortes, A. (2007) 'Cascade fuzzy logic controller for an anaerobic digester', in *Proceedings of the Fourth Congress of Electronics, Robotics and Automotive Mechanics*, CERMA, 25–28 September 2007, IEEE Xplore.

Mata-Alvarez, J. (1987) 'A dynamic simulation of a two-phase anaerobic digestion system for solid wastes', *Biotechnology and Bioengineering*, vol. 30, pp. 844–851.

Mata-Alvarez, J., Macé, S. and Llabres, P. (2000) 'Anaerobic digestion of organic solid wastes. An overview of research achievements and perspectives', *Bioresource Technology*, vol. 74, pp 3–16.

Mattocks, R., Kintzer, B. and Wilkie, A. (2008) *AgSTAR handbook: A manual for developing biogas systems at commercial farms in the united states*, http://epa.gov/agstar/resources/handbook.html [accessed May 2012].

McEniry, J., O'Kiely, P., Clipson, N.J.W., Forristal, P.D., and Doyle, E.M. (2006) 'The microbiological and chemical composition of baled and precision-chop silages on a sample of farms in County Meath', *Irish Journal of Agricultural and Food Research*, vol. 45, pp. 73–83.

Meher, K.K., Murthy, M.V.S. and Gollakota, K.G. (1994) 'Psychrophilic anaerobic digestion of human waste', *Bioresource Technology*, vol. 50, no 2, pp. 103–106.

Mishima, D., Tateda, M., Ike, M. and Fujita, M. (2006) 'Comparative study on chemical pretreatments to accelerate enzymatic hydrolysis of aquatic macrophyte biomass used in water purification processes', *Bioresource Technology*, vol. 97, pp. 2166–2172.

Moghtaderi, B., Sheng, C., and Wall, T.F. (2006) 'An overview of the Australian biomass resources and utilization technologies', *BioResources*, vol. 1, no 1, pp. 93–115.

Moller, H.B., Sommer, S.G. and Ahring, B.K. (2004) 'Methane productivity of manure, straw and solid fractions of manure', *Biomass and Bioenergy*, vol. 26, pp. 485–495.

Molnar, L. and Bartha, I. (1988) 'High solids anaerobic fermentation for biogas and compost production', *Biomass*, vol 16, no 3, pp. 173–182.

Morris, D. (2006) 'The next economy: From dead to living carbon', *Journal of the Science of Food and Agriculture*, vol.86, pp. 1743–1746.

Murphy, J.D., and Power, N.M. (2009) 'An argument for using biomethane generated from grass as a biofuel in Ireland', *Biomass Bioenergy*, vol. 33, pp. 504–512.

Murphy, J., Braun, R., Weiland, P. and Wellinger, A. (2011) 'Biogas from crop digestion', IEA Bioenergy, Task 37, Energy from Biogas, September 2011.

Neves, L., Oliveira, R. and Alves, M.M. (2006a) 'Anaerobic co-digestion of coffee waste and sewage sludge', *Waste Management*, vol. 26, no 2, pp 176–181.

Neves, L., Ribeiro, R., Oliveira, R. and Alves, M.M. (2006b) 'Enhancement of methane production from barley waste', *Biomass and Bioenergy*, vol. 30, pp. 599–603.

Nges, I.A. and Liu, J. (2010) 'Effects of solid retention time on anaerobic digestion of dewatered-sewage sludge in mesophilic and thermophilic conditions', *Renewable Energy*, vol. 35, pp. 2200–2206.

Nizami, A.S. and Murphy, J. (2010) 'What type of digester configurations should be employed to produce biomethane from grass silage?', *Renewable and Sustainable Energy Reviews* vol. 14, pp. 1558–1568.

Nizami, A.S. and Murphy, J.D. (2011) 'Optimizing the operation of a two-phase anaerobic digestion system digesting grass silage', *Environmental Science & Technology*, vol. 45, no 17, pp 7561–7569.

Nizami, A.S., Korres, N.E, and Murphy, J.D. (2009) 'Review of the integrated process for the production of grass biomethane', *Environmental Science and Technology*, vol. 43, no 22, pp. 8496–8508.

Nizami, A.S., Singh, A. and Murphy, J.D. (2011) 'Design, commissioning, and start-up of a sequentially fed leach bed reactor complete with an upflow anaerobic sludge blanket digesting grass silage', *Energy and Fuels*, vol. 25, no 2, pp 823–834.

Norberg, A. (2004) Uppsala presentation to the California Delegation on the Swedish biogas tour. JTI, June 11.

Krizsan, S.J. and Randby, A.T. (2007) 'The effect of fermentation quality on the voluntary intake of grass silage by growing cattle fed silage and the sole feed', *Journal Animal Science*, vol 85, pp. 984–996.

Kung, L., Carmean, B.R. and Tung, R.S. (1990) 'Microbial inoculation or cellulase enzyme treatment of barley and vetch silage harvested at three maturities', *Journal of Dairy Science*, vol. 73, pp. 1304–1311.

Kurakake, M., Ide, N. and Komaki, T. (2007) 'Biological pretreatment with two bacterial strains for enzymatic hydrolysis of office paper', *Current Microbiology*, vol. 54, no 6, pp. 424–428.

Laura, R.D. and Idnani, M.A. (1971) 'Increased production of biogas from cowdung by adding other agricultural waste materials', *Journal of the Science of Food and Agriculture*, vol. 2, 164–167.

Lee, M.R.F., Evan, P.R., Nute, G.R., Richardson, R.I. and Scollan, N.D. (2009) 'A comparison between red clover silage and grass silage feeding on fatty acid composition, meat stability and sensory quality of the M. *longissimus* muscle of dairy cull cows', *Meat Science*, vol. 81, pp. 738–744.

Lehtomaki, A. (2006) 'Biogas production from energy crops and crop residues', PhD dissertation, Jyvaskyla studies in biological and environmental science. University of Jyvaskyla, Jyvaskyla.

Lehtomaki, A., Huttunen, S., Lehtinen, T.M. and Rintala, J.A. (2008) 'Anaerobic digestion of grass silage in batch leach bed processes for methane production', *Bioresource Technology*, vol. 99, no 8, pp 3267–3278.

Li, D., Yuan, Z., and Sun, Y. (2010) 'Semi-dry mesophilic anaerobic digestion of water sorted organic fraction of municipal solid waste (WS-OFMSW)', *Bioresource Technology*, vol 101, no 8, pp 2722–2728.

Li, J., Kumar Jha, A., He, J., Ban, Q., Chang, S. and Wang, P. (2011) 'Assessment of the effects of dry anaerobic codigestion of cow dung with waste water sludge on biogas yield and biodegradability', *International Journal of the Physical Sciences*, vol 6, no 15, pp. 3679–3688.

Linder, G. (2009a) 'Failure Analysis and Smart Grid Control Protocols for Anaerobic Digesters', MS Thesis, Dept. Elect. Eng, Clarkson University, Potsdam, New York.

Linder, G. (2009b) 'The importance of standard SCADA protocols to the reliable operation of distributed farm-scale anaerobic digesters', Presented at IEEE PSCE 2009 Seattle, Washington.

Liqian, W. (2011) 'Different pretreatments to enhance biogas production. A comparison of thermal, chemical and ultrasonic methods', MSc Thesis, Applied and Environmental Sciences, Halmstad University.

Lissens, G., Vandevivere, P., de Baere, L., Biey, E.M. and Verstraete, W. (2001) 'Solid waste digesters: process performance and practice for municipal solid waste digestion', *Water Science and Technology*, vol. 44, pp. 91–102.

Liu, C.F., Yuan, X.Z., Zeng, G.M., Li, W.W. and Li, G. (2008). 'Prediction of methane yield at optimum pH for anaerobic digestion of organic fraction of municipal solid waste', *Bioresource Technology*, vol. 99, no 4, pp. 882–888.

Liu, H.W., Walter, H.K., Vogt, G.M., and Holbein, B.E. (2002) 'Steam pressure disruption of municipal solid waste enhances anaerobic digestion kinetics and biogas yield', *Biotechnology and Bioengineering*, vol. 77, pp. 121–130.

Liu, J.G. and Olsson, B.M. (2004) 'Control of an anerobic reactor towards maximum biogas production', *Water Science Technology*, vol. 50, no 11, pp. 189–198.

Lusk, P., Wheeler, P. and Rivard, C. (1996) 'Deploying anaerobic digesters: Current status and future possibilities', National Renewable Energy Laboratory, U.S. Department of Energy, Prepared under Task No. WM513231, January 1996.

Madlener, R., Antunes, C.H. and Dias, L.C. (2009) 'Assessing the performance of biogas plants with multi-criteria and data envelopment analysis', *European Journal of Operational Research*, vol. 197, no3, pp. 1084–1094.

Mahnert, P., Heiermann, M. and Linke, B. (2005) 'Batch and Semi-continuous biogas production from different grass species', *Agricultural Engineering International. The CIGRE Journal*, manuscript EE 05010, vol. VII.

Mais, U., Esteghlalian, A.R., Saddler, J.N. and Mansfield, S.D. (2002) 'Enhancing the enzymatic hydrolysis of cellulosic materials using simultaneous ball milling', *Applied Biochemistry and Biotechnology*, vol 98, pp. 815–832.

Hamzawi, N., Kennedy, K.J. and McLean, D.D. (1998) 'Anaerobic digestion of co-mingled municipal solid waste and sewage sludge', *Water Science and Technology*, vol. 38, no 2, pp. 127–132.

Hashimoto, A.G., Varel, V. and Chen, Y.R. (1981) 'Ultimate methane yield from beef cattle manure: effect of temperature, ration constituents, antibiotics and manure age', *Journal of Agricultural Wastes*, vol. 3, no 4, pp. 241–256.

Hee, R.B. (1995) 'Knowing the basics of PLCs', *EC & M Magazine*, Penton Business Media, http://www.ecmweb.com/mag/electric_knowing_basics_plcs/index.html [accessed May 2012]

Herren, H.R., Bassi, A.M., Tan, Z. and Binns, W.P. (2012) 'Green jobs for a revitalized food and agriculture sector', Natural Resources Management and Environment Department, Food and Agriculture Organization of the United Nations.

Hill, D.T. and Jenkins, S.R. (1989) 'Measuring alkalinity accurately in aqueous systems containing high organic acid concentrations', *Transactions of the American Society of Agricultural Engineers*, vol. 32, pp. 2175–2178.

Holm-Nielsen, J.B., Al Seadi, T. and Oleskowicz-Popiel, P. (2009) 'The future of anaerobic digestion and biogas utilization' *Bioresource Technology*, vol 100, no 22, pp. 5478–5484.

Idler, C., Heckel, M., Herrmann, C. and Heiermann, M. (2007) 'Influence of biological additives in grass silages on the biogas yield', *Research Papers in IAg Eng & LU Ag.*, vol. 39, no. 4, pp. 69–82.

Ileleji, K.E., Martin, C. and Jones, D. (2008) 'Basics of energy production through anaerobic digestion of livestock manure', in *Bioenergy. Fuelling America through renewable resources*, IB-406-W, Purdue extension, Purdue University.

Jain, S.R., Mattiasson, B. (1998) 'Acclimatization of methanogenic consortia for low pH biomethanation process', *Biotechnology Letters*, vol. 20, no 8, pp. 771–775.

Jha, A.K., Li, J., Nies, L. and Zhang, L. (2011) 'Research advances in dry anaerobic digestion process of solid organic wastes', *African Journal of Biotechnology*, vol 10, no 65, pp. 14242–14253.

Karagiannidis, A. and Perkoulidis, G. (2009) 'A multi-criteria ranking of different technologies for the anaerobic digestion for energy recovery of the organic fraction of municipal solid wastes' *Bioresource Technology*, vol 100, pp. 2355–2360.

Karellas, S., Boukis, I. and Kontopoulos, G. (2010) 'Development of an investment decision tool for biogas production from agricultural waste', *Renewable and Sustainable Energy Reviews*, vol 14, pp. 1273–1282.

Kayhanian, M. (1994) 'Performance of a high-solids anaerobic digestion process under various ammonia concentrations', *Journal of Chemical Technology and Biotechnology*, vol. 59, pp. 349–352.

Kayhanian, M. (1999) 'Ammonia inhibition in high-solids biogasification: an overview and practical solutions', *Environmental Technology*, vol. 20, pp. 355–365.

Keady, T.W.J., Mayne, C.S. and Fitzpatrick, D.A. (2000) 'Prediction of silage feeding value from the analysis of the herbage at ensiling and effects of nitrogen fertilizer, date of harvest and additive treatment on grass silage composition', *Journal of Agricultural Science*, vol. 134, pp. 353–368.

Kelleher, M. (2007) 'Anaerobic digestion outlook for MSW streams', *Biocycle Energy*, August, pp. 51–55.

Kim, J., Park, C., Kim, T.H., Lee, M., Kim, S., Kim, S.W. and Lee, J. (2003) 'Effects of various pretreatments for enhanced anaerobic digestion with waste activated sludge', *Journal of Bioscience and Bioengineering*, vol. 95, no 3, pp. 271–275.

Korres, N. E., Singh, A., Nizami, A. S. and Murphy, J. D. (2010) 'Is grass biomethane a sustainable transport biofuel?', *Biofuels, Bioproducts and Biorefining*, 4, 310–325.

Korres, N.E., Thamsiriroj, T., Smyth, B., Nizami, A.S., Singh, A. and Murphy, J.D. (2011) 'Grass biomethane for agriculture and energy' in E. Lichtfouse (ed.) 'Genetics, Biofuels and Local Farming Systems', *Sustainable Agriculture Reviews* vol. 7, pp. 5–50, Springer Science & Business Media.

Krich, K., Augenstein, D., Batmale, J., Benemann, J., Rutledge, B. and Salour, D. (2005) 'Biomethane from dairy waste – a sourcebook for the production and use of renewable natural gas in California', A study funded through the Value-Added Agricultural Product Market Development Grant Program, Administered by USDA Rural Development, July 2005, http://www.suscon.org/cowpower/biomethaneSourcebook/biomethanesourcebook.php [Accessed February 2012].

Demirbas, A. and Demirbas, H.A. (2004), 'Estimating the calorific values of lignocellulosic fuels', *Journal Energy, Exploration & Exploitation*, vol. 20, no 1, pp. 105–111.

Desai, M., and Madamwar, D. (1994) 'Anaerobic digestion of a mixture of cheese whey, poultry waste and cattle dung: a study of the use of adsorbents to improve digester performance', *Environmental Pollution*, vol. 86, no 3, pp. 337–340.

Desai, M., Patel, V. and Madamwar, D. (1994) 'Effect of temperature and retention time on biomethanation of cheese whey poultry waste–cattle dung', *Environmental Pollution*, vol. 83, no 3, pp. 311–315.

Deublein, D. and Steinhauser, A. (2008) *Biogas from waste and renewable resources*, Weinheim: Wiley-VCH.

EC (2006) 'Biofuels in the European Union—a vision for 2030 and beyond', Final report of the Biofuels Research Advisory Council, EUR 22066, Directorate-General for Research Sustainable Energy Systems.

Elaiyaraju, P. and Partha, N. (2012) 'Biogas production from co-digestion of orange peel waste and jatropha de-oiled cake in an anaerobic batch reactor', *African Journal of Biotechnology*, vol 11, no 14, pp. 3339–3345.

Erol, M., Haykiri-Acma, H., and Kuçukbayrak, S. (2010) 'Calorific value estimation of biomass from their proximate analyses data', *Renewable Energy*, vol 35, no 1, pp. 170–173.

EU Waste Catalogue. (2001) *EU Waste Catalogue and Hazardous Waste List*.

Eze, J.I. and Ojike, O. (2012) 'Anaerobic production of biogas from maize wastes', *International Journal of the Physical Sciences*, vol 7, no 6, pp. 982–987.

Fan, L.T., Gharpury, M.M. and Lee, Y.H. (1981) 'Evaluation of pre-treatments for enzymatic conversion of agricultural residues', *Biotechnology and Bioengineering Symposium*, vol 11, pp. 29–45.

Ferrer, I., Campas, E., Palatsi, J., Porras, S., and Flotats, X. (2004) 'Effect of a thermal pretreatment and the temperature range on the anaerobic digestion of water hyacinth (*Eichornia crassipes*)', In *Proceedings of Anaerobic Digestion 10th World Congress*, vol.4, pp 2107–2109, Montreal, Canada.

FNR (2010) 'Biogas- Messprogramm II', 61 Biogasanlagen im Vergleich, Fachagentur Nachwachsende Rohstoffe e. V., Gulzow, Germany.

Fountoulakis, M.S. and Manios, T. (2009) 'Enhanced methane and hydrogen production from municipal solid waste and agro-industrial by-products co-digested with crude glycerol', *Bioresource Technology*, vol 100, no 12, pp. 3043–3047.

Franco, A., Mosquera-Corral, A., Campos, J.L. and Roca, E. (2007a) 'Learning to operate anaerobic bioreactors', in A. Méndez-Vilas (ed.) *Communicating current research and educational topics and trends in applied microbiology*, vol. 2, Formatex 2007, pp. 618–627, http://www.formatex.org/microbio/index.htm [accessed August 2012].

Franco, A., Roca, E. and Lema, J.M. (2007b) 'Enhanced start-up of upflow anaerobic filters by pulsation', *Journal of Environmental Engineering*, vol. 133, no. 2, pp. 186–190.

Fricke, K., Santen, H., Wallmann, R., Huttner, A., and Dichtl, N. (2007) 'Operating problems in anaerobic digestion plants from nitrogen in MSW', *Waste Management*, vol. 27, pp. 30–43.

Fry, L.J. and Merill (1973) 'Methane digesters for fuel gas and fertilizer', Newsletter No.3, Santa Cruz, CA: New Alchemy Institute.

Fulford, D. (1988). *Running of biogas program handbook*, London: Intermediate Technology Publications.

Garba, B. (1996) 'Effect of temperature and retention period on biogas production from ligrocellulosic material', *Renewable Energy*, vol. 9 no 1–4, pp. 938–941.

Gebrezgabher, S.A., Meuwissen, M.P.M., Oude Lansink, A.G.J.M. and Prins, B.A.M. (2009) 'Economic analysis of anaerobic digestion—A case of green power biogas plant in the Netherlands', *17th International Farm Management Congress*, Bloomington/Normal, IL, pp. 231–244.

Gerardi, M.H. (2003) *The microbiology of anaerobic digesters*, Hoboken, NJ: John Wiley & Sons.

Ghaly A. (1996) 'A comparative study of anaerobic digestion of acid cheese whey and dairy manure in a two-stage reactor', *Bioresource Technology*, 58, 61–72.

Gunaseelan, V. and Nallathambi, S. (1997) 'Anaerobic digestion of biomass for methane production: a review', *Biomass Bioenergy*, vol. 13, pp. 83–114.

Buekens, A. (2005) 'Energy recovery from residual waste by means of anaerobic digestion technologies', in *Proceedings of the future of residual waste management in Europe*, 17–18 November, Luxembourg.

Burke, D.A. (2001) 'Dairy waste anaerobic digestion handbook. Options for recovering beneficial products from dairy manure', Environmental Energy Company, www.makingenergy.com [Accessed July 2012].

Buxton, D.R. and O'Kiely, P. (2005) 'Preharvest plant factors affecting ensiling. Silage science and technology', *Agronomy Monograph* vol 42, pp. 199–250.

Callander, I.J. and Barford, J.P. (1983) 'Recent advances in anaerobic digestion technology', *Process Biochemistry*, vol. 18, pp. 24–30.

Cecchi, F., Pavan, P., Musacco, A., Mata-Alvarez, J. and Vallini, G. (1993) 'Digesting the organic fraction of municipal solid waste: Moving from mesophilic (37°C) to thermophilic (55°C) conditions', *Waste Management Research*, vol. 11, no 5, pp. 403–414.

Cecchi, F., Bolzonella, D., Pavan, P., Macé, S., and Mata-Alvarez, J. (2011) 'Anaerobic digestion of the organic fraction of municipal solid waste for methane production: research and industrial application', *Comprehensive Biotechnology* (2nd edn), vol. 6, pp. 463–472.

Chae, K.J., Yim, S.K., Choi, K.H., Park, W.K. and Lim, D.K. (2008) 'Anaerobic digestion of swine manure: Sung-Hwan farm-scale biogas plant in Korea', http://www.bvsde.paho.org/bvsacd/unam7/swine.pdf [accessed July 2012].

Chaudhary, B.K. (2008) 'Dry continuous anaerobic digestion of municipal solid waste in thermophilic conditions', MEng thesis in Environmental Engineering and Management, Asian Institute of Technology, School of Environment, Resources and Development, Thailand.

Chen, Y., Cheng, J.J., and Creamer, K.S. (2008) 'Inhibition of anaerobic digestion process: A review', *Bioresource Technology*, vol. 99, pp. 4044–4064.

Cho, J.K. Park, S.C. and Chang, H.N. (1995) 'Biochemical methane potential and solid state anaerobic digestion of Korean food wastes', *Bioresource Technology*, vol 52, no 3, pp 245–253.

Chynoweth, D.P., Turick, C.E., Owens, J.M., Jerger, D.E. and Peck, M.W. (1993) 'Biochemical methane potential of biomass and waste feedstocks', *Biomass and Bioenergy*, vol. 5, pp. 95–111.

Ciborowski, P. (2001) 'Anaerobic digestion of livestock manure: a feasibility assessment', Report prepared in completion of USEPA Grant CX 825639-01-0, Minnesota Pollution Control Agency, March 2001.

Clarke, G. and Deon, R. (2004) *Practical modern SCADA protocols: DNP3, IEC 60870.5 and related systems*. Oxford: Newnes.

Clavero, T. and Razz, R. (2002) 'Effects of biological additives on silage composition of mott dwarf elephantgrass and animal performance', *Revista Científica FCV-LUZ*, vol.12, no. 4, pp. 313–316.

Clemens, J., Trimborn, M., Weiland, P., and Amon, B. (2006) 'Mitigation of greenhouse gas emissions by anaerobic digestion of cattle slurry', *Agriculture, Ecosystems & Environment*, vol 112, no 2–3, pp. 171–177.

Colmenarejo, M.F., Sanchez, E., Bustos, A., Garcia, G. and Borja, R. (2004) 'A pilot-scale study of total volatile fatty acids production by anaerobic fermentation of sewage in fixed-bed and suspended biomass reactors', *Process Biochemistry*, vol. 39, no 10, pp. 1257–1267.

Dawson, L.E.R., Ferris, C.P., Steen, R.W.J., Gordon, F.J. and Kilpatrick, D.J. (1999) 'The effects of wilting grass before ensiling on silage intake', *Grass and Forage Science*, vol.54, pp. 237–247.

de Baere, L.A., Devocht, M., van Assche, P., and Verstraete, W. (1984) 'Influence of high NaCl and NH_4Cl salt levels on methanogenic associations', *Water Research*, vol. 18, pp. 543–548.

Dearman, B. and Bentham, R.H. (2007) 'Anaerobic digestion of food waste: comparing leachate exchange rates in sequential batch systems digesting food waste and biosolids', *Waste Management*, vol 27, pp. 1792–1799.

Delgenes, J.P., Penaud, V., and Moletta, R. (2003) 'Pre-treatment for the enhancement of anaerobic digestion of solid wastes', in Mata-Alvarez, J. (ed.) *Biomethanization of the organic fraction of municipal solid waste*, pp. 201–228. London: IWA Publishing.

Demirbas, A. (2004) 'Combustion characteristics of different biomass fuels', *Progress in Energy and Combustion Science*, vol 30, no 2, pp. 219–230.

Anonymous (2009) 'User manual for the automatic monitoring, management and early-warning system', in *European Biogas Initiative to improve the yield of agricultural biogas plants*, Project no. 513949, EU-Agro-Biogas.

APHA (American Public Health Association) (1998) *Standard methods for the examination of water and wastewater*, , 20th edn, Washington DC: American Water Works Association and Water Environment Federation, .

Appels, L., Baeyens, J., Jan Degreve, J. and Dewil, R. (2008) 'Principles and potential of the anaerobic digestion of waste-activated sludge', *Progress in Energy and Combustion Science*, vol. 34, pp. 755–781.

Arsova, L. (2010) 'Anaerobic digestion of food waste: Current status, problems and an alternative product', M.S. Thesis, Department of Earth and Environmental Engineering, Fu Foundation of Engineering and Applied Science, Columbia University, May 2010.

Arvanitoyannis, I.S. and Varzakas, T.H. (2008) 'Vegetable waste treatment: comparison and critical presentation of methodologies' *Critical Reviews in Food Science and Nutrition*, vol. 48, no 3, pp. 205–247.

Asia, I.O., Oladoja, N.A. and Bamuza-Pemu, E.E. (2006) 'Treatment of textile sludge using anaerobic technology', *African Journal of Biotechnology*, vol. 5 no 18, pp. 1678–1683.

Azaizeh, H. and Jadoun, J. (2010) 'Co-digestion of olive mill wastewater and swine manure using up-flow anaerobic sludge blanket reactor for biogas production', *Journal of Water Resource and Protection*, vol. 2, pp. 314–321.

Bach-Knudsen, K.E. (1997) 'Carbohydrate and lignin contents of plant materials used in animal feeding', *Animal Feed Science Technology*, vol. 67, pp. 319–338.

Bardiya, N. and Gaur, A.C. (1997) 'Effects of carbon and nitrogen ratio on rice straw biomethanation', *Journal of Rural Energy*, vol. 4, no 144, pp. 1–16.

Barnett, A. (1978) *Biogas technology in the third world: a multidisciplinary review*, Ottawa: IDRC.

Baserga, U. (1998) *Vergarung von Extensogras-Silage in einer Feststoff-Pilotanlage und einer landwirtschaftlichen Co-Vergarungs-Biogasanlage* [Anaerobic digestion of silage from extensive grassland in a solid state pilot plant and a farm-scale co-fermentation biogas plant].Tanikon: Eidg. Forschungsanstalt fur Agrarwirtschaft und Landtechnik, Forschungsprogramm Biomasse.

Batstone, D.J., Gernaey, K.V. and Steyer, J.-Ph. (2004) 'Instrumentation and control in anaerobic digestion', in M. van Loosdrecht and J. Clement (eds), *Water and Environmental Management Series*, 2nd IWA Leading Edge Conf. on Water and Wastewater Treatment Technologies, pp. 173–182. London: IWA Publishing.

Beno, Z., Boran, J., Houdkova, L., Dlabaja, T. and Sponar, J. (2009) 'Cofermentation of kitchen waste with sewage sludge', *Chemical Engineering Transactions*, vol. 18, pp. 677–682.

Bernet, N. and Beline, F. (2009) 'Challenges and innovations on biological treatment of livestock effluents', *Bioresource Technology*, vol. 100, pp. 5431–5436.

Bjornsson, L., Mattiasson, B. and Henrysson, T. (1997) 'Effects of support material on the pattern of volatile fatty acid accumulation at overload in anaerobic digestion of semi-solid waste', *Applied Microbiology and Biotechnology*, vol. 47, pp. 640–644.

Blanc, F.C. and Molof, A.H. (1973) 'Electrode potential monitoring and electrolytic control in anaerobic digestion', *Water Pollution Control Federation*, vol. 45, no 4, pp 655–667.

Braun, R. (1992) 'Verwertung und Entsorgung organischer Nebenprodukte und Reststoffe der Industrie', *Müll und Abfall H.* 12, 841–851.

Braun, R. (2007) 'Anaerobic digestion: a multi faceted process for energy, environmental management and rural development', in P. Ranalli (ed.) *Improvement of crop plants for industrial end users*, pp. 335–416. Amsterdam: Springer.

Braun, R. and Wellinger, A. (2003) 'Potential co-digestion' in *Energy from biogas and landfill gas*, IEA Bioenergy Report, Task 37.

Braun, R., Weiland, P. and Wellinger, A. (2009) 'Bioenergy from energy crop digestion', in *Energy from biogas and landfill gas*, IEA Bioenergy Report, Task 37.

Bruni, E., Jensen, A.P., Pedersen, E.S. and Angelidaki, I. (2010) 'Anaerobic digestion of maize focusing on variety, harvest time and pretreatment', *Applied Energy*, vol.87, pp. 2212–2217.

Conclusions

The rationale for monitoring biogas production is that biogas plants which operate without monitoring may be underperforming. As has been shown, variability which is sometimes inherited, as in the case of lignocellulosic feedstock, can easily act as a disturbing factor for the process of feedstock degradation resulting in lower biogas production. Furthermore, for economic reasons more plants are operating in a critical range where there is a risk of a digester failure. In the case of larger plants, the cost of online monitoring is only a small fraction of the total costs. Controlling the biogas production process can be difficult because different feedstocks are used hence they have different process requirements. There are also different responses with the same equipment but in combination with the application of data warehouse and mining techniques, as described in the next chapter, monitoring can be proved an invaluable tool for the biogas production industry.

References

Ahring, B.K., Sandberg, M. and Angelidaki, I. (1995) 'Volatile fatty acids as indicators of process imbalance in anaerobic digesters', *Applied Microbiology and Biotechnology*, vol. 43, pp. 559–565.

Akkaya, A.V. (2009) 'Proximate analysis based multiple regression models for higher heating value estimation of low rank coals', *Fuel Processing Technology*, vol. 90, no 2, pp. 165–394.

Al Seadi, T. (undated) 'Good practice in Quality Management of AD Residues from biogas production', Task 24 Energy from Biological Conversion of Organic Waste, IEA Bioenergy.

Al Seadi, T., Rutz, D., Prassl, H., Köttner, M., Finsterwalder, T., Volk, S. and Janssen, R. (2008) *Biogas handbook*, Esbjerg: University of Southern Denmark.

Alvarez-Ramirez, J.M. (2002) 'Feedback control design for an anaerobic digestion process', *Journal of Chemical Technology and Biotechnology*, vol. 77, pp. 725–734.

Amon, T., Kryvoruchko, V., Amon, B., Moitzi, G., Lyson, D., Hackl, E., Jeremic, D., Zollitsch, W. and Pötsch, E. (2003) 'Optimierung der Biogaserzeugung aus den Energiepflanzen Mais und Kleegras', Biogas production from the energy crops maize and clover grass', *Endbericht*, no 1249.

Amon, T., Amon, B., Krvvoruchko, V., Machmuller, A., Hopfner-Sixt, K., Bodiroza, V., Hrbek, R., Friedel, J., Potsch, E., Wagentristl, H., Schreiner, M. and Zollitsch, W. (2007) 'Methane production through anaerobic digestion of various energy crops grown in sustainable crop rotations', *Bioresource Technology*, vol. 98, no 17, pp. 3204–3212.

Angelidaki, I. and Ahring, B.K. (1994) 'Thermophilic anaerobic digestion of livestock waste: the effect of ammonia', *Applied Microbiology and Biotechnoly*, vol.38, no 4, pp 560–564.

Angelidaki, I. and Ellegaard, L. (2002) 'Anaerobic digestion in Denmark: past, present and future' in *Anaerobic digestion for sustainability in waste (water) treatment and reuse*. Proceedings of 7th FAO/SREN-Workshop, 19–22 May 2002, pp. 129–138.

Angelidaki, I., Hendriksen, H.V., Mathrani, I.M. and Sorensen, A.H. (1996). *The biogas process. Lecture notes for energy from biomass*, 1st edn, Lungby: Institute of Environmental Science and Engineering.

Angelidaki, I., Ellegaard, L. and Ahring, B. (2003) 'Applications of the anaerobic digestion process', in B. Ahring (eds), *Biomethanation II*, pp. 1–33. Heidelberg: Springer Berlin

Anonymous (2001) 'Working document: Biological treatment of biowaste', 2nd draft. EC/DG ENV.A. Brussels February 2001.

Anonymous (2006) *Biogas as a road transport fuel. An assessment of the potential role of biogas as a renewable transport fuel*, Brighton: National Society for Clean Air and Environmental Protection.

Anonymous (2008) 'Benchmarking Report on Critical Points and Influential Factors at Agricultural Biogas Plants', in *European Biogas Initiative to improve the yield of agricultural biogas plants*, EU-Agrobiogas, Project no. 513949, partner No2, IGER.

frequency, and the flow and quantity of biogas. The pH, temperature and gas flow of each digester were measured on a minute basis by sensors and meters. Averages of 60 minute values were generated automatically and stored in PLC memory as an hour reading. They found that the PLC and SCADA system was less helpful when dealing with process stability and up-scaling issues. However, after fixing the commissioning problems, the process control system makes digester operation much simpler.

Human–machine interface (HMI) panel

All of the system process, components and variables are displayed on a colour control panel or a touch screen control called a human–machine interface (HMI) panel. The HMI panel includes a menu structure as well as data display. The local system debugging and configuration of system parameters is made easy by the HMI panel (Linder, 2009b). There are many colours on the screen that fluctuate with changing digester parameters including temperature, sprinkling rates, and others. These facilities will help farmers to operate the system if it is upscaled for farm use (Nizami et al., 2009).

SCADA (supervisory control and data acquisition) program

The SCADA monitors and controls the entire system that is spread out through the digester by interacting with the PLC (Clarke and Deon, 2004). Thus, with the automatic control actions and feedback control loop of the PLC, the SCADA will monitor the overall digester performance (Figure 12.1). In leach bed digesters the PLC keeps the leach bed heated but with SCADA the temperature can be altered (Nizami et al., 2011). PLC passes the feedback control loop, which is monitored for overall performance by SCADA. The program is further transferred to the PLC to write the program into a removable chip such as electrically erasable programmable read-only memory (EEPROM) or erasable programmable read-only memory (EPROM) through a programming board (Hee, 1995). Under SCADA, Liu and Olsson (2004) presented a cascade control idea (proportional–proportional) for an anaerobic digester; cascade is a multi-loop feedback configuration control system (Alvarez-Ramirez, 2002) that bypasses the time-consuming task of manually turning controllers (Martínez-Sibaja et al., 2007). Among various digesters the high rate reactors have advantages including an increased net energy production and lower sludge production. They are also considered complex and sensitive systems (Shuiwen et al., 2007). In Mexico only 74 per cent of the installed volume of AD plants include UASB type reactors. Cascade is a multi-loop feedback configuration control system among those in use (Alvarez-Ramirez, 2002).

Software

Those variables that are not measureable are estimated on-line by software sensors. Software models the data and information of measurements provided by in-line methods (Spanjers and Van Lier, 2006). For example, the COD and alkalinity are estimated on the basis of the actual observation of the loading rate, the gaseous flow rate, VFA, gas composition, and other factors. The PLC and HMI are programmed using different application software such as Unitronics DataXport. The entry and editing of the ladder-style logic, debugging and troubleshooting the PLC software are carried out using a software application (Hee, 1995).

Table 12.9 Variables in AD and their measuring techniques.

Variables	Importance in anaerobic digestion	Measuring principles/ Instrumentation
Alkalinity	Indicate anaerobic activity and process performance	Titrimetric methods Non-titrimetric methods Software sensor
Biochemical oxygen demand (BOD)	Measure of organic concentration and biodegradability of substrate leachate	Biochemical fuel cell Non-biochemical techniques
Biomass	Knowing the anaerobic digester capacity	Total suspended solids (TSS) Capacitance measurement Software sensor
Chemical oxygen demand (COD)	Indicate concentration of substrate organic compounds	Titrimetric or spectrometry methods IR-spectrometry Software sensor
Flow of liquid and gas	Provide information on digester feeding rate and loading rate	Ultrasonic or electromagnetic principle Rotameters Thermal mass flow or gas flow meter
Hydrogen	Indicate imbalance among essential microbial groups	Gas chromatography Palladium sensor Hydrogen/air fuel cell detector Trace reduction gas analyser Membrane inlet mass spectroscopy Amperometric dissolved hydrogen probe Dissolved hydrogen monitor Pd metal oxide semiconductor sensor
Hydrogen sulphide (H_2S)	Indicate the accumulation of toxic compounds like hydrogen sulphate	Dissolved hydrogen analyser Chalcogenide glass sensor Visible light spectroscopy UV-spectrometry Gold film sensor
Methane (CH_4)	Indicate the anaerobic activity of the substrate	IR spectrometry Gas sensitive semiconductors Dissolved gas sensor
pH	Provide information on process performance	Electrochemical sensor Titrimetry VFA and alkalinity analyser
Total suspended solid (TSS)	Indicate process performance and provide information on reactor type and design	Spectrometric methods Submersible UV/VIS spectrometer Turbidity meter
Total organic carbon (TOC)	Knowing the digester performance	Automated analyser UV absorption Fourier transform IR-spectrometry
Toxicity	Indicate imbalance of particular biological process	Automated instruments pH-stat titration biosensor Toxicity meter
Volatile fatty acids (VFA)	Indicate any disturbance or imbalance of methanogens	Gas chromatography Titrimetry methods IR-spectrometry Membrane inlet mass spectrometry

Note: based on Spanjers and Van Lier, 2006; Steyer, 2002; Batstone et al., 2004

Instrumentation

Instruments send data or information to a data acquisition system or a computer called an on-line instrument. In off-line instrumentation the data or information are not readily available to a computer, but it still measures, stores and displays data. Such instruments send data naturally and are placed in the process stream (Spanjers and Van Lier, 2006). In the example of chromatography, substances are separated into stationary and mobile phases based on their affinity. A liquid or a gas phase represents the mobile phase and a solid represents the stationary phase. Chromatography is employed to measure the organic components of substrates even when they are present in very low concentrations. The VFA of anaerobic digesters is mostly measured by chromatography. In electrochemistry the measuring components present in a liquid are measured by current and electric potential using electrodes. There are always new developments in electrochemistry-based instrumentation. In this way many different variables are observed simultaneously using multiple probes but all electrodes are within one body. An example of this includes a solid-state sensor (a Pd metal oxide semiconductor sensor) designed to measure dissolved gases in the digester. Spectrometric-based instrumentations use ultraviolet, visible and infrared range light to measure the transmission, absorbance, diffusion and fluorescence (Batstone et al., 2004). Important variables such as total organic carbon (TOC), COD, and biological oxygen demand (BOD) are measured by spectrometric procedures and UV absorbance. New developments in spectrometry include mid-IR spectrometry, which measures VFA, alkalinity, COD, TOC, change in feeding composition and digestion disturbances. The qualitative identification of various compounds in a substrate is measured by mass spectrometry. In titrimetric-based methods the measuring components react with a unit volume of reagent solution to measure the amount of reagent. Variables such as VFA, alkalinity, concentrations of ammonium and bicarbonate are all estimated by titrimetric methods (Spanjers and Van Lier, 2006). Table 12.9 gives a summary of the variables in AD and their measuring techniques.

PLC (programmable logic controller) program

The PLC takes the information associated with the activity within a certain time period and provides output in the form of actions. The PLC works for multiple input and output arrangements and with an extended temperature range (Linder, 2009a). The PLC system of the digester operates continuously by reading the program and following the given instructions (Figure 12.4). The PLC receives analogue signals and generates a 4–20 mA numerical signal after reading the analogue input. When these numbers are received they are converted into a variable signal (volts or mA). The discrete output signals are in the form of current and work at 4–20 mA with 4 being 0 per cent and 20 being 100 per cent of output (Hee, 1995). The PC is connected to the PLC through an Ethernet connection to enable data uploading. Thus, the measurement of all variables is logged on an internal flash drive and further uploaded to a PC. With this facility the optimisation of the system becomes much easier. The PLC controls the electric pump, solenoids, and temperature in a digester. Nizami and Murphy (2011) and Nizami et al. (2011) used PLC and SCADA in sequential leach bed reactor with upflow anaerobic sludge blanket (SLBR-UASB) and continuously stirred tank reactor (CSTR) type digesters to monitor and control various process parameters including temperature, pH, flow of circulating leachate, pump rates, leachate holding and discharge

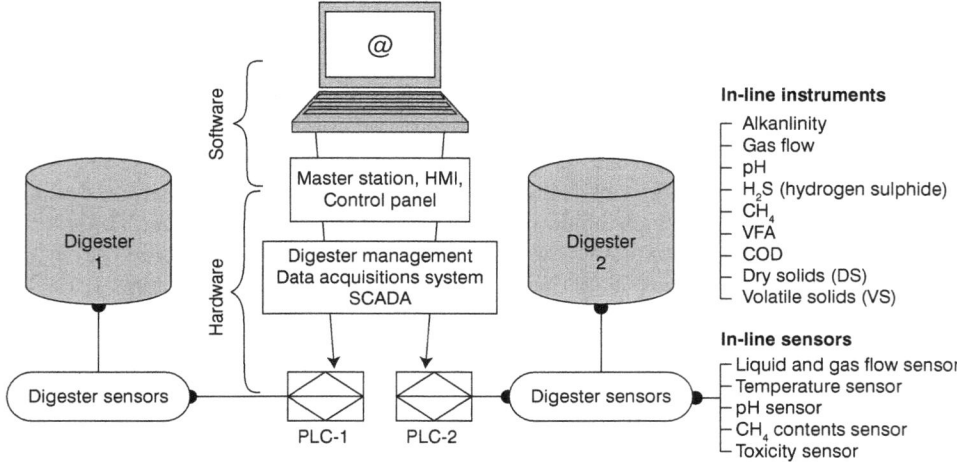

Figure 12.4 Process control system of a digester

Box 12.1 Summary of process control system components

- *Instrumentation*

The physical, chemical, biological techniques or combination of these for the measurement variables.

- *PLC (Programmable logic controller)*

The digester operation and functions are controlled by the PLC, which works as the digester brain.

- *HMI (Human machine interface)*

A control panel in the form of an LCD touch screen, which allows humans to intervene with digester operation.

- *SCADA (Supervisory control and data acquisition)*

This is incorporated into the PLC, where it logs an array of stations for providing information about digester health, e.g. different thermocouples.

- *Software*

Variables which are not measureable on-line are estimated by software sensors.

that even small pH value variations are sufficient to affect the inhibiting effect. Despite these observations Schnurer and Jarvis (2010) stated that generally, ammonium is harmless for the anaerobic microorganisms.

Volatile fatty acids

The stability of the AD process is reflected by the concentration of intermediate products such as VFA (e.g. acetate, propionate, butyrate), produced during acidogenesis, with a carbon chain of up to six atoms (Al Seadi et al., 2008). The concentration of these compounds is one of the most important parameters in the monitoring of the AD process (Parawira, 2004) since it is commonly accepted that an increase in their concentration over the acceptable range which can lead to pH value reductions (Al Seadi et al., 2008) indicates an unbalanced digestion process ((Ahring et al., 1995; Bjornsson et al., 1997). However, the accumulation of VFAs will not always result in pH value reductions due to the buffer capacity in the digester. The same concentration of VFA can be optimal for one type of digester but inhibitory for another type, hence the VFA concentration cannot be recommended as a stand-alone process monitoring parameter (Al Seadi et al., 2008).

Process control systems

As a result of ineffective monitoring and control (Ward et al., 2008), inhibitory compounds to methanogens (e.g. fatty acids) are formed resulting in inefficient substrate digestion and biogas production (Pullammanappallil et al., 1998). The use of buffers or an equalisation container can manage the disturbance caused by substrate variation particularly in a commercial scale digester. Furthermore, the application of remote control, in conjunction with robust communication monitoring and early handling of problems can significantly contribute to reduction of the variation as described above (Linder, 2009a). Thus, a process control system is required to operate digesters at their optimum operating conditions and to overcome the disturbances of feed overloading, underloading, inhibitor entering and temperature disturbance (Pullammanappallil et al., 1998). Modern practices and protocols for monitoring and control of digesters are known as 'smart grid' technologies (Linder, 2009a). These technologies are reducing the perception that digesters are unreliable and expensive as they reduce the maintenance of an off-site facility, hence rural communities can economically benefit due to the growth of small companies for the application of small grid AD technologies (Linder, 2009b). In a process control system, software, hardware and other instruments are connected with a digester to record and control different parameters and variables (Figure 12.4). These parameters and variables include alkalinity, gas flow, temperature, pH, redox potential, hydrogen, hydrogen sulphide, methane and VFA. Most of these variables are measured by in-line procedures and the instruments are based on physical, chemical, and biological techniques or any combination necessary (Spanjers and Van Lier, 2006). For example, the COD can be measured either by titrimetric or spectrometric methods (Spanjers and Van Lier, 2006). The hardware is in the form of a PLC (programmable logic controller) and SCADA (supervisory control and data acquisition) program. A short description of each unit of the process control system is given in Box 12.1.

Redox potential

Redox or reduction oxidising potential has been successfully used as a control and/or monitoring factor in many AD systems (Colmenarejo et al., 2004; Wang et al., 2006) because organic materials in an anaerobic environment are subjected to degradation by redox reaction catalyzed enzymes. AD as a complex process requires strict anaerobic conditions (i.e. redox potential < −200 mV) to proceed (Appels et al., 2008). According to Zupancic and Grilc (2011) a low redox potential is necessary for optimum performance in the AD in agreement with Blanc and Molof (1973) who observed a correlation between increasing redox potential and lower digester performance. Redox potential for methanogenic bacteria should be lower than −250 mV, preferably between −300 and −330 mV (Table 12.8). Vasilian and Trchunian (2008) stated that the regulation of anaerobic bacterial metabolism is accomplished through changes in redox potential. They assumed that bacteria under anaerobic conditions are sensitive to changes in the redox potential and therefore could have a redox taxis. A low redox potential can be achieved and/or maintained if the addition of any oxidising agent such as oxygen, sulphates, nitrates or nitrites into the digester is prohibited (Deublein and Steinhauser, 2008).

Ammonia/ammonium

Ammonia is produced by the decomposition of nitrogenous matter, mostly in the form of proteins and urea (Kayhanian, 1999) and primarily inhibits the methane-producing organisms (Schnurer and Jarvis, 2010). Ammonium ion (NH_4^+) and free ammonia (NH_3) are the two principal forms in the equilibrium of inorganic ammonia nitrogen in aqueous solution and the predominating form depends mainly on the process, temperature and pH (when the temperature rises, for example, this equilibrium steadily shifts towards free ammonia) (Schnurer and Jarvis, 2010). It has been reported that among anaerobes, the methanogens are the least tolerant and the most likely to cease growth due to NH_3 inhibition (Kayhanian, 1994) of which the ability of its molecules to diffuse into the cell cause proton imbalance and/or potassium deficiency (de Baere et al., 1984). Nevertheless, based on results of methane production and growth rate, there is contradictory information in the literature about the sensitivity of aceticlastic and hydrogenotrophic methanogens towards ammonia concentration (Chen et al., 2008). According to these authors some results showed that the inhibitory effect was stronger for the aceticlastic than for the hydrogenotrophic methanogens while others observed a relatively high resistance of acetate consuming methanogens to high levels of the total ammonia nitrogen in comparison to hydrogen utilising methanogens. Additionally, the role of ammonium considering its toxicity to anaerobes is not clear. As has been stated, ammonium, although it represents an ideal form of nitrogen for microbial cell growth, is toxic to mesophilic methanogenic microorganisms at concentrations over 3000 mgL^{-1} and pH over 7.4 whereas thermophilic methanogenic microorganisms are generally more sensitive to ammonium concentration since inhibition occurs at 2200 mgL^{-1} of ammonium nitrogen (Zupancic and Grilc, 2011). A study of ammonium inhibition in thermophilic digestion shows an inhibiting concentration to be over 4900 mgL^{-1} when using non-fat waste milk as substrate (Sung and Liu, 2003). All the above indicate that ammonium inhibition varies with substrate type. Furthermore as Deublein and Steinhauser (2008) stated the inhibition by ammonium increases with rising pH value at constant concentration, e.g. the ratio ammonium to ammonia is 99/1 at pH = 7 and 70/30 at pH = 9. They also reported relationship exists between the hydraulic retention time and the volatile solids converted to gas (Burke, 2001). The effects of HTR alterations on biogas (Table 12.4) or methane production are still unclear. Clemens et al. (2006), for example, recorded a small effect on the CH_4 production by extending the HRT from 29 to 56 days in a study when potato starch was used as an additive during the digestion of cattle slurry. In contrast, Sanchez et al. (1992) found that increased HTR resulted in the improvement of organic matter digestion while anaerobically treating cattle dung. Shorter retention time is likely to face the risk of washout of an active bacterial population while longer retention time requires a large volume digester and hence more capital cost (Yadvika et al., 2004). Easily degradable feedstock such as food, industrial waste or manure can be digested at short (15–30 days) HRT while processes fed with lignocellulosic feedstocks, i.e. energy crops, are operated at longer HRT of over 100 days (Braun, 2007; FNR, 2010). In tropical countries like India, HRT varies from 30–50 days while in countries with colder climates it may go up to 100 days. The optimal value varies according to technology and process details (for a specific digester, it may for example even change from day to day with changing feedstock or from season to season with changing temperatures), temperature and waste composition (Buekens, 2005; Romano and Zhang, 2008). Buekens (2005) stated that research into reducing retention time in the AD process has focused on i) separating the consecutive stages of digestion to assist bacteria to optimise their performance, ii) improvements of mixing patterns and internal circulation, iii) introducing surfaces, such as foam or other porous carriers, on which bacteria can be colonised and thrive permanently due to reduced bacteria washing out with the effluent, iv) determination of appropriate feedstock pre-treatment to increase its digestibility, and finally v) control of operating conditions.

Organic loading rate (OLR)

The organic loading rate and the HRT are two major parameters used for planning the size of digesters. Their optimum values are specific to the substrate and the operating temperature of digesters (Romano and Zhang, 2008). Additionally, there is an optimum feed rate for a particular size of plant, which will produce maximum gas output and beyond which a further increase in the quantity of substrate will not proportionately produce more gas. Nevertheless, research on organic loading rate is contradictory. Xie et al. (2002), for example, recorded that OLR affected the digester performance significantly when the co-digestion of the solid fraction of separated pig manure was evaluated with dried grass silage. When OLR increased three-fold the volumetric methane yield increased by 88 per cent whereas the production of the specific methane (i.e. the amount of methane that is produced per kg of volatile solids) decreased by 38 per cent. In addition when OLR of manure increased from 346 kg VS day^{-1} to 1030 kg VS day^{-1}, biogas yield increased from 67 to 202 m^3 day^{-1} (Yadvika et al., 2004). In contrast, methane yield was found to increase with reduction in loading rate (Vartak et al., 1997) although as these authors stated a reactor subject to decreasing OLR is inefficient in terms of feedstock treatment and energy production. The OLR determines how much volatile solids are put to the digester (Ciborowski, 2001), i.e. the quantity of biomass fed per unit volume of the digester per unit time usually as total solids or volatile solids (kg (m^3 of digester volume d)$^{-1}$) (Buekens, 2005). According to Tchobanoglous et al. (2003) a typical OLR in CSTRs ranges from 1 to 5 kg VS (m^3·d)$^{-1}$. OLR is a useful criterion for assessing the performance of reactors (Rajeshwari et al., 2000; Lissens et al., 2001).

operation (Yardika, 2004) should be attained. Correction of C/N ratio can be achieved by adding materials high in N where the C/N ratio is high (Fry and Merill, 1973; Barnett, 1978) or manure, sewage sludge (Buekens, 2005) nitrogen in farm cattle urine or by fitting latrine to the plant (Fulford, 1988) where the C/N ratio is low. It is obvious that maintaining a balanced C/N ratio (Deublein and Steinhauser, 2008) is necessary for optimum biogas production. The C/N ratio may be either monitored explicitly or simply by keeping track of the waste types entering the facility, and knowing the relative make-up of each (Buekens, 2005).

Alkalinity

The AD industry calls for a higher tolerance under extreme conditions hence anaerobic treatment under acidic or alkaline conditions may be important (van Lier et al., 2001). Nevertheless, as stated earlier in this chapter pH fluctuations below or above certain limits significantly affect the growth rate and performance of anaerobes, and hence alkalinity, which is the ability of a solution to neutralise acids, can serve as a buffer to prevent rapid change in pH (Schnurer and Jarvis, 2010). As has been stated by various authors (Callander and Barford, 1983; Parawira, 2004; Parawira et al., 2006) it is essential that the reactor contents provide enough buffering capacity to neutralise any possible VFA accumulation and maintain pH within acceptable levels for stable AD operation. In AD processes alkalinity is influenced by carbon dioxide and carbonate ions since the ammonia released by the decomposition of nitrogen-rich feedstocks with high proportions of proteins and amino acids can react with dissolved carbon dioxide (i.e. carbonate ions) to form ammonium bicarbonate (Schnurer and Jarvis, 2010). When the pH is around 7.0, as is often the case in anaerobic reactors, the alkalinity is practically equivalent to the concentration of bicarbonate, hydrogen sulphide, dihydrogen phosphate or ammonia which buffers the system in the optimum pH range for the process to run efficiently (Parawira, 2004; Spanjers and van Lier, 2006). VFA buffers the system at a low pH that is inhibitory to the biomass matrix in the digester (Zaher et al., undated). Alkalinity is generally determined by titration of a sample at two pH endpoints, i.e. titration of a sample at pH 4.3 (total alkalinity) or titration of a sample at pH 5.75 (partial alkalinity, a measure of bicarbonate concentration) (Hill and Jenkins, 1989; APHA, 1998). The difference between total and partial alkalinity is called intermediate alkalinity (a measure of the concentration of VFA) (Franco et al., 2007). A total alkalinity of 1.5 g $CaCO_3$/L is recommended for an adequate performance of the anaerobic systems (Franco et al., 2007) although the true monitoring parameter used is the ratio between intermediate alkalinity to total alkalinity which must be maintained at a level lower than 0.3–0.4 (Ripley et al., 1986; Franco et al., 2007).

Hydraulic retention time (HRT)

The HRT is an important design parameter influencing the economics of digestion and methane yield (Anonymous, 2009). HRT is the average time spent by the input feedstock inside the digester before it comes out (Yadvika et al., 2004) and it is defined as digester volume divided by the volume of daily feedstock input. It is dependent on the type of digester, for example, vertical digesters require a slightly higher hydraulic retention time than horizontal digesters. The HRT is important since it establishes the quantity of time available for bacterial growth and subsequent conversion of the organic material to gas. A direct increase in VFA concentration (Svoboda, 2003). This can be a consequence of overloading the biomass with organic material or due to the presence of inhibitors like antibiotics or disinfectants (Svoboda, 2003). Conversely, according to Yadvika et al. (2004), prolific methanogenesis may result in a higher concentration of ammonia which is accompanied by pH values above 8.0, a fact that impedes acidogenesis. Stabilisation of pH can be achieved by feeding the digester at an optimum loading rate (Yadvika et al., 2004) or by adding chemicals such as calcium carbonate or lime to the system when pH values drop due an increased loading rate, increase in VFAs and decreased methanogenic activity (Buekens, 2005). It has been reported by Chae et al. (2008) that stabilisation of pH was achieved when alkalinity was maintained in the range 2966–6606 mgL^{-1} as $CaCO_3$. The pH value is a function of VFA concentration, bicarbonate concentration and alkalinity of the system as well as the fraction of CO_2 in digester gas (Liu et al., 2007; Chae et al., 2008) along with the retention time and loading rate particularly in batch reactors (Arsova, 2010). In order to agree on a constant pH value it is crucial to adjust the relationship between the VFA and bicarbonate concentrations (Liu et al., 2007). In general, pH values above 5.0 increase the efficiency of CH_4 production more than 75 per cent (Jain and Mattiasson, 1998). Even though the optimal range of pH for obtaining maximal biogas yield in anaerobic digestion is 6.5–7.5 its range in AD plants is relatively wide and the optimal value of pH varies with substrate and digestion technique (Liu et al., 2008).

The carbon/nitrogen (C/N) ratio

The C/N ratio represents the relationship between the amount of carbon and nitrogen contained in organic materials and it is an indicative measure of the nutrient balance that anaerobic organisms require for growth (Verma, 2002). Materials with different C/N ratios differ widely in their yield of biogas (Table 12.8). It is generally found that during anaerobic digestion microorganisms utilise carbon 25–30 times faster than nitrogen (Marchaim, 1992). The C/N ratios in AD to ensure sufficient nitrogen supply for cell production and the degradation of the carbon present in the feedstock (Fricke et al., 2007) vary between 10:1 and 30:1 (Malik et al., 1987; Marchaim, 1992; Bardiya and Gaur, 1997; Buekens, 2005; Schnurer and Jarvis, 2010; Liqian, 2011) with an optimum, in most cases, C/N ratio between 15:1 and 25:1 (Yen and Brune 2007; Liu et al., 2008; Schnurer and Jarvis, 2010). Excess amounts of some nutrients may however also become inhibitory to the AD process (Chen et al., 2008). Hence, a high C/N ratio will lead to a rapid consumption of nitrogen, in other words a reduced protein formation and a decline in the energy and structural metabolism of the anaerobes (Deublein and Steinhauser 2008; Liqian, 2011) resulting in lower substrate degradation efficiency (Zupancic and Grilc, 2011) and consequently in lower gas production rates (Fulford, 1988; Buekens, 2005). On the contrary, a low C/N ratio, or too much nitrogen, can cause ammonia to accumulate which would lead to pH values above 8.5 (Verma, 2002; Chaudhary, 2008; Liqian, 2011) resulting in reduced anaerobic efficiency as described in the previous paragraph. Additionally, the quality of the digestate declines with ammonia production (Buekens, 2005). As stated above, increased nitrogen concentration and pH values are the most important factors affecting the C/N ratio. It has also been reported that C/N ratio varies with temperature (Laura and Idnani, 1971), low levels of phosphorus and trace elements, process decomposition efficiency and the composition of the substrate (Schnurer and Jarvis, 2010). The proper composition of the feedstock in which the C/N ratio remains within the desired range for efficient AD plant

Table 12.8 Comparison of the performance of mesophilic and thermophilic anaerobic digestion.

Performance characteristics	Mesophilic digestion	Thermophilic digestion
Gas production rate	Contradictory reports	Contradictory reports
Pathogen reduction	Lower	Higher
Effluent VFAs	Lower	Higher (contradictory)
Dewaterability	Contradictory reports	Contradictory reports
Process stability	Higher	Lower (contradictory)
Methane content	Higher	Lower
Energy requirement	Lower	Higher
Odour	Lower	Higher
Product/substrate inhibition	Lower	Higher

Note: based on Ahring, 1994; Ghosh, 1998; Ahn and Forster, 2000, cited in Parawira, 2004.

(2011) reported that temperature fluctuations under thermophilic conditions should be kept within ±1 °C because a departure from a normal temperature range even of ±2 °C can result in 30 per cent less biogas production (Zupancic and Jemec, 2010). However, the same authors stated that microorganisms under mesophilic conditions can tolerate temperature fluctuations of ±3 °C. On the other hand, Meher et al. (1994) and Nozhevnikova et al. (1999) stated that long-term adaptation of active psychrophilic microbial communities can be essential for efficient treatment of cattle dung at low temperatures. Finally, based on the literature, the relative performances of anaerobic digesters at thermophilic and mesophilic conditions do not show a fully clear-cut and consistent difference (Table 12.8).

pH

pH is an important operating parameter because it affects the growth of anaerobes during all stages of AD. However, it has been stated by Chae et al. (2008) to be a very poor indicator in highly buffered substrates such as swine manure. In terms of biochemistry, the early stages of feedstock breakdown require an acidic environment, whereas in the later stages, when the methane is actually produced, a neutral pH environment is advantageous (Anonymous, 2008). Hence, the pH value for hydrolysis and acidogenesis stages should lie between 4.0 and 6.5 whereas during the methanogenesis stage it should be around 7.0 (Kim et al., 2003). At early stages of digestion, i.e. hydrolysis and acidogenesis, a reduction in pH as a result of VFA and COD increase was recorded (Nizami et al., 2009). According to Siegert and Banks (2005) pH changes occur when VFA concentration exceeds 4000 mgL^{-1} whereas Yadvika et al. (2004) stated that the highest limit of VFA concentration above which disturbances of the AD equilibrium could be observed did not exceed 2000 mgL^{-1}. Amon et al. (2007) reported that VFA concentrations greater than 1000–1500 mgL^{-1} during acetogenesis and methanogenesis stages inhibit the process. In addition, acid accumulation that occurs when volatile solids in the digester are overloaded can cause a reduction in the pH value resulting in an increase of acidogenic bacteria which produce more organic acids and consequently even lower, to a lethal level for methanogenic bacteria, pH values (Buekens, 2005). Since the acetogenic phase of the digestion has a higher reaction rate than the methanogenic, accumulation of VFA can occur in the reactor causing a decrease in pH and a further

Table 12.7 Range of process parameters affecting optimum AD performance.

Factors	Hydrolysis/Acidogenesis	Methanogenesis
Temperature	25–35 °C	Mesophilic: 30–40 °C Thermophilic: 50–60 °C
pH	4–6.3	6.7–7.5
C/N ratio	10–45	20–30
Alkalinity (buffering capacity)	$\geq 4{,}000$ mg L^{-1} bicarbonate	
HRT energy crops	60–120 days	
HRT manure and food wastes	10–30 days	
Organic loading rate	Various rates	
Redox potential	+400 to –300 mV	< –250 mV
Ammonium/Ammonia	$\leq 1{,}500$ mg/L NH$_4^+$ -N/ < 80 mg/L NH$_3$	
VFA or organic acids from acetic (C2) to caproic (C6)	1000–4000 mg/l	

Note: Based on Svoboda, 2003; Anonymous, 2008; Deublein and Steinhauser, 2008

possible fluctuations in biogas production. A large HRT, for example, is used in conventional biogas plants due to a large volume of the digester which, on the other hand, is associated with a high production cost (Yadvika et al., 2004). The performance of a biogas plant can be controlled and improved by studying and monitoring the variation in parameters like pH, temperature, loading rate, etc. Any drastic change in these can adversely affect the biogas production. So these parameters should be varied within a desirable range to operate the biogas plant efficiently (Yadvika et al., 2004). The most important AD parameters with their current functional range are presented in Table 12.7.

Temperature

Operational temperature or its variations inside the digester has a major role on the biogas production process (Buekens, 2005) because it affects greatly the methane-forming and volatile acid-forming bacteria (Gerardi, 2003). According to different temperature ranges the anaerobic digestion can be classified as psychrophilic (12–30 °C), mesophilic (30–40 °C, preferably 35 °C) and thermophilic (50–60 °C, preferably 55 °C) with anaerobes performing best at either mesophilic or thermophilic temperatures (Desai and Madamwar, 1994; Zennaki et al., 1996; Song et al., 2004; Nges and Liu, 2010). Thermophilic digestion has been reported to produce higher gas yield in comparison to mesophilic digesters (Kelleher, 2007) due to higher digestion rate and loading rate, better degradation efficiency of the substrate, improved solids settling and lower retention time (Varel et al., 1980; Hashimoto et al., 1981; Buekens, 2005; Kelleher, 2007) although according to Takashima et al. (2011) microorganisms in mesophilic digestion have less demand on nutrients. Desai et al. (1994) reported that maximum gas production with a methane content of 62 per cent was obtained with a retention time of ten days at 40 °C in agreement with Angelidaki and Ahring (1994) who observed that a reduction in temperature below 55 °C resulted in an increase of biogas yield. In addition, Garba (1996) showed that maximum production of biogas from lignocellulosic material was achieved at thermophilic temperatures. Also, Zupancic and Grilc

design, equipments and feedstock utilisation). They can be operated under psychro-, meso- and thermophilic conditions. Based on moisture content of the feedstock, digesters can be classified as dry with high solids concentration (i.e. > 20–50 per cent) and wet with low solids concentration (i.e. < 12–15 per cent) (Buekens, 2005; Karagiannidis and Perkoulidis, 2009; Nizami and Murphy, 2010). Besides these standard configurations, recent advances focus on the improvement of reactor volume by utilising the dry digestion by continuous flow stirred tank reactors (CSTRs) for the digestion of high concentration of total solid (TS) feedstocks as in the case of energy crops, fodder crops, organic fraction of municipal solid wastes (OFMSW), and separated organics (e.g. kitchen waste or biowaste) (Karellas et al., 2010). Despite the utilisation of dry digestion for a wide range of feedstocks, the nature (e.g. volatile solids) of this is also an important factor which affects digester design (Murphy and Power, 2009) and increases variation of the methane outcome. Digesters, for example, optimised for OFMSW may not be ideal for grass silage because the volatile solids content of grass silage is significantly higher than that of OFMSW, i.e. up to 92 per cent compared with values as low as 60 per cent. Thus the digestate from grass may be quite liquid in nature (solids content of less than 5 per cent) as opposed to digestate from OFMSW which may have a solids content of over 20 per cent. Two-phase digester configuration supports high growth rates of hydrolytic and methanogenic bacteria (Gunaseelan and Nallathambi 1997; Verrier et al., 1987; Mata-Alvarez 1987) hence this design/configuration is most appropriate for high solid feedstocks such as grass silage. Furthermore, work conducted by Lehtomaki (2006) and Yu et al. (2002) suggests that incorporation of a high rate reactor, such as an upflow anaerobic sludge blanket (UASB), with a CSTR is an efficient digester configuration for feedstocks originated from fodder crops. Additionally, soluble organic wastes are treated using high-rate biofilm systems (e.g. anaerobic filters, fluidised bed reactors and UASB reactors) whereas slurry and high TS feedstocks are mainly treated in CSTRs (Angelidaki and Ellegaard, 2002). A noticeable variation in biogas and biomethane yield ranging from 0.16 to 0.72 m^3 kg^{-1} VS and from 0.37 to 0.6 m^3 kg^{-1} VS respectively for various digester types (i.e. semi-continuous, batch, continuous and leach-UASB) under various feedstocks (i.e. fresh and ensiled grasses, food and municipal solid wastes) has been reported (Cho et al., 1995; Baserga, 1998; Amon et al., 2003; Mahnert et al., 2005; Prochnow et al., 2005; Lehtomaki et al., 2008; Li et al., 2010; Jha et al., 2011). As Buekens (2005) stated, changes in feed or operating conditions can either disturb the equilibrium or result in intermediaries that may inhibit the overall process or shut it down altogether. It is crucial to use adequate designs, enhancing stability, as well as control technologies to continually monitor and adjust the environment to prevent such occurrence.

Variation due to process parameters

In a world where carbon trading and carbon neutral are the current buzz words, AD seems quite promising to many stakeholders (Kelleher, 2007) but as Ward et al. (2008) reported, many industrial anaerobic digesters are currently operating below their optimal capacity. In addition variation associated with the biogas production due to dynamics of the microorganisms involved and various process factors such as temperature, pH, carbon to nitrogen (C/N) ratio, hydraulic retention time (HRT), the lack of process stability, loading rates and specific requirements for waste composition (Van der Berg and Kennedy, 1983; Angelidaki et al., 2003; Ward et al., 2008; Weiland, 2010) lowers even more the production potential of AD. Consequently, adjustments, sometimes feasible but not optimal, are employed to equalise the bacterial strains *Sphingomonas paucimobilis* and *Bacillus circulans* improved enzymatic hydrolysis of office paper from municipal wastes. Nevertheless, as stated by Verma (2002) lignocellulosic organic materials such as paper and cardboard do not readily degrade under anaerobic conditions and they are better suited for waste-to-energy plants. Additionally, Palatsi et al. (2011) found that specialised microbial populations, e.g. b-oxidising/proteolitic bacteria (*Syntrophomonas* sp., *Coprothermobacter* sp. and *Anaerobaculum* sp.), and syntrophic methanogens (*Methanosarcina* sp.) could limit the efficiency of fresh slaughter waste AD.

Fungi inoculants

Depending on the feedstock the effectiveness of fungi use as a biological pre-treatment can be achieved by different species or combination of them such as *Aspergillus niger*, *A. awamori*, *A. oryzae* and *A. terreus* (Taherzadeh and Karimi, 2008). As stated by Sun and Cheng (2002) the most effective species for biological pre-treatment of lignocelluloses amongst brown-, white- and soft-rot was found to be the white-rot fungi most probably due to its potential to degrade lignin (Sanchez, 2009). Taniguchi et al. (2005) evaluated biological pre-treatment of rice straw using four white-rot fungi (i.e. *Phanerochaete chrysosporium*, *Trametes versicolor*, *Ceriporiopsis subvermispora*, and *Pleurotus ostreatus*) and found that the use of the latter resulted in selective degradation of the lignin and increased the susceptibility of rice straw to enzymatic hydrolysis.

Enzymes

Treatment with cellulase enzymes during silage preparation caused an increased degradation of plant cell wall constituents that were more susceptible to bacterial decomposition to silo (Clavero and Razz, 2002). Additionally, Ridla and Vehida (1998) and Sajko et al. (1997) reported that the addition of cellulosic enzymes facilitated the breakdown of a component of structural carbohydrates during ensiling which resulted in improved degradation during silage fermentation and most probably increased methane production during AD. Additional benefits include the reduction in quantity of digestate and reduced energy consumption during AD (Zimbardi et al., 1999).

Variation due to digester type and configuration

The design of the digester is an important factor that influences the organic loading rate and consequently the methane yield and the performance of AD (Ward et al., 2008). The properties of the feedstock (Arvanitoyannis and Varzakas, 2008) and operational parameters (Vandevivere et al., 2003) form the basis of criteria to design a digester. Anaerobic digesters, based on the way of feeding substrate, are classified as batch, semi- and continuous digesters (i.e. in batch digesters, for example, the feedstock is inserted once into the digester for a certain period of time to complete the digestion activity, whereas in continuous digesters the feedstock is constantly or regularly fed into the digester). Digesters according to the phases of digestion activities can be also be classified as continuous single stage, double or multiple stage (i.e. in single-phase digesters, for example, all phases of anaerobic digestion occur within the same compartment in the digester, whereas in two-phase digesters, hydrolysis, acidogenesis and acetogenesis are separated from methanogenesis). Digesters can be vertical or horizontal treatment units (i.e. based on plant size with different dimensions, not enhance the AD of water hyacinth (*Eichhorinia crassipes*) because its water solubility increased only slightly under the specific experimental conditions. In contrast, AD of pasteurised slaughterhouse waste resulted in a fourfold increase of CH_4 yields after thermal treatment at 70 °C for 1 hour (Norberg, 2004, cited in Krich et al., 2005).

Ultrasonic pre-treatment has been shown to be effective in disintegrating sewage sludge, resulting in greatly improved fermentation rates (Vera et al., 2004) and reduced retention time (Verma, 2002). The destruction of volatile solids increases according to the degree of cell disintegration resulting in increased biogas production (Krich et al., 2005). However, the application of this technology to manure solids is untried and its success is uncertain due to the lignocellulosic character of manure (Krich et al., 2005).

Chemical pre-treatment

Chemical pre-treatments include treatments with alkalis, acids, solvents or oxidants and result in changes of the feedstock composition by reducing the particulate organic matter to soluble form (i.e. proteins, fats, carbohydrates or other compounds of lower molecular weight) (Verma, 2002). More particularly, alkaline, sulphuric acid, sodium hydroxide and hydrogen peroxide treatments have been used to facilitate hydrolysis (Pavlostathis and Gossett, 1985; Prasad et al., 2007) and delignification of lignocellulosic materials as in the case of wheat straw, cotton stalks, water hyacinth and water lettuce (Sun and Cheng, 2002; Sun et al., 2005; Mishima et al., 2006; Silverstein et al., 2007). Nevertheless, scientific evidences concerning the efficiency of chemical pre-treatment on biomethane production are contradictory. Us and Perendeci (2012) have recorded an increase in methane potential when H_2SO_4 was used to optimise thermal pre-treatment of vegetable residues compared with untreated samples. Additionally, pre-treatment of wastes by alkaline hydrolysis before co-digestion with activated sludge resulted in biomethane yield from 25 L to 222 L CH_4 kg^{-1} VS accompanied with 67 per cent and 84 per cent reduction of total and volatile solids respectively (Neves et al., 2006). On the contrary, Shahriari et al. (2012) recorded no differences in biomethane production when hydrogen peroxide (H_2O_2) was used to enhance the AD of the organic fraction of municipal solid waste.

Biological pre-treatment

Biological pre-treatment, in which either microbes and/or enzymes are used for partial degradation of lignocellulose, offers a cost-effective solution in comparison with other pre-treatments but it requires a specific environment to work efficiently (Ward et al., 2008).

Bacteria inoculants

It has been reported that strains of some bacteria and some fungi species can enhance biogas production by stimulating the activity of particular enzymes (Santosh et al., 2004). As such, Tirumale and Nand (1994) stated that some cellulolytic strains of *Actinomycetes* and mixed bacteria consortia have been found to improve biogas production from cattle dung by up to 44 per cent. In addition, Woolford (1984) and Idler et al. (2007) during ensilage reported that heterofermentative bacteria (as compared with homofermentative bacteria) facilitate the production of intermediates for later use by methanogens hence they could be more beneficial for efficient anaerobic digestion. In addition, Kurakake et al. (2007) found that

The main aim of pre-treatment is to convert the feedstock being digested into a suitable form for higher or faster biogas production (Petersson et al., 2007). The feedstock determines the need and type of pre-treatment; in the case of the organic fraction of municipal solid wastes for example the pre-treatment is usually a fundamental operational process of the AD plant (Arsova, 2010). Pre-treatment can also facilitate removing unwanted material from the feedstock, some of which could be input to recycling schemes. It has been stated that methane production from AD of plant residues can be increased by pre-treatment of the feedstock because of the disruption of the polymer chains to soluble compounds more easily accessible to bacteria populations (Lehtomaki, 2006). This is of particular importance in the case of biomass, the components of which vary in their accessibility to AD. Hydrolysis of lignocellulose, for example, has been found to be a rate limiting step in the AD of solid biomass (Cecchi et al., 1993; Mata-Alvarez et al., 2000). Thermal and chemical pre-treatments can improve hydrolysis and promote solubilisation (Verma, 2002) thereby enhancing the potential for methane production. An ideal pre-treatment process for feedstock with high concentration of organic materials such as lignocellulose should increase its homogenisation and the surface area by breaking up the lignocellulose complex, decrease the lignin content and reduce the crystallinity of cellulose (Fan et al., 1981; Verma, 2002). Major pre-treatments include physical (mechanical and thermal), chemical, biological treatments and their combinations. According to Qi et al. (2005) a combination of different pre-treatment approaches with different operational procedures is required for optimum cost-effective pre-treatment manipulation accompanied by minimum energy consumption.

Physical pre-treatment

The most important effect of physical pre-treatment on agricultural biomass is particle size reduction (Palmowski and Müller, 1999) mainly through chipping (size of biomass particles usually 10–30 mm), milling, gridding (size of biomass particles 0.2–2 mm) (Mais et al., 2002; Peltola et al., 2004; Stamatelatou et al., 2012), or hydrothermal, high-pressure steaming, expansion, extrusion, pyrolysis, irradiance (Taherzadeh and Karimi, 2008), and ultrasonic (Vera et al., 2004) treatments. Nevertheless, results reported in the literature concerning biomethane production increase by the employment of milling or gridding are controversial, a fact that increases the overall variation of the system. Zeng et al. (2007) reported that reduced size of maize stover resulted in increased biomethane production (e.g. large maize stover particles of size 425–710 μm were 1.5 times less productive than these with size 53–75 μm). In addition, Bruni et al. (2010) reported that mechanical pre-treatment of maize silage (whole plant) increased the methane yield by 0.04 ± 0.01 m^3 CH$_4$ kg^{-1} VS whereas Delgenes et al. (2003) stated that milling enhances methane production from 5 to 25 per cent. In contrast, Peltola et al. (2004) showed that although impact grinding can increase the soluble COD content of the organic fraction of municipal solid waste by approximately 2.5 times due to partial disintegration of plant cells of the organic fraction of municipal solid waste, it did not increase biogas production. However, they reported an increase in the AD pace as a result of the pre-treatment. Mechanical pre-treatment leads to a lower quality digestate since the removal of all contaminants is not possible especially for the smaller fraction particles such as heavy metals (RISE-AT, 1998).

Thermal pre-treatment can increase the CH$_4$ yield of certain substrates (Krich et al., 2005) although it is not an effective technique for the AD of all substrates such as those of plant origin (Ferrer et al., 2004). They found that thermal pre-treatment at 80 °C did

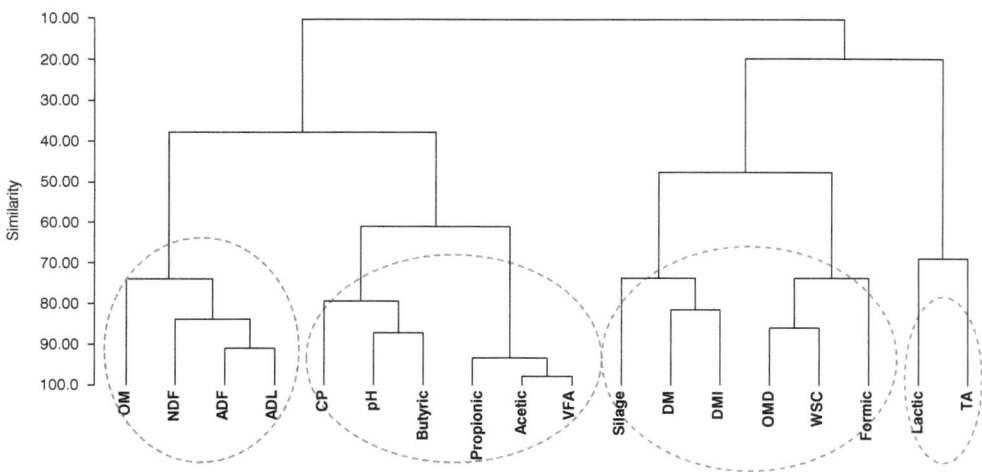

Figure 12.3 Dendrogram based on hierarchical cluster analysis for the characterisation of the similarity of various physico-chemical characteristics of grass silages

OM = Organic matter; NDF = Neutral detergent fibres; ADF = Acid detergent fibres; ADL = Acid detergent lignins; CP = Crude protein; VFA = Volatile fatty acids; Silage = Silage type (i.e. wilted vs. unwilted or first, second cut etc.); DM = Dry matter; DMI = Dry matter intake; OMD = Organic dry matter; WSC = Water-soluble carbohydrates; TA = Total acids

2006) and post-harvest treatment (Kung et al., 1990; Dawson et al., 1999) were analysed by multivariate analysis (i.e. hierarchical cluster analysis) to identify, based on various physico-chemical characteristics, how similar these silages are.

In Figure 12.3 four major clusters (within circles) and six sub-clusters (e.g. propionic, acetic and VFA) were formed consisting of two to six grass silage traits (i.e. OM, NDF, ADF and ADL; CP, pH, butyric acid; propionic acid, acetic acid and VFA; silage, DM and DMI; OMD, WSC and formic acid and finally lactic acid and TA). The similarity between the silage variables as presented indicates that these traits describe the same variation pattern hence measurements in one of these, VFA for example, will most probably reveal the same information as as measurements in acetic acid whereas any of these characteristics can be approximately estimated by measurements of propionic acid concentration.

Variation due to pre-treatment

The difficulty of releasing all the substances useful in AD from the lignocellulosic matrix of the agricultural biomass and agro-industrial residues is an issue that adds up to total variation in biogas production (Sarath et al., 2008; Zhang, 2008; Vandamme, 2009). This necessitates even more the use of pre-treatment or pre-processing despite the cost increase in the overall economics of 'second generation' digestion processes. A large number of potentially interesting pre-treatment methods have been developed to solve problems such as bottlenecks and variation on the biomethane production potential of various feedstocks. Hence, an evaluation of the need for pre-treatment methods is of interest. Consequently, if these methods can be optimised in the near future, they will relieve current societal tensions regarding the use of biomass and crops for food or for platform chemicals and biofuels (Morris, 2006).

Table 12.6 Effects of grassland type on biogas and methane production[a] (m^3 kg^{-1} VS).

Grassland type/utilisation	Biogas	Biomethane
Intensive (monoculture, fresh grass and silage)	0.70–0.72	–
Extensive (fresh grass and silage)	0.54–0.58	–
Extensive (fresh grass and hay)	0.50–0.60	–
Extensive (silage)	0.50–0.55	0.22
Mixed (fresh grass and silage)	0.65–0.86	0.31–0.36
Mixed (silage)	0.56–0.61	0.30–0.32
Grass and clover (mixed, silage)	0.532–0.427	0.37–0.297
Intensive grassland (monoculture, silage)	–	0.39

[a]Digester operating conditions (e.g. retention time, temperature) and digester type are not mentioned. Based on Prochnow et al., 2005

0.86 m^3 kg^{-1} VS) for perennial ryegrass compared with other grasses (both fresh and ensiled) such as cocksfoot (*Dactylis glomerata*) in which a biogas yield of 0.65–0.72 m^3 kg^{-1} VS added was recorded (Mahnert et al., 2005). Biogas composition depends on the feedstock, e.g. biogas from grass contains around 55 per cent CH_4 (Smyth et al., 2009), where as biogas from chicken slurry contains 60–80 per cent CH_4 (Steffen et al., 1998).

Physico-chemical characteristics as potential monitoring variables for both biogas and animal feed production: a grass silage example

As mentioned above the major problem in using agricultural biomass is their variable dimensions and compositions due to the different production systems that generated them. Traits of grass and grass silage, for example, have been identified by many researchers based on quantitative (e.g. dry matter, dry matter concentration, particle size, milling resistance, leaf-to-stem ratio), qualitative (e.g. voluntary intake, intake potential, dry matter intake, digestibility, D-value, palatability etc.), and chemical (e.g. neutral detergent fibre, acid detergent fibre, acid detergent lignin, crude fibre, crude protein, water soluble carbohydrates, organic matter, butyric, acetic, propionic and lactic acids, buffering capacity, pH, etc.) characteristics along with ensiling techniques (e.g. wilted, use of additives) (Kung et al., 1990; Chynoweth et al., 1993; Bach-Knudsen, 1997; Peyraud et al., 1997; Dawson et al., 1999; Keady et al., 2000; O'Kiely et al., 2002; McEniry et al., 2006; Van Dorland et al., 2006; Pakarinen et al., 2008; Lee et al., 2009; Nordheim-Viken and Volden, 2009) that affect their feeding value and intake potential and most probably their biogas/biomethane production (Korres et al., 2011). The question arises about how the most appropriate characteristics of feedstock for both animal feed and biomethane production can be distinguished. In other words, how repetitions in measurements which reveal the same piece of information can be avoided. The application of various statistical techniques such as multivariate analysis of variance (e.g. cluster, principal component analysis) as part of data mining techniques in conjunction with data warehouse and mining (Chapter 13) can be used for this purpose. Data on silages' physico-chemical characteristics obtained from various studies in which traits and feed factors that could possibly affect voluntary intake of grass silage (and possibly biogas production potential) based on fermentation quality (Krizsan and Randby, 2007), husbandry factors (Nordheim-Viken and Volden, 2009), harvesting method (McEniry et al.,

Table 12.5 Examples of biomethane yield (m³ kg⁻¹ VS) from digestion of various plant species and plant material.

Crop	Biomethane	Crop/crop residues	Biomethane
Zea mays (Maize*)	0.205–0.45	Hordeum vulgaris (Barley)	0.353–0.658
Triticum aestivum (Wheat**)	0.384–0.426	Triticale	0.337–0.555
Avena sativa (Oat**)	0.250–0.295	Sorghum bicolour (Sorghum)	0.295–0.372
Secale cereale (Rye**)	0.283–0.492	Lathyrus spp. (Peas)	0.390
Grass	0.298–0.467	Medicago sativa (Alfalfa)	0.340–0.500
Clover grass	0.290–0.390	Brassica oleracea (Kale)	0.240–0.334
Trifolium pratense (Red clover)	0.300–0.350	Phalaris arundinacea (Reed canary)	0.340–0.430
Trifolium spp (Clover)	0.345–0.350	Lolium perenne (Ryegrass)	0.390–0.410
Cannabis sativa (Hemp)	0.355–0.409	Urtica spp (Nettle)	0.120–0.420
Linum usitatissium (Flax)	0.212	Miscanthus × Giganteus (Miscanthus)	0.179–0.218
Helianthus annus (Sunflower)	0.154–0.400	Rheum rhabarbarum (Rhubarb)	0.320–0.490
Brassica napus (Oilseed rape)	0.240–0.340	Brassica rapa (Turnip)	0.314
Helianthus tuberosus (Jerusalem artichoke)	0.300–0.370	Sorghum × Drummondii (Sudan grass)	0.213–0.303
Solanum tuberosum (Potato)	0.276–0.400	Chaff	0.270–0.316
Beta vulgaris (Sugar beet)	0.236–0.381	Straw	0.242–0.324
Fodder beet	0.420–0.500	Leaves	0.417–0.453

Note: *whole crop; **grain; common names of plant species are provided within parentheses. Based on Braun, 2007; Murphy et al., 2011

certainly will influence the energetic yield of a material. High-density materials present a higher mass per volume and they have an advantage of resulting in a higher combustion yield. Another consideration related to the feedstock is its moisture content. The wetter the material the more suitable it will be for handling with standard pumps instead of energy intensive concrete pumps and physical means of movement (Verma, 2002). In contrast, the wetter the material, the more volume and area it takes up relative to the levels of gas that are produced. The moisture content of the target feedstock will also affect what type of system is applied for its treatment. Even when the feedstock is produced exclusively by sources of vegetative origin, i.e. plant species or crop residues, the variation in biomethane yield is evident (Table 12.5).

Furthermore, variation in biogas/biomethane production when the feedstock is of common origin, e.g. grassland, has been reported (Prochnow et al., 2005). As these authors stated, various grassland types (e.g. mixed vs. monoculture, intensively vs. extensively managed) can affect biogas and biomethane output (Table 12.6).

Utilisation intensity of grassland can affect biogas production potential. Usually, in temperate climates, perennial ryegrass (*Lolium perenne*) is the dominant species in intensively managed grasslands because of its high water soluble and non-structural carbohydrates content (Smith et al., 2002; Buxton and O'Kiely, 2005) and reduced concentration of crude fibre (Nizami et al., 2009). It has been reported that there is a greater biogas yield (0.83–

Table 12.4 Variation of chemical characteristics, retention time and biogas production for various feedstocks.

Feedstock	Total solids (%)	Volatile solids (% of TS)	C/N ratio	Biogas yield (m^3 kg^{-1} VS)	Retention time (days)
Pig slurry/manure	3–8	70–80	3–18	0.25–0.60[a]	20–40
Cow slurry/manure	5–12	75–85	6–24	0.20–0.30	20–30
Poultry dung	10–30	70–80	3–10	0.35–0.68[a] (0.41)	> 30
Goat/sheep dung			12/19		
Stomach/intestine content	15	80		0.40–0.68	
Waste fat				1.00	
Slaughterhouse waste			22–37	0.7	
Whey	1–12	80–95		0.35–0.95	3–10
Concentrated whey	20–25	90		0.80–0.95	
Flotation sludge/animal fat				0.775	
Fermented slops	1–5	80–95	4–10	0.35–0.78	3–10
Stillage from breweries				0.60	
Waste from sawmills	20–80		511		
Wood wastes	60–70	99.6	723		
Garden wastes	60–70	90	100–150	0.20–0.50	8–30
Fruit wastes	15–20	75	7–35	0.20–0.50	8–20
Food waste	10	80	15–32	0.40–0.60	10–20
Households wastes	40–60		18		
Green waste (markets)				0.55	
Biowastes (source separated)				0.45	
Garden wastes	60–70	90		0.20–0.50	
Paper	85–95		173		
Sewage sludge	0.5–5		6–8	0.16–0.30	
Water hyacinth			25		
Flower bulbs	10	80		0.80[a] (0.50)	
Leaves	80	90	30–80	0.10–0.30	8–20
Wood shavings	80	95	511		
Straw	70	90	90	0.35–0.45	10–50
Maize straw			60		
Energy maize	35–39	35		0.65[a] (0.39)	
Grass	20–25	90	12–26	0.20–0.64	10
Grass silage	15–25	90	10–25	0.35–0.56	10
Straw/stems	70–90	80–90	50–150	0.10–0.375	

[a]Assuming biomethane content 60%; numbers in parentheses indicate methane yield (m^3 kg^{-1} VS).
Note: Data from Webb and Hawkest, 1985; Braun, 1992; Angelidaki et al., 1996; RISE-AT, 1998; Nordberg, 1999; Ortenblad, 2000; Weilard, 2000; Braun and Wellinger, 2003; Moller et al., 2004; Amon et al., 2007; Ileleji et al., 2008; Braun et al., 2009; Al Seadi, undated; Gebrezgabher et al., 2009; Schnurer and Jarvis, 2010.

Table 12.3 Important economic and technical considerations for the selection of the feedstock in AD.

Economic parameters	Technical parameters
Availability of feedstock	Consistency of physicochemical characteristics
Cost per unit of feedstock	Fitness in application
Transportation cost	Rheological characteristics
Pre-treatment cost	Surface tension factors
Storage cost	Product recovery impact
Stabilisation cost	Process yield
Price fluctuations	Product concentration Product type
Safety factors	Overall productivity

feedstock, whilst parameters such as solids, elemental and organic analyses are important for digester design and operation. Anaerobes can break down organic material quickly in the case of short chain hydrocarbons such as sugars, but over longer periods of time in the case of cellulose and hemicellulose. Anaerobic microorganisms are unable to significantly break down long chain woody molecules such as lignin (Wastesum, 2006). Various feedstock parameters directly affect the yield and quality of the biogas product (Lusk et al., 1996). The content of fixed carbon, heating value, volatile substances, ash, carbon to nitrogen ratio (C/N ratio) and cellulose/lignin contents are the most important factors to be considered for energy analyses of lignocellulosic and organic waste materials (Demirbas, 2004; Demirbas and Demirbas 2004; Moghtaderi et al., 2006; Akkaya, 2009; Erol et al., 2010). According to Owen et al. (1979) the theoretical methane yield for a given feedstock depends on the chemical oxygen demand (COD) value. In general, this relationship states that for 100 per cent conversion efficiency one gram of substrate COD will result in 350 mL of methane at standard temperature and pressure. Another key consideration in AD is the C/N ratio of the initial substrate that is subjected to anaerobic decomposition (see below). Potential feedstocks for AD can be highly variable depending on their sources and biogas potential. Some of their chemical characteristics are shown in Table 12.4. Despite some similarities of volatile solids as a percentage of total solids (VS %TS) there is large variation in TS% (i.e. 1 per cent to 90 per cent) and biogas yield (i.e. 0.15 to 0.95 $m^3 kg^{-1}$ VS). Based on the results shown in Table 12.4, animal originated wastes (e.g. whey, intestine content and manures/slurries) show great potential for biogas yield followed by food wastes and lignocellulosic feedstocks.

Physical characteristics

Feedstock size or feedstock size reduction is another factor that can improve the hydrolysis rate for polymeric feedstocks. The size of the feedstock should not be too large otherwise it would result in the clogging of the digester and also it would be difficult for microbes to carry out their digestion (Yadvika et al., 2004). Smaller particle sizes between 0.088 to 0.4 mm (Sharma et al., 1988) would provide large surface area for adsorbing the substrate and increased microbial activity (Lusk et al., 1996; Yadvika et al., 2004). The density of lignocellulosic material is also very important for bioenergy production, since it establishes a relation between the mass and volume of the by-products considering the voids that

Table 12.2 Biowastes suitable for biological treatment according to the European Waste Catalogue.

Waste category	Category constituents
Waste from agriculture, horticulture, aquaculture, forestry, hunting and fishing, food preparation and processing (Ghaly, 1996; Neves et al., 2006a; Beno et al., 2009; Sagagi et al., 2009; Al Seadi, undated; Herren et al., 2012; Elaiyaraju and Partha, 2012; Chapter 8 in this book)	Waste from agriculture, horticulture, aquaculture, forestry, hunting and fishing
	Waste from the preparation and processing of meat, fish and other foods of animal origin
	Wastes from the preparation, processing and conservation production of fruit, vegetables, cereals, edible oils, cocoa, tea and tobacco; production of yeast, preparation and fermentation of molasses
	Wastes from sugar processing
	Wastes from the dairy products industry
	Wastes from the baking and confectionery industry
	Wastes from the production of alcoholic and non-alcoholic beverages (coffee, tea and cocoa excluded)
Wastes from wood processing, furniture production of panels and furniture, pulp, paper and cardboard (Kurakake et al., 2007; Kelleher, 2007)	Wastes from wood processing and the production of panels. Wastes from wood processing and furniture
	Wastes from pulp, paper and cardboard production and processing
Waste from the leather, fur and textile industries (Asia et al., 2006)	Wastes from the leather and fur industry
	Wastes from the textile industry
Waste packing, absorbents, wiping cloths, filter materials and protective clothing not otherwise specified	Packaging (including separately collected municipal packaging waste)
Waste from waste management facilities, off-site waste water treatment plants and the preparation of water intended for human consumption and water for industrial use (Azaizeh and Jadoun, 2010; Li et al., 2011; Elaiyaraju and Partha, 2012)	Wastes from anaerobic treatment of waste
	Wastes from waste water treatment plants not otherwise specified
	Wastes from the preparation of water intended for human consumption or water for industrial use
Municipal wastes (household waste and similar commercial, industrial and institutional wastes) including separately collected fractions (Cecchi et al., 2011)	Separately collected fractions
	Garden and park wastes
	Other municipal wastes

Note: Modified by Al Seadi (undated); EU (2001); Anonymous (2001)

Chemical characteristics

Anaerobic digesters typically can accept any biodegradable material, although the level of biodegradability is the key factor for its successful application. Substrate composition is a major factor in determining the methane yield and methane production rates from the digestion of biomass. Techniques are available to determine the compositional characteristics of the application of AD systems is highly dependent on feedstock since this determines the rate of the anaerobic degradation and must be taken into consideration in the process technology and process operation. Some of the factors affecting the performance of the AD process are related to the choice of the feedstock since its physical and chemical characteristics affect the design and operation of anaerobic digesters along with the biogas production and process stability (Zhang et al., 2007). It has been shown, for example, that the hydrolysis of crops and crop residues prior to AD could significantly reduce the hydraulic retention time (HRT) of some digesters to below 100 days whereas the type and/or mixture of feedstock would also influence the biogas yield (Anonymous, 2008). Feedstock for AD includes a wide range of substances, easily convertible to biomethane by anaerobic bacteria, as it can range from degradable organic wastes to complex high-solid waste (e.g. grass, maize, solid municipal organic wastes etc.). The most common feedstock categories for biogas production are listed below.

- Animal manure and slurry (Clemens et al., 2006; Holm-Nielsen et al., 2009; Li et al., 2011; Provenzano et al., 2011; Riano et al., 2011; Xie et al., 2012) ;
- Agricultural residues and by-products (Molnar and Bartha, 1988; Nuri et al., 2008; Parawira et al., 2008; Holm-Nielsen et al., 2009; Rao et al., 2010; Eze and Ojike, 2012);
- Organic wastes from food and agro industries (vegetable and animal origin) (Salminen and Rintala, 2002; Dearman and Bentham, 2007; Zhang et al., 2007; Fountoulakis and Manios, 2009);
- Organic fraction of municipal waste (Zhang et al., 2007; Beno et al., 2009; Chapter 7 in this book);
- Sewage sludge (Sharma et al., 1997; Hamzawi et al., 1998; Neves et al., 2006; Beno et al., 2009);
- Edible and non-edible energy crops such as maize, grass, miscanthus, sorghum, clover, sunflower and others (Vandevivere 1999; Mahnert et al., 2005; Stinner et al., 2008; Smith et al., 2008; Uzodinma and Ofoefule, 2009; Madlener et al., 2009; Korres et al., 2010; Chapter 4 and Chapter 5 in this book);
- Seaweeds and algae (Chapter 6 in this book).

The list of the potential feedstock for biogas production can be expanded if the biological materials suitable for bio-processing, according to EU Legislation (EU, 2001), as listed in Table 12.2, are added to it.

Variation in biogas/biomethane production due to feedstock

The ultimate choice of feedstock type for a given digestion process is a complex decision, based on imperatives given by the microbial strain involved, the nature of the end product and on technical and economic considerations such as these presented in the Table 12.3.

Additionally, the physical and chemical characteristics need to be considered if the maximum output is to be achieved. The major problem in using plant residues and organic wastes is their variable dimensions and compositions due to the different production systems that generated them, high moisture content, and great volume. For this reason, they present different potentials for energy and must be studied for bettering decision-making procedures and if possible the classification of these feedstocks based on various common characteristics.

Table 12.1 Typical composition of biogas.

Substance	Volume content (%)
Methane CH_4	55–75
Carbon dioxide CO_2	25–45
NH_3, N_2, Urea	1–5
Carbon monoxide CO	0–0.3
H_2	0–3
Hydrosulphide H_2S	0.1–0.5
O_2 and H_2O	Traces

Note: Adopted from Karellas et al. (2010)

limiting, consequently a thermophilic catabolism of fat is preferred. The biogas produced during anaerobic digestion is composed of CH_4 (55–75 per cent), CO_2 (25–45 per cent) and trace elements (Table 12.1), and can be used as an energy source, i.e. heat, electricity or as a transport fuel (Figure 12.1).

Each phase in the biogas production chain as described in Figure 12.1 could affect significantly the production of biogas and increase variation in the final product within and between AD production systems. The process of AD has been widely applied to treat a range of feedstocks (Chapters 4–8). The potential of each feedstock for optimum biogas production varies significantly due to various qualitative and quantitative characteristics along with pre-processing, digester design and configuration (Chapter 9) and AD process parameters. According to Anonymous (2008) the methane output per kg of volatile solids (VS) from different biogas plants has been recorded to vary widely (up to three times) due to feedstock. These plants have shown a variation (based on standard deviation) of the specific methane yield as low as 7 per cent while others could be considered unstable with values over 100 per cent of their mean values. Homogeneity of the feedstock for maximum biogas production, through the application of an appropriate pre-treatment technique, is required in most AD systems (Anonymous, 2006). Nizami et al. (2009) reported that economically viable and operationally efficient pre-treatment options include physical/mechanical, chemical, thermal and biological techniques have been widely used especially in the case of the lignocellulosic feedstocks. The variation due to feedstock and feedstock pre-treatment can affect hydraulic retention time (from 30 to more than 100 days) (Anonymous, 2008). Therefore the intention of this chapter is to describe those generic factors and process control parameters which affect overall biogas production variation and can be considered for all biogas plants. Particular attention is given to the identification of the key parameters for AD process monitoring at the most representative stage of the whole biogas production chain, aiming to highlight the real possibility of plant breakdown and aid process diagnosis.

Feedstock

The technology for the production of biogas/biomethane by AD of organic materials which are abundant, low-cost and renewable in nature, is readily available. In fact, several thousand biogas plants are already in operation in many developed (Korres et al., 2011) and developing countries such as China, India, Pakistan, Sri Lanka and others (Chapter 19). The

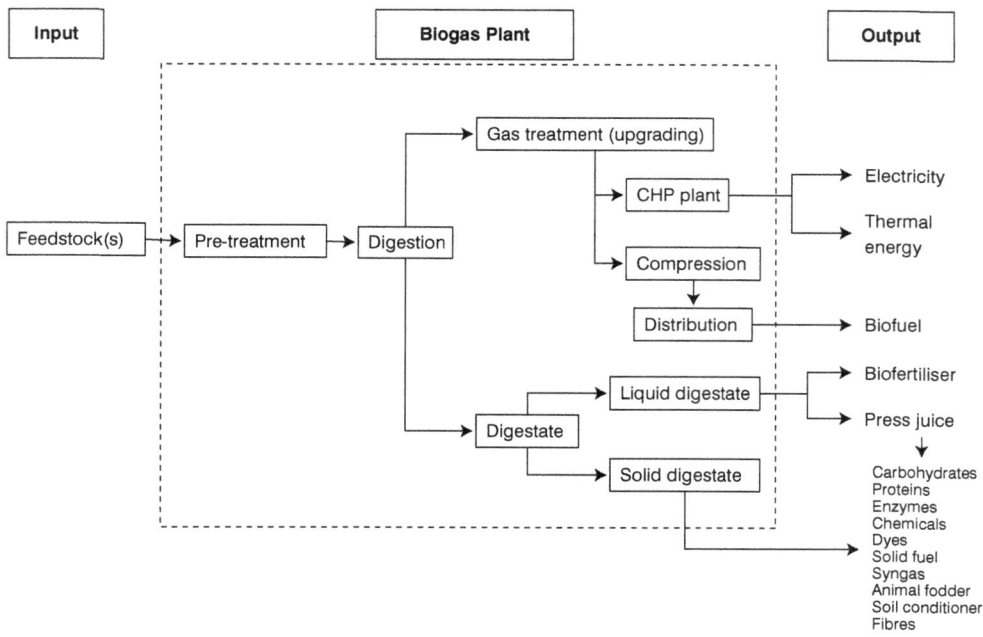

Figure 12.1 Biogas production chain. Based on Raven and Gregersen (2005); Karellas et al. (2010); Korres et al. (2011); CHP = combined heat and power

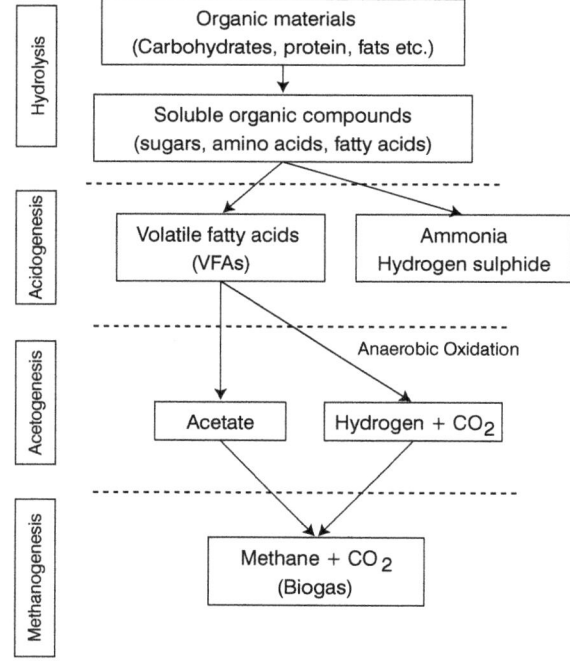

Figure 12.2 Schematic representation of methane production by AD. Based on Reith et al. (2003) and Weiland (2010)

Chapter 12

Variation in anaerobic digestion
Need for process monitoring

Nicholas E. Korres[1]* and Abdul Satar Nizami[2]

[1]26 Grigoroviou str, Patisia, GR-11141, Athens, Greece;
[2]Department of Chemical Engineering and Applied Chemistry, University of Toronto, ON, Canada
*Corresponding author email: nkorres@yahoo.co.uk

Variation in biogas production chain

The EU policies concerning renewable energy systems have put forward a fixed goal of supplying 20 per cent of the European energy demands from renewable energy systems by year 2020 (EC, 2006). A major part of the renewable energy will originate from European farming and forestry whereas at least 25 per cent of all bioenergy in the future can originate from biogas, produced from wet organic materials such as animal manure, whole crop silages, wet food and feed wastes (Holm-Nielsen et al., 2009). Production of biogas is an integrated process in which numerous stages and permutations are involved. Generally the overall biogas production can be divided into three phases, namely the input phase (i.e. production/collection of the feedstock, transportation and storage), the biogas plant/processing phase (i.e. pre-treatment, anaerobic digestion *per se*, gas treatment and digestate treatment) and the output phase (i.e. production of various goods and value-added products as in biorefineries) (Figure 12.1).

Through the whole biogas production chain the driving biological process is anaerobic digestion. Anaerobic digestion (AD) involves the degradation of organic materials, under anaerobic conditions, by micro-organisms that leads to the formation of biogas, the renewable energy source, and digestate, a residual that can be applied to land as a substitute for manufactured fertilisers (Chapter 17). Four successive biological stages are involved in the anaerobic degradation of organic matter, namely hydrolysis, acidogenesis, acetogenesis and methanogenesis (Figure 12.2). The conversion of complex polymers into monomers by extra-cellular enzymes during hydrolysis is followed by the transformation of the latter into volatile fatty acids (i.e. acetic, propionic and butyric acids) and hydrogen (H_2) during acidogenesis. Acetate, carbon dioxide (CO_2) and H_2 are produced from volatile fatty acids during acetogenesis and finally converted into methane (CH_4) during methanogenesis (Bernet and Beline, 2009; Korres et al., 2011).

It has been stated that priority should be given to methanogenic bacteria due to their lower growth rate and sensitivity in environmental fluctuations. Nevertheless, as mentioned by Deublein and Steinhauser (2008), exceptions to this rule due to feedstock characteristics should be considered. For example, with lignocellulosic feedstocks the hydrolysis limits the process and therefore it is the stage that merits consideration whereas with proteinaceous feedstocks, a single-stage plant is quite sufficient because the pH optima are the same in hydrolytic and methanogenic stages of AD. With fats, the hydrolysis proceeds more rapidly with increasing emulsification, therefore acetogenesis is

Gornal, L. (undated) 'Hemispherical dome gas bags'. Available: http://ds.dial.pipex.com/town/terrace/ae198/GasBags.html [Accessed February 2012].

Karellas, S., Boukis, I. and Kontopoulos, G. (2010) 'Development of an investment decision tool for biogas production from agricultural waste', *Renewable and Sustainable Energy Reviews*, vol 14, pp. 1273–1282.

Krich, K., Augenstein, D., Batmale, J., Benemann, J., Rutledge, B. and Salour, D. (2005) 'Biomethane from dairy waste – a sourcebook for the production and use of renewable natural gas in California', A study funded through the Value-Added Agricultural Product Market Development Grant Program, administered by USDA Rural Development, July 2005. http://www.suscon.org/cowpower/biomethaneSourcebook/biomethanesourcebook.php [Accessed February 2012].

Mangione, T. (2011), 'Cover system helps store biogas for on-site power applications'. PennWell Corporation. Available: http://www.waterworld.com/index/display/article-display/293884/articles/waterworld/volume-23/issue-6/editorial-feature/cover-system-helps-store-biogas-for-on-site-power-applications.html) [AccessedMarch 2012].

Mathiasson A (2008) 'Vehicle gas utilization in Sweden – today and tomorrow'. In *2nd Nordic Biogas Conference*. The Swedish Gas Association, Malmo.

Murphy, J.D. and Power, N.M. (2009) 'An argument for using biomethane generated from grass as a biofuel in Ireland', *Biomass and Bioenergy*, vol 33, pp. 504–512.

Patterson, T., Esteves, S., Dinsdale, R. and Guwy, A. (2011) 'Life cycle assessment of biogas infrastructure options on a regional scale', *Bioresource Technology*, vol 102, pp. 7313–7323.

Persson, M., Jonsson, O. and Wellinger, A. (2006) 'Biogas upgrading to vehicle fuel standards and grid injection', International Energy Agency Task 37 Bioenergy.

Robertson, G. and Grace, P. (2004) 'Greenhouse gas fluxes in tropical and temperate agriculture: The need for a full-cost accounting of global warming potentials', *Environment, Development and Sustainability*, vol 6, pp. 51–63.

Ross, C.C, Drake, T.J. and Walsh, J.L. (1996) *Handbook on biogas utilization*, 2nd edn, US Dept of Energy, South-eastern Regional Biomass Energy Program, July 1996, Tennessee Valley Authority.

Siemens AG Water Technologies. (2011). 'Digester gas holder systems'. Available: http://www.water.siemens.com/en/products/sludge_biosolids_processing/digesters/Pages/envirex_product_dystor_gas_holder.aspx [Accessed January 2012].

Smyth, B.M., Smyth, H. and Murphy, J.D. (2010) 'Can grass biomethane be an economically viable biofuel for the farmer and the consumer?', *Biofuels, Bioproducts and Biorefining*, vol 4, pp. 519–537.

Urban, W., Girod, K. and Lohman, H. (2008a), *Ergebnisse der Markterhebung 2007–2008*, Obeerhausen: Fraunhofer UMSICHT.

Urban, W., Girod, K. and Lohman, H. (2008b), *Technologien und Kosten der Biogasaufbereitung und Einspeisung in das Erdgasnetz. Ergebnisse der Markterhebung 2007–2008*, Oberhausen: Fraunhofer UMSICHT.

cent investment subsidy for network connection, which biomethane suppliers are entitled to under German regulations (GasNZV).

If biomethane is to be used as a vehicle fuel, a filling station is required. Filling stations can incorporate fast and slow fill systems; the latter is used for captive fleet overnight slow fill, while the former can be used by cars with a fill time of approximately 10 mins. The cost of building a medium-scale biomethane/BioCNG (Biogas-based CNG) service station is estimated at €500,000 (AEBIOM, 2009).

Conclusions

The provision of sufficient gas storage is an important design consideration which can have a major effect on the efficiency and viability of a biogas plant. Biogas storage systems can be used to maintain operating pressure for low-pressure on-site boilers and gas engines.

For biomethane distribution, the use of biogas pipelines or gas grid injection at larger scale offers the most efficient mode of distribution from an economic and environmental impact viewpoint. It allows for greater energy efficiency and flexibility of end use. In countries where AD is a mature technology (e.g. Germany), the latest trends show that injecting biomethane into the gas grid is an attractive method for distribution. As petroleum transport fuels continue to rise, many public and private enterprises are looking towards alternative fuels. CNG vehicles are increasing across the globe. With the expanding CNG market, the opportunity (AEBIOM, 2009) for using biomethane as a transport fuel is also increasing. Efficient biomethane distribution networks will facilitate the change from petroleum based transport to biomethane and/or natural gas. By using biomethane as a transport fuel, better profits can be made, therefore making investment in AD capital projects more attractive, e.g. gas transport via pressurised pipeline is considerably more energy efficient than electrical transport via high voltage cables.

References

AEBIOM (2009) *A biogas road map for Europe*, Brussels: European Biomass Association.

Anonymous (2011) 'Opportunities for combined heat and power at wastewater treatment facilities: market analysis and lessons from the field', U.S. Environmental Protection Agency, Combined Heat and Power Partnership, October 2011, available: http://epa.gov/chp/documents/wwtf_opportunities.pdf [Accessed May 2012].

Argini, M., Baroni, M., Corbeau, A.S., Cozzi, L. et al. (2011) 'Are we entering a golden age of gas? Special Report', *World energy outlook* 2011, OECD/IEA, available: http://www.iea.org/weo/docs/weo2011/WEO2011_GoldenAgeofGasReport.pdf, [Accessed May 2012].

Birkmose, T., Lyngsø Foged, H. and Hinge, J. (2007) 'State of biogas plants in European agriculture', Negotiated procedure No. IP/B/AGRI/IC/2007-020, Prepared for European Parliament, Directorate General Internal Policies of the Union, Directorate B – Structural and Cohesion Policies, Danish Agricultural Advisory Service.

Browne, J., Nizami, A.S., Thamsiriroj, T. and Murphy, J.D. (2011) 'Assessing the cost of biofuel production with increasing penetration of the transport fuel market: A case study of gaseous biomethane in Ireland', *Renewable and Sustainable Energy Reviews*, vol 15, pp. 4537–4547.

Budzianowski, W.M. (2012) 'Benefits of biogas upgrading to biomethane by high-pressure reactive solvent scrubbing'. *Biofuels Bioproduct & Biorefining*, vol 6, pp. 12–20.

Eisentraut, A. (2010) 'Sustainable production of second generation biofuels', Information paper, International Energy Agency.

from about 21 MJ/m³ to 37.7 MJ/m³. Upgrading methods such as HPWS require compression of the biogas to about 3 bars. Economies of scale mean biogas upgrading is best suited to large-scale AD systems; 500 kWe biogas plant requiring 250 m³ raw biogas per hour is typical (Urban et al., 2008b; Budzianowski, 2012).

Biogas and biomethane transport and distribution

Biogas or biomethane distribution is usually the final major design consideration in a renewable gas system. The distribution options largely depend on plant capacity and the gas end use. Traditionally biogas was used on-site or piped over short distances to the points of utilisation. In large-scale biogas plants, purifying biogas to biomethane is becoming more common. By utilising the existing natural gas distribution grid, the distribution of biomethane can be made more economically and environmentally advantageous. Other concepts for biogas transport and distribution include a number of small-scale biogas plants linked into a dedicated biogas pipeline which can then be used to distribute the biogas to a large CHP or upgrading plant. This system may work well for a number of small-scale biogas plants in close vicinity, which seek to optimise the energy output from the biogas.

The utilisation of biomethane as a transport fuel in compressed natural gas (CNG) vehicles is seen as an attractive outlet for biomethane as a renewable fuel. With increasing petroleum prices and the finite resources available coupled with increasing global demand, locally produced biofuel is a major goal for many countries. More particularly, the use of biogas as a transport fuel, via anaerobic digestion of various feedstocks, after upgrading to biomethane, has recently started to gain attention in many European countries, such as Sweden, Austria, France and Switzerland (Mathiasson, 2008). It is worth mentioning that the potential for biogas production based on manure in the EU is 827 PJ, whereas today only about 50 PJ is produced from agricultural biogas plants with added animal manure, energy crops and organic waste added to these (Birkmose et al., 2007). Additionally, biogas production from biomass has been strongly promoted in many developing regions including Asia, Latin America and some regions of West Africa (Eisentraut, 2010).

There are many avenues for biomethane distribution for the transport market; biomethane can be compressed to approximately 250 bar, stored on-site in high pressure gas cylinders. These are collected from biomethane producers and are transported to a centralised gas filling station. However, this system of collection may not be the most efficient from an energy viewpoint as the transportation vehicles will also use fuel. Additionally, compression of biomethane to 250 bar which costs approximately €0.11/m_n^3 biomethane also requires significant electrical input, a value of 0.35 kWeh/m³ biomethane (Murphy and Power, 2009).

In countries with extensive natural gas networks, injection of biomethane into the existing natural gas grid could be a more advantageous supply route with respect to energy efficiency and associated environmental benefits of displacing transport vehicles (Smyth et al., 2010). The gas distribution grid operates at approximately 4.2 bar so no additional compression is needed for injection to the distribution grid as HPWS upgrading plants typically pressurise biomethane up to 7–9 bar (Smyth et al., 2010). There are a few European countries, most notably Germany, where injection of biomethane into the national grid is becoming more common. From discussions with the industry, the cost of connection to the gas grid can vary widely and depends on distance to the network, ground conditions and the type of pipe etc. Smyth et al. (2010) estimated the capital cost of grid connection at €200,000 while Urban et al. (2008a) estimated a high cost of connection of approximately €300,000 including a 50 per

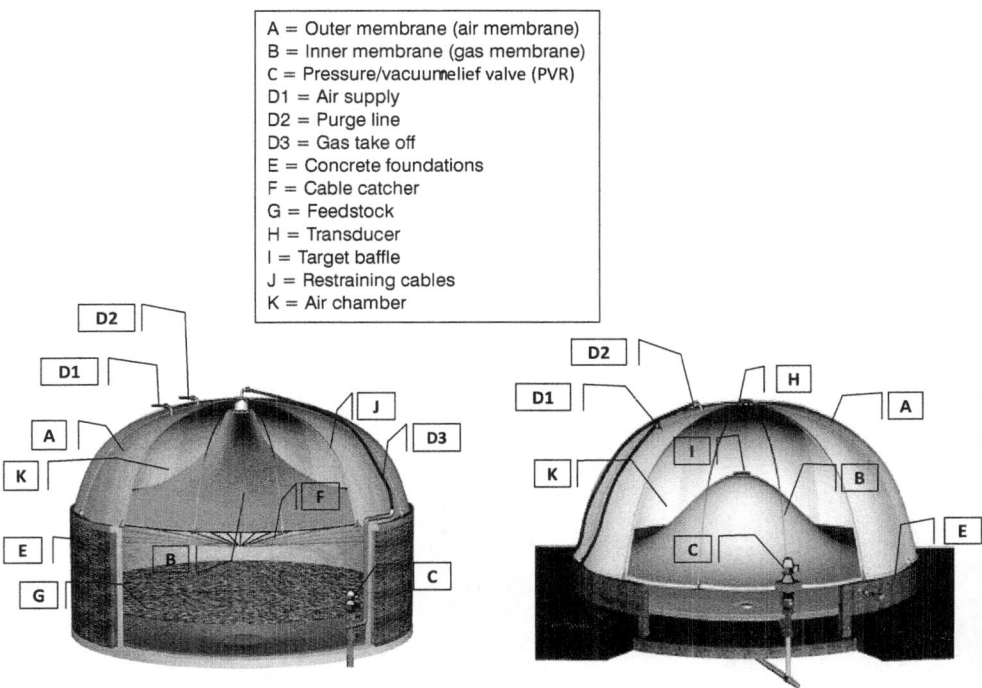

Figure 11.4 Cross-section of flexible membrane covers (based on Mangione, 2011) (courtesy of Siemens Industry, Inc.)

Medium-pressure biogas storage

If a system requires a gas pressure greater than several mbars but less than 15 bars (approx. 200 psi), clean biogas (H_2S removed) may be compressed and stored in tanks such as propane gas tanks (Table 11.1). At these higher pressures, the use of safety codes, protocols and devices (e.g. insurance investigations, pressure vessel codes, pressure safety devices) are required (most of the times by law). Medium-pressure storage tanks are less expensive than their low-pressure counterparts but the requirements for compression and gas clean-up make medium-pressure storage more expensive. In exchange for the higher cost, the same volume of gas can be stored in a smaller vessel and the stored biogas possesses a higher energy density than that in a low-pressure system.

High-pressure biomethane storage

When biogas is purified and refined to about 98 per cent methane by removing carbon dioxide, hydrogen sulphide and water vapour, the resulting product is biomethane which can be used in the same manner as natural gas. Since biomethane is usually for off-site use (e.g. as fuel in compressed natural gas vehicles) it is either compressed and stored in cylinders or is injected into a gas distribution pipeline where access to such a network is available. Using available technology such as high-pressure water scrubbing (HPWS) or pressure swing adsorption, the non-methane fractions removed from the biogas bring its energy content

Storage and distribution of biomethane 189

Figure 11.3 Weather-proof outer layer of flexible membrane cover

process. A pre-set operating pressure is continuously maintained between the two membranes. Such systems can hold up to three times as much gas as conventional digester covers, thereby reducing the cost of gas storage. Flexible membrane covers can provide an economic storage option for low-pressure gas systems. Depending on the substrates used in the digester and their corresponding biogas yield, the gas storage can hold up to a day's supply of biogas or more. By storing biogas while performing routine maintenance on gas processing equipment, no biogas is wasted.

The biogas storage membranes shown in Figures 11.3 and 11.4 are composed of a high quality rubber and are characterised by good UV and ozone stability. The gas is efficiently processed by means of a large seeding surface in combination with a timber frame roof. The roof and gas storage can be opened and closed quickly, facilitating quick repair. Over the lifetime of a project, the ability to store gas while allowing easy access to the tanks for inspection or maintenance purposes should significantly increase plant efficiency. A small air pump keeps the outer members' shape while maintaining a constant pressure on the inner gas store as shown in Figure 11.4.

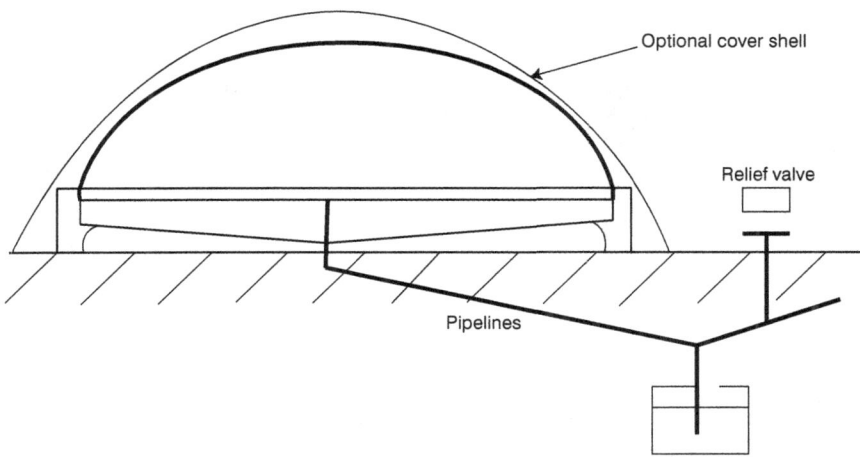

Figure 11.2 Biogas low-pressure gas bag storage facility

traps should be built above any winter water table and below frost levels in soil. Gas bags may need protection from the elements as bags may be punctured or lost in strong winds. They should be made from a material that is resistant to hydrogen sulphide on the inside layer, provide high flexural strength and be resistant to UV degradation on the outer layer.

Rigid digester cover

In the rigid cover design, gas storage is simply incorporated in the headspace of the digester which is constructed of a ridged material such as mild or stainless steel, aluminium, fibreglass or reinforced concrete. It is important that the gas holding section of the digester be gas tight, which may require the application of a sealant to prevent gas escaping. It has been reported that storage of the biogas can be less effective than other gas holding systems as the gas pressure can fluctuate substantially within the digester. Without gas pressure regulation even simple appliances cannot be set in an optimal way, therefore a gas pressure regulator or a floating gas holder is necessary if the biogas is required at constant pressure for downstream gas loads.

Flexible membrane cover

Many commercial digesters now use flexible membrane covers which incorporate gas storage and facilitate constant gas flow to energy conversion units. The flexible membrane cover should not react with hydrogen sulphide in biogas. A variety of flexible membrane covers are commercially available, either as an integral component of the digester, or as part of an auxiliary unit. These types of covers are often used with plug flow and complete-mix digesters. Flexible membrane materials commonly used for these gas holders include high-density polyethylene, low-density polyethylene, linear low-density polyethylene and chlorosulfonated polyethylene covered polyester (Krich et al., 2005).

Usually two durable membranes are used; the outer membrane is cable restrained and remains inflated in a fixed position (Siemens AG Water Technologies, 2011); the inner membrane moves freely as it stores or releases gas generated from the anaerobic digestion

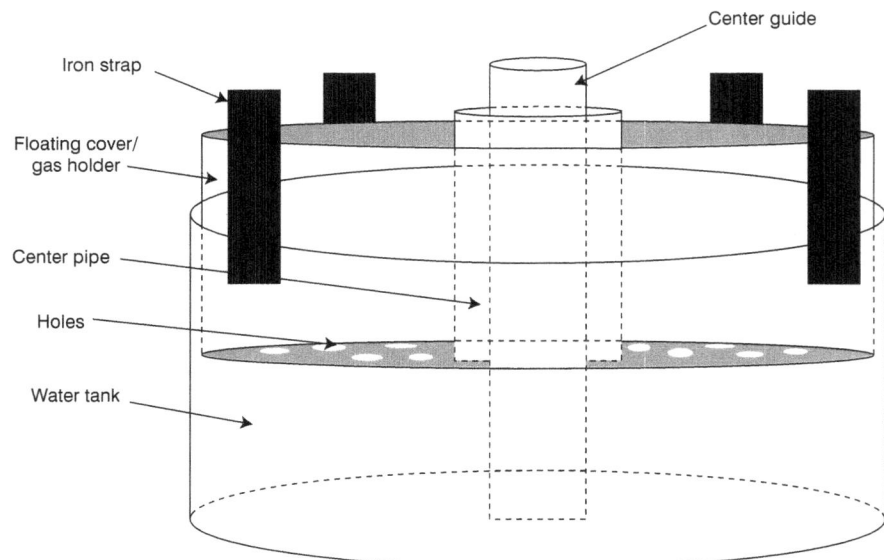

Figure 11.1 Floating tank cover

water jacket. The gas is collected in the gas holding tank, which rises or falls according to the amount of gas stored (i.e. biogas from the digester fills the gas tank, displacing water and pushing the gas tank up; when the biogas is used, water replaces the consumed gas and the gas tank falls). The gas drum is prevented from tilting by a guiding frame along the outside walls of the gas holding tank. In larger biogas plants a separate floating gas holding tank may be required to fulfil the desired storage and pressure conditions. Such gas storage floating tanks are simple in principle and typically operate at low pressures; however storage pressure can be increased by placing weights on top of the floating tank.

Gas bags

Gas storage bags are a good storage option for small-scale temporary gas storage. A large variety of gas bags are available on the market in many shapes and sizes. Flexibility and relative ease of installation are the main advantages of gas bag storage. One example of gas bag storage is the hemispherical dome gas bag (Figure 11.2), which come in sizes of 25–500 m^3 (Gornall, undated). Gas bags can provide overnight biogas storage when there is no need for prior gas scrubbing or gas compression. Typically gas bags have no moving parts except the variable volume membrane. Generally there are no motors to compress the gas and the low-pressure system provides very low stress on components. The gas is normally extracted from the storage dome by means of a small gas fan or pump to provide pipeline pressures between the bag and the boiler or CHP unit. This enables the digester and gas bag to be operated at very low gas pressures (Table 11.1).

Gas bags are often mounted on level hard bases that exceed the diameter of the gas bag, with a deep trench to the circumference of the base from the centre of the bag to take the outlet pipe below frost level. They should be located such that gas pipes are never horizontal but fall to a water trap or away from a water trap to avoid blockage by standing water. Water

Table 11.1 On-farm storage options for biogas and biomethane.

Purpose of storage	Pressure (mbar)	Storage device	Material	Size (m³)
Short and intermediate storage for on-farm use	< 7	Floating cover	Reinforced and non-reinforced plastics, rubbers	Variable volume usually less than one day's production
	< 137	Gas bag	Reinforced and non-reinforced plastics, rubbers	14–1,025
	137–413	Water sealed gas holder	Steel	325
		Weighted gas bag	Reinforced and non-reinforced plastics, rubbers	82–2,600
		Floating roof	Plastic, reinforced plastic	Variable volume, usually less than one day's production
Storage for later on- or off-farm use	700–200,000	Propane or butane tanks	Steel	186
	> 200,000	Commercial gas cylinders	Alloy steel	32.5

Note: Based on Ross et al., (1996).

the flexible membrane cover, which allows more efficient gas storage than older steel tanks, since it does not react with H_2S in the biogas (Krich et al., 2005).

In small-scale systems, the need for biogas storage is required for shorter time periods, e.g. during plant maintenance or at peak production. Therefore the storage volume required is relatively small and may allow for one to two days' gas production. In general for smaller biogas systems, low-pressure gas storage is sufficient to meet the demands of the system.

In Table 11.1, the basic on-farm storage options are presented along with specifications of the size, materials and storage pressure.

The major advantage of a digester with an integral storage component has to be the reduced capital cost of the system. However, such a design features several areas that require special attention. The roof of the digester must be insulated. Non-insulated covers are susceptible to large temperature fluctuations which will cause operating problems in the digester. Floating and flexible covers present a second problem. The cover will most probably deflate as the gas is used, so the digester will need to be sheltered to some degree to protect its cover from wind damage.

Floating cover

In smaller AD applications a common and simple gas storage design is the floating gas top system, in which a floating tank with holes in its bottom is directly installed on top of an open-top tank digester filled with water or slurry, for gas storage (Figure 11.1). The floating cover or gas holder where the gas is stored floats either directly on the slurry or in a concentric systems may be required to satisfied biomethane distribution demands. The operational requirements of high pressure systems are more expensive due to the higher maintenance involved with compressors and health and safety requirements. Installation cost premiums arising from more robust and high specification equipment, adds to high storage expenses in these systems. The additional costs are often justifiable as the end product has a higher energy content and is more valuable than raw biogas as in the biomethane for transport fuel model (Krich et al., 2005; Browne et al., 2011). The extent and cost of biogas storage is influenced by the gas end use and scale of operation.

Biogas end uses and storage options prerequisites

Biogas can be utilised in a wide range of applications from small household scale gas stoves and lamps (see Appendix) to industrial scale CHP units or biogas upgrading plants where biogas is purified to biomethane and can be used off-site in the same manner as natural gas. Each of these applications requires biogas storage to a lesser or greater degree according to specifications and uses.

As biogas in its raw form has lower energy content (expressed by the Wobbe index as described in Chapter 10) and different chemical properties to natural gas, domestic appliances such as gas stoves and boilers need to be designed or modified to facilitate efficient combustion of biogas. Due to its chemical composition (i.e. biogas usually contains on average 60 per cent methane and 40 per cent carbon dioxide (CO_2) and some other minor constituents), it will not burn clean in a standard natural gas burner because the high CO_2 content indicates that the fuel/air ratio is increased and also the flame speed is slower, making flames less stable.

The removal of hydrogen sulphide (H_2S) and other contaminants such as ammonia gas may be necessary to maintain engine performance and reduce the risk of failure from corrosion of mechanical components. In this case, biogas storage must be resistant to any potential chemical corrosion from such gas components. Hydrogen sulphide is a very corrosive component of biogas. In boilers it oxidises to form sulphuric acid which can dissolve the metal parts of the heat exchangers and chimneys. In internal combustion engines, hydrogen sulphide reacts with copper alloys and rapidly destroys the bearings and other engine parts. Removal of H_2S is therefore a prerequisite for safe biogas processing (i.e. cleaning) in biogas upgrading plants which tend to be relatively large facilities in order to take advantage of economies of scale, biogas storage and distribution is an important aspect of the process. It is fair to suggest that the larger the scale of operation, the greater the need for gas storage.

Low-pressure biogas storage

Low-pressure storage systems are the simplest and least expensive methods for on-site utilisation or intermediate storage of biogas. Biogas can be stored between the liquid level of the digester and the digester cap as the roof can float (i.e. rise as more gas is stored) or can be made of flexible material. Floating gas holders on the digester form a low-pressure storage option for biogas systems and typically operate at pressures below 140 mbars (approximately 2 psi). In current industry practice there are a number of different low-pressure storage options for biogas, which can be incorporated into the digester itself or can be separate from the main digester. These include gas bag, fixed digester cover, floating top cover and flexible membrane cover systems (Table 11.1). In recent years there has been a trend towards using

Biogas storage options

Biogas storage is an integral part of digester design which is necessary to compensate for fluctuations in the production and consumption of biogas. In addition, storage is an important operational option because it secures adequacy of the product for later on-site use and/or after transportation to off-site distribution points or systems. Hence, ensuring sufficient gas storage will dampen market volatility and improve energy security (Argini et al., 2011). The correct selection and dimensioning of the biogas storage system can make a substantial contribution to the efficiency and safety of the biogas plant. Suitable biogas storage assures a continuous supply of biogas and reduces gas losses, enhancing digester safety and reliability. As biogas is not produced at a constant rate, its storage is required to ensure downstream equipment operates at a constant pressure. Gas storage systems are employed to stabilise variations in gas production, quality and consumption.

A wide variety of materials have been used to make biogas storage vessels. Medium- and high-pressure storage vessels are usually constructed of steel while low-pressure storage vessels can be made of steel, concrete, plastics and polymers. Each material possesses advantages and disadvantages that must be considered individually. In recent years major advances have been made in synthetic membrane designs. A variety of engineered materials such as reinforced plastics, rubber, nylon and polyester fabrics are used to create flexible digester covers.

Certain criteria affect the choice of the storage system, such as digester volume, gas production rate and working gas pressure in the system along with the required downstream pressure to maintain gas supply to external loads, i.e. boilers or CHP units are used for the assessment of the appropriate gas storage selection. In general, large gas storage allows for high CHP utilisation and increased efficiencies (Anonymous, 2011). The delivery pressure required for the final biogas conversion system will affect the choice of biogas storage. The operating gas pressure for most anaerobic digesters rarely exceeds 60 millibars (mbars), which is only adequate to supply simple direct combustion devices such as flares or simple boilers without compression. However, more often than not, there is a need for a fan, blower or compressor to overcome the pressure drop along delivery piping and in the biogas cleaning processes. The use of biogas in gas engines often requires compression to high pressures to achieve minimal storage volume.

Gas storage can be as simple as a floating top digester operating at low gas pressures of a few mbars (e.g. small-scale biogas applications such as a biogas stove or boiler require low operating pressures of about 10–60 mbars and gas storage may be simply incorporated into the head space of the digester) or as elaborate as storage in pressurised gas cylinders up to 300 bars. In general, low-pressure gas storage systems offer the least expensive and least complex gas storage systems for AD facilities. In many small-scale AD processes, biogas is combusted directly in simple biogas boilers without compression and the need for biogas storage is usually of a temporary nature, such as when production exceeds consumption or during maintenance of digester equipment.

However, in larger biogas plants supplying gas to CHP units, much larger operating gas pressures of up to 3 bars are required. For undisturbed electricity and heat production, biogas storage and compression is required, although the biogas may be stored for later use when applications require variable power or when production is greater than consumption.

For larger scale AD facilities designed for high energy output (e.g. 500 kWe), the gas storage requirements may be substantial. In biogas upgrading plants, high-pressure storage

Chapter 11

Storage and distribution of biomethane

Nicholas E. Korres
26 Grigoroviou str, Patisia, GR-11141, Athens, Greece
Email: nkorres@yahoo.co.uk

Introduction

The conversion of biomass to biogas, which typically contains 55–75 per cent methane (Karellas et al., 2010) depending on the feedstock and anaerobic digestion (AD) technology used, is one of the major advantages of AD (Persson et al., 2006). However the rate of biogas production is a function of the biochemical process and is prone to fluctuations during long-term continuous digestion. Depending on the nature and role of the AD process (i.e. bioenergy production or waste treatment with biogas production as co-product) gas storage and distribution has a significant role in the overall process scheme, hence, the associated investment costs for the project.

For commercial biogas plants, where income comes from the sale of renewable electricity, heat or biomethane as transport fuel, sufficient storage is required to allow constant energy output and minimise the effect of fluctuations in biogas production which will invariably happen over time. The capital investment costs for combined heat and power (CHP) utility or biogas upgrading equipment can be relatively large, making high efficiency a priority, e.g. achieving 8000 hours of operation per annum of production is desired to reduce the unit cost of energy production. The ultimate goal of energy conversion systems is to achieve consistent energy output at the highest possible conversion rate, although in practice this is often difficult due to the complexity of the AD process.

Where AD is utilised in a waste treatment process based on gate-fee income rather than energy output, the role of biogas storage and distribution may play a less critical role as the objective of the plant is to maximise waste throughput to a certain environmental standard. However because methane is a potent greenhouse gas with high global warming potential (Robertson and Grace, 2004) uncontrolled methane emissions are to be avoided. Biogas storage is required to prevent methane being discharged or flared off unnecessarily, particularly in periods of low demand (Anonymous, 2011). Flaring of biogas due to insufficient biogas storage, results in energy wastage. It has been demonstrated that methane losses during biogas upgrading can negatively impact on the economics of the upgrading system (Patterson et al., 2011). With increasing environmental awareness and the introduction of emission offsets mechanisms such as CAP-and-Trade, waste treatment companies are now taking advantage of the stored energy in biogas via gas storage and utilisation.

FAIR-CT 96-2083 (DG12-SSMI) 31 Nijmegen: AD-NETT. www.agrienvarchive.ca/bioenergy/download/AD_techsum_biogas_AD-NETT.pdf (accessed July 2012).

Schweigkofler, M. and Niessner, R. (2001) 'Removal of siloxanes in biogases', *Journal of Hazardous Materials*, vol 83, pp183–196.

Strevett, K.A., Vieth, R.F. and Grasso, D. (1995) 'Chemo-autotrophic biogas purification for methane enrichment: mechanism and kinetics', *Chemical Engineering Journal and Biochemistry Engineering Journal*, vol 58, pp 71–79.

Tower, P. (2003) 'New technology for removal siloxones in digester gas. Results in lower maintenance costs and air quality benefits in power generation equipment', WEFTEC 03 78th Annual Technical Exhibition and Conference, 11–15 October. www.appliedfiltertechnology.com/page4813.asp (accessed 23 May 2011).

Ward, A.J., Hobbs, P.J., Holliman, P.J. and Jones, D.L. (2008) 'Optimisation of the anaerobic digestion of agricultural resources', *Bioresource Technology*, vol 99, pp 7928–7940.

Wheless, E. and Pierce, J. (2004). 'Siloxanes in landfill and digester gas update', presentation at SWANA's 27th Annual Landfill Gas Symposium, March 2004, San Antonio, TX.

Zhao, Q., Leonhardt, E., MacConnell, C., Frear, C. and Chen, S. (2010) 'Purification technologies for biogas generated by anaerobic digestion', *Climate friendly farming*, CSANR Research Report 2010. http://csanr.wsu.edu/publications/researchreports/CFF%20Report/CSANR2010-001.Ch09.pdf (accessed April 2011).

Kohl, A. and Nielsen, R. (1997) *Gas purification*, 5th edition, Houston, TX: Gulf Publishing Company.

Krich, K., Augenstein, A., Batmale, J., Benemann, J., Rutledge, B. and Salour, D. (2005) 'Upgrading dairy biogas to biomethane and other fuels', in K. Andrews (ed.). *Biomethane from dairy waste – A sourcebook for the production and use of renewable natural gas in California*, pp 47–69, San Francisco, CA: Clear Concepts.

Kumar, P.S., Hogendoorn, J.A., Feron, P.H.M. and Versteegh, G.F. (2002) 'New absorption liquids for the removal of CO_2 from dilute gas streams using membrane contactors', *Chemical Engineering*, vol 57, pp 1639–1651.

Lekvam, K. and Bishnoi, P.R. (1997) 'Dissolution of methane in water at low temperature and intermediate pressure,' *Fluid Phase Equilibria*, vol 131, 297–309.

Lems, R. and Dirkse, E.H.M. (2010) 'Small scale biogas upgrading: Green gas with the DMT Carborex-MS® system', 15th European Biosolids and Organic Resources Conference, organised by Aqua Enviro Technology Transfer, www.europeanbiosolids.com (accessed 12 Mar 2011).

Miltner, M., Makaruk, A. and Harasek, M. (2008) 'Application of gas permeation for biogas upgrade— Operational experiences of feeding biomethane into the Austrian gas grid', 16th European Biomass Conference and Exhibition, Valencia, Spain.

Patterson, T., Esteves, S., Dinsdale, R. and Guwy, A. (2011) 'An evaluation of the policy and techno-economic factors affecting the potential for biogas upgrading for transport fuel use in the UK', *Energy Policy*, vol. 39, no 3, pp1806–1816.

Perry, R.H. and Green, D.W. (1997) *Perry's chemical engineers' handbook* (7th edn), New York: McGraw-Hill.

Persson, M. (2003) 'Evaluation of upgrading techniques for biogas' ['Utvärdering av uppgraderingstekniker för biogas'], Rapport SGC 142 Malmö: Svenskt Gastekniskt Center AB, www.sgc.se/dokument/SGC142.pdf (accessed 15 May 2011).

Persson, M. (2007) 'Biogas upgrading and utilisation as a vehicle fuel', in *Proceedings of The Future of Biogas in Europe III*, University of Southern Denmark. www.websrv5.sdu.dk/bio/Probiogas/down/work07/Proceedings.pdf (accessed17 July 2011).

Persson, M., Wellinger, A., Rehnlund, B. and Rahm, L. (2007) 'Report on technological applicability of existing biogas upgrading processes.' Biogasmax. www.biogasmax.co.uk/media/report_on_technological_2007__041639600_1025_22052007.pdf (accessed 21 July 2011).

Pierantozzi, R. (1993) 'Carbon dioxide,' in J. Kroschwitz and M. Howe-Grant (eds), *Encyclopedia of chemical technology*, vol 5, pp 35–53, New York: John Wiley & Sons.

Prabucki, M.J., Doczyck, W. and Asmus, D. (2001) 'Removal of organic silicon compounds from landfill and sewer gas'. in *Proceedings Sardinia 2001, 8th International Waste Management and Landfill Symposium*, pp 631–639, Cagliari, Italy, vol 2. Cagliari: Cisa.

Rasi, S. (2009) 'Biogas composition and upgrading to biomethane', PhD thesis, University of Jyväskylä, Finland.

Rasi, S., Veijanen, A. and Rintala, J. (2007) 'Trace compounds of biogas from different biogas production plants', *Energy*, vol. 32, pp 1375–1380.

Rasi, S., Lehtinen, J. and Rintala, J. (2010) 'Determination of organic silicon compounds in biogas from wastewater treatments plants, landfills, and co-digestion plants', *Renewable Energy*, vol 35, pp 2666–2673.

Rasi, S., Läntelä, J. and Rintala, J. (2011) 'Trace compounds affecting biogas energy utilisation – A review', *Energy Conversion and Management*, vol 52, pp 3369–3375.

Rutledge, B. (2005) 'California biogas industry assessment,' white paper, Pasadena, TX: WestStart-Calstart.

Ryckebosch, E., Drouillon, M. and Vervaeren, H. (2011) 'Techniques for transformation of biogas to biomethane,' *Biomass and Bioenergy*, vol 35, 1633–1645.

Schomaker, A.H.H.M., Boerboom, A.A.M., Visser, A. and Pfeifer, A.E. (2000) 'Anaerobic digestion of agro-industrial wastes: information networks – technical summary on gas treatment', Report No.:

EEG. (2009) 'Technology bonus for biogas upgrading', Act on granting priority to renewable energy sources (Renewable Energy Sources Act, EEG), Germany, http://www.bmu.de/files/english/pdf/application/pdf/eeg_2009_en_bf.pdf (accessed 5 July 2011).

Fegan, B. (2001) *ASHRAE handbook fundamentals*, Michigan Public Service Commission, Energy Information Administration, MichCon.

Franco, A., Mosquera-Corral, A., Campos, J.L. and Roca. E. (2007) 'Learning to operate anaerobic bioreactors', In A. Mendez-Vilas (ed.) *Communicating Current Research and Educational Topics and Trends in Applied Microbiology*, vol 1, pp 618–627 www.formatex.org/microbio/pdf/contents.pdf (accessed 14 June 2012).

Gomes, V.G. and Hassan, M.M. (2001) 'Coalseam methane recovery by vacuum swing adsorption', *Separaration and Purification Technology* vol 24, pp 189–196.

Günther, L. (2007) 'DGE GmbH presentation: Purification of biomethane using pressure less purification for the production of biomethane and carbon dioxide', INNOGAS, www.dge-wittenberg.com/english/vortraege/DGE%20Fachtagung%20WB%202006%20teil1-EN.pdf (accessed 12 July 2011).

Hagen, M., Polman, E., Jensen, J., Myken, A., Jönsson, O. and Dahl, A. (2001) 'Adding gas from biomass to the gas grid', Report SCG 118. Malmö, Sweden: Swedish Gas Center. www.sgc.se/dokument/sgc118.pdf (accessed 12 June 2012).

Hagmann, M., Heimbrand, E. and Hentschel, P. (1999) 'Determination of siloxanes in biogas from landfills and sewage treatment plants', *Proceedings Sardinia 1999, 7th International Waste Management and Landfill Symposium*, Cagliari, Italy.

Hagmann, M., Hesse, E., Hentschel, P. and Bauer, T. (2001) 'Purification of biogas – removal of volatile silicones', In T.H. Christensen, R. Cossu and R. Stegmann (eds), *7th International Waste Management and Landfill Symposium*, vol. II, pp. 641–644. Cagliari, Italy.

Hammer, M., Pettersson, K. and Svensen, B. (2007) 'Guide to developing a local biogas strategy for sustainable large-scale consumption and production in collaboration between town and countryside', BiogasMax Report Biogas as vehicle fuel—Market expansion to 2020 air quality, www.medcoast.org/download/18.726a005111ed95650fa80002283/Guide+to+developing+a+local+biogas+strategy.pdf (accessed 14 September 2011).

Heguy D. and Bogner, J. (2004) 'Cost-effective hydrogen sulfide treatment strategies for commercial landfill gas recovery: role of increasing C&D (construction and demolition) waste', *MSW Management* March/April, http://waste.environmental-expert.com (accessed May 2012).

Holm-Nielsen and Al Seadi (2004) 'Manure-based biogas systems – Danish Experience', in P. Lens, B. Hamelers, H. Hoitink and W. Bidlingmaier (eds) *Resource Recovery and Reuse in Organic Solid Waste Management*, London: IWA Publishing.

IEA Bioenergy. (2006) 'Biogas upgrading to vehicle fuel standards and grid injection' IEA Bioenergy, Task 37, http://www.seai.ie/Renewables/Bioenergy/Biogas_upgrading_to_vehicle_fuel_standards_and_grid_connection_IEA.pdf (accessed November 2011).

IEA Bioenergy. (2009) 'Biogas upgrading technologies –developments and innovations' IEA Bioenergy Task 37, www.iea-biogas.net/_download/publi-task37/upgrading_rz_low_final.pdf (accessed 12 May 2011).

IPCC. (2007) 'Technical Summary', in S. Solomon., D. Qin., M. Manning., Z. Chen., M. Marquis., K.B. Averyt., M. Tignor, and H.L. Miller (eds) *Climate change 2007: The physical science basis. contribution of Working Group I to the Fourth Assess ment Report of the Intergovernmental Panel on Climate Change*, Cambridge: Cambridge University Press.

ISET. (2008) 'Biogas upgrading to biomethane', in *Proceedings of the European Biomethane Fuel Conference*, 21 Feb 2008, Kassel, Germany.

Jönsson, O. (2009) 'Biogas upgrading – Technologies, framework and experience', Microphilox Project Workshop, Barcelona, Spain, 26 March, www.microphilox.com/pdf/OJonsson%20Natural%20gas%20grid%20injection.pdf (accessed 14 September 2011).

Khanal, S.K. (2008) *Anaerobic biotechnology for bioenergy production: principles and applications*, New York: John Wiley & Sons.

Biomethane is less corrosive than biogas and also is potentially more valuable as a fuel. For these reasons, it may be both possible and desirable to store biomethane for on- or off-site uses. Compressed biomethane is generally stored in order to increase the energy density, save storage space and facilitate transportation. Depending on the purpose, biogas can be compressed to high pressure, medium pressure or low pressure prior to storage.

References

Accettola, F. and Guebitz, G.M. and Schoeftner, R. (2008) 'Siloxane removal from biogas by biofiltration: biodegradation studies', *Clean Technology and Environmental Policy*, vol 10, pp 211–218.

Agency for Renewable Resources (2012). 'Biomethane'. Gülzow: Fachagentur Nachwachsende Rohstoffe e. V. (FNR) www.mediathek.fnr.de/broschuren/bioenergie/biogas/biomethan.html.

Ajhar, M. and Melin, T. (2006) 'Siloxane removal with gas permeation membranes', *Desalination*, vol 200, pp 234–235.

Allen, M.R., Braithwaite, A. and Hills, C.C. (1997) 'Trace organic compounds in landfill gas at seven U.K. waste disposal sites', *Environmental Science and Technology*, vol 31, pp 1054–1061.

Angelidaki, I. and Sanders, W. (2004) 'Assessment of the anaerobic biodegradability of macropollutants', *Re/Views in Environmental Science and Bio/Technology*, vol 3, pp 117–129.

Appels, L., Baeyens, J. and Dewil, R. (2008) 'Siloxane removal from biosolids by peroxidation' *Energy Conversion and Management*, vol 49, pp 2859–2864.

Arnold, M. (2009) 'Reduction and monitoring of biogas trace compounds', Research Notes 2496, Espoo: VTT Tiedotteita.

Beil, M. (2009) 'Over-view on biogas upgrading technologies', European Biomethane Fuel Conference, Goteborg, Sweden, Biogasmax, www. biogasmax.eu/media/3t3_overview_on_upgrading_iset__062510600_0654_3009-2009.pdfs (accessed 06 July 2011).

Benjaminsson, J. (2006) 'NYA Renings och uppgraderingstekniker for biogas', Report163, Swedish Gas Centre, www.sgc.se/rapporter/resources/SGC163.pdf (accessed 24 July 2011).

Berndt, A. (2006) 'CarboTech Engineering GmbH, Intelligent utilisation of biogas—Upgrading and adding to the grid', Jonkoping, May, www.german-renewable-energy.com/Renewables/Redaktion/PDF/es/es-World-Bioenergy-2006-Carbotech-schulte-biogas,property=pdf,bereich=renewables,sprache=es,rwb=true.pdf (accessed 28 June 2011).

Berndt, A. (2012) 'CarboTech Engineering GmbH', personal communication.

Betterton, E.A. (1992) 'Henry's law constants of soluble and moderately soluble organic gases: Effects on aqueous phase chemistry', *Advance in Environmental Science and Technology*, vol 24, pp 1–50.

Bourque, H. (2006) 'Use of liquefied biogas in transport sector', Conference sur les credits CO_2 et la valorisation du biogas, www.apcas.qc.ca (accessed May 2012).

Buswell, E. and Neave, S. (1930) *Laboratory studies of sludge digestion*. Illinois Div. of State Water Survey 30.

de Hullu, J., Maassen, J., vanMeel, P., Siamak, S., Vaessen, J., Bini, L. and Reijenga, J. (2008) 'Biogas upgrading: Comparing different techniques', Eindhoven University of Technology, www.students.chem.tue.nl/ifp24/Final%20prese-nation.ppss, accessed 10 May 2011.

DENA. (2009) 'Executive Report: The German market for biomethane', Deutsche Energie-Agentur GmbH (DENA), German Energy Agency, www.exportinitiative.de/fileadmin/user_upload/Table_of_Contents_v3_Biomethan.pdf (accessed 13 June 2011).

Deublein, D. and Steinhauser, A. (2008) *Biogas from waste and renewable resources. An introduction*, Weinheim: Wiley-VCH.

Dewil, R., Appels, L. and Baeyens, J. (2006) 'Energy use of biogas hampered by the presence of siloxanes', *Energy Conversion and Management*, vol 47, pp 1711–1722.

Dortmund, D., and Doshi, K. (1999) 'Recent developments in CO_2 removal membrane technology', http://www.uop.com/gasprocessing/TechPapers/CO$_2$RemovalMembrane.pdf (accessed 24 May 2011).

Table 10.11 Most commonly used biogas and biomethane storage systems. HDPE: high-density polyethylene; LDPE: low-density polyethylene.

Pressure (kPa)	Storage device	Material
Low (< 14)	Water sealed gas holder or gas bag	Steel, rubber, plastic, vinyl, HDPE, LDPE
Medium (150–1,500)	Propane or butane tanks	Steel
High (20,000–35,000)	Commercial gas cylinders	Alloy

Source: Modified after Appels et al. (2008).

Gas compressors suitable for flammable gas are generally used for compressing the biomethane. These differ from regular compressors in several respects, e.g. the cylinder is located further from the crankcase, higher quality packing, hardened connecting rods, passageways are provided to vent leaks away from the crankcase and to prevent explosions, inlet and exhaust ports are designed to let contaminants pass through instead of collect in the compressor, and explosion-proof motors and electrical connections are used on all equipment.

The compressed biomethane is usually stored in steel cylinders such as those typically used for storage of other commercial gases (Table 10.11). Storage facilities must be adequately fitted with safety devices such as rupture disks and pressure relief valves. The energy required for compression, assuming the biogas contains 60 per cent methane, is approximately 10 per cent of the energy content of the stored biogas.

Conclusions

Biogas is mainly composed of CH_4 (45–70 per cent) and CO_2 (30–55 per cent) with smaller amounts of H_2S (0–2000 ppm), some trace compounds and water vapour. Biogas can be used for the production of heat and/or electricity, or it can be transformed to vehicle fuel or natural gas substitute. However, biogas has to be cleaned and upgraded in order to reach the quality requirements for vehicle fuel or grid injection.

Biogas upgrading includes adsorption (pressure swing adsorption), absorption (water scrubbing and chemical amine), membrane separation and cryogenic purification. The selection of the technology is dependent on the amount and composition of the raw biogas, the quality of biomethane desired, and economic and environmental issues. In addition, the biomethane yield and purity are also taken into account. In many cases, not the technique itself, but the specific situation on-site and/or the availability of a technology supplier in that country, will determine which installation is most feasible. Therefore, there is no 'standard' solution that is best practice, but feasibility studies are needed prior to installation. Along with investment and maintenance costs, operational costs on the use of chemicals, heat, electricity and/or water determine the choice of the technology. Moreover, a technique that requires heat or water can be an economically appropriate choice if excess heat or wastewater is available in the nearby vicinity. Among the leading biogas upgrading countries in Europe, water scrubbing is the most popular technology in Sweden while Germany has the highest installed capacity of PSA technology. In the Netherlands all three technologies, water scrubbing, PSA and membrane technology, are popular. Methane losses generally occur during biogas upgrading. Any CH_4 lost in the process is not only a revenue loss but also poses a potential environmental threat as the GWP of CH_4 is much greater than that of CO_2.

Chemical absorption has two real waste streams, namely a stream containing CO_2 and a catalyst stream. The catalyst stream can be either amine for the absorption of CO_2 or Fe/EDTA complexes for the absorption of H_2S (Hagen et al., 2001; Persson, 2003; Krich et al., 2005). The spent amine solution must be replaced a few times a year and thus is also a waste. This solution can be separated into a water phase and the amines using a membrane. The clean water phase can then be purged to a river. The CO_2 is emitted to the atmosphere. The off-gas from the membrane process still contains CH_4 which needs treatment. Part of it can be fed back into the inlet of a multistage set-up and thus improve the yield (Rutledge, 2005). On the other hand, the cryogenic process produces a gas stream with a high percentage of CO_2, which probably could be recovered and used for industrial purposes (Zhao et al., 2010).

Biomethane compression and storage

Biogas cannot be stored easily as it does not liquefy under pressure at ambient temperature (critical temperature and pressure required are 82.5°C and 4750 kPa, respectively). Biomethane is generally compressed to reduce the storage requirements, increase its energy density, and increase pressure to facilitate the gas flow in pipelines. Furthermore, compression is sometimes needed to match the pressure requirements of the downstream gas application or utilisation equipment. Both H_2S and water should be removed prior to compression. Removal of H_2S will prevent corrosion, while drying of biomethane will avoid condensation of water vapour during compression.

Biomethane compression for vehicle fuel and for natural gas grid injection

Biogas can be used without any compression in combustion devices such as flares and simple boilers. For vehicle fuel use, biomethane is generally compressed to high pressures of 20,000–25,000 kPa (IEA Bioenergy, 2009). In addition, biomethane is tanked and transported at 20,000–25,000 kPa or higher pressures in order to minimise the need for additional compression at the refuelling station (Persson, 2003). Finally, three different gas pressures are maintained in the storage tank. For instance, when a car is refuelling, first the low pressure tank is opened, depending on its remaining pressure in the tank. When the gas flow is limited due to low pressure differences, the second tank opens and the first one closes and so on. The tank is usually filled to a maximum of 25,000 kPa. Mostly, pressure in the high pressure storage tank is around 30,000 kPa. This way of operation limits the energy use for compression.

Biomethane is increasingly injected into the natural gas grid (IEA Bioenergy, 2009). In the natural gas grid, there are typically three different pressure levels: high (HP), medium (MP) and low (LP). For instance, in the main transport (long distance and regional) pipelines in Belgium, the gas is transported under high pressure (6000–8000 kPa) to the actual gas distributors. Then it is transported in medium (50–500 kPa) to low pressure (< 50 kPa) lines to the regional and local gas companies. Finally, the gas is delivered to the end consumers in low pressure pipes. The pressure levels are slightly different in different countries. It must be noted that a biogas fed into a high pressure grid may no longer be sensible with regard to energy and economic aspects. As the amount of energy required for compression increases quadratically with the pressure, a gas conditioning process is no longer sensible.

Table 10.10 Summary of CH_4 losses associated with biogas upgrade technologies.

Technology	Methane losses (%)				
	Berndt, 2006	Günther, 2007	Persson (2007)	Benjaminsson (2006)	Agency for Renewable Resources (2012)
PSA	Medium	5.5	< 2[a]	N.A.	1–5
Water scrubbing	Medium	4.7	< 2[a]	N.A.	0.5–2
Physical absorption	High	13.75	< 2	N.A.	1–4
Chemical (amine) scrubbing	Low	0.03	< 0.1	N.A.	0.1
Membrane (low pressure)	N.A.	N.A.	N.A.	< 1.5	2–8
Membrane (high pressure)	High	N.A.	N.A.	N.A.	N.A.
Cryogenic	N.A.	N.A.	< 2–10[b]	< 2	N.A.

Source: Adapted after Patterson et al. (2011) and Agency for Renewable Resources (2012)
[a] Manufacturers figures—higher losses have been noted at some plants (Persson, 2007); [b] Manufacturers figures—losses of 10–18% have been noted at some plants (Persson, 2007); N.A.: Not Available.

to Table 10.10 only chemical scrubbing as such appears to meet that biogas upgrading bonus requirement. In all upgrading technologies, CH_4 losses can be managed by burning it. This means an additional unit which increases both investment and operation cost should be taken into account when evaluating the upgrading technology. Nevertheless, it offers the opportunity to recuperate heat or electricity as an auxiliary energy source for the upgrading plant.

WASTE STREAMS

The process waste stream mainly consists of process wash water/liquids or exhaust gases that contain pollutants. These streams need to be either treated or regenerated before final disposal or use. The waste gas produced with PSA and membrane separation can be used in a gas engine to generate energy, and thus emissions can be reduced. The PSA process creates a liquid stream, which contains the adsorbed material from the molecular sieves, and also some amounts of CH_4. Normally, the liquid waste stream is recycled back through the adsorption process, which reduces the amount of CH_4 in the waste stream and increases the content of CH_4 in the product gas (Persson, 2003). In a water scrubbing process, liquid and gas waste streams are generated. The first waste stream is the off-gas. As air is used in the final stage of the process to strip CO_2 and regenerated water, this stream mainly consists of air enriched in CO_2 along with traces of H_2S and CH_4. Because of H_2S, the off-gas needs to be treated. However, high dilution of CH_4 in the off-gas clearly jeopardises the off-gas treatment (Schomaker et al., 2000; Krich et al., 2005). The second waste stream consists of wash water which is purged and replaced with clean water to keep dissolubility as high as possible and avoid accumulations of CO_2 and H_2S.

Table 10.9 Cost estimates of upgrading biogas to biomethane from studies undertaken during 2007–2009.

Technology	Cost per m³ of biomethane (€/Nm³)			
	de Hullu et al. (2008)	Persson et al. (2007)	Jönsson (2009)	Hammer et al. (2007)
PSA	0.25	0.11–0.16	0.11–0.22	N.A.
Water scrubbing	0.13	N.A.	N.A.	0.11[c]
Physical adsorption	N.A.	N.A.	N.A.	N.A.
Chemical (amine) scrubbing	0.17[a] –0.28[b]	N.A.	N.A.	N.A.
Membrane (low pres.)	0.12[a]–0.22[b]	N.A.	N.A.	N.A.
Cryogenic	0.44	N.A.	N.A.	N.A.

Source: Adapted after Patterson et al. (2011)
[a] Costs without H_2S removal; [b] Costs with H_2S removal; [c] Calculated from reported capital and operational costs of the Falköping upgrading plant assuming 5% interest rate and 10 year depreciation period using the method described in de Hullu et al. (2008); N.A.: Not Available.

Impact on the environment

The environmental impact of the upgrading processes is an important factor that can be used to compare the feasibility of different upgrading technologies. If the pollutants, especially CH_4, that were removed from biogas during upgrading are emitted to the atmosphere or to other environment, the contamination of the environment will jeopardise the goal of producing sustainable fuels (biomethane). Pollution from the upgrading can be classified into CH_4 losses (CH_4 slip) and process waste streams. A brief description of various CH_4 losses and waste streams from different upgrading technologies follows below.

METHANE LOSSES

The most important parameter in evaluating the environmental performance of upgrading technologies is the amount of CH_4 lost during the upgrading process. This is also called CH_4 slip. In general, any CH_4 loss in the process is not only a revenue loss but also poses an environmental impact as CH_4 has a global warming potential (GWP) 25 times greater than CO_2 (IPCC, 2007). CH_4 losses generally occur during biogas upgrading and are dependent on the technology used (Table 10.10). Among the current upgrading technologies, chemical scrubbing was considered to have the lowest CH_4 losses of 0.03 per cent. PSA, water scrubbing and membrane technology have CH_4 losses in the medium range (1–2 per cent), although losses from 8 to 10 per cent have been measured at some plants, whilst cryogenic separation has the highest methane loss (2–10 per cent). As a trade-off, plant operators sometimes allow a certain percentage of CH_4 to remain in the off-gas in favour of high product gas quality and the associated upgrading costs. In such cases, the off-gas may contain 1–4 per cent CH_4 and can be blended with higher CH_4 content gases for combustion in a CHP or boiler plant, or captured within a catalytic converter. However, to limit the environmental impact from CH_4 losses, Germany has introduced a biogas upgrading bonus in the feed-in tariff system (in 2008) if the CH_4 losses within the upgrading process are less than 0.5 per cent (EEG, 2009). According

Table 10.8 Investment and maintenance costs for the five different upgrading technologies.

Technology	Plant capacity (1,000 m³/h of raw biogas) (Berndt, 2012)			
	Investment cost (€/yr)	Utility cost (€/yr)	Maintenance cost (€/yr)	Methane losses (m³/yr)
PSA	270,000	202,000	62,000	54,000
Water scrubbing	255,000	233,000	54,000	52,000
Physical absorption	234,000	218,000	45,000	108,000
Chemical (amine) scrubbing	213,000	415,000	75,000	4,500
Membrane separation	255,000	270,000	66,000	43,300
Cryogenic separation	N.A.	N.A.	N.A.	N.A.

Source: Berndt (2012)
Assumptions: 13 € ct/kWh (electrical), 5 € ct/kWh (heat), investment: cost 10 years depreciation, 5% interest rate
N.A.: Not Available.

COST PRICE PER NM³ BIOMETHANE

The cost price per Nm³ biomethane is considered as the most important economic parameter while evaluating the economic performance of upgrading technologies. It is calculated by using (Eq. 10.10), in which the interest rate on the investment is also considered.

$$\text{Price per Nm}^3 = \frac{\frac{\text{investment}}{\text{depreciation period}} + \text{investment} \times \text{interest rate} + \text{annual cost}}{\text{Nm}^3 \text{ produced biomethane per year}} \quad \text{(Eq. 10.10)}$$

Table 10.9 presents the cost estimates of upgrading biogas to biomethane during the study period 2007–2009. Among the studied technologies, high pressure water scrubbing was considered the cheapest due to its lowest investment costs while cryogenic separation has both the highest investment cost and the highest cost price per Nm³ of biomethane. The costs for cryogenic technology without H_2S removal was €0.12 and €0.44/Nm³ biomethane if H_2S removal was included (de Hullu et al., 2008). Similarly, the costs for chemical scrubbing were €0.17 and €0.28/Nm³ of biomethane without and with H_2S removal, respectively, as the final H_2S concentration in upgraded biogas should be less than 5 mg/Nm³ (de Hullu et al., 2008). As the investment costs of PSA are also quite high, the cost price of €0.25/Nm³ biomethane was higher than the other four technologies. The costs of the water scrubbing and membrane technology were in a lower range of €0.12–0.13/Nm³ of biomethane (de Hullu et al., 2008). Although this cost is low in comparison with other technologies, difficulties with yield and purity as well as the potential for fouling membranes (requiring membrane replacement) raises operating costs and strongly impacts on project economics. Moreover, costs for widely used technologies such as PSA and water scrubbing can vary by up to 100 per cent. As there is a lack of data relating to the costs associated with less widely used technologies, the lowest costs associated with water scrubbing should be considered as indicative only.

0.6 kWh/Nm³ biomethane (Patterson et al., 2011). This wide variation in energy requirement is attributed to the inclusion or omission of the thermal energy required to regenerate the amine absorbent. Data relating to the physical absorption process was more consistent than membrane or chemical scrubbing technologies. The high fluctuation in energy requirements for the membrane separation technology is mostly due to the variations in membrane types and operating pressures. This high variation in energy use between and within the upgrading technologies indicates an opportunity to optimise the process and energy requirements. However, it should be noted that different factors may be included in the energy estimation in different studies, e.g. in some studies energy needed for pre-treatments or off-gas treatment is probably included and in others it is not.

Economic assessment of upgrading technologies

The total cost for cleaning and upgrading biogas includes the cost of investment as well as of operation of the plant and maintenance of the equipment. CO_2 removal is considered as the most expensive part when biogas is upgraded to biomethane.

INVESTMENT AND MAINTENANCE COSTS

Investment and maintenance costs are dependent on upgrading technology and plant capacity and in turn on the raw biogas quality and its variation, and the desired quality of the upgraded gas. For instance, inclusion of the H_2S removal step in chemical scrubbing or membrane separation could double these costs. Generally, investment costs increase with increased capacity but at the same time investment per unit of installed capacity decreases for larger plants (Beil, 2009). Although plants are operated automatically, they still need manpower for daily supervision, repair and maintenance. The need for staff is not proportional to the size of the plant, which leads to higher costs for upgrading biogas at smaller plants (de Hullu et al., 2008; Beil, 2009).

Typical investment costs are €1.2–1.5 million for a plant treating 1000 Nm³/h of raw gas and €0.5–0.8 million for a plant treating 250 Nm³/h of raw biogas (Beil, 2009). The major operation costs are electricity, personnel, and depending on technology, cost of water or chemicals. Investment as well as operation and maintenance costs should be divided per kWh produced in the plant to estimate the cost price per Nm³ biogas upgraded. Typical operational costs can range from €7–13/MWh for a 1000 Nm³/h unit and €13–17/MWh for a 250 Nm³/h of raw biogas unit (Beil, 2009).

Table 10.8 presents the investment and maintenance costs for the five different upgrading technologies. Data available from various upgrading technologies for a 1000 m³/h (raw biogas) upgrading plant indicate that the lowest maintenance costs are associated with water scrubbing and membrane separation (Beil, 2009). According to a report published by German Energy Agency, annual maintenance costs may be 10 per cent lower for amine scrubbing, 42 per cent lower for PSA and 200 per cent higher for water scrubbing than the value presented here (DENA, 2009).

Table 10.7 Summary of energetic requirements of biogas upgrading technologies.

Technology	Energy requirement (kWh$_{el}$/Nm^3 biomethane)					
	Persson (2007)[a]	Agency for Renewable Resources (2012)[b]	Berndt (2006)[f]	Günther (2007)[c]	Benjaminsson (2006)	Miltner et al. (2008)
PSA	0.5–0.6 (0.3–1.0)	0.20–0.25	0.335	0.285	N.A.	N.A.
Water scrubbing	0.3 (0.45–0.9)	0.20–0.30	0.43	0.391	N.A.	N.A.
Physical absorption	0.4	0.23–0.33 (el) 0.3 (thermal)	0.49	0.511	N.A.	N.A.
Chemical (amine) scrubbing	(0.15)	0.06–0.15 (el) 0.5–0.8 (thermal)	0.646[d]	0.126[e]	N.A.	N.A.
Membrane separation	N.A.	0.18–0.25	0.769[d]	N.A.	0.27 (low pressure)	0.378
Cryogenic	N.A.	0.18–0.33	N.A.	N.A.	0.42	N.A.

Source: Adapted after Patterson et al. (2011) and Agency for Renewable Resources (2012).
[a] Figures reported are from operational plants. Figures in parentheses are from manufacturers. [b] Figures were based on data obtained from a 1,000m^3/h raw biogas plant and energy requirements were calculated for raw biogas; [c] CarboTech manufacture carbon molecular sieves used in PSA plants and DGE GmbH design and build PSA plants; [d] The value is the sum of electric and thermal energy. [e] This figure is related to raw gas and also accounts for methane losses and regeneration energy; N.A.: not available.

Table 10.6 Technical information along with the advantages and disadvantages of different biogas upgrading techniques.

Upgrading technology	Description	Working pressure (bars)	Technical availability per year (%)	Maximum achievable methane yield (%)	Maximum achievable methane purity (%)	Advantages	Disadvantages
Pressure swing adsorption	CO_2, higher C_xH_y, H_2S, Si-, Fl-, Cl-components, odours will be removed by activated carbon/carbon molecular sieve. Pre-treatment for water vapour and H_2S removal required. Regeneration: vacuum	4–7	94	91	98	More than 97% CH_4 enrichment. Low power demand. Low level of emissions. Adsorption of N_2 and O_2	Additional complex H_2S removal step needed
High pressure water scrubbing	CO_2 and H_2S are absorbed by means of water by high pressure. No pre-treatment for water vapour and H_2S removal. Water regeneration by air stripping	7–10	96	94	98	Removal gases and particulate matter. High purity, good yield. Simple technique, no special chemicals or equipment required. Neutralisation of corrosive gases	Limitation of H_2S absorption due to changing pH. H_2S damages equipment. Requires a lot of water, even with the regeneration process
Chemical absorption	CO_2 and H_2S are absorbed by means of scrubbing fluid (e.g. amines, polyethyleneglycol (PEG) etc.). Regeneration: air stripping for PEG; heating for amine. Pre-treatment: water vapour and H_2S removal for PEG; H_2S removal for amine	PEG: 7–10; Amine: Atmp	91	90	98	Almost complete H_2S removal	Depends on selected amine as only one component is removed. Expensive catalyst
Membrane process	CO_2 is separated due to different permeation rates at a membrane. Pre-treatment for water vapour and H_2S removal required	8–10	98	78	89.5	Compact and light in weight. Low maintenance. Low energy requirements. Easy process	Relatively low CH_4 yield. H_2S removal step needed. Membranes can be expensive
Cryogenic separation	CO_2 is liquefied by high pressure and low temperatures and separated by rectification column	25–40	98	98	91	Can produce large quantities with high purity. Easy scaling up. No chemicals used in the process	A lot of equipment is required

Source: Adapted after Beil (2009); Berndt (2006)

Biological methane enrichment

Biological methane enrichment is also known as chemo-autotrophic biogas upgrading (Strevett et al., 1995). In this technology, different methanogens utilise CO_2 as a carbon source and H_2 as an energy source. The raw biogas is fed to the hollow fibres packed with methanogens and operated at mesophilic and thermophilic temperatures. Thermophilic methanogens exhibit rapid methanogenesis, while mesophilic methanogens give more complete conversion of the available CO_2 (Strevett et al., 1995). Use of the thermophilic methanogen *Methanobacterium thermoautotrophicum* can consume H_2S, thus both unwanted components are removed simultaneously. The main advantage of biological enrichment is that CO_2 and H_2S are effectively removed, while approximately doubling the original CH_4 mass. In contrast, physico-chemical treatment methods only remove the contaminating gas components such as CO_2, H_2S, H_2O etc., without changing CH_4 mass. Furthermore, physico-chemical treatment generates additional waste, unwanted end products and off-gas, which either need a further treatment or result in CH_4 losses. The purified biogas in biological enrichments process contains about 96 per cent CH_4 and 4 per cent CO_2, while H_2 and H_2S are completely removed (Strevett et al., 1995).

Comparison of biogas upgrading techniques

In this section, the five most common upgrading technologies are compared. Although these upgrading technologies have their own advantages and disadvantages, some common factors such as technical specifications, energy requirements, economics (price per Nm^3 upgraded biogas, investment costs and operating costs) and the impact on the environment (CH_4 losses, off-gas and waste streams) were compared and are presented in Table 10.6.

Technical availability

All plants have a high technical availability around 95 per cent (de Hullu et al., 2008). The plants are generally stopped only for repair or maintenance. Data available from various upgrading technologies for a 1000 m^3/h (raw biogas) upgrading plant indicate that the highest technical availability is for membrane separation, water scrubbing and physical absorption systems (Beil, 2009). Water scrubbing, PSA and chemical scrubbing have higher CH_4 enrichment capacity compared with membrane and cryogenic technologies (Table 10.6).

Energetic performance

One of the key parameter when considering the economic and environmental performance of upgrading technologies is the energy required for upgrading a unit amount of biogas to biomethane. Table 10.7 summarises the amount of energy required to upgrade raw biogas to biomethane. The energy requirements presented in Table 10.7 are for the produced biomethane and not for raw biogas. Thus, the raw biogas composition can affect the energy demand. Upgrading technologies with a lower energetic requirement for upgrading will result in more net energy being available for end use.

The energy requirements for water scrubbing (0.20–0.43 kWh/Nm^3 biomethane) seems to be more consistent than for PSA technology (0.24–0.6 kWh/Nm^3 biomethane). The energy requirement for chemical (amine) scrubbing ranged from 0.12 kWh/Nm^3 to

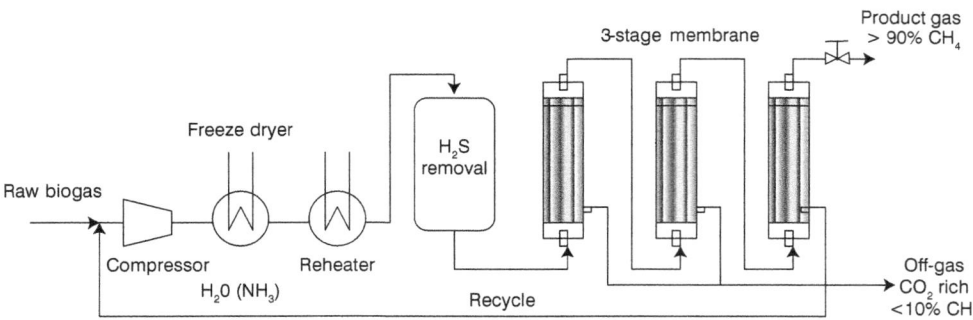

Figure 10.5 Flow chart of membrane biogas purification process (source: adapted after Miltner et al., 2008)

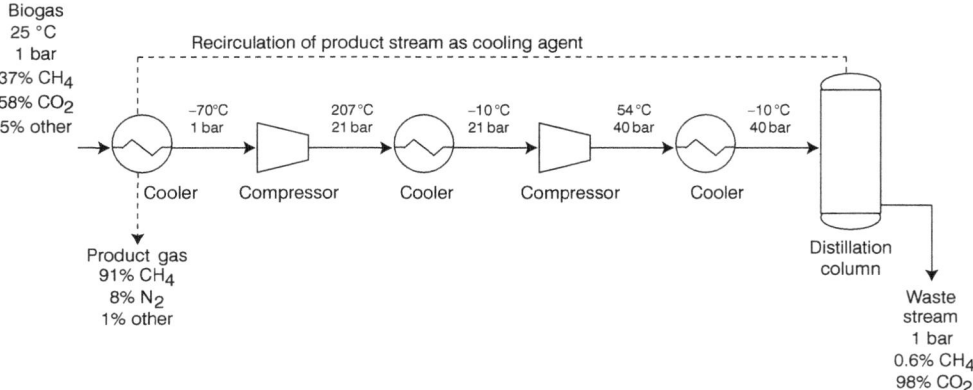

Figure 10.6 Flow chart of cryogenic biogas purification process (courtesy: ISET, 2008)

through the first heat exchanger which cools the gas down to –70 °C (Zhao et al., 2010). The heat recovered by the heat exchanger is used to preheat the produced gas before leaving the plant (Hagen et al., 2001). The first cooling step is followed by a cascade of compressors and heat exchangers which cool the inlet gas progressively to –100 °C and compress it to 4000 kPa. In these conditions, a gas–solid phase balance exists, with CO_2 being the solid phase and the gaseous phase containing > 97 per cent CH_4 (Deublein and Steinhauser, 2008). Finally, the gas enters the distillation column, where CH_4 is separated from the other contaminants, mainly H_2S and CO_2. Sometimes, CO_2 (target contaminant) is allowed to condense in an expansion tank. High purity CO_2 is produced and is sold as an industrial product (Zhao et al., 2010).

The main advantage of cryogenic separation is the high purity of the upgraded biogas (99 per cent CH_4) and high yield (Zhao et al., 2010). If needed, CH_4 can also be liquefied as compressed biomethane gas. Furthermore, siloxanes are removed without an additional removal unit. The main limitation of this technology is the need for complex process equipment, mainly compressors, turbines and heat exchangers, and energy demand. This raises both the capital and operating costs relative to other technologies (de Hullu et al., 2008).

Two basic membrane separation systems are in use: (1) gas–gas separation and (2) gas–liquid absorption separation. In gas–gas separation, the gases are present on each side of the membrane and separation is undertaken at high pressure (> 2000 kPa), although some systems can operate at 800–1000 kPa (Lems and Dirkse, 2010). Biogas is generally upgraded to a maximum of 92 per cent CH_4 in one-step, but with a multiple stage process, a final CH_4 concentration of > 96 per cent can be achieved (Miltner et al., 2008). The off-gases from the first stages are recycled within the process to enhance CH_4 capture, whilst the off-gas from the final stage (which may still contain 10–20 per cent CH_4) is flared, used for heat production or captured catalytically. This technology has been applied for some time for the upgrading of natural gas. Membranes made of acetate-cellulose separate small polar molecules such as CO_2, H_2O and the remaining H_2S, but are not effective in separating N_2 from CH_4. On the other hand, some membranes are very specific for some molecules, i.e. H_2S and CO_2, and can be used to separate different modules (IEA Bioenergy, 2009).

In a gas–liquid absorption system, a liquid adsorbent is present on one side of the membrane wall and operated close to atmospheric pressure. A micro-porous hydrophobic membrane separates the raw gas stream from a liquid phase absorbent. Absorbents such as NaOH (e.g. for H_2S separation) or heat regenerative amine solutions (e.g. for CO_2 separation) are used. The molecules from the gas stream, flowing in one direction, diffuse through the membrane, and will be absorbed on the other side by the liquid flowing in a counter current. Liquid is prevented from flowing to the gas side due to slight pressurisation of the gas. With amine solution, biogas with 55 per cent CH_4 can be upgraded to > 96 per cent CH_4 in one step. The amine solution can be regenerated by heating. The process can yield high purity CO_2 and is sold as an industrial product (IEA Bioenergy, 2009).

The advantages of membrane separation are that the process is compact, light in weight, has low energy and maintenance requirements, involves easy processing and has almost no start-up time. The disadvantages of the system are the relatively low CH_4 yield and high membrane cost. The concentrations of H_2S and H_2O are generally reduced prior to separation of CH_4 and CO_2. Due to imperfect separation, multiple stages may be required (IEA Bioenergy, 2009) and may increase CH_4 loss. This can be partly avoided by recirculation (Rutledge, 2005). Therefore, CH_4 recovery rate is an important criterion for membrane selection. Membranes separate the feed gas into a CH_4-rich stream and a CO_2-rich stream, which also contains some CH_4 (30 per cent v/v). In order to improve the CH_4 recovery, multiple membranes are used and CO_2-rich gas from the second membrane is fed into the feed of the first membrane system. Various configurations (serial, parallel and combinations) for multiple stage membrane systems are possible. The most common configuration is shown in Figure 10.5.

Cryogenic technique

Cryogenic technology is based on the principle that CH_4 and all other biogas components have different boiling points and thus liquefy at a different temperature–pressure range. The boiling points of CH_4 and CO_2 are -160 °C and -78 °C at atmospheric pressure, respectively (Zhao et al., 2010). By progressively cooling, different components in the biogas will condense at different temperatures and thus can be separated. Compression is done in a series of stadia with interim cooling (Figure 10.6). Raw biogas is first compressed and dried to prevent freezing in the following cooling step. The dried and compressed gas is then streamed

The main advantages of PSA technology are its reliability in producing a constant high-quality product gas (> 97 per cent CH_4), low power demand and simultaneous removal of N_2 and O_2 (IEA Bioenergy, 2009; Ryckebosch et al., 2011). The main disadvantage of PSA technology is an additional H_2S removal step needed before PSA as H_2S, adsorbing irreversibly, may damage the molecular sieve functionality. Similarly, the PSA process also requires dry gas before the upgrading process (Hagen et al., 2001; IEA Bioenergy, 2009). Furthermore, the process is relatively more expensive than other upgrading technologies and post-treatment of off-gasses, which contain significant amounts of CH_4, is required (Persson, 2003; IEA Bioenergy, 2009).

Membrane separation

Membranes can be used for separation of CH_4 and CO_2. In membrane separation, the components of a gas mixture are separated by the difference in particle size or affinity with certain specific molecules passing through a membrane while retaining others. The driving force behind this process is a difference in partial pressure between the gases (de Hullu et al., 2008). The efficiency of separation is determined by the flux of CO_2 through the membrane which is given by Fick's law (Dortmund and Doshi, 1999):

$$J = (k.D.\, \Delta p)\, /\, l \qquad \text{(Eq. 10.9)}$$

where J = flux
 k = solubility of CO_2 in the polymer
 D = diffusion coefficient of CO_2 through the polymer
 Δp = pressure difference over the membrane
 l = thickness of the membrane.

Permeability (P) is a function of $k.D$, and is affected by the operating pressure and temperature. The selectivity (α) of the membrane is dependent on the characteristics of the polymer used for the membrane and gives information about the permeability of various components like CO_2, H_2O and H_2S compared with CH_4 (Lems and Dirkse, 2010). In Figure 10.4, a relative diffusion rate of various components of biogas is illustrated. In general, permeability and selectivity work opposite to each other. Gases with high permeability require a lower membrane area, whilst membranes with higher selectivity for CH_4 can result in a better CH_4 recovery. Therefore, a trade-off between high permeability and high selectivity should be made in order to optimise separation. Note: Pf and Pp (Pf > Pp) in Figure 10.4 are the partial pressures of a gas at the feed and permeate sides, respectively.

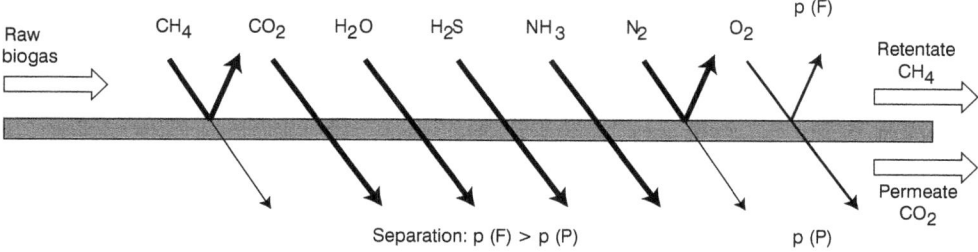

Figure 10.4 Relative permeation rate of various biogas components (source: adapted after Miltner et al., 2008; Lems and Dirkse, 2010)

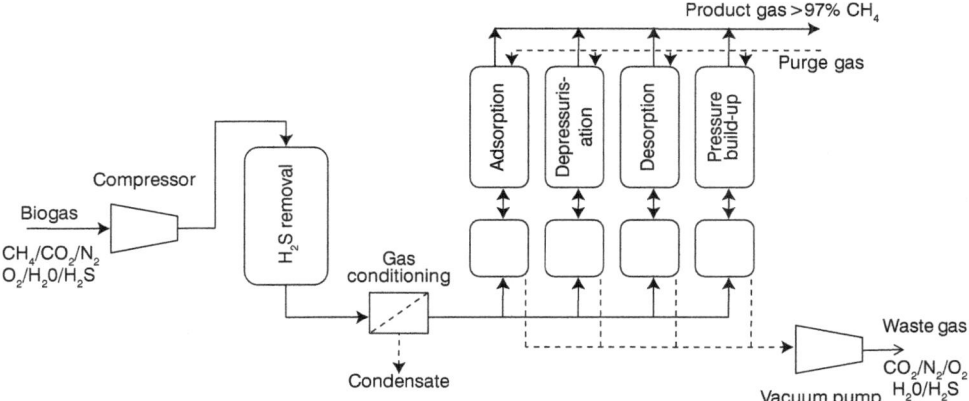

Figure 10.3 Pressure-swing adsorption schematic process (Source: Adapted after ISET, 2008)

passes through interstitial spaces (IEA Bioenergy, 2009). In addition to CO_2, PSA can also separate O_2 and N_2 from CH_4 (Ryckebosch et al., 2011). Selectivity of adsorption is achieved by different mesh sizes and/or application of different gas pressures. Sometimes different adsorbents are integrated into the PSA process to improve the process efficiency. For instance, activated carbon impregnated with potassium iodide can catalytically react with O_2 and H_2S to form water and S, which facilitates the removal of S compounds in addition to CO_2 (Zhao et al., 2010).

The PSA process is operated on the basis of equilibrium or kinetic selectivity, depending on the residence time in the column (Ryckebosch et al., 2011). Based on equilibrium selectivity, separation is achieved by retaining the strongly adsorbed components of a gas mixture within the column, while the less strongly adsorbed species is concentrated in the off-gas. In the case of separation based on kinetic selectivity, the faster diffusing species is retained on the absorbent, while the slower diffusing component is concentrated in the high pressure product gas (Gomes and Hassan, 2001).

A PSA unit usually consists of three or four columns connected in series. The columns are packed with the absorbents and operated in a cyclic batch process consisting of different stadia: adsorption, depressurisation, desorption and pressurisation (Figure 10.3) (ISET, 2008). The number of columns and operational sequence is dependent on the raw gas composition or the required quality of output gas.

In a cyclic-batch PSA process, each column is successively pressurised (700–800 kPa) with biogas and depressurised to lower pressure (Gomes and Hassan, 2001). The energy requirement for gas compression is reduced by coupling the pressure released from one column to pressurise the subsequent column. During the adsorption step, CO_2 is allowed to be adsorbed on to the absorbent bed until saturation, while CH_4 passes through the bed. If the breakthrough point is reached, the adsorption process will be switched to the next column. Upon saturation, the bed is depressurised, releasing a CH_4/CO_2 gas mixture with a high CH_4 content. The CH_4/CO_2 mixture is then recycled to the inlet of the PSA unit. During the desorption stage, the pressure is further dropped, releasing a gas containing mainly CO_2 with little CH_4 and thus allowing the regeneration and re-use of the adsorbent material. Care must be taken not to release too much CH_4 in the off-gas as CH_4 is a 25 times stronger greenhouse gas than CO_2 (IPCC, 2007)

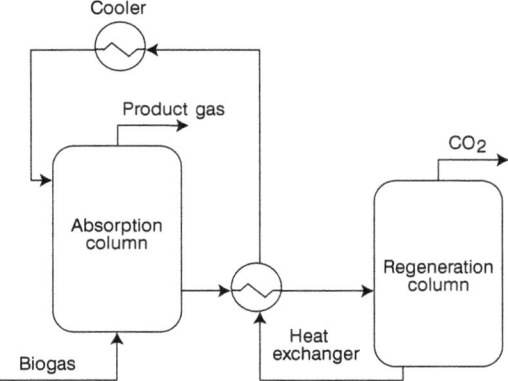

Figure 10.2 Flow chart of chemical absorption process (source: adapted after ISET, 2008)

Organic amines are highly selective in absorbing CO_2 and can dissolve significantly more CO_2 per unit volume when compared with water scrubbing, leading to smaller volumes and plant sizes. CO_2 is absorbed in the amine solution and reacts at quasi-atmospheric pressure with the chemical in the absorption column (Figure 10.2). The advantage of chemical amine scrubbing is complete H_2S removal, high efficiency and reaction rates and it is also effective at lower pressures compared with water and Selexol scrubbing, thereby leading to reduced compression energy requirements (Hagen et al., 2001; Persson, 2003; Krich et al., 2005). However, H_2S removal prior to chemical scrubbing is recommended to prevent irreversible damage of the chemical. After the reaction, the chemical is regenerated with steam or heat and CO_2 can be recovered. Some heat input is required to regenerate the amine solution prior to recirculation. However, 75 per cent of that heat needed in the regeneration phase can be re-used, e.g. in biogas plants. Due to the highly selective nature of the amine solvent, less CH_4 absorption occurs and overall CH_4 losses are extremely low. The CH_4 content in the produced gas can reach as high as 99.5 per cent (Hagen et al., 2001; Persson, 2003; Krich et al., 2005). Because of these advantages, the process is commonly used in industrial applications, including natural gas purification. The disadvantages are the additional chemical inputs needed and the need to treat waste chemicals from the process.

Pressure swing adsorption

Pressure swing adsorption (PSA) is the most versatile and widely used gas separation and purification technology due to its flexibility over a wide range of temperatures and pressures, low energy requirements and high efficiency. The PSA process is based on the ability of various adsorbent materials to selectively retain one or more components of a gas mixture under varying pressure conditions. Adsorbent materials such as activated carbon, natural zeolites (alumina silicates), synthetic zeolites, activated alumina, silica gels and polymeric sorbents are used (Krich et al., 2005; Bourque, 2006). These adsorbent materials are highly porous and separate gas components under high pressure according to molecular size, thus acting as a molecular sieve. For instance, CH_4 (molecular size of 3.8 Å) and CO_2 (molecular size 3.4 Å) separation is achieved by using an adsorbent with a pore size of 3.7 Å. Therefore, CO_2 is allowed to enter into the matrix of the absorbent material and thus is retained in the cavities of the absorbent and this is not irreversible, whilst CH_4

treatment plants where treated wastewater can be used (IEA Bioenergy, 2009). On the other hand, in the regenerative absorption process the wash water is regenerated after the biogas scrubbing process. The main advantage of this technique is that the total amount of water required is much lower compared with single-pass scrubbing.

Physical absorption (organic solvents)

A non-reactive organic solvent such as polyethylene glycol is used to physically absorb the unwanted component of the gas stream. Polyethylene scrubbing relies on the same principle as that of water scrubbing. Both CO_2 and H_2S are more soluble than CH_4 in the organic solvent. Selexol® and Genosorb® are the trade names for the polyethylene glycol used in this process. The main difference between water and Selexol is that CO_2 and H_2S are more soluble in Selexol which results in a lower solvent demand and reduced pumping, thus requiring a smaller installation. Furthermore, Selexol has a low vapour pressure, so the losses of chemicals are very low (Ryckebosch et al., 2011). In addition, water and halogenated hydrocarbons (contaminants in biogas from landfills) are removed when biogas is scrubbed with Selexol and thus an additional drying of the upgraded gas is not required (Hagen et al., 2001; Rutledge, 2005). During the absorption process, some of CH_4 may also dissolve into the absorption liquid. In many plants, a flash tank is used to recover CH_4 from the absorbent at an intermediate pressure (200–400 kPa) before the desorption unit, and thus decreases CH_4 losses. Spent absorbents are then regenerated by depressurising and/or heating. Selexol scrubbing is always designed with recirculation. Due to formation of elementary S, stripping the Selexol solvent is normally done with steam or inert gas rather than with air. Removal of H_2S prior to Selexol treatment is highly recommended. The Selexol process has been used successfully for upgrading landfill gas at several sites in the US and Europe (IEA Bioenergy, 2009).

Chemical (amine) scrubbing

In the chemical scrubbing process, reversible chemical bonds are formed between the solute and the solvent. Aqueous solutions of amines (i.e. mono-, di- or tri-ethanolamine) or aqueous solution of alkaline salts (i.e. sodium, potassium and calcium hydroxides) are used as solvents. The chemical reaction between an aqueous amine solution and CO_2 is shown below (Kumar et al., 2002).

Absorption of CO_2 : (Eq. 10.7)

$$2\ RNH_2 + H_2O + CO_2 \leftrightarrow RNH_3^+ + HCO_3$$

Desorption of CO_2 : (Eq. 10.8)

$$RNH_3^+ + HCO_3 \leftrightarrow RNH_2 + H_2O + CO_2$$

where R is the remaining organic component and is not specific in this equation. In (Eq. 10.7), the reaction of CO_2 with an amino acid (RNH_2) can be seen. The conversion of CO_2 to HCO_3^- is not significant as not many OH^- ions from the conversion of $H_2O \leftrightarrow H^+ + OH^-$ are present in the solution due to very low pH. Since the OH^- ions are in equilibrium with the amine molecules, the conversion of H_2O to OH^- and CO_2 to HCO_3^- has to be taken into account.

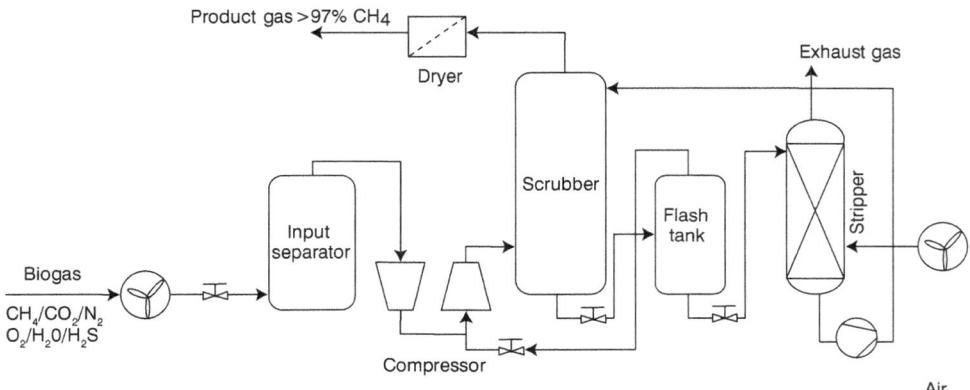

Figure 10.1 Flow chart of water scrubbing technology (Source: Adapted after ISET, 2008)

A schematic diagram of water scrubbing technology is presented in Figure 10.1. Raw biogas usually is pressurised to around 900–1200 kPa and fed to the bottom of a packed scrubbing tower. Water is fed in at the top of the tower so that the absorption process is operated counter-currently. The scrubbing tower (height-to-diameter ratio 20:1) is packed with a high surface area media (e.g. pall rings) to provide a high contact area between the gas and water. As the raw biogas moves up the column against the flow of water, CO_2 and H_2S are dissolved in the liquid stream. The concentration of CO_2 decreases in the water as the gas becomes more and more concentrated with CH_4. H_2S is also absorbed by the water. Upgraded gas leaves from the top of the column. If any CH_4 is dissolved in the water, it is usually captured by depressurising the water to 200–400 kPa within a flash tank. Gases released are then returned to the bottom of the scrubbing column, thus recirculating the released CH_4 in the flash tank. Finally, the gas is completely depressurised in the stripping tank and CO_2 is released by stripping with air. The released CO_2 is discharged into the atmosphere. The CH_4-rich product gas is then dried and compressed to around 20,000 kPa for storage. The CH_4 content in dried product gas (biomethane) can reach 95 per cent (Schomaker et al., 2000).

After scrubbing, the wash water can be regenerated by using a desorption column. Air, steam or an inert gas is used to release the CO_2 in the wash water. The vent gas is either released to the atmosphere or treated in a gas filter to remove odour-causing compounds. Scrubbing water can be used once in a single-pass system, or re-circulated following removal of dissolved gases. Stripping with air is generally not recommended when high levels of H_2S are handled since the water quickly becomes contaminated with elementary S which causes operational problems.

Water scrubbing requires a large amount of water. For example, the regenerative absorption process that washes 330 Nm^3/h of raw biogas requires approximately 50 l/h of water (de Hullu et al., 2008). Depending on whether wash water is regenerated or not, water scrubbing is designed as single-pass and regenerative absorption technology. In single-pass scrubbing, the wash water is used only once. No contamination in the water occurs with H_2S and CO_2. The main advantage of this technology is its simplicity and it is relatively inexpensive. Also, the loss of CH_4 is relatively small (less than 2 per cent) because of the large difference in solubility of CO_2 and CH_4 (Krich et al., 2005). The disadvantage of this technique is that it requires a large amount of water. Single-pass scrubbing is used, for example, in sewage

Table 10.5 Solubility of CO^2 in water at different temperatures and pressures.

Pressure (kPa)	Solubility of CO_2 g/100 g water			
	0 °C	10 °C	20 °C	30 °C
100	0.40	0.25	0.15	0.10
2000	3.15	2.15	1.30	0.90
5000	7.70	6.95	6.00	4.80

Source: Adapted after Appels et al. (2008)

$$P_A = X_A H_A \quad \text{(Eq. 10.5)}$$

where P_A = vapour pressure of component A above the liquid mixture, X_A = mole fraction of A in the liquid mixture, and H_A = Henry's law constant. At higher pressures, Henry's law is no longer valid (Lekvam and Bishnoi, 1997), and temperature becomes a more important factor than pressure for gas solubility (Pierantozzi, 1993). Table 10.5 illustrates how solubility of CO_2 in water is governed by pressure and temperature. As the pressure increases, the solubility of CO_2 in water increases; but decreases as the temperature increases.

According to Dalton's law, the total pressure is the sum of all partial pressures. Thus, if the total pressure is increased, the partial pressure of the components increases by the same factor. This is mathematically expressed as:

$$Pt = PP1 + PP2 + PP3, \text{etc.} \quad \text{(Eq. 10.6)}$$

where Pt = Total pressure, PP1, etc. = partial pressure of the first gas, etc.

This means that the saturation concentration of the components rises as well. However, when higher pressures are reached, the dissolubility of the components will no longer increase linearly with the pressure. The maximum pressure is up to 2000 kPa where the dissolubility can be described according to Henry's law (Perry and Green, 1997).

The pH also affects the solubility of the components in water. The pH of water is dependent on the amount of H_2S and CO_2 that has been dissolved into water, and becomes lower when more H_2S and CO_2 are dissolved. However, the dissolubility of CO_2 and H_2S decrease with decrease in pH. At higher pH, sulphur and carbonate ions will precipitate. Thus, optimal pH for water scrubbing is considered to be around 7.

Water scrubbing

Water scrubbing or absorption in water is the most widely used gas upgrading technology in Europe with 48 upgrading units in 2011 (IEA Bioenergy, 2009). Both CO_2 and H_2S are simultaneously removed due to the difference in binding forces of the polar CO_2 and H_2S and the non-polar CH_4. Removal of CO_2 from the biogas increases the CH_4 content in the upgraded gas. This method also removes H_2S since H_2S is more soluble than CO_2 in water. However, condensed moisture or particulates present within the raw gas stream have to be removed prior to the water scrubbing unit. The advantages of water scrubbing are that no special chemicals are required and both CO_2 and H_2S are removed in the process. Furthermore, the cost of water can be reduced by using effluent from a wastewater treatment plant. The disadvantages of water scrubbing are that it requires a lot of water even with regeneration, as well as limitations on H_2S removal, because the CO_2 decreases the pH of the solution and corrosion caused by H_2S is a risk.

acids such as diluted nitric or sulphuric acids and also with activated carbon (Hagen et al., 2001; Ryckebosch et al., 2011).

Oxygen and nitrogen

The presence of O_2 and N_2 in biogas lowers the heating value of the gas and can also cause corrosion. O_2 and N_2 are typically absent in all biogases except in landfill gas. Landfill gas contains relatively high concentrations of O_2 (1–3 per cent vol) and N_2 (1–17 per cent vol) (Rasi et al., 2007; IEA Bioenergy, 2009) as air may penetrate in the landfill gas collection systems, which are operated by using vacuum, through landfill cover. O_2 and N_2 may occasionally be present in biogas due to leakage or as air/O_2 is injected into the biogas reactor to control H_2S. Removal of O_2 and N_2 – if present in large quantities – can be costly and impede the use of biogas for vehicle or grid injection (IEA Bioenergy, 2009).

Carbon dioxide

The CO_2 content in the biogas ranges from 30 to 50 per cent depending on the feedstock composition and process conditions. CO_2 is generally removed to increase the energy content (Wobbe index) of the gas to produce biomethane, e.g. for vehicle fuel use or gas grid injection As the CO_2 of the gas is removed, its relative density is decreased and the calorific value is increased, thus increasing the Wobbe index. CO_2 removal is usually performed after the removal of water (vapour), H_2S, siloxanes and NH_3. Some CO_2 scrubbing technologies may also remove the above-mentioned contaminants.

The following techniques are employed for removal of CO_2 from the biogas: (1) physical and chemical CO_2 absorption, (2) pressure swing adsorption (PSA), (3) membrane separation, (4) cryogenic separation and (5) biological methane enrichment (Rasi et al., 2011; Ryckebosch et al., 2011). The selection of the CO_2 removal techniques is however dependent on the raw biogas composition and flow rate, biomethane use, methane losses, techno-economical and environmental issues associated with the technology. Examples of case studies with different technologies can be found e.g. in IEA Bioenergy Task 37 (http://www.iea-biogas.net/).

Absorption technologies

In the absorption process, compounds from the gas stream are dissolved into a solvent liquid stream by mass transfer (physical absorption). The driving force for mass transfer is the difference in the solute concentration between the gaseous and liquid phases. Absorption technologies are classified into physical and chemical processes. In physical absorption processes, a non-reactive fluid is used to physically absorb the unwanted component of the gas stream. In chemical processes, a chemical reaction between the compound from the gas stream and the component of the liquid phase occurs. The chemical reaction can be reversible or irreversible (Kohl and Nielsen, 1997).

The solubility of gases into water is dependent on several factors, such as pressure, temperature, pH, liquid/gas ratio etc. According to Henry's law there is a linear relationship at equilibrium between the partial pressure of a gas and its concentration in dilute solution (Betterton, 1992):

valves, cylinder walls and liner surfaces and result in overheating of sensitive engine parts (thermal insulator) and affect the function of spark plugs (Tower, 2003). Furthermore, the catalytic gas exhaust treatment is affected (Hagmann et al., 1999). In gas engines, siloxanes deposit is usually noticed on the nozzles and blades, and in lubricant oil causing erosion of the turbine blades and subsequently lowering operating efficiency (Tower, 2003). To prevent premature engine failure due to silicate deposition, manufacturers of gas engines introduced a limit value of 1 mg/l of silicon in the oil of gas engines (Prabucki et al., 2001) or a maximum total siloxane concentration ranging from 0.03 mg/Nm^3 of biogas (Capstone Microturbines) to 28 mg/Nm^3 (Caterpillar) (Dewil et al., 2006).

Part of the siloxanes present in the biogas is removed during the treatment of other contaminants such as H_2S and water vapour. For instance, cooling biogas to 5 °C with a gas dryer removes some water and approximately 30–40 per cent of the siloxanes (Khanal, 2008), while the remaining siloxanes can be removed by using a scrubber system. The two major methods for removal of siloxanes are physical or chemical absorption (Ryckebosch et al., 2011). In physical absorption, long carbon chain organic solvents are used either in a spraying device or a packed column. The major disadvantage of the physical adsorption method is its inability to achieve complete removal of siloxanes since the highly volatile siloxanes are easily stripped from the solvent at elevated gas flow rates (Appels et al., 2008).

Siloxanes are chemically absorbed in materials such as activated carbon, polymer beads, molecular sieves and silica gel (Hagmann et al., 2001; Schweigkofler and Niessner, 2001; Ajhar and Melin, 2006; Deublein and Steinhauser 2008). Schweigkofler and Niessner (2001) made a comparative study of adsorption materials using polymer beads, silica gel and activated carbon and reported that all exhibited large adsorption capacities for the siloxane D5. Although activated carbon is the most commonly used absorbent, silica gel is considered as a highly cost-effective absorbent for simultaneous biogas drying and siloxane removal. Thermal regeneration of silica gel and activated carbon is usually performed at 250 °C for about 20 min. More than 95 per cent of silica gel can be regenerated while the regeneration capacity of activated carbon is lower (Schweigkofler and Niessner, 2001). It should be noted that due to a competitive adsorption of siloxanes and other biogas contaminants (e.g. moisture in the siliga gel) the performance of the adsorption process may need large adsorption capacities (Schweigkofler and Niessner, 2001).

Furthermore, cryogenic condensation of the siloxanes from biogas has also been reported, but is considered expensive (e.g. Hagmann et al., 2001). Schweigkofler and Niessner (2001) reported that over 88 per cent of the initial siloxane concentrations were still present in both landfill gas and AD-plant biogas after cryogenic treatment.

Ammonia

Ammonia is formed during the anaerobic degradation of feedstocks rich in protein. Ammonia concentration in biogas is generally very low, not exceeding 0.1 mg/Nm^3 (Rasi et al., 2007). A high concentration of ammonia in the biogas can lead to the formation of nitrous oxide (NO_x) during combustion. Most of the engines can accept an ammonia concentration of 100 mg/Nm^3 (IEA Bioenergy, 2006). In practice, ammonia formation can be avoided by controlling the biogas process, as ammonia is formed at high pH and temperature. Adjusting the C/N ratio of the feedstock can also avoid ammonia formation. Some ammonia is removed during CO_2-removal processes like adsorption and absorption processes with water or organic

Table 10.4 Some common siloxanes occurring in biogas.

Siloxane	Abbreviation	Molecular weight (g/mol)	Boiling point (°C)	Water solubility at 25 °C (mg/l)
Hexamethyldisiloxane	L2	162	106.9	0.93
Hexamethylcyclotrisiloxane	D3	223	135.2	1.56
Octamethyltrisiloxane	L3	237	153	0.034
Octamethylcyclotetrasiloxane	D4	297	175.7	0.056
Hexamethyltetrasiloxane	L4	311	194	0.00674
Decamethylcyclopentasiloxane	D5	371	211.2	0.017
Dodecamethylpentasiloxane	L5	385	232	0.000309
Dodecamethylcyclohexasiloxane	D6	444	245.1	0.005
Trimethylsilanol	–	90	–	4.26E+4

Source: Adapted after Rasi et al. (2007, 2009); Arnold (2009)

Siloxane compounds are widely used in many industrial processes and are frequently added to consumer products, such as hygiene products, cosmetics and biopharmaceuticals, fuel additives, car waxes, detergents and antifoams. Most of these siloxanes reach the wastewater and are not degraded in a conventional-activated sludge wastewater treatment process. Generally, a large part of siloxanes is volatilised to the atmosphere during the sludge treatment. However, a significant amount of siloxanes still remain adsorbed to the sludge flocs (Appels et al., 2008). During the AD of the sludge, the adsorbed siloxanes are released and volatilised due to the breakdown of the organic material and the high temperature in the reactor. Thus, the biogas is enriched with siloxanes. Typical siloxane concentrations found in biogas are between 30 and 50 mg/Nm^3 with peaks up to 400 mg/Nm^3 in some wastewater treatment plants (Dewil et al., 2006).

Siloxanes are commonly present in landfill and sewage gases but are absent or present only in low amounts in biogases produced during AD of agricultural wastes, energy crops or food waste (Rasi, 2009). In comparison to landfill gas, sewage gas usually has a higher siloxane concentration. The concentration of siloxanes in landfill gas can range from 1 to 400 mg/Nm^3 (Rasi et al., 2010).

The major siloxanes identified in landfill and sewage gases are octamethylcyclotetrasiloxane (D4) and decamethylcyclopentasiloxane (D5) (Schweigkofler and Niessner 2001; Wheless and Pierce, 2004; Arnold, 2009). These two often make up > 90 per cent of the total siloxane concentration (Rasi, 2009). Landfill gas may contain significant quantities of other siloxanes, such as hexamethylcyclotrisiloxane (D3) and dodecamethylcyclohexasiloxane (D6), as well as hexamethyldisiloxane (L2), octamethyltrisiloxane (L3), hexamethyltetrasiloxane (L4) and dodecamethylpentasiloxane (L5) (Wheless and Pierce 2004). In addition, sewage gas may also contain other organic silicon compounds such as methoxytrimethyl silane, tetramethylsilane, trimethylfluorosilane, and trimethylpropoxysilane that cause similar detrimental effects in the combustion process (Rasi, 2009).

During biogas combustion, siloxanes are converted to microcrystalline silicon dioxide (SiO_2), a residue with chemical and physical properties similar to glass (Accettola et al., 2008; Rasi, 2009). These crystalline deposits have a surface thickness of several millimetres. Extensive deposits of SiO_2 can lead to serious engine damage by abrasion of gas engine

The produced FeS is precipitated and thus H_2S is removed from the biogas. This method is efficient in reducing high concentrations of H_2S in the biogas but it is not efficient in achieving a low and stable level of H_2S (Krich et al., 2005).

H_2S present in the biogas can be removed after the reactor using different physical and chemical methods as well as using biofilters (Rasi et al., 2011; Ryckebosch et al., 2011). H_2S is also removed in the same processes as CO_2 is removed, e.g. in water scrubbing (chemical) and activated carbon treatment. The description for these processes is given later in this chapter. Other H_2S removal methods include e.g. adsorption using iron oxide or hydroxide in a rust-covered steel wool reaction bed, and chemical absorption with NaOH, $FeCl_2$ or $Fe(OH)_3$, and biological filters (for details see review by Ryckebosch et al., 2011). Biofilters are based on similar bioprocesses as in the reactor. Biofilters are typically filled with plastic carriers. Biological filtration is commonly applied in removing odours (oxygen rich situation) (Krich et al., 2005), and H_2S and NH_3 from the biogas. Biological filtration can reduce H_2S content from 2000–3000 to 50–100 cm^3/Nm^3 (Schomaker et al., 2000; Persson, 2003).

Halogenated compounds

Halogenated compounds found in biogas are chlorine, bromine and fluorine containing substances (e.g. carbon tetrachloride, chlorobenzene, chloroform and trifluoromethane). In AD plants, the concentration of halogenated compounds in the biogas is generally very low. Higher concentration of halogenated compounds is generally noticed in biogases from industrial landfills and sewage treatment plants (Rasi et al., 2007). Most halogenated volatile compounds are directly released from the waste, and their quantity depends on their presence in the feedstock as well as process conditions such as the pressure and the temperature.

Halogenated compounds break down in the engine during combustion. In the water phase, combustion products form the corrosive and acidic hydrogen chloride, hydrogen bromide and hydrogen fluoride. If biogas is used for energy production, compounds containing organochloride contribute to corrosion in vehicle or combustion engines. Incinerating halogens at a low temperature < 400 °C may also lead to the formation of dioxides and furans such as polychlorinated dibenzo-*p*-dioxins and polychlorinated dibenzofurans (Allen et al., 1997). Halogens also dissolve in motor oil. Depending upon the halogen concentration, more frequent oil changes may be necessary.

Halogenated compounds can be removed with activated carbon. Small molecules like CH_4, CO_2, N_2 and O_2 can pass through the pores but the larger molecules are absorbed. Generally, two columns are used in parallel: one for absorption and the other one for regeneration. Regeneration is done by heating the activated carbon to 200 °C, thus evaporating the adsorbed components which are thereafter removed by an inert gas (IEA Bioenergy, 2009).

Siloxanes

Siloxanes (C_2H_6OSi) are volatile compounds containing Si–O bonds with organic radicals bonded to the Si. The organic radicals can include methyl, ethyl, and other organic functional groups. Table 10.4 presents the most common siloxanes present in the biogas (Rasi et al., 2007, 2010, 2011; Arnold, 2009). Siloxanes are either linear or cyclic (Dewil et al., 2006). Linear siloxanes are designated with the letter L or M nomenclature while cyclic compounds are designated with the letter D (Wheless and Pierce, 2004). The amount of silicon in the molecules is referred to by the number following the letter (Arnold, 2009).

generally takes place in a dryer, which is filled with the adsorption dryer, e.g. silica, aluminum oxide or magnesium oxide. The system usually consists of two columns packed with drying agent and operated in parallel. One column is used for absorption while the other one is used for regeneration of the drying agent. During the absorption cycle, pressurised gas is led through the absorption column and water is adsorbed on the drying agent. Regeneration is achieved by evaporating the water through decompression and heating to approximately 200 °C.

Hydrogen sulphide

Most of the sulphur in biogas is in the form of hydrogen sulphide (H_2S). In addition, varying levels of carbonyl sulphide (COS), mercaptans (thiols) and disulphides are also present in the biogas (Rasi, 2009). H_2S is produced from the degradation of proteins and other sulphur containing compounds present in the organic feedstock by sulphur reducing bacteria. For instance, methanethiol and dimethylsulphide (DMS) are formed when sulphur containing amino acids, e.g. in manure, are degraded. DMS is then reduced to CH_4 and methanethiol, which further decomposes into CH_4, CO_2 and H_2S (Rasi et al., 2007). The sulphur concentration in biogas varies depending on the feed and the process used. For instance, typical H_2S concentrations in agricultural biogas plants (manure as main feedstock) vary between 500 and 3000 ppmv and <100 ppmv (Heguy and Bogner, 2004) in landfills, depending on landfill waste characteristics.

H_2S is toxic and strongly corrosive to many kinds of steel. H_2S needs to be removed in order to avoid corrosion in compressors, gas storage tanks and engines. When raw biogas is combusted, H_2S is converted into sulphur oxides (sulphur dioxide and sulphur trioxide). Sulphur oxide then reacts with water to form sulphuric acid (H_2SO_4), which corrodes metallic components of the engine. The H_2S content at 300–500 ppmv or more decreases the energy conversion efficiency and damages the catalytic converter (Holm-Nielsen and Al Seadi, 2004). Therefore, H_2S must be removed from the biogas or at least reduced to less than 1000 ppmv (0.1 per cent by volume) to prevent corrosion of boilers or to 100–500 ppmv H_2S (equal to 0.01–0.05 per cent by volume) to prevent damage to the CHP plant and other equipment, e.g., heat exchangers and catalysts.

H_2S can be removed in the reactor by oxidising it to elemental sulphur (S) or by forming insoluble metal sulphides (Hagen et al., 2001; Persson, 2003). Techniques involved are air/O_2 dosing and addition of iron chloride. In air/O_2 dosing, a small amount of O_2 (2–6 per cent) is introduced into the reactor by an air pump. H_2S is oxidised biologically to elemental S by a group of specialised sulphide oxidising micro-organisms (*Thiobacillus* spp.). These autotrophic bacteria are usually present in the reactor and utilise CO_2 from the biogas as a carbon source:

$$H_2S + O_2 \rightarrow 2\,S + 2\,H_2O \qquad (Eq.\ 10.2)$$

H_2S concentrations of 20–100 cm³/Nm³ in treated biogas with H_2S removal efficiencies of 80–99 per cent can be achieved by this technique (Krich et al., 2005; IEA Bioenergy, 2009).

H_2S can also be removed in the reactor by adding iron chloride (FeCl) into the reactor or dosed through the feed tank. The Fe^{3+} ions react with the sulphide (S^{-2}) ions in H_2S to form FeS (particles) and elemental S:

$$2\,Fe^{3+} + 3\,S^{2-} \rightarrow 2\,FeS + S \qquad (Eq.\ 10.3)$$

$$Fe^{2+} + S^{2-} \rightarrow FeS \qquad (Eq.\ 10.4)$$

Table 10.3 Biogas contaminants and possible impact on the downstream application.

Contaminant	Possible impact
Dust, foam and solid particles	Clogging of gas pipelines, compressors, gas storage tanks
Water	Corrosion in gas pipelines, compressors, gas storage tanks and engines due to reaction with H_2S and CO_2 to form acids Accumulation of water in pipes Condensation and/or freezing due to high pressure
H_2S	Corrosion in compressors, gas storage tanks and engines Toxic concentrations of H_2S (> 5 cm^3m^{-3}) remain in the biogas SO_2 and SO_3 are formed due to combustion, which are more toxic than H_2S and cause corrosion with water
CO_2	Low calorific value
Siloxanes	Formation of SiO_2 and microcrystalline quartz due to combustion; deposition at spark plugs, valves and cylinder heads abrading the surface
Hydrocarbons	Corrosion in engines due to combustion
NH_3	Corrosion when dissolved in water
O_2/air	Explosive mixtures due to high concentrations of O_2 in biogas
Halogenated compounds	Corrosion in combustion engines

Source: Adapted after review by Ryckebosch et al. (2011)

Solid particles and water

Biogas leaving the reactor contains typically dust particles and foam and is normally saturated with water vapour. In most biogas utilisation applications these are removed in order to prevent e.g. mechanical abrasion and corrosion in downstream equipment. A wide variety of methods are available to deal with this, as recently presented with detailed descriptions in a review by Ryckebosch et al. (2011). Dust particles are normally removed with 2–5 micro mesh dust collector filters while cyclones are used to remove foam.

Biogas also contains water vapour. The amount of water vapour in biogas depends on the AD process temperature, feed characteristics, and process (dry versus wet fermentation). The lower the temperature, the lower the water vapour content in the raw biogas and vice versa. The moisture reduces the calorific value of biogas, and it may condense, e.g. in the gas pipelines. The condensate can then react with H_2S to form ionic hydrogen and/or H_2SO_4, which can cause corrosion of steel (Table 10.3). Water can be removed by physical separation of condensed water or by chemical drying of the biogas.

A commonly used method to remove excess water vapour from biogas is through sensible pipe-works, which condense and remove the moisture as water. Condensed water can be removed by physical separation using moisture traps, water taps, demisters or cyclone separators (Ryckebosch et al., 2011). Also refrigeration can be used to promote condensation of water vapour from biogas or from compressed biogas.

Chemical drying methods involve adsorption and absorption techniques usually applied at elevated pressures. Adsorption using alumina or zeolites/molecular sieves is the most common chemical drying technique (Ryckebosch et al., 2011). Other chemical methods to remove water from gas are absorption by triethylene glycol and hygroscopic salts. The drying process

Table 10.2 Theoretical methane percentage in biogas calculated for different major substrates using Buswell's equation.

Substrate	Composition	Methane yield at STP (Nm³/tonne volatile solids)	Methane percentage in biogas (v/v, %)
Carbohydrate	$(C_6H_{10}O_5)n$	415	50
Protein	$C_5H_7O_2N$	496	69
Lipids	$C_{57}H_{104}O_6$	1014	70

Source: Adapted after Angelidaki and Sanders (2004).

in the biogas is always lower than the theoretical value. As CH_4 is practically insoluble in water, it is always present in the gas phase only. At low temperatures, CO_2 is found dissolved in the liquid phase only, e.g. in some AD wastewater treatment processes as well as in low temperature landfill leachates.

During the AD process, most of the utilised carbon from the substrate is converted into CH_4, while only a small part (5–10 per cent) is used to synthesise new biomass, indicating high energy efficiency. In the biogas, carbon is present in its most oxidised form (CO_2) and its most reduced form (CH_4). The ratio between CH_4 and CO_2 is however dependent on the oxidation state of the carbon compounds present in the feedstock, i.e. the more reduced the organic carbon content is, the more CH_4 will be produced (Angelidaki and Sanders, 2004). If the chemical composition of feedstock is known and biodegradation is complete, the theoretical methane yield can be calculated by means of Buswell's equation (Buswell and Neave, 1930). Table 10.2 presents the theoretical methane percentage in the biogas for different substrates during the AD process.

Biogas contaminants and treatment

In addition to CH_4 and CO_2, biogas contains moisture, dust, sulphur compounds, and organic trace compounds. More than 500 different trace contaminants have been detected in landfill gas alone (Rasi et al., 2007). In this chapter, the presence of these contaminants in biogas is presented as well as their effects on the downstream energy conversion, and their treatment. The main emphasis is on CO_2 removal techniques as they are under intensive development due to increasing interest to upgrade biogas to biomethane.

Table 10.3 presents the major biogas contaminants and possible impact on the downstream application. Biogas can be utilised as such, e.g. in cooking, but for most applications it is processed to facilitate the function and maintenance of the utilisation equipment as well as to increase the energy intensity of the gas. For almost all applications (CHP) dust, moisture and sulphur compounds are removed, e.g. to avoid corrosion. In addition, removal of contaminants like halogenated hydrocarbons and siloxanes is required for both vehicle fuel use and gas grid injection applications, while removal of CO_2 is needed to increase the energy value of the gas. In general, the end use of the biogas sets its quality demands and thus the need for its processing (IEA Bioenergy, 2009).

Typically biogas cleaning and upgrading may contain several unit operations, which remove different contaminants. However, many of the processes remove several different types of contaminants at the same time.

Biogas can be compared with natural gas as they are used in similar energy systems. However, it must be noted that natural gas composition varies greatly among different production sites. When compared with natural gases, biogas typically contains lower or similar amounts of CH_4 and higher amounts of CO_2 and lower levels of hydrocarbons other than CH_4. Natural gas, used as domestic or industrial fuel, typically contains 55–98 per cent of CH_4 and higher hydrocarbons such as ethane (1–14 per cent), propane (0–4 per cent), butane (0–2 per cent), pentane (0–0.5 per cent) and hexane (0–2 per cent) and non-combustible gases: CO_2 (0–2 per cent); O_2 (0–1.2 per cent) and N_2 (0.4–17 per cent) (Fegan, 2001). Thus, the energy content of biogas per unit volume is typically lower compared with natural gas (Table 10.1). By separating CO_2 from the biogas in an upgrading process, the energy content of upgraded biogas is increased.

The heating value (energy value) of a fuel is the amount of heat released during the combustion of a known quantity of the fuel. The heating value is expressed as higher heating value (HHV) or lower heating value (LHV). The HHV of biogas is the energy that is released when 1 Nm^3 of biogas is combusted and the water vapour formed in the combustion is condensed, thus releasing its condensation energy in surplus. On the other hand, the LHV is the energy formed when the water vapour is not condensed. For example, the LHV and HHV of CH_4 are 50 and 55.5 MJ/kg, respectively.

The Wobbe index (WI) is commonly used to define the domestic gas quality to the end-user (Eq. 10.1). In this expression, relative density is the quotient of gas over air density at standard temperature and pressure (STP). As gas fuels are mixtures of gas and air, this quotient is a measure of the amount of gas fuel over air that is used in a combustion system. WI is therefore proportional to the energy flux of a given gas mixture through an orifice.

$$WI(MJ/Nm^3) = \frac{\text{calorific value}(MJ/Nm^3)}{\sqrt{\text{relative density}}} \qquad \text{(Eq. 10.1)}$$

Gases with the same WI have the same combustion energy output in a given combustion appliance, e.g., gas engines or burners. For biogas to have similar combustion behaviour as that of natural gas, the WI should therefore also be similar. The minimum amount of CH_4 required as well as the maximum amount of N_2 (air) that could be present in the gas is dependent on the WI. Similar to LHV and HHV, there are also a lower and a higher WI.

Factors affecting biogas composition

The amount and the composition of biogas produced depends on several factors such as the characteristics of the feedstock used (e.g. organic content, fibre content, lignin content, total solids content, calorific value), type of reactor (mixed tank, plug flow, garage system etc.) and process used (batch, semi-continuous, continuous), process parameters implied (e.g. temperature, hydraulic retention time and organic loading rate, mixing regime, degassing system used, sludge retention time), and the presence (e.g. antibiotics) or formation (e.g. volatile fatty acids, ammonia, H_2S, H_2) of inhibitory compounds (see e.g. Ward et al., 2008; Ryckebosch et al., 2011).

The biogas composition is also dependent on the process pH (Franco et al., 2007) as CO_2 is highly soluble in water. Depending on the pH value, CO_2 is either partially dissolved in the liquid phase as carbonic acid or converted to bicarbonate. Thus, CO_2 concentration

Table 10.1 Composition of biogas from dedicated AD plants, sewage treatment plants and landfills and as a reference composition of natural gas in the Netherlands.

Parameter	Unit	Farm AD plant	Centralised AD plant	Landfill	Sewage treatment plant	Natural gas (The Netherlands)
CH_4	vol-%	55–60	60–70	30–65	60–65	81–89
Other hydrocarbons	vol-%	0	0	0	0	3.5–9.4
H_2	vol-%	0	0	0–3	0	N/A
CO_2	vol-%	35–40	30–40	25–45	35–40	0.67–1
N_2	vol-%	<1–2	2–6	<1–17	<1–2	0.28–14
O_2	vol-%	<1	0.5–1.6	<1–3	<0.05–0.7	0
H_2S	ppmv	25–2000	0–2000	30–500	0.5–6800	0–2.9
NH_3	ppmv	≈100	≈100	≈5	<1–7	0
Halogenated compounds (as Cl^-)	mg/Nm^3	<0.01	<0.25	0.3–225	0–2	N/A
Siloxanes	mg/Nm^3	<0.03–<0.2	<0.08–<0.5	<0.3–36	<1–400	N/A
Wobbe index	MJ/Nm^3	24–33	24–33	20–25	25–30	44–55
Lower heating value	MJ/Nm^3	23	23	16	22	31–40
Density	kg/Nm^3	1.2	1.2	1.3	1.2	0.8

Source: adapted after Rasi et al. (2007); IEA Bioenergy (2009)

Note: 1 kiloWatt hour (kWh)/Nm^3 = 3.6 MJ/Nm^3; a normal cubic metre (Nm^3) is the metric expression of gas volume at standard conditions and it is usually defined as being measured at 0 °C and 1 atmosphere of pressure. N/A = Not available

Chapter 10

Biogas upgrading and compression

Prasad Kaparaju,[1] Saija Rasi[2] and Jukka Rintala[3]*

[1]Department of Biological & Environmental Science, University of Jyväskylä, P.O. Box 35, FI-40014, Jyväskylä, Finland; [2]MTT Agrifood Research Finland, Bioenergy and Environment, FI-31600 Jokioinen, Finland; [3]Department of Chemistry and Bioengineering, Tampere University of Technology, P.O. Box 541, FI-33101 Tampere, Finland
*Corresponding author email: prasad.kaparaju@jyu.fi

Introduction

In the anaerobic digestion (AD) process microorganisms convert organic material into biogas under anaerobic conditions. In man-made engineered systems, biogas is produced at sewage treatment plants, in AD plants from industrial and municipal biowastes as well as in farm AD plants from agricultural feedstocks such as manure, crop residues and energy crops. Biogas is also produced in landfills. The biogas composition varies in different production sites with methane (CH_4) and carbon dioxide (CO_2) as the main components while sulphide compounds are typically present as well as various trace compounds including siloxanes. CH_4 is the main energy component in the biogas, while it is also a strong greenhouse gas (GHG). Biogas, similar to natural gas, is typically used for combined heat and power (CHP) generation and increasingly as vehicle fuel and for distribution through the natural gas grid. Moisture, dust and sulphur (S) compounds are typically removed before any energy use, for example to prevent corrosion. For vehicle use and natural gas grid injection as well as for some CHP applications, removal of CO_2 and other contaminants is needed to produce high value product, referred to as biomethane. This chapter presents biogas composition and its upgrading for biomethane production with emphases on CO_2 removal.

Biogas composition and energy value

Biogas is composed primarily of CH_4 (45–70 per cent) and CO_2 (30–45 per cent) with smaller amounts of hydrogen sulphide (H_2S; 0–2000 ppm) and ammonia (NH_3; 0–590 ppm). In addition, trace amounts of hydrogen (H_2), nitrogen (N_2), carbon monoxide (CO), saturated or halogenated hydrocarbons and oxygen (O_2) are occasionally present in the biogas. The biogas is usually saturated with water vapour (H_2O) (Rasi et al., 2007).

The biogas composition varies from production site to site, e.g. depending on the type of feedstock and process technology and conditions used for the biogas production. Table 10.1 presents a typical biogas composition as an average from a range of production sites. Biogases from farm (typically using manure and crops) and centralised (typically using biowastes) AD plants as well as sewage treatment plants contain relatively high CH_4 content with low concentrations of trace compounds (Rasi et al., 2007). The CH_4 content can be much lower in the landfill gas while it may also contain some O_2 and N_2 (Rasi et al., 2007).

Nizami, A.S., Thamsiriroj, T., Singh, A. and Murphy, J.D. (2010) 'The role of leaching and hydrolysis in a two phase grass digestion system', *Energy and Fuels*, vol 24, no 8, pp 4549–4559.

Nizami, A.S., Singh, A. and Murphy, J.D. (2011) 'Design, commissioning, and start-up of a sequentially fed leach bed reactor complete with an upflow anaerobic sludge blanket digesting grass silage', *Energy and Fuels*, vol 25, no 2, pp 823–834.

Nizami, A.S., Orozcoc, A., Groom, E., Dieterich, B. and Murphy, J.D. (2012) 'How much gas can we get from grass?', *Applied Energy*, vol 92, pp 783–790.

Oregon Department of Energy. (2009) 'Biogas technology', Oregon Department of Energy, US http://www.oregon.gov/energy/renew/biomass/biogas.shtml [accessed: January 2012].

Paula, J.D.R. and Foresti, E. (1992) 'Kinetic studies on a UASB reactor subjected to increasing COD concentration', *Water Science Technology*, vol 25, no 7, pp 103–111.

Prochnow, A., Heiermann, M., Drenckhan, A. and Schelle, H. (2005) 'Seasonal pattern of biomethanisation of grass from landscape management', *Agricultural Engineering International The CIGR E Journal*, vol 7, manuscript EE 05 011.

Qi, B.C., Aldrich, C., Lorenzen, L. and Wolfaardt, G.W. (2005) 'Acidogenic fermentation of lignocellulosic substrate with activated sludge', *Chemical Engineering Communications*, vol 192, no 9, pp 1221–1242.

Rafique, R., Poulsen, T.G., Nizami, A.S., Asam, Z.Z., Murphy, J.D. and Kiely, G. (2010) 'Effect of thermal, chemical and thermo-chemical pre-treatments to enhance methane production', *Energy*, vol 35, no 12, pp 4556–4561.

Reddy. (2012) 'Biogas to renewable natural gas – Ontario & BC case studies', Canadian Farm and Food Biogas Conference and Exhibition. March 5–7. London Convention Centre, London, Ontario, Canada, www.gtmconference.ca/site/index.php [accessed: April 2012].

Richard, M., Barry, K. and Ann, W. (2008) *AgSTAR handbook: A manual for developing biogas systems at commercial farms in the United States* www.epa.gov/agstar/documents/AgSTAR-handbook.pdf [accessed: November 2011].

Saville, B.A., Duff, B. and Porter, S. (2008) *Market assessment of agricultural and industrial anaerobic digestion potential in Canada*, Technical Report for Natural Resources Canada, Report no. NRC-07-00096.

Singh, A., Nizami, A.S., Korres, N.E. and Murphy, J.D. (2011) 'The effect of reactor design on the sustainability of grass biomethane', *Renewable and Sustainable Energy Reviews*, vol 15, no 3, pp 1567–1574.

US-DOE. (2011) Alternative and advanced fuels. What is biogas? US DOE http://www.afdc.energy.gov/afdc/fuels/emerging_biogas_what_is.html [accessed: February 2012].

US-EPA (2012) 'Landfill Methane Outreach Program, Basic information'. US EPA http://www.epa.gov/lmop/basic-info/index.html [accessed: April 2012].

Vandevivere, P., De Baere, L. and Verstraete, W. (2003) 'Types of anaerobic digester for solid wastes', in J. Mata-Alvarez (ed.), *Biomethanization of the organic fraction of municipal solid wastes*. London: IWA Press.

Ward, A.J., Hobbs, P.J., Holliman, P.J. and Jones, D.L. (2008) 'Optimization of the anaerobic digestion of agricultural resources', *Bioresource Technology*, vol 99, no 17, pp 7928–7940.

Weiland, P. and Rozzi, A. (1991) 'The start up, operation and monitoring of high rate anaerobic treatment systems', discussers report, *Water Science Technology*, vol 24, no 8, pp 257–277.

Yu, H.W., Samani, Z., Hanson, A. and Smith, G. (2002) 'Energy recovery from grass using two-phase anaerobic digestion', *Waste Management*, vol 22, no 1, pp 1–5.

DSIRE (Database of State Incentives for Renewable Energy). (2011b) 'California – net metering', Database of State Incentives for Renewable Energy, US, www.dsireusa.org [accessed: January 2012].

Green J. (2012) 'Ontario biogas outlook', Canadian Farm and Food Biogas Conference and Exhibition. March 5–7. London Convention Centre, London, Ontario, Canada, www.gtmconference.ca/site/index.php [accessed: April 2012].

IEA (International Energy Agency). (2005) 'Biogas production and utilisation', IEA Bioenergy Task 37, International Energy Agency, www.iea-biogas.net [accessed: April 2012].

IEA (International Energy Agency). (2011) 'Country reports of member countries, Istanbul,' IEA Bioenergy Task 37, International Energy Agency, www.iea-biogas.net [accessed: April 2012].

IEA (International Energy Agency). (2012) 'Focus Canada' Newsletter: 2/2012, IEA Bioenergy Task 37, www.iea-biogas.net [accessed: May 2012].

Igoni, A.H., Ayotamuno, M.J., Eze, C.L., Ogaji, S.O.T. and Probert, S.D. (2008) 'Design of anaerobic digesters for producing biogas from municipal solid-waste', *Applied Energy*, vol 85, no 6, pp 430–438.

Jha, A.K., Li, J., Nies, L. and Zhang, L. (2011) 'Research advances in dry anaerobic digestion process of solid organic wastes', *African Journal of Biotechnology*, vol 10, no 65, pp 14242–14253.

Juanga, J.P. (2005) 'Optimizing dry anaerobic digestion of organic fraction of municipal solid waste', M.E. Thesis, Asian Institute of Technology, Bangkok, Thailand.

Korres, N.E., Singh, A., Nizami, A.S. and Murphy, J.D. (2010) 'Is grass biomethane a sustainable transport biofuel?', *Biofuels, Bioproducts, Biorefinery*, vol 4, no 3, pp 310–325.

Korres, N.E., Thamsiriroj, T., Smyth, B.M., Nizami, A.S., Singh, A. and Murphy J.D. (2011) 'Grass biomethane for agriculture and energy', in E. Lichtfouse (Ed.) *Sustainable Agriculture Reviews* vol 7, 5–49.

Lehtomäki, A., Huttunen, S., Lehtinen, T.M. and Rintala, J.A. (2008) 'Anaerobic digestion of grass silage in batch leach bed processes for methane production', *Bioresource Technology*, vol 99, no 8, pp 3267–3278.

Lemmer, A. and Oechsner, H. (2001) 'Kofermentation von Gras und Silomais', *Landtechnik*, vol 56, no 6, pp 412–413.

Lettinga, G. (1995) 'Anaerobic digestion and wastewater treatment systems', *Antonie van Leeuwenhoek*, vol 67, no 1, pp 3–28.

Li, D., Yuan, Z. and Sun, Y. (2010) 'Semi-dry mesophilic anaerobic digestion of water sorted organic fraction of municipal solid waste (WS-OFMSW)', *Bioresource Technology*, vol 101, no 8, pp 2722–2728.

Linder, G. (2009) 'Failure analysis and smart grid control protocols for anaerobic digesters', MS Thesis, Dept. Elect. Eng, Clarkson, University, Potsdam, New York.

Mähnert, P., Heiermann, M. and Linke, B. (2005) 'Batch- and semi-continuous biogas production from different grass species', *Agricultural Engineering International, the CIGR Ejournal*, vol 7, manuscript EE 05 010.

Marchaim, U. (1992) *Biogas processes for sustainable development*, Rome: FAO Agricultural Services Bulletin.

Milner, A. (2012) 'RNG end use in transportation – what's next in Canada', Canadian Farm and Food Biogas Conference and Exhibition. March 5–7. London Convention Centre, London, Ontario, Canada, www.gtmconference.ca/site/index.php [accessed: May 2012].

Mullins, P.A. and Tikalsky, S.M. (2006) *Anaerobic digester implementation issues, Phase II – A survey of California farmers* (Dairy Power Production Program). California Energy Commission.

Nizami, A.S. and Murphy, J.D. (2010) 'What type of digester configurations should be employed to produce biomethane from grass silage?' *Renewable and Sustainable Energy Reviews*, vol 14, no 6, pp 1558–1568.

Nizami, A.S. and Murphy, J.D. (2011) 'Optimizing the operation of a two-phase anaerobic digestion system digesting grass silage', *Environmental Science & Technology*, vol 45, no 17, pp 7561–7569.

Nizami, A.S., Korres, N.E. and Murphy, J.D. (2009) 'A review of the integrated process for the production of grass biomethane', *Environmental Science and Technology*, vol 43, no 22, pp 8496–8508.

fee for wastes otherwise destined for landfill. The goal of organic waste diversion can be achieved with efficient management of organic 'by-products' from municipalities and the food processing industry. The continuity of the various incentive systems is nonetheless critical for maintaining long-term growth and investment into the AD sector. Furthermore, grant programmes and loan guarantees for digester development will continue to encourage anaerobic digester installation and development. There is great potential for the use of enriched biomethane as an indigenous gaseous biofuel within many jurisdictions, because of the network of natural gas grids already available. The fertile land base and livestock activities required are already at hand.

References

AgSTAR: An outreach programme of US EPA (2002) 'Managing manure with biogas recovery systems improved performance at competitive costs', The Agstar Program. US EPA, http://www.epa.gov/nscep/ [accessed: November 2011].

AgSTAR: An outreach programme of US EPA (2010) 'Farm anaerobic digestion systems: A 2010 snapshot'. The Agstar Program. US EPA, www.epa.gov/agstar/documents/2010_digester_update.pdf [accessed: December 2011].

Amon, T., Kryvoruchko, V., Amon, B., Moitzi, G., Lyson, D., Hackl, E., Jeremic, D., Zollitsch, W. and Pötsch, E. (2003) 'Optimierung der Biogaserzeugung aus den Energiepflanzen Mais und Kleegras' ['Biogas production from the energy crops maize and clover grass'], *Endbericht*, no 1249.

Anderson R. (2012) 'The economics of anaerobic digester technology for Ontario farmers', Canadian Farm and Food Biogas Conference and Exhibition. March 5–7. London Convention Centre, London, Ontario, Canada, www.gtmconference.ca/site/index.php [accessed: April 2012].

Asam, Z., Poulsen, T.G., Nizami, A.S., Rafique, R., Kiely, G. and Murphy, J.D. (2011) 'How can we improve biomethane production per unit of feedstock in biogas plants?', *Applied Energy*, vol 88, no 6, pp 2013–2018.

Bal, A.S. and Dhagat, N.N. (2001) 'Upflow anaerobic sludge blanket—a review', *Indian Journal of Environmental Health*, vol 43, no 2, pp 1–82.

Baserga, U. (1998) *Vergarung von Extensogras-Silage in einer Feststoff-Pilotanlage und einer landwirtschaftlichen Co-Vergarungs-Biogasanlage* [Anaerobic digestion of silage from extensive grassland in a solid state pilot plant and a farm-scale cofermentation biogas plant], Tanikon: Eidg. Forschungsanstalt fur Agrarwirtschaft und Landtechnik, Forschungsprogramm Biomasse.

Baserga, U. and Egger, K. (1997) *VergarungvonEnergiegras zurBiogasgewinnung* [Anaerobic digestion of grass energy crops for biogas production]. Tanikon: Bundesamt fur Energiewirtschaft, Forschungsprogramm Biomasse.

C2ES (Center for Climate and Energy Solutions). (2011) 'Anaerobic digeste', *Climate tech book*, Publication Library, Center for Climate and Energy Solutions, www.c2es.org/_taxonomy/term/3142 [accessed: February 2012].

Cho J.K., Park, S.C. and Chang, H.N. (1995) 'Biochemical methane potential and solid state anaerobic digestion of Korean food wastes', *Bioresource Technology*, vol 52, no 3, pp 245–253.

Crolla, A., Duke, C., Kinsley, C. and Sauvé, T. (2012) 'Advantages and limitations with using various substrates in manure biogas plants', Canadian Farm and Food Biogas Conference and Exhibition. March 5–7. London Convention Centre, London, Ontario, Canada, www.gtmconference.ca/site/index.php [accessed: May 2012].

De Baere, L.D., and Mattheeuws, B. (2008) 'State-of-the-art 2008 – anaerobic digestion of solid waste', *Waste Management World*, vol 9, no 4, pp 1–8.

Demirbas, A. and Ozturk, T. (2005) 'Anaerobic digestion of agricultural solid residues', *International Journal of Green Energy*, vol 1, no 4, pp 483–494.

DSIRE (Database of State Incentives for Renewable Energy). (2011a) 'Net metering map', Database of State Incentives for Renewable Energy, US, www.dsireusa.org [accessed: February 2012].

where raw biogas of 750 m³ h⁻¹ is upgraded to pipeline quality gas. The substrate used in the digester is agricultural waste and manure. In September 2012, the plant began its first injection (Reddy, 2012).

Future research

- There is potential to further increase methane production by co-digesting lignocellulosic substrates with other substrates. The benefits of increased buffering capacity, microbiological stability and hydrolysis, and the decline of H_2S and ammonia were all reported in co-digestion of various substrates. Hydrolysis through use of slurry should also be investigated.
- The application of nanotechnology in AD in the form of chips and sensory equipment will provide new ways of monitoring and controlling the inhibitory effects of volatile fatty acids (VFA), COD, pH and temperature. This may also allow more rapid/effective adjustment of reactor operating conditions to better control biogas production rates.
- The value-added products of lignocellulosic substrates and the production of biofuel will assist with reducing GHG (greenhouse gases) emissions. However, it is important to calculate the GHG emissions of these value-added products as they are shaped into final marketable products at an additional energy and monetary cost. This measurement/calculation will guide any attempt to scale up AD processes from the laboratory scale (biogas production) to the farm scale (enriched biomethane production) and on to the industrial scale.
- Carbon dioxide is removed during the upgrading of biogas into enriched biomethane; this carbon dioxide removal provides an additional greenhouse gas (GHG) emission that can be minimized by reincorporation into the biogas production process. Dissolved carbon dioxide forms carbonic acid in water, and thus increases the CH_4 content of biogas. If this water is mixed with leachate, it could partially dissolve hemicellulose without harming the AD process.
- The continuity of incentive systems is critical for maintaining long-term growth, innovation and investment in the AD sector. The grant programmes and loan guarantees for digester development will encourage the anaerobic digester installation and development.

Conclusions

A high-rate reactor such as UASB converts volatizable organics into biogas with great efficiency and has a short retention time. This type of reactor is well suited to substrates with solids content between < 4 to 15 per cent and can process a wide range of substrates when attached with leach bed reactors. The BMP assays provide varying results, and therefore, one cannot assume that the biogas rate observed in a batch test will be realized in a larger scale operation, even for essentially the same feedstock. The chosen digester type will lead to different biogas production rates even for the 'same' feedstock due to differences in seasonal growth, harvesting regimes, pretreatment, digester operating conditions and procedures. Managing this variation is a key element of AD system operation. Anaerobic digesters can be cost competitive if the price of carbon rises and subsequently increases the cost of fossil fuels, or if other digester attributes are incentivized, e.g., by receipt of a tipping due to uncertainty about long-term tax incentives and subsidies (C2ES, 2011). Thus, cost is one of the important barriers limiting widespread use of anaerobic digesters. There are a number of factors affecting the yield and net cost of biogas plants. These include substrate methane potential, digester configuration, quantity of substrate, HRT, digester operating and capital cost, potential use of biogas and value-added products of digestate. The cost-effectiveness of a digester mainly depends on its size, type and biogas use. Additionally, the geographic location of the project and fuel type to be displaced by the digester also have an effect on the digester's cost-effectiveness. For example, in a community where electricity is less costly but thermal energy is expensive, it would be advantageous for a digester to upgrade biogas to biomethane to displace natural gas. Sometimes, however, the digester benefits are neglected in the cost analysis. For example, when waste is diverted to an anaerobic digester, the cost for municipal waste disposal is reduced. Furthermore, the costs associated with waste odour control and water protection in landfills are reduced when wastes are diverted to anaerobic digestion (C2ES, 2011). The US-EPA has estimated a digester capital cost based on farm and animal size. These costs are expressed per animal unit (AU), which is equal to a live animal weight of 454 kg (C2ES, 2011). For a covered lagoon type digester, the capital cost is $150–400 per AU. The capital cost of a complete mix or plug flow digester is $200–400 per AU. These calculations do not include digester operating costs (Agstar, 2002).

In some jurisdictions, there are restrictions on the quantity of electricity that can be sold to the grid, which can introduce another impediment to the widespread use of anaerobic digesters (DSIRE, 2011a). The Energy Policy Act of 2005 requires net metering for consumers upon request, in every state of the US. Net metering is the ability for electricity clients to sell electricity produced on-site back to the utility (C2ES, 2011). Utilities will gain experience as more units are connected, thus reducing potential issues with connections and grid management. Electricity providers will increase their experience during incorporation of anaerobic digesters into the electrical grid and so the current situation has room for improvement. Moreover, the electricity rates and policy implementation have limited the use of anaerobic digesters in the agriculture sector. For example, utility providers in the state of California cannot apply for standby charges, interconnection fees and minimum monthly charges (DSIRE, 2011b). Thus, utility providers do not buy excess electricity. In such situations, many farmers flare their excess gas rather than providing free energy to the grid (Mullins and Tikalsky, 2006).

Many European countries have specific targets for the use of biogas as a vehicular fuel. In Germany and Austria, natural-gas-powered vehicles require natural gas with 20 per cent biogas. This mandate dovetails well with the development of anaerobic digesters in Germany, which is also expanding due to feed-in tariffs for the biogas. The feed-in-tariffs require the purchase of energy from certain energy production facilities at a favorable rate. Therefore, it mandates the utilities to purchase biogas from anaerobic digesters and provides a financial return to AD projects (C2ES, 2011). In 2009, there were 4000 agricultural digesters in Germany and there were 6800 agricultural digesters in 2011 (IEA, 2011). Sweden also estimates that biogas will fulfill half of their need for fuelling its nearly 11,500 natural gas vehicles. On a global scale in 2010, there were 70,000 vehicles that operate on biogas (US-DOE, 2011). Within Ontario, the current share of natural gas operated vehicles is < 0.1 per cent. The estimated number of natural gas vehicles on the road is 9600 and there are 41 public and 18 private refuelling stations in Ontario (Milner, 2012). The first biogas upgrading plant to inject biogas into the gas grid in Canada is in Abbotsford BC by Catalyst Power,

Table 9.6 Challenges of the AD industry.

Technical	Biological	Financial	Environmental and regulatory
Reactor design	Digestate stabilization	Capital cost and financing	Feedstock collection
Process control system	Inoculum stabilization	Incentives	Utility connections
Downstream products and processing	Feedstock characteristics	Co-products and markets	Permits and regulatory approvals
Digester operations	Mono or co-digestion	On-farm versus centralized digesters	Nutrient management issues
Temperature operating regime	Inhibitory substances		
Feedstock type			
Process scale			

Source: Nizami et al., 2011; Saville et al., 2008

Table 9.7 Conditions of a stabilized anaerobic digester.

Parameters	Optimal range
Alkalinity	1,500–4,000 mg $CaCO_3$ L^{-1}
pH	6.8–7.2
VS/TS (volatile solids / total solids)	> 45%
NH_4^+-N (ammonium)	< 1,500 mg L^{-1}
C/N/P (carbon to nitrogen to phosphorus proportion)	100–120:5:1
C/N (carbon to nitrogen ratio)	20–30:1
IA/TA (intermediate alkalinity to total alkalinity)	0.1–0.2 (< 0.4)

Source: Crolla et al., 2012

not been studied in the depth required to upscale these approaches to the commercial and industrial scale; further optimization of both operational and process parameters is required (Nizami et al., 2012).

Commercial-scale anaerobic digesters are often run below their ORL due to ineffective monitoring and control measures. This can cause instability in the digestion process by forming inhibitory compounds, such as high VFA, which effect methanogens (Nizami et al., 2009). In the US, many installed digesters failed during 1970s due to inadequate operating expertise, technical support and monitoring procedures (Richard et al., 2008). This rate of failure reached 70 per cent with plug flow and continuously mixed type digesters (Linder, 2009). A process control system of online close-loop monitoring with sensory equipment, data-reading software and programmed control are beneficial for optimal digester performance (Nizami et al., 2009).

Although the current generation of digesters, including dry batch, dry continuous and high-rate reactors, have been improved, there are lingering reliability and cost issues with respect to long-term use and digester operation and installation (C2ES, 2011). When US-generated biogas is utilized for energy, agricultural digesters have a payback period that ranges between 3 and 7 years (Agstar, 2002). The payback period for a WWTP (wastewater treatment plant) digester is less than 3 years. This can be further reduced if food waste is accepted as a co-substrate (C2ES, 2011). The financial incentives play a positive role in the expansion of digesters. The commercialization of digester projects has declined, mainly

Table 9.5 Comparison of different anaerobic digesters applied for high solid substrates.

Digester type	Substrate type	Operating conditions	HRT (days)	Biogas yield (m^3 kg^{-1} VS added)	Methane yield (m^3 kg^{-1} VS added)	Studies
Semi-continuous digester	Three ensiled grass species	Mesophilic, mono-digestion	28	0.56–0.61	0.3–0.32	Mähnert et al. (2005)
	Grass silage	Mesophilic, co-digestion	25–60	0.08–0.39	–	Lemmer and Oechsner (2001)
Batch digester	Fresh and ensiled grass	Mesophilic, co-digestion	25	0.5–0.72	–	Baserga and Egger (1997)
	(MSW) Municipal solid waste	Mesophilic	60	–	0.314	Li et al. (2010)
	Fresh and ensiled grass species	Mesophilic, mono-digestion	28	0.65–0.86	0.31–0.36	Mähnert et al. (2005)
	Food wastes	Mesophilic	120	–	0.373	Cho et al. (1995)
	Ensiled mixture of grass and clover	Mesophilic, mono-digestion	59	0.42–0.53	0.29–0.37	Amon et al. (2003)
	MSW (municipal solid waste)	Thermophilic	28	–	0.32	Juanga (2005)
Continuous digester	Grass silage and hay	Mesophilic, co-digestion	18–36	0.5–0.55	–	Prochnow et al. (2005)
	MSW	Thermophilic	25	–	0.278	Jha et al. (2011)
	Grass silage	Mesophilic, co-digestion	20	0.5–0.55	–	Baserga (1998)
	Grass silage	Mesophilic, mono-digestion	50	–	0.451	Nizami et al. (2012)
Batch leach-UASB digester	Grass silage	Mesophilic, mono-digestion	55	0.16–0.39	0.06–0.197	Lehtomäki et al (2008)
	Grass silage	Mesophilic, mono-digestion	30	–	0.341	Nizami et al. (2012)

Table 9.4 Factors affecting anaerobic digester inputs and outputs.

Digester inputs	Digester outputs	
Digester operation	Pre-treatment	Post-treatment
Temperature	Co-digestion	Biogas
pH	Temperature	Biomethane
Recirculation	Physical, chemical, biological	Fertilizer
DS and VS content	and thermal treatments	By-products
OLR	Inoculum/nutrition addition,	
Retention time	C:N ratio	
Particle size		
Mixing		

cannot be directly applied to these other substrates (Nizami and Murphy, 2010). Various BMP tests can provide different predictions of biomethane yield for the same feedstock, due to different methodologies, including drying the substrate before the test, different ratios of inoculum to substrate, different sizes of the vessel, the degree of mixing of the substrate, or addition of micro minerals and trace metals (Nizami et al., 2012; Asam et al., 2011). These factors and others that affect the performance of anaerobic digesters are grouped in Table 9.4. Nizami et al. (2012) have studied three different types of BMP (i.e., micro, small and large BMP), and determined the potential of methane production for grass silage. The results vary with each type of BMP assay. Thus, one cannot assume that a batch test will provide an exact indication of biogas production for a given feedstock; small reactor vessels with inoculum from a different feedstock may not yield the same production of biogas as a continuous digester. Furthermore, even for the same feedstock, the biogas production rate will be affected by the type of digester. This has clear implications for scale-up; translating lab data and BMP test results into commercial operation can be a challenge. The biogas yields of different anaerobic digesters applied to high solid substrates are summarized in Table 9.5.

Challenges in the AD industry

There are various challenges in the development of the AD industry pertaining to biological, technical, economic and environmental criteria (Table 9.6). Controlling AD requires maintaining delicate microbial ecosystems. Both the temperature and physical input into the digesters (e.g., inorganic and non-digestible waste) must be closely regulated to enable optimal gas production and to avoid process failure. The optimum ranges of parameters that influence the stability of the digestion process are shown in Table 9.7. Performance problems in the US throughout the 1980s significantly impeded the commercialization of agricultural digesters (Oregon, 2009). There is less work carried out to analyse lignocellulosic substrates with an aim to improve biogas yield. Thus, the research literature on best digester configurations for lignocellulosic substrates is arguably scarce (Nizami and Murphy, 2010). Digesters can be evaluated for optimal reactor configuration while at the same time using equal quantities of the same substrate loaded at the same rate. As the substrate is a significant factor, the digester configuration needs to be optimized for different feedstocks (e.g., OFMSW, green waste), in order to determine the best configuration for a specific high solids substrate. Solid content and volatile content in the feedstock are also important aspects to be considered (Nizami et al., 2009). Procedures and technologies adopted at the laboratory/pilot scale have

Table 9.3 Comparison of different digester characteristics.

Digester type	DS content (%)	VS removal (%)	HRT (days)	OLR (kg VS m^{-3} d^{-1})
One-stage	10–40	40–80	10–60	0.7–15
Two-stage	2–40	40–75	10–15	10–15 for second stage
Dry	20–50	40–70	14–60	12–15
Wet	2–12	40–75	25–60	< 5
Batch	25–40	40–70	30–60	12–15
Continuous	2–15	40–75	30–60	0.7–1.4
High rate	< 4–15%	75–98	0.5–12	10–15

Source: Nizami and Murphy, 2009; Vandevivere et al., 2003

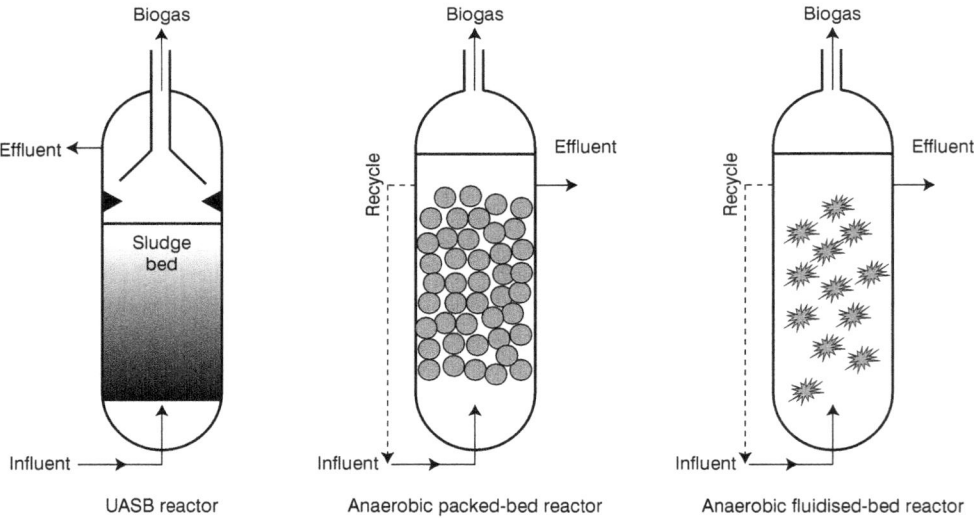

Figure 9.2 Types of high rate digesters (Vandevivere et al., 2003)

This property is attained by attachment of the biomass to denser carrier particles and results in the formation of settleable granules (Lettinga, 1995). For small and commercial scale use, upflow anaerobic filters, UASB, anaerobic packed-bed reactors and fluidized bed reactor systems are all employed for high rate AD (Figure 9.2). UASB is preferred amongst high rate systems (Weiland and Rozzi, 1991) with DS contents of less than 4 per cent (Marchaim, 1992). UASB reactors are also preferred when a high organic loading rate (OLR) is required (Bal and Dhagat, 2001; Paula and Foresti, 1992). High rate digesters incorporated with leach beds (Lehtomäki et al., 2008; Yu et al., 2002) or CSTRs in two-stage or multi-stage arrangements can be useful for high solids substrate digestion (Nizami and Murphy, 2010, 2011).

A variety of digesters are fed with a variety of high solid substrates (OFMSW, grass silage, maize silage, green waste and others). These substrates have different chemical compositions and VS (volatile solid) contents, even though they have similar DS contents. Thus, the data from the literature on optimum digester types using lignocellulosic substrates

Table 9.2 Comparison of advantages and disadvantages of various digester types.

Digester	Disadvantages	Advantages
One-stage	• Higher retention time • Foam and scum formation	• Simpler design with less technical failure
Two-stage	• Complex and expensive to build and maintain • Solid particles need to be removed from the feedstock in the second stage	• Increased overall degradation due to recirculation • Constant feeding rate to methanogenic stage • More robust and less susceptible to failure
Dry digester	• Complex and expensive transport and handling of waste • Only structured material can be used • Material handling and mixing is difficult	• Higher biomass retention • Controlled feeding and spatial niches • Pre-treatment is simpler
Wet digester	• Scum formation during crop digestion • High consumption of water and energy • Short-circuiting • Sensitive to shock loads	• Dilution of inhibitors with fresh water
Batch digester	• Channelling and clogging • Larger volume • Lower biogas yield	• No mixing, stirring and pumping required • Low input in terms of process and mechanical demands • Low capital cost
Continuous digester	• Rapid acidification and larger VFA (volatile fatty acids) production	• Simplicity in design and operation • Low capital costs
High-rate digester	• Longer start-up times • Channelling at low feeding rates	• Higher biomass retention • Controlled feeding • Lower investment cost • No supporting material required

Source: Nizami and Murphy, 2010; Vandevivere et al., 2003

problem of substrate flotation (Nizami et al., 2011). However, the retention time is relatively longer – typically greater than 50 days for lignocellulosic substrates (Table 9.3). Recently, dry batch digestion systems have come into vogue; a feedstock is inserted and gas production begins, increases, peaks, decreases and then ceases. Half of the feedstock remains in the batch for a second cycle to serve as inoculum for the next batch (Nizami et al., 2010). Coupling dry batch digestion with a high-rate reactor packed with methanogens allows the batch reactor to be fully emptied at the end of each cycle, thus reducing retention time. A dry batch digester is optimized for hydrolysis and acidogensis as it converts volatile solids found in the substrates into a high strength COD (chemical oxygen demand) leachate.

Factors affecting digester design and configuration

A high rate reactor like UASB converts the volatizable organics into biogas with great efficiency and has only a short retention time (Nizami and Murphy, 2011; Singh et al., 2011).

Table 9.1 Various types of anaerobic digesters.

Classification basis	Digester types
Substrate feeding	Batch and continuous digesters
Operating temperature	Mesophilic, thermophilic and psychrophilic digesters
Substrate DS contents	Dry and wet digesters
Substrate type	High solids (20–40% DS) and low solids (< 20% DS) digesters
AD process complexity	Single stage/phase and multistage/multiphase digesters
Scale of digester	Farm-based, food processing and centralized digesters

Perspectives on anaerobic digesters

Anaerobic digesters are available in a number of designs and configurations that are classified based on the DS (dry solid) content of the feedstock, the number of phases or stages, the operating temperature, and the method of feeding the substrate (Table 9.1). Examples include dry batch and dry continuous processes for high solid substrates, sequential batch processes for OFMSW (organic fractions of municipal solid waste) and UASB (upflow anaerobic sludge blanket) for slurries (Qi et al., 2005). Each type of digester has its own merits and drawbacks with respect to the type of substrate, operating conditions, pre-treatment needs and potential outcomes (Nizami et al., 2012; Rafique et al., 2010). A detailed comparative analysis of the advantages and disadvantages of various digester types is outlined in Table 9.2.

In AD, the design of the digester is critical to ensure a high biogas production rate. The design is determined by operational parameters including HRT (hydraulic retention time), mixing, volume of the reactor, temperature (Vandevivere et al., 2003) and other specific properties of the substrates (Igoni et al., 2008). A reduced HRT, higher OLR (organic loading rate) and methane yield are characteristics of an optimized digester design (Ward et al., 2008). The criteria for designing a digester include biological, technical, economic and environmental disciplines. Biologists look at the rate, stability and completion of the biochemical reactions. For an engineer, the point of interest is the wear and tear of machine parts and the maintenance of electro-mechanical devices. From an economic point of view, a good reactor design should be fixed in its operational costs. An environmentalist will consider the energy and material recovery and the related gases emissions (Nizami et al., 2011; Vandevivere et al., 2003). The DS content, solubility time, hydrolysis rate and C/N (carbon to nitrogen) ratio each play a key role in a lignocellulosic feedstock digester due to the substrate characteristics (Qi et al., 2005). Anaerobic digesters are built with different materials including steel, brick, plastic or concrete. The digester's shape varies, from a pond, silo or basin to a trough. The digester operation can either be on the surface or underground, with components such as a premixing tank/area, a digestion vessel, a collection and storage area for biogas and a system with which to spread and distribute the digestate (Demirbas and Ozturk, 2005).

The yield of biogas from high solid substrates varies depending on the type of digester and the properties of the substrate. Substrate properties can change from season to season, country to country and site to site (Nizami et al., 2009). A comparison and review of anaerobic digesters suitable for various substrates is described by De Baere and Mattheeuws (2008), Vandevivere et al. (2003) and Nizami and Murphy (2010). Wet continuous digesters give good yields of biogas, provided an effective agitation system is installed to overcome the

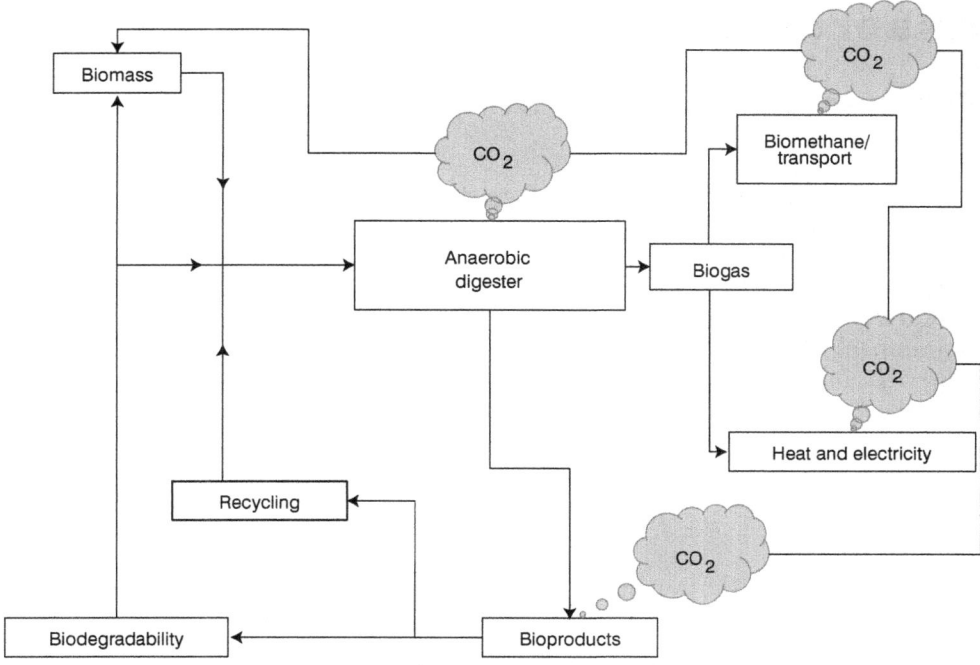

Figure 9.1 Anaerobic digestion as a means to produce energy and bioproducts

anaerobic sludge blanket) and covered lagoon digesters. According to the US Environmental Protection Agency (US-EPA), 510 MSW (municipal solid waste) facilities use landfill biogas to generate electricity and heat (US-EPA, 2012). Wastewater treatment plants (WWTPs) now frequently include anaerobic digesters for waste management and energy production.

After the enactment of the Canadian Green Energy and Green Economy Act in 2009, Ontario was the first province to incentivize production of biogas using industrial and agricultural wastes as well as landfill sites (Anderson, 2012). A biogas financial assistance programme was introduced in Ontario in 2008 by OMAFRA (Ontario Ministry of Agriculture, Food and Rural Affairs) (IEA, 2012) and farmers and agri-food businesses were allocated $11.2 million under this programme (Anderson, 2012). This was followed by the FIT (Feed-in-Tariff) and the MFIT (Micro-Feed-in-Tariff) programmes in 2009, administered by the OPA (Ontario Power Authority). There are currently 10 operating biogas plants, each capable of producing 5000 kW of electricity in Ontario. Moreover, 20 additional plants are in the planning and construction phases (Green, 2012). Some of these facilities upgrade and inject biogas from farm anaerobic digesters into the natural gas grid. Recently, British Columbia and Quebec have also promoted AD initiatives (IEA, 2012). The target set by the OPA is that 12 per cent of all its energy will be renewable by 2025. This is attainable as there is a similar number of livestock in Ontario as in Germany, and thus the German success with AD can be read with optimism (Green, 2012).

The aim of this chapter is to seek new perspectives in anaerobic digesters and highlight the challenges in development of the AD industry. The new perspectives include the digester design, substrate properties and variable biogas yields from similar digester types and (BMP) biomethane potential assays. The challenges discussed include the process scale and control system and digester costs and financing.

Chapter 9

Anaerobic digesters

Perspectives and challenges

Abdul-Sattar Nizami,[1,2*] *Bradley A. Saville*[1] *and Heather L. MacLean*[1,2]

[1]Department of Chemical Engineering and Applied Chemistry, University of Toronto, Toronto, ON; [2]Department of Civil Engineering, University of Toronto, Toronto, ON
*Corresponding author email: abdulsattar.nizami@utoronto.ca; nizami_pk@yahoo.com

Introduction

Anaerobic digestion (AD) is a well recognized process in the treatment of wastewater and industrial organic waste; it is also being used for bioenergy production, often using animal manure and agricultural residues. An enormous amount of organic matter can be reduced using the AD process. Such biodegradable organic materials include waste paper, residual food, grass clippings, sewage and animal waste. Lignocellulosic substrates such as agricultural residues, energy crops and silages have recently been used as digester substrates. Biogas is produced as the result of this process and can be used in a CHP (combined heat and power) system or to produce electricity (Figure 9.1). Biogas contains methane (55–80 per cent), carbon dioxide (20–45 per cent) and traces of other gases, including nitrogen, hydrogen and hydrogen sulfide. If the biogas is upgraded to remove contaminants and CO_2, the resulting 'biomethane' can be distributed via the natural gas grid for heating purposes or compressed for use as a vehicular fuel (Korres et al., 2010).

The use of biogas in developing countries permits heating and cooking in rural areas. Among Asian countries, China and Nepal have 8 million and 50,000 anaerobic digesters respectively (IEA, 2005). In Europe, AD is used to produce an upgraded biogas (97 per cent methane) and can be used as a natural gas alternative or to generate electricity. The substrates used in such cases are agricultural, industrial and municipal wastes. Germany is the European leader with 6800 anaerobic digesters; Austria follows with just 551 digesters (IEA, 2011). In the US, digesters first saw widespread use in the 1970s, primarily in the processing of manure. The decade between 1970 and 1980 saw approximately 120 agricultural digesters commercialized (C2ES, 2011). The development was limited due to economics and digester performance (Oregon, 2009). The number of agricultural digesters has increased due to recent government incentives and policies (Korres et al., 2011; Singh et al., 2011). Factors like funding, guarantees for loans and electricity renewable portfolio standard (RPS) policies have helped to stimulate growth. Today, a large portion of the 162 existing agricultural anaerobic digester systems in the US are new. In 2010, these units generated about 453 million kWh of energy, sufficient to power 25,000 average-sized homes (AgSTAR, 2010). Depending on the waste management system, different types of digesters are employed, including complete mix, plug flow, UASB (upflow

Part III

Anaerobic digestion technology

Voigt, J., Haeffner, B., Schieder, D., Ellenrieder, J., and Faulstich, M. (2009) 'Biogas aus Resten der Nahrungsmittel- und Getränkeindustrie'. *Brauwelt*. vol. 46. pp. 1384–1388.

Von Nordenskjöld, R. (2008). 'Prozesswasser- und Reststoffaufbereitung durch Anaerob/Aerob-Fermentation'. *Brauwelt*. vol. 45. pp. 1334–1338.

Von Nordenskjöld, R. and Stippler, K. (2003) 'Prozesswasser- und Reststoffaufbereitung in Brauereien durch Anaerob/Aerob-Fermentation'. *Brauwelt*. vol. 6/7. pp. 136–141.

Walla, C., and Schneeberger, W. (2008). 'The optimal size for biogas plants'. *Biomass and Bioenergy*. vol. 32. pp. 551–557.

Weiland, P. (1993) 'One- and two-step anaerobic digestion of solid agroindustrial residues'. *Water Science and Technology*, vol. 27, no. 2, pp. 145–151.

Wellinger, A., Murphy, J., and Baxter, D. (2012) *The biogas handbook: science, production, applications*. London: Woodhead Publishing.

Yeoh, B.G. (1997) 'Two-phase anaerobic treatment of cane-molasses alcohol stillage'. *Water Science and Technology* vol. 36, no. 6–7, pp. 441–448.

Mussatto, S.I., and Roberto, I.C. (2005) 'Acid hydrolysis and fermentation of brewers' spent grain to produce xylitol'. *Journal of Food and Agriculture.* vol. 85, no.14, pp. 2453–2460.

Narziß, L. (1995) *Abriß der Bierbrauerei.* Stuttgart: Ferdinand Enke Verlag.

Nguyen, T.L.T., Hermansen, J.E., and Sagisaka, M. (2009) 'Fossil energy savings potential of sugar cane bio-energy systems'. *Applied Energy* vol. 86, pp. 132–139.

Noike, T., Endo, G., Chang, J.-E., Yaguch, J.-I. and Matsumoto, J.-I. (1985), 'Characteristics of carbohydrate degradation and the rate-limiting step in anaerobic digestion'. *Biotechnology and Bioengineering,* vol. 27, pp. 1482–1489.

Ortner, M. (2010) unpublished data.

Paraskeva, P., and Diamadopolus, E. (2006) 'Technologies for olive mill wastewater (OMW) treatment: a review', *Journal of Chemical Technology and Biotechnology,* vol. 81, pp. 1475–1485.

Parawira, W., Murto, M., Zvauya, R., and Mattiasson, B. (2004) 'Anaerobic batch digestion of solid potato waste alone and in combination with sugar beet leaves'. *Renewable Energy,* vol. 29, pp. 1811–1823.

Paredes, C., Cegarra, A., Roig, A., Sánchez-Monedero, M.A., and Bernal, M.P. (1999) 'Characterization of olive mill wastewater (alpechin) and its sludge for agricultural purposes'. *Bioresource Technology,* vol. 67, pp. 111–115.

Pesta, G. (2009) 'Alternative Verwertung'. *Recycling Magazin,* vol. 10.

Pesta, G., Faulstich, M., Meyer-Pittroff, R., and Bochmann, G. (2006) 'Der Einsatz von Rest- und Abfallstoffen aus der Brauerei als Sekundärrohstoffe'. In *Handbuch zum 39. Technologischen Seminar Weihenstephan,* pp. 37/1–37/9, 26 Jamuary–15 February.

Pollitt, M. (2011). 'British Sugar's flagship factory to make beet biogas' *Norfolk Eastern Daily Press.* http://anaerobic-digestion-news.blogspot.com/2011/06/british-sugar-flagship-factory-to-make.html.

Renewable Fuels Association (2011) http://www.ethanolrfa.org.(accessed 13 December 2011).

Rincón, B., Raposo, F., Borja, R., Gonzalez, J.M., Portillo, M.C., and Saiz-Jimenez, C. (2006) 'Performance and microbial communities of a continuous stirred tank anaerobic reactor treating two-phases olive mill solid wastes at low organic loading rates', *Journal of Biotechnology,* vol. 121, pp. 534–543.

Rincón, B., Borja, R., González, J.M., Portillo, M.C., and Sáiz-Jiménez (2008) 'Influence of organic loading rate and hydraulic retention time on the performance, stability and microbial communities of one-stage anaerobic digestion of two-phase olive mill solid residue', *Biochemical Engineering Journal,* vol. 40, no. 2, pp. 253–261.

Roig, A., Cayuela, M.A., and Sánchez-Monedero (2006) 'An overview on olive mill wastes and their valorisation methods', *Waste Management,* vol. 26, pp. 960–969.

Rosentrater, K.A., Hall, H.R. and Hansen, C.L. (2006) 'Anaerobic digestion potential for ethanol processing residues'. *Proceedings ASABE Annual International Meeting,* July 9–12, Portland, OR.

Salminen, E., Rintala, J., Lokshina, L.Y. and Vavilin, V.A. (1995) 'Anaerobic batch degradation of solid poultry slaughterhouse waste'. *Water, Science and Technology,* pp. 33–41.

Schnürer, A. and Nordberg, A. (2008). 'Ammonia, a selective agent for methane production by syntrophic acetate oxidation at mesophilic temperature'. *Water Science and Technology,* pp. 735–740. doi:10.2166/wst.2008.097.

Senn, T. and Pieper, H.J. (2001) 'Classical methods'. In Roehr, M. (ed.) *The biotechnology of ethanol* pp. 7–84, Weinheim: Wiley-VCH.

Sezun, M., Grilc, V., and Marinsek-Logar, R. (2010) 'Anaerobic digestion of mechanically and chemically pretreated lignocellulosic substrate'. *Proceedings Venice 2010, Third International Symposium on Energy from Biomass and Waste* Venice, Italy, November.

Speece, R.E. (1996) *Anaerobic Biotechnology for Industrial Wastewaters.* Nashville, TN: Archae Press.

Tiejun, G. and Xiaomei, L. (2010) 'Using thermophilic anaerobic digestate effluent to replace freshwater for bioethanol production'. *Bioresource Technology* vol. 102, pp. 2126–2129.

Gannoun, H., Othman, N.B., Hassib, B., and Moktar, H. (2007) 'Mesophilic and thermophilic anaerobic co-digestion of olive mill wastewaters and abbatoir wastewaters in an upflow anaerobic filter', *Industrial & Engineering Chemistry Research*, vol. 46, pp.6737–6743.

Gavala, H.N., Skiadas, I.V., Ahring, B.K., and Lyberatos, G. (2005) 'Potential for biohydrogen and methane production from olive pulp'. *Water Science and Technology*, vol. 52, no. 1–2, pp. 209–215.

Gelegenis, J., Georgakakis, D., Angelidaki, I., Christopoulou, N., and Goumenaki, M. (2007) 'Optimization of biogas production from olive-oil mill wastewater, by codigesting with diluted poultry-manure', *Applied Energy*, vol. 84, pp. 646–663.

Genie, G.V. (1982) *Juice extraction in the beet sugar factory*, Amsterdam: Elsevier.

Goberna, M., Gadermaier, M., García, C., Wett, B., and Insam, H. (2010) 'Methanogenic communities fermenting cattle excreta and olive mill wastes into biogas monitored using the AnaeroChip and quantitative PCR'. *Applied Environmental Microbiology* vol. 76, pp. 6564–6571.

Goberna, M., Podmirseg, S.M., Waldhuber, S., Knapp, B.A., García, C., and Insam, H. (2011) 'Pathogenic bacteria and mineral N in soils following the land spreading of biogas digestates and fresh manure'. *Applied Soil Ecology* vol. 49, pp. 18–25.

Hamdi, M. (1996) 'Anaerobic digestion of olive mill wastewaters', *Review. Process Biochemistry*, vol. 31 no. 2, pp. 105–110.

Harada, H., Uemura, S., Chen, A., and Jayadevan, J. (1996) 'Anaerobic treatment of a recalcitrant distillery wastewater by a thermophilic UASB reactor'. *Bioresource Technology* vol. 55, no. 3, pp. 215–221.

Herfellner, T., Bochmann, G., and Meyer-Pittroff, R. (2006). ‚Wirtschaftlich sinnvolle Verfahren?' *Brauindustrie*. vol. 8. pp. 42–45.

Hutnan, M., Drtil, M., and Mrafkova, L. (2000) 'Anaerobic biodegradation of sugar beet pulp'. *Biodegradation*, vol. 11, pp. 203–211.

Hutnan, M., Drtil, M., Derco, J., Mrafkova, L., Hornak, M., and Mico, S. (2001) 'Two-step pilot-scale anaerobic treatment of sugar beet pulp'. *Polish Journal of Environmental Studies*, vol. 10, no. 4, pp. 237–243.

Jarrell, K.F., Saulnier, M., and Ley, A. (1987) 'Inhibition of methanogenesis in pure cultures by ammonia, fatty acids, and heavy metals, and protection against heavy metal toxicity by sewage sludge'. *Canadian Journal of Microbiology* vol. 33, pp. 551–555.

Kanaucho, O., Fujiyama, Y., Mitsuyama, K., and Araki, Y. (2001) 'Development of a functional germinated barley foodstuff from brewers' spent grain for the treatment of ulcerative colitis'. *Journal of the American Society of Brewing Chemists* vol. 59, pp. 59–62.

Kayhanian, M. (1994) 'Performance of a high-solids anaerobic digestion process under various ammonia concentrations'. *Journal of Chemical Techology and Biotechnology* vol. 59, pp. 349–352.

Klenk, I., Landquist, B., and de Imaña, O.R. (2012) 'The product carbon footprint of EU beet sugar'. *Sugar Industry*, vol. 137, no. 3, pp. 169–177.

Koppar, A., and Pullammanappallil, P. (2008) 'Single-stage, batch, leach-bed, thermophilic anaerobic digestion of spent sugar beet pulp'. *Bioresource Technology*, vol. 99, pp. 2831–2839.

Koutrouli, E.C., Kalfas, H., Gavala, H.N., Skiadas, I.V., Stamatelatou, K., and Lyberatos, G. (2009) 'Hydrogen and methane production through two-stage mesophilic anerobic digestion of olive pulp', *Bioresource Technology*, vol. 100, pp. 3718–3723.

Kryvoruchko, V., Machmüller, A., Bodiroza, V., Amon, B., and Amon, A. (2009) 'Anaerobic digestion of by-products of sugar beet and starch potato processing'. *Biomass and Bioenergy*, vol. 33, pp. 620–627.

Licht, F.O. (2012) http://www.agra-net.com/portal2 (accessed 4 April 2012).

Lurgi GmbH (2006) 'Bioethanol' (Company information). http://gep-france.com/biocarb/Bioethanol-Lurgi.pdf (accessed 3 January 2012).

Micard, V., Renard, C.M.G.C., and Thibault, J.-F. (1996) 'Enzymatic saccharification of sugar beet pulp'. *Enzyme and Microbial Technology*, vol. 19, pp. 162–170.

Murphy, J.D. and Power, N.M. (2008) 'How can we improve the energy balance of ethanol production from wheat?' *Fuel* vol. 87, pp. 1799–1806.

Boubaker, F., and Chekih, R.B. (2010) 'Two-phase anaerobic co-digestion of olive mill wastes in semi-continuous digesters at mesophilic temperature'. *Bioresource Technology*, vol. 101, pp. 1628–1634.

Braun, R. (1982) *Biogas – Methangärung organischer Abfallstoffe* [Biogas – methane fermentation of organic wastes[. Vienna: Springer.

Braunegg, G. (2006) 'Tierreststoffverwertung: Verwertung von Reststoffen aus Schlachtung und Fleischverarbeitung mit hoher Wertschöpfung'. *Berichte aus Energie- und Umweltforschung* 60.

Brooks, L., Parravicini, V., Svardal, K., Kroiss, H., and Prendl, L. (2008) 'Biogas from sugar beet press pulp as substitute of fossil fuel in sugar beet factories'. *Water Science and Technology*. vol. 58 no. 7, pp. 1497–1504.

Buckland, G., and Gonzales, A.C. (2010) *Trends in olive oil production, supply and consumption in Mediterranean countries from 1961 to the present day. Olives and olive oil in health and disease prevention*. New York: Elsevier.

Cail, R.G., and Barford, J.P. (1985) 'A comparison of an upflow floc (tower) digester and UASB system treating cane juice stillage'. *Agricultural Wastes* vol. 14, no, 4, pp. 291–299.

Cassidy, D.P., Hirl, P.J. and Belia, E. (2008) 'Methane production from ethanol co-products in anaerobic SBRs'. *Water Science Techology* vol. 58, no. 4, pp. 789–793.

CEFS (Comité Européen des Fabricants de Sucre) (2011) 'Statistics 2011', www.cefs.org, (accessed 5 April 2012).

Chemie Report (2007) http://www.alsa.at/chemiereport/stories/6913 (accessed 6.4.2012).

Chen, Y., Cheng, J.J. and Creamer, K.S. (2008) 'Inhibition of anaerobic digestion process : A review'. *Bioresource Technology*, vol. 99, pp. 4044–4064. doi:10.1016/j.biortech.2007.01.057.

Coughlan, M.P., Mehra, R.K., Considine, P.J., O'Rorke, A., and Puls, J. (1985) 'Saccharification of agricultural residues by combined cellulolytic and pectinolytic enzyme systems'. *Biotechnology and Bioengineering Symposium*, 15, 447–458.

Demirel, B., and Scherer, P. (2009) 'Bio-methanization of energy crops through mono-digestion for continuous production of renewable biogas'. *Renewable Energy*, 34, 2940–2945.

Drosg, B., Wirthensohn, T., Konrad, G., Hornbachner, D., Resch, C., Wäger, F., Loderer, C., Waltenberger, R., Kirchmayr, R. and Braun, R. (2008) 'Comparing centralised and decentralised anaerobic digestion of stillage from a large-scale bioethanol plant to animal feed production'. *Water Science and Technology* vol. 58, no. 7, pp. 1483–1489.

Drosg, B., Fuchs, W., Meixner, K., Waltenberger, R., Kirchmayr, R., Braun, R. and Bochmann, G. (2013) 'Anaerobic digestion of stillage fractions – estimation of the potential for energy recovery in bioethanol plants'. *Water Science and Technology*, vol. 67, no. 3, pp. 494–505.

EU (2006). *The European Sugar Sector*, Brussels: European Commission, DG Agriculture and Rural Development.

European Commission (2000) Decision 2000/766/EC.

European Commission (2002) Regulation (EC) 1774/2002.

European Commission (2009) Regulation (EC) 1069/2009.

Fang, C., Boe, K., and Angelidaki, I. (2011) 'Anaerobic co-digestion of by-products from sugar production with cow manure'. *Water Research*, vol. 45, pp. 3473–3480.

FAO (2012) 'FAOSTAT database', Food and Agriculture Organization of the United Nations. http://faostat.fao.org/ (accessed at 31 January 2012).

Farhadian, M., Borghei, M., and Umrania, V. (2007) 'Treatment of beet sugar wastewater by UAFB bioprocess'. *Bioresource Technology*, vol. 98, pp. 3080–3083.

Freudenreich, P., and Bach, H. (1993) 'Anfall und Verwertung von Schlachtnebenprodukten'. In *Beiträge zur Erzeugung und Vermarktung von Fleisch*. Kulmbacher Reihe Band 12. Kulmbach: Institut für Fleischerzeugung und Vermarktung der Bundesanstalt für Fleischforschung.

Fuchs, W., and Drosg, B. (2010) *Technologiebewertung von Gärrestbehandlungs- und Verwertungskonzepten*. Vienna: Eigenverlag der Universität für Bodenkultur Wien.

Fuchs, W. and Drosg, B. (2013) 'Assessment of the state of the art of technologies for the processing of digestate residue from anaerobic digesters', *Water Science and Technology*, vol. 67, no. 9, pp. 1984–1993.

The main bottlenecks in the anaerobic digestion of industrial feedstocks, are ammonia inhibition and foaming (slaughterhouse waste, bioethanol residues), management of digestate (bioethanol residues), lignocellulose containing compounds (brewers' spent grains) or other inhibiting substances (e.g. polyphenols in olive oil waste). The five examples described in this chapter represent these typical bottlenecks and challenges, which can be overcome by use of appropriate technologies and adequate process control. Moreover there has been much progress especially in the field of efficiency of biomass usage through cascading (bio-refinery concept) of organic residual materials.

To summarise, anaerobic digestion (AD) is a well-known technology to treat industrial organic residues almost regardless of their consistency. The utilisation of industrial organic residues by AD is an appropriate way to improve both the process and the economic efficiency of an industrial factory. Furthermore, it enables a controlled stabilisation of the organic material, reduces greenhouse gas emissions and contributes to the closing of nutrient cycles.

References

AAT Abwasser- und Abfalltechnik GmbH (n.d.) 'Biogas aus Zuckerrüben' http://www.aat-biogas.at/aktuell/news-detail/article/eridania-it (accessed 4 April 2012).

Alburquierque, J.A., Gonzálvez, J. García, and Cefarra, J (2004) 'Agrochemical characterisation of "alperujo", a solid by-product of the two-phase centrifugation method for olive oil extraction', *Bioresource Technology*, vol. 91, pp. 195–200.

Alkaya, E. and Demirer, G. (2011a) 'Anaerobic mesophilic co-digestion of sugar-beet processing wastewater and beet-pulp in batch reactors', *Renewable Energy*, vol. 36, pp. 971–975.

Alkaya, E. and Demirer, G.N. (2011b) 'Anaerobic acidification of sugar beet processing wastes: Effect of operational parameters', *Biomass and Bioenergy*, vol. 35, pp. 32–39.

Angelidaki, I. and Ahring, B.K. (1993) 'Thermophilic anaerobic digestion of livestock waste: The effect of ammonia', *Applied Microbiology and Biotechnology*, vol.38, no 4, pp. 560–564.

Anonymous (2012) 'Vereinsnachrichten, Biogas in Dinterloord', *Sugar Industry*, vol. 136, no 3, p. 125.

Azaizeh, H. and Jadoun, J. (2010) 'Co-digestion of olive mill wastewater and swine manure using up-flow anaerobic sludge blanket reactor for biogas production', *Journal of Water Resource and Protection*, vol. 2, pp. 314–321.

Barth Report (2010) *Hops 2009/2010*. Nuremberg: Barth.

Bischofsberger, W., Rosenwinkel, K.H., Dichtl, N., Seyfried, C.F. and Böhnke, B. (2005) *Anaerobtechnik* [*Anaerobic Technology*], Berlin and Heidelberg: Springer, .

Bochmann, G., Herfellner, H., Susanto, F., Kreuter, F. and Pesta, G. (2007) 'Application of enzymes in anaerobic digestion', *Water Science and Technology*, vol.56, no 10, pp. 29–35.

Bochmann, G., Drosg, B., Ortner, M., Schönlieb, M., Andres-Lainez, S., Kirchmayr, R. and Braun, R. (2010) 'Influence of thermal pre-treatment to increase digestibility of brewers' spent grains', *Proceedings IWA AD 12 Guadalajara, Mexico*.

Böchzelt, H.G., Graf, N., Habel, R.W., Lomsek, J., Wagner, S., and Schnitzer, H. (2002) Möglichkeiten der Wertschöpfungs-steigerung durch Abfallvermeidung (biogener Reststoffe) und Nebenproduktnutzung – Feasibility Study. Hrsg.: Joanneum Research. Frohnleiten. Im Auftrag des BMVIT und Land Steiermark.

Borja, R., Rincón, B., Raposo, F., Alba, J., and Martín, A. (2002) 'A study of anaerobic digestability of two-phases olive mill solid waste (OMSW) at mesophilic temperature', *Process Biochemistry*, vol. 38, pp. 733–742.

Bothast, R.J. and Schlicher, M.A. (2005) 'Biotechnological processes for conversion of corn into ethanol'. *Applied Microbial Biotechnology* vol. 67, pp. 19–25.

Boubaker, F., and Chekih, R.B. (2007) 'Thermophilic anaerobic co-digestion of olive mill wastewater with olive mill solid wastes in a tubular digester'. *Chemical Engineering Journal*, vol. 132, pp. 195–203.

similar biogas production in two-stage as in one-stage anaerobic digestion, but a better COD removal. The optimal HRT in the first stage is two days; the optimal waste to inoculum ratio is 1:1 for sugar beet pulp and wastewater at mesophilic temperature (Alkaya & Demirer, 2011b). Brooks et al. (2008) achieved a stable biogas production of 530 Nm³/t COD or 610 Nm³/t VS at standard temperature and pressure conditions (STP) at an OLR of 10 kg COD/(m³·d). Methane content was 50–53 per cent and COD removal was 72 per cent. Single-stage, batch, unmixed, leach-bed, laboratory scale thermophilic anaerobic digestion of spent sugar beet pulp resulted in 0.336 Nm³ CH_4 /kg VS and 95 per cent of the methane yield was achieved after 8 days (Koppar & Pullammanappallil, 2008). This result confirms the fast degradability of sugar beet pulp. Biochemical methane potential (BMP) of sugar beet pulp tests resulted in 430 Nm³ CH_4/ kg VS (Kryvoruchko et al., 2009). Thermophilic co-digestion of sugar beet pulp, desugared molasses and cow manure showed a decrease in the inhibiting potential of the desugared molasses, mainly due to the dilution with manure, which provides a buffer capacity and nutrients (Fang et al., 2011). Sugar beet molasses were digested in an upflow anaerobic fixed bed reactor at mesophilic temperature, with 20 h HRT and influent COD ranging from 7.8 to 9.6 kg COD/ (m³·d). The COD removal ranged from 75 to 93 per cent (Farhadian et al., 2007). Sugar beet leaves improved the methane yield from potato waste by up to 62 per cent (0.32 Nm³ CH_4 /kg $VS_{degraded}$) by optimising the C/N ratio by batch fermentation (Parawira et al., 2004). The wastes from the sugar industry have a high energy potential, so this should be used to substitute fossil energy and for making sugar production more profitable.

Conclusions

The residues of five typical food, beverage and biofuel processes with high potential for biomethane production were described in detail. This is by no means a complete list. Information on, for example, the biomethane potential or chemical characterisation of other industrial by-products and residues can be found in standard works such as Braun (1982), Speece (1996), and Bischofsberger et al. (2005). Although most sectors of the food, beverage and biofuel industry are able to generate revenue by selling their organic residues as an animal feed, some of them have to pay for their disposal (e.g. abattoirs). The cost of disposal depends on the waste composition and varies from country to country. In Austria, the disposal costs range between 25 and 30 €/t for blood and between 40 and 45 €/t for other residues coming from the pig slaughter process. Economic factors in recent years such as constantly increasing costs of both energy and chemical fertilisers, have led to a paradigm shift, especially within the food and beverage industry.

More and more companies are optimising their energy balance in terms of utilising their organic residues to become energy self-sufficient by using integrated AD technology and combined heat and power plants. Among the industries presented in this chapter, the sugar industry has the strongest interest in integrating AD technology to its production process. There are already a handful of AD plants which are fully integrated into the production process. Abattoirs also have great interest due to the lack of alternative utilisation. The successful integration of AD technology in the production process of an abattoir in Austria may lead to the construction of further process-integrated biogas plants. AD plants have already been implemented in the bioethanol industry. However, no large-scale mono-fermentation AD plants for breweries or olive mills are known, although there is considerable potential.

Table 8.9 Chemical characterisation of the wastes from sugar production from sugar beet – also for comparison sugar beet silage is shown (Weiland 1993; Brooks et al., 2008; Demirel & Scherer, 2009; Fang et al., 2011; Alkaya & Demirer, 2011a).

Substrate	pH	TS (%)	VS (%)	COD (g/kg)	TKN (g/kg)
Sugar beet pressed pulp	3.9–4.0	15–18	14–17	180–260	1.2–3.1
Sugar beet silage	3.3	20	19	265	3.1
Desugared molasses	n.a.	49.8	32.6	49.8	6.7
Wastewater	6.8	6	2.8	6.62	0.01

Table 8.10 Composition of sugar beet (Micard et al., 1996).

Component	% TS
Rhamnose	2.4
Fucose	0.2
Arabinose	20.9
Xylose	1.7
Mannose	1.1
Galactose	5.1
Glucose	21.1
Galacturonic acid	21.1
Ferulic acid	0.8
Diferulic acid	0.04
Protein (N x 6.25)	11.3

cent of the dry matter. The lignocellulosic fraction of the dried pulp is: cellulose 22–30 per cent; hemicellulose 24–32 per cent; pectic substances 24–32 per cent; and lignin 3–4 per cent (Coughlan et al., 1985). The pulp fragments have a cylindrical shape with 6–9 mm diameter and 20–40 mm length. The desugared molasses contain high amounts of ions: potassium 160 g/l; sodium 36 g/l; and calcium 5 g/l – which can inhibit the biogas process (Fang et al., 2011). The interest in the anaerobic digestion of residues from the sugar industry has increased over the last two decades.

Experience in AD of sugar production residues

The digestibility of sugar industry residues has been examined at different conditions, concerning the organic loading rate (OLR), temperature, hydraulic retention time (HRT), mono- and co-digestion, and one- and two-stage fermentation. In 1993, Weiland focused on the effect of the C/N ratio on the anaerobic digestion of agro-industrial residues, including SBP, and the one- and two-step fermentation. The C/N ratio of sugar beet pulp was determined to be between 35 and 40, which is optimal for the biogas process. Therefore, Weiland (1993) found no interest in investigating a two-step (also called stage or phase) process. Nevertheless, according to the literature, other advantages of the two-stage fermentation of sugar beet pulp were expected later. Hutnan et al. (2000) reported

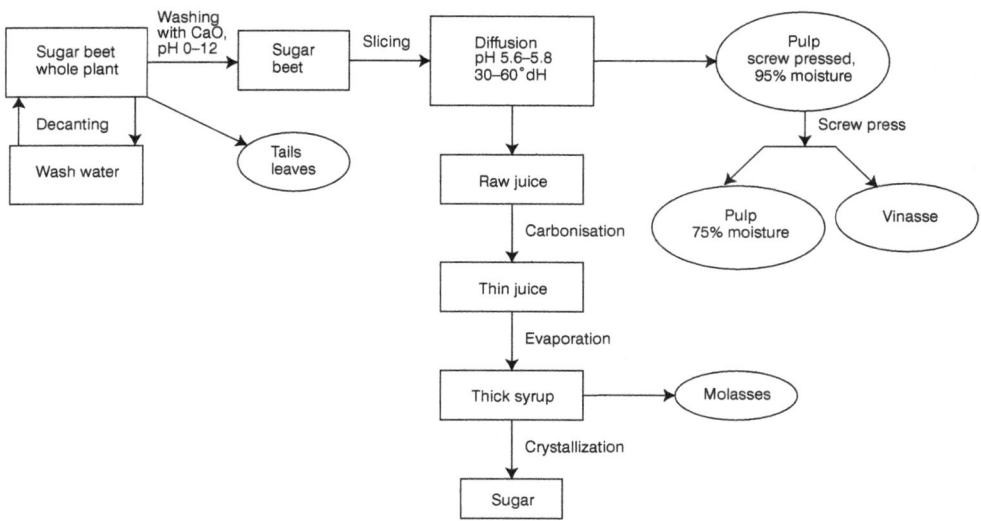

Figure 8.3 Overview of the accumulation of by-products in sugar beet processing

In summer 2013, a biogas plant on the site of the largest European sugar factory, British Sugar, in Wissington, England, should start its operation (Pollitt 2011).

Sugar beet refinery process

The sugar beet is first washed with a dilute CaO solution at pH 10–12 (Figure 8.3). The liquid is recirculated after sedimentation. The next step is slicing of the sugar beet to pulp. In the subsequent diffusion step, the extraction of the sugar into an aqueous solution at pH 5.6–5.8, 30–60 °dH (German degrees) in a countercurrent exchanger, first for 5 minutes at 70–78°C to denature the cells, then for 70–85 minutes at 69–73°C. To avoid microorganism activity, formaldehyde is added. The liquid phase is the so-called raw juice. The solid phase is the pulp with 95 per cent moisture which is then screw pressed to 75 per cent moisture.

Impurities from the raw juice are removed by carbonisation with milk of lime at 60–70°C, pH 10.8–11.9 for 20 minutes, followed by 30 minutes at 80–85°C. At this stage, impurities such as sulfate, phosphate, citrate and oxalate are precipitated as their calcium salts, and proteins, saponins and pectins also aggregate in the presence of Ca^{2+} ions. By multiple-effect evaporation, usually five stages, the raw juice is converted to thick syrup. The last step is crystallisation. The remaining molasses contain some remaining sugar, and also impurities (Genie 1982).

Characterisation of the residues from sugar production

The sugar beet residues most commonly used for biogas production are sugar beet pulp (SBP) and desugared molasses. The characteristics of wastes from the sugar industry are shown in Table 8.9.

The sugar beet pressed pulp composition offers good possibilities for biological treatment (Tables 8.9, 8.10). The main components of the sugar beet pulp are sugars – about 74 per

and OMSW alone or with a nitrogen-rich substrate seems to be appropriate for achieving high OLR and short HRT in order to dispose of the polluting waste as fast as possible in an environmentally friendly way – as digestate – while gaining energy at the same time.

Wastes from sugar beet processing facilities

Another organic residue, whose potential for anaerobic digestion should be extended, is the waste from the sugar industry: sugar beet pulp, molasses and sugar beet leaves. These by-products can be used as animal feed, for paper, yeast and amino acid production, for the generation of alcohol including ethanol, and as a soil conditioner (EU 2006). Nevertheless, there are two reasons to implement the anaerobic technology in the sugar-producing industry: to reduce the ecological impact in terms of CO_2 emissions and to reduce operational costs.

World sugar production rose from 166.6 million tons in 2007/08 to 174.1 million tons in 2011/12. In Europe, where sugar is produced from sugar beet, production increased from 25.7 million tons in 2007/08 to 29.4 million tons in 2011/12 (Licht, 2012). In the European Union and Switzerland, 2.16 million tons of molasses and 6.52 million tons of fresh pulp (22 per cent dry matter) were accumulated in 2010/11 (CEFS Statistics, 2011). For the processing of one ton of sugar beet, excluding drying of sugar beet pulp, about 170 to 330 kWh are needed (Brooks et al., 2008). The product carbon footprint (PCF) of sugar from sugar beet produced and refined in the EU, and from sugar cane imported and refined in the EU has been compared (Klenk et al., 2012). The PCF range for EU refined cane sugar is on average 642–771 kg CO_{2eq}/t sugar, which is similar if not higher than the one for the EU beet sugar: 242–771 kg CO_{2eq}/t sugar. The overseas transport and refining of sugar cane also adds a significant amount of emissions to the PCF. The land-use efficiency of beet sugar is higher than that of sugar cane; 51 per cent more land is required by cane systems to produce an equivalent set of products (sugar and co-products) with an equivalent amount of greenhouse gas emissions. The impact of the emissions from land-use change for sugar cane is also significant, but is rarely taken into account (as discussed for other crops in Chapter 1).

Another advantage of converting the energy from factory residues into biogas to replace energy production from fossil fuels is the decrease in costs for sugar production to make it competitive on the world sugar market. In 2006, the Common Market Organisation (CMO) for sugar production in the EU was reformed. The quota for sugar production was reduced by 30 per cent to 13.3 million tons. Prices sank by 36 per cent to 404.4 €/ton in 2009/10 and the minimum sugar price is not guaranteed by intervention mechanisms any longer(CEFS Statistics, 2011). This reform and the high amount of fossil energy needed for the production of sugar beet could give good reason for the use of energy from renewable sources, e.g. biogas from factory residues.

There are already examples of application of this concept on a large scale. Since 2008, the biogas plant in Kaposvár, Hungary (Magyar Cukor ZRT, Agrana) has used almost the half of the daily sugar beet pulp produced, and provides 110,000 m³ biogas (55 per cent methane), which covers 40 per cent of the energy need of the factory (Chemie Report 2007). In the Netherlands, two biogas plants have been built at factories of Suiker Unie, Dinterloord (2011) and Vierverlaten (2012). The substrates used are sugar beet pulp, sugar beet tails, residues from the potato industry and other agricultural products (Anonymous 2012). The biomethane will be injected into the national grid. Since 2011, a biogas plant has been in operation near Parma, Italy, with co-digestion of sugar beet pulp and maize silage (AAT n.d.).

potential of about 1 GJ per ton of olives (Gelegenis et al., 2007) and COD removal up to 90 per cent (Rincón et al., 2006). The anaerobic process should be optimised for fast conversion of the wastes and for phenol and fat degradation.

A large number of laboratory studies over two decades demonstrated the up-flow anaerobic sludge bed reactor or anaerobic filters to be suitable for OMW digestion (Roig et al., 2006). Dilution during the start-up of the reactor is recommended to keep the concentration of inhibitors low in order to provide adaptation time for the archaea. The upflow anaerobic filter (UAF) operating system offers more process stability and shorter start-up time (Hamdi 1996). Later studies confirmed the effectiveness of the UASB reactor and reported operation parameters like an HRT of up to 25 days, COD removal up to 90 per cent, average organic loading rate (OLR) of 5 kg COD /(m^3·d) and 0.30–0.35 Nm^3 CH$_4$/kgCOD$_{removed}$ (Paraskeva & Diamadopolus, 2006).

The low nitrogen content and buffer capacity, and high content of inhibiting compounds, makes the OMW unsuitable for mono-digestion. Co-digestion fermentation with OMSW at thermophilic conditions showed that an HRT of 36 days and an OLR of 3.62 kg COD /(m^3·d) are optimal for obtaining 69 per cent soluble COD removal and methane production of 46 Nm^3/m^3 OMW per day (Boubaker & Chekih, 2007). Two-phase anaerobic co-digestion under mesophilic conditions has also been explored (Boubaker & Chekih, 2010). The optimal HRT value for the first stage was determined to be 24 days, and for the second stage 36 days, with 82 per cent COD removal and 70–78 per cent polyphenol removal. Thermophilic co-digestion with abattoir wastewaters in a UAF reduces the main problems encountered during their mono-digestion by optimising the C/N ratio and decreasing the polyphenol concentration (Gannoun et al., 2007). Poultry and swine manure have also been reported as suitable co-substrates for OMW fermentation. Addition of 70 per cent (v/v) poultry manure with high TKN (4.9 g/l) and alkalinity (20.2 g CaCO$_3$/l) to the OMW resulted in a stable process HRT of 18 days and a biogas yield of 1.53 Nm^3/kg COD per day with a methane content of 65 per cent (Gelegenis et al., 2007). The swine manure and OMW optimal mix ratio was reported to be 33–67 per cent (v/v). In this case 85–95 per cent COD removal with 0.55 Nm^3 biogas/kg COD per day was reached (Azaizeh & Jadoun, 2010).

Pre-treatment methods like addition of soluble calcium salts for precipitating the lipids and pH adjustment with CaCO$_3$ and NH$_4^+$ led to COD and polyphenol removal of 78–88 per cent and 12 per cent, respectively. Targeted polyphenol removal was tested with coagulation, extraction or oxidation, and led to 40 per cent COD reduction and up to 13 per cent phenol removal. About 80 per cent of the polyphenols with molecular weight lower than 500 Da were degraded during the methanogenic anaerobic stage. For the phenols with a molecular weight more than 1000 Da, adsorption on betonite was successful. The final aerobic treatment stage achieved up to 96 per cent COD removal (Paraskeva & Diamadopolus, 2006).

Despite its low pH and high organic matter and phenol concentration, there have been several studies on mono-digestion of OMSW. The maximum methane production was found to be 0.244 Nm^3 CH$_4$ /kg COD$_{removed}$ at standard temperature and pressure conditions (STP) at an OLR of 9.2 kg COD /(m^3·d) and an HRT of 17 days (Rincón et al., 2008). An interesting application of OMSW could be for hydrogen production in two-stage anaerobic fermentation (Koutrouli et al., 2009). The bio-hydrogen potential was estimated at 1.6 mmol H$_2$ per g total solids in two-stage fermentation (Gavala et al., 2005).

At a large scale, olive mill wastes are applied as a co-substrate in anaerobic digesters (biogas plant in Foggia, Southern Italy). The overview of studies on anaerobic digestion of olive mill residues shows rising interest on its disposal possibilities. The co-digestion of OMW

Extraction process

Olive oil extraction can be carried out in a number of ways. In the traditional pressing process, olive emulsion is decanted from wastewater after pressing and a solid fraction (the olive husk) remains. However, this process is no longer common in industry. Instead, centrifuges are used for olive oil extraction nowadays. Depending on the centrifugation system, two- and three-phase systems are common. The three-phase system can be completely automated, produces high-quality oil and is a compact process. The huge amount of wastewater – 1 to 1.6 m³ per ton of olives – led to the development of the two-phase extraction system, which provides 0.2 m³ wastewater per ton of olives and is used by roughly 90 per cent of the olive mills. In the three-phase system, 550 kg olive cake accumulates per ton of olives; in the two-phase system the solid phase amounts to 800 kg olive wet cake per ton of olives (Roig et al., 2006).

Characterisation of olive mill waste

The characteristics of olive oil waste (Table 8.8) vary depending on geographical, seasonal, varietal or methodological factors. Two-phase olive pomasse is an acid effluent consisting of water 60–70 per cent, lignin 13–15 per cent, cellulose and hemicellulose 18–20 per cent, mineral solids 2.5 per cent (Borja et al., 2002). The antimicrobial and the phytotoxic effects, observed when applying the olive mill wastes directly to the soil are due to the phenolic and long chain fatty acid content and led to the development of other valorisation methods: evaporation, physico-chemical treatment, including coagulation, precipitation, microbiological treatment, composting, extraction of valuable products, solid state fermentation of the solid phase and, last but not least, anaerobic digestion (Roig et al., 2006).

Beside its low pH and high polyphenol content (Table 8.8), the OMW has low alkalinity (3.8 g $CaCO_3$/l) and a low amount of ammonium nitrogen (750 mg/l) and total Kjeldahl nitrogen (TKN) (1.65 g/l), which contributes to the instability of the anaerobic process (Boubaker & Chekih, 2007).

Limitations and bottlenecks

Olive oil production is a seasonal process. The treatment process for safe disposal of such an amount of waste should be flexible in terms of continuity on the one hand, and effective and robust to avoid the necessity of waste storage over a year, on the other hand. The main advantage of anaerobic digestion of olive mill waste is the high energy potential of the substrate due to its high COD, and the possibility to use the digestate as a fertiliser. The production of biomethane can be up to 25 Nm^3CH_4 per ton of olives, with a heat-production

Table 8.8 Chemical characterisation of the waste from olive oil production residues (Alburquierque et al., 2004; Paredes et al., 1999; Gelegenis et al., 2007).

Substrate	pH	TS (%)	COD (g/kg)	Fat (%)	Polyphenols (% TS)
Olive mill wastewater (OMW)	4.80–5.50	4.12–16.38	150	0.55–11.37	1.32–3.99
Olive mill solid waste (OMSW)	4.86–6.45	23.92	183–280	2.5–3	0.62–2.39

Eggen, R.I.E. (1994) 'Regulated gene expression in methanogens', *FEMS Microbiology Reviews*, vol 15, pp. 251–260.

Ehlers, C., Weidenbach, K., Veit, K., Deppenmeier, U., Metcalf, W.W. and Schmitz, R.A. (2005) 'Development of genetic methods and construction of a chromosomal $glnK_1$ mutant in *Methanosarcina mazei* strain Göl', *Molecular Genetics and Genomics*, vol 273, no 4, pp. 290–298.

Ferry, J.G. (1999) 'Enzymology of one-carbon metabolism in methanogenic pathways', *FEMS Microbiology Ecology*, vol. 23, pp. 13–38.

Fonknechten, N., Perret, A., Perchat, N., Tricot, S., Lechaplais, C., Vallenet, D., Vergne, C., Zaparucha, A., Le Paslier, D., Weissenbach, J. and Salanoubat, M. (2009) 'A conserved gene cluster rules anaerobic oxidative degradation of L-ornithine', *Journal of Bacteriology*, vol 191, pp. 3162–3167.

Franke-Whittle, I.H., Goberna, M., Pfister, V. and Insam, H. (2009a) 'Design and development of the ANAEROCHIP microarray for investigation of methanogenic communities', *Journal of Microbiological Methods*, vol 79, pp. 279–288.

Franke-Whittle, I.H, Knapp, B.A, Fuchs, J., Kaufmann R. and Insam H. (2009b) 'Application of COMPOCHIP microarray to investigate the bacterial communities of different composts', *Microbial Ecology*, vol 57, pp. 510–521.

Gernhardt, P., Possot, O., Foglino, M., Sibold, L. and Klein, A. (1990) 'Construction of an integration vector for use in the archaebacterium *Methanococcus voltae* and expression of a eubacterial resistance gene', *Molecular Genetics and Genomics*, vol 221, pp. 273–279.

Goyal, A., Ghosh, B. and Eveleigh, D. (1991) 'Characteristics of fungal cellulases', *Bioresource Technology*, vol 36, pp. 37–50.

Grabber J.H., Schatz, P.F., Kim, H., Lu, F. and Ralph, J. (2010) 'Identifying new lignin bioengineering targets: 1. Monolignol-substitute impacts on lignin formation and cell wall fermentability', *BMC Plant Biology*, vol 10, article 114.

Graham, D.E. and White, R.H. (2002) 'Elucidation of methanogenic coenzyme biosyntheses: from spectroscopy to genomics', *Natural Products Report*, vol 19, pp. 133–147.

Himmel, M.E., Ding, S.Y., Johnson, D.K., Adney, W.S., Nimlos, M.R., Brady, J.W. and Foust, T.D. (2007) 'Biomass recalcitrance: Engineering plants and enzymes for biofuels production' *Science*, vol 315, pp. 804–807.

Howard, R.L., Abotsi, E., Jansen van Rensburg, E.L. and Howard, S. (2003) 'Lignocellulose biotechnology: issues of bioconversion and enzyme production', *African Journal of Biotechnology*, vol 2, pp. 602–619.

Hu, S., Zheng, H., Gu, Y., Zhao, J., Zhang, W., Yang, Y., Wang, S., Zhao, G., Yang, S. and Jiang, W. (2011) 'Comparative genomic and transcriptomic analysis revealed genetic characteristics related to solvent formation and xylose utilization in *Clostridium acetobutylicum* EA 2018', *BMC Genomics*, vol 12, article 93.

Hu, W.J., Harding, S.A., Lung, J., Popko, J.L., Ralph, J., Stokke, D.D., Tsai, C.J. and Chiang, V.L. (1999) 'Repression of lignin biosynthesis promotes cellulose accumulation and growth in transgenic trees', *Nature Biotechnology*, vol 17, pp. 808–812.

Illanes, A., Cauerhff, A., Wilson, L. and Castro, G.R. (2012) 'Recent trends in biocatalysis engineering', *Bioresource Technology*, vol 115, pp. 48-57.

Ilmberger, N., Meske, D., Juergensen, J., Schulte, M., Barthen, P., Rabausch, U., Angelov, A., Mientus, M., Liebl, W., Schmitz, R.A. and Streit, W.R. (2012) 'Metagenomic cellulases highly tolerant towards the presence of ionic liquids-linking thermostability and halotolerance', *Applied Microbiology and Biotechnology*, vol 95, pp. 135–146.

Khalid, A., Arshad, M., Anjum, M., Mahmood, T. and Dawson L. (2011) 'The anaerobic digestion of solid organic waste', *Waste Management*, vol. 31, no 8, pp 1737–1744.

Kohler, P.R.A. and Metcalf, W.W. (2012) 'Genetic manipulation of *Methanosarcina* spp.', *Frontiers in microbiology*, vol 3, article 259.

Kotsyurbenko, O.R., Glagolev, M.V., Nozhevnikova, A.N. and Conrad, R. (2001) 'Competition between homoacetogenic bacteria and methanogenic archaea for hydrogen at low temperature', *FEMS Microbiology and Ecology*, vol 38, pp. 153–159.

Krakat, N., Schmidt, S. and Scherer, P. (2011) 'Potential impact of process parameters upon the bacterial diversity in the mesophilic anaerobic digestion of beet silage', *Bioresource Technology*, vol 102, pp. 5692–5701.

Krause, L., Diaz, N.N., Bartels, D., Edwards, R.A., Puhler, A., Rohwer, F., Meyer, F. and Stoye, J. (2006) 'Finding novel genes in bacterial communities isolated from the environment', *Bioinformatics*, vol 22, pp. 281–289.

Kumar, R., Singh, S. and Singh, O.V. (2008) 'Bioconversion of lignocellulosic biomass: biochemical and molecular perspectives', *Journal of Industrial Microbiology and Biotechnology*, vol 35, no 5, pp. 377–391.

Kumar, S., Dagar, S.S., Mohanty, A.K., Sirohi, S.K., Puniya, M.C., Kuhad, R.C., Sangu, K.P.S., Griffith G.W. and Puniya, A.K. (2011) 'Enumeration of methanogens with a focus on fluorescence in situ hybridization', *Naturwissenschaften*, vol 98, pp. 457–472.

Lange, M. and Ahring, B.K. (2001) 'A comprehensive study into the molecular methodology and molecular biology of methanogenic Archaea', *FEMS Microbiology Reviews*, vol 25, no 5, pp. 553–571.

Lapado, J. and Whitman, W.B. (1990) 'Method for isolation of auxotrophs in the methanogenic archaebacteria: role of the acetyl-CoA pathway of autotrophic CO_2 fixation in *Methanococcus maripauludis*', *Proceedings of the National Academy of Science of the United States of America*, vol 87, pp. 5598–5602.

Lee, C.K., Herbold, C.W., Polson, S.W., Wommack, K.E., Williamson, S.J., McDonald, I.R., and Cary, S.C. (2012a) 'Groundtruthing Next-Gen sequencing for microbial ecology – biases and errors in community structure estimates from PCR amplicon pyrosequencing', *PLoS ONE*, vol 7, e44224.

Lee, S.-H., Kang, H.-J., Lee, Y.H., Lee, T.J., Han, K., Choi, Y. and Park, H.-D. (2012b) 'Monitoring bacterial community structure and variability in time scale in full-scale anaerobic digesters', *Journal of Environmental Monitoring*, vol 14, pp. 1893–1905.

Lessner, D.J., Lhu, L., Wahal, C.S. and Ferry, J.G. (2010) 'An engineered methanogenic pathway derived from the domains *Bacteria* and *Archaea*', *mBio*, vol 1, e00243-10.

Malherbe, S., and Cloete, T.E. (2003) 'Lignocellulose biodegradation: fundamentals and applications: A review', *Environmental Science and Biotechnology*, vol 1, pp. 105–114.

McCarty, P.L. (1982) 'One hundred years of anaerobic treatment', In D.E. Hughes, D.A. Stafford, B.I. Wheatley, W. Baader, G. Lettinga, E.J. Nyns, W. Verstraete, and R.L. Wentworth (Eds.), *Anaerobic Digestion, 1981*, pp. 3–21, Amsterdam :Elsevier Biomedical Press.

McInerney, M.J., Bryant, M.P., Hespell, R.B. and Costerton, J.W. (1981) 'Svntrophmonnas wolfei gen. nov. sp. nov., anaerobic, syntrophic, fatty acid-oxidizing bacterium', *Applied and Environmental Microbiology*, vol 41, pp. 1029–1039.

Menon, V. and Rao, M. (2012) 'Trends in bioconversion of lignocellulose: Biofuels, platform chemicals & biorefinery concept', *Progress in Energy and Combustion Science*, vol 38, pp. 522–550.

Mevarech, M. and Werczberger, R. (1985) 'Genetic transfer in *Halobacterium volcanii*', *Journal of Bacteriology*, vol 162, pp. 461–462.

Miyamoto K. (ed.) (1997) 'Renewable biological systems for alternative sustainable energy production', *FAO Agricultural Services Bulletin* 128, http://www.fao.org/docrep/W7241E/W7241E00.htm, accessed April 2012.

Narihiro, T. and Sekiguchi, Y. (2011) 'Oligonucleotide primers, probes and molecular methods for the environmental monitoring of methanogenic archaea', *Microbial Biotechnology*, vol 4, pp. 585–602.

Nelson, M.C., Morrison, M., Schanbacher, F., and Yu, Z. (2012) 'Shifts in microbial community structure of granular and liquid biomass in response to changes to infeed and digester design in anaerobic digesters receiving food-processing wastes', *Bioresource Technology*, vol 107, pp. 135–143.

Nelson, M.C., Morrison, M. and Yu, Z. (2011) 'A meta-analysis of the microbial diversity observed in anaerobic digesters', *Bioresource Technology*, vol 102, pp. 3730–3739.

Pagani, I., Liolios, K., Jansson, J., Chen, I.-M.A., Smirnova, T., Nosrat, B., Markowitz, V.M. and Kyrpides, N.C. (2012) 'The Genomes OnLine Database (GOLD) v.4: status of genomic and metagenomic projects and their associated metadata', *Nucleic Acids Research*, vol. 40, Database issue pp. D571–D579.

Papoutsakis, E.T. (2008) 'Engineering solventogenic clostridia'. *Current Opinion in Biotechnology* vol 19, pp. 420–429.

Parawira, W. (2012) 'Enzyme research and applications in biotechnological intensification of biogas production', *Critical Reviews in Biotechnology*, vol 32, pp. 172–186.

Perret, A., Lechaplais, C., Tricot, S., Perchat N., Vergne, C., Pellé, C., Bastard, K., Kreimeyer, A., Vallenet, D., Zaparucha, A., Weissenbach, J. and Salanbout, M. (2011) 'A novel Acyl-CoA beta-transaminase characterized from a metagenome', *PLoS ONE*, vol 6, no 8, e22918.

Pinto, A.J. and Raskin, L. (2012) 'PCR biases distort bacterial and archaeal community structure in pyrosequencing datasets,' *PLoS ONE*, vol 7, no 8, e43093.

Porat, I. and Whitman, W.B. (2009) 'Tryptophan auxotrophs were obtained by random transposon insertions in the *Methanococcus maripaludis* tryptophan operon', *FEMS Microbiology Letters*, vol 297, pp. 250–254.

Ragsdale, S.W. and Pierce, E. (2008) 'Acetogenesis and the Wood-Ljungdahl pathway of CO2 fixation', *Biochimica et Biophysica Acta*, vol 1784, pp. 1873–1898.

Reeve, J.N., Nölling, J., Morgan, R.M., and Smith, D.R. (1997) 'Methanogenesis: genes, genomes, and who's on first?' *Journal of Bacteriology*, vol 179, pp. 5975–5986.

Rittmann, B.E., Krajmalnik-Brown, R. and Halden, R.U. (2008) 'Pre-genomic, genomic and postgenomic study of microbial communities involved in bioenergy', *Nature Reviews in Microbiology*, vol 6, pp. 604–612.

Rother, M. and Metcalf, W.W. (2005). 'Genetic technologies for Archaea', *Current Opinion in Microbiology*, vol 8, pp. 745–751.

Saha, B.C. (2000) 'Alpha-L-arabinofuranosidases biochemistry, molecular biology and application in biotechnology', *Biotechnology Advances*, vol 18, pp. 403–423.

Sakai, S., Takaki, Y., Shimamura, S., Sekine, M., Tajima, T., Kosugi, H., Ichikawa, N., Tasumi, E., Hiraki, A.T., Shimizu, A., Kato, Y., Nishiko, R., Mori, K., Fujita, N., Imachi, H. and Takai, K.. (2011) 'Genome sequence of a mesophilic hydrogenotrophic methanogen *Methanocella paludicola*, the first cultivated representative of the order *Methanocellales*', *PLoS ONE* 6(7): e22898.

Sánchez, C. (2009) 'Lignocellulosic residues: Biodegradation and bioconversion by fungi'. *Biotechnology Advances*, vol 27, pp. 185–194.

Sanz, J.L., and Köchling, T. (2007) 'Molecular biology techniques used in wastewater treatment: An overview.' *Process Biochemistry*, vol 42, pp. 119–133.

Sarmiento, F.B., Leigh, J.A. and Whitman, W.B. (2011) 'Genetic systems for hydrogenotrophic methanogens' *Methods in Enzymology*, vol 494, pp. 43–73..

Schlüter, A., Bekel, T., Diaz, N.N., Dondrup, M., Eichenlaub, R., Gartemann, K.H., Krahn, I., Krause, L., Kromeke, H., Kruse, O., Mussgnug, J.H., Neuweger, H., Niehaus, K., Puhler, A., Runte, K.J., Szczepanowski, R., Tauch, A., Tilker, A., Viehover, P. and Goesmann, A. (2008) 'The metagenome of a biogas-producing microbial community of a production-scale biogas plant fermenter analysed by the 454-pyrosequencing technology'. *Journal of Biotechnology*, vol 136, pp. 77–90.

Sharma, R., Katoch, M., Srivastava, P.S. and Quazi, G.N. (2009) 'Approaches for refining heterologous protein production in filamentous fungi', *World Journal of Microbiology and Biotechnology*, vol 25, pp. 2083–2094.

Shen, B., Sun, X., Zuo, X., Shilling, T., Apgar, J., Ross, M., Bougri, O., Samoylov, V., Parker, M., Hancock, E., Lucero, H., Gray, B., Ekborg, N.A., Zhang, D., Johnson, J.C.S., Lazar, G. and Raab, R.M. (2012) 'Engineering a thermoregulated intein-modified xylanase into maize for consolidated lignocellulosic biomass processing', *Nature Biotechnology*, vol 30, pp. 1131–1136.

Simon, C. and Daniel, R. (2009) 'Achievements and new knowledge unraveled by metagenomic approaches', *Applied Microbiology and Biotechnology*, vol 85, pp. 265–276.

Simon, C. and Daniel, R. (2011) 'Metagenomic analyses: Past and future trends', *Applied Environmental Microbiology*, vol 77, pp. 1153–1161.

Talbot, G., Topp, E., Palin, M.F., and Massé, D.I. (2008) 'Evaluation of molecular methods used for establishing the interactions and functions of microorganisms in anaerobic bioreactors,' *Water Research*, vol 42, pp. 513–537.

Tokura, M., Ohkuma, M. and Kudo, T. (2000) 'Molecular phylogeny of methanogens associated with flagellated protists in the gut and with the gut epithelium of termites', *FEMS Microbiology Ecology*, vol 33, no 3, pp. 233–240.

Tracy, B.P., Jones, S.W., Fast, A.G., Indurthi, D.C. and Papoutsakis, E.T. (2012) 'Clostridia: the importance of their exceptional substrate and metabolite diversity for biofuel and biorefinery applications', *Current Opinion in Biotechnology*, vol 23, pp. 364–381.

Uchiyama, T. and Miyazaki, K. (2009) 'Functional metagenomics for enzyme discovery: challenges to efficient screening', *Current Opinion in Biotechnology*, vol 20, pp. 616–622.

Van Soest, P.J., Robertson, J.B. and Lewis, B.A. (1991) 'Methods for dietary fiber, neutral detergent fiber and nonstarch polysaccharides in relation to animal nutrition. Symposium: Carbohydrate methodology, metabolism and nutritional implications in dairy cattle', *Journal of Dairy Science* vol 74, pp. 3583–3597.

Vitikainen, M., Arvas, M., Pakula, T., Oja, M., Penttilä M. and Saloheim, M. (2010) 'Array comparative genomic hybridization analysis of *Trichoderma reesei* strains with enhanced cellulase production properties', *BMC Genomics*, vol 11, article 441.

Whitman, W.B., Bowen, T.L. and Boone, D.R. (2006) 'The methanogenic bacteria', in M. Dworkin, S. Falkow, E. Rosenberg, K.H. Schleifer and E. Stackebrandt (eds) *The Prokaryotes Volume 3: Archaea. Bacteria: Firmicutes, Actinomycetes*, Heidelberg: Springer-Verlag.

Wirth, R., Kovács, E., Maróti, G., Bagi, Z., Rákhely, G., and Kovác, K.L. (2012) 'Characterization of a biogas-producing microbial community by short-read next generation DNA sequencing,' *Biotechnology for Biofuels*, vol 5, article 41.

Wright, A.G. and Pimm, C. (2003) 'Improved strategy for presumptive identification of methanogens using 16S riboprinting', *Journal of Microbiological Methods*, vol 55, pp. 337–349.

Wu, D., Hugenholtz, P., Mavromatis, K., and Pukall, R.Dalin, E., Ivanova, N.N., Kunin, V., Goodwin, L., Wu, M., Tindall, B.J., Hooper, S.D., Pati, A., Lykidis, A., Spring, S., Anderson, I.J., D'haeseleer, P., Zemla, A., Singer, M., Lapidus, A., Nolan, M., Copeland, A., Han, C., Chen, F., Cheng, J.F., Lucas, S., Kerfeld, C., Lang, E., Gronow, S., Chain, P., Bruce, D., Rubin, E.M., Kyrpides, N.C., Klenk H.P. and Eisen J.A.(2009) 'A phylogeny-driven genomic encyclopaedia of Bacteria and Archaea', *Nature*, vol 462, pp. 1056–1060.

Zakrzewski, M., Goesmann, A., Jaenicke, S., Jünemann, S., Eikmeyer, F., Szczepanowski, R., Al-Soud, W.A., Sørensen, S., Pühler, A. and Schlüter, A. (2012) 'Profiling of the metabolically active community from a production-scale biogas plant by means of high-throughput metatranscriptome sequencing', *Journal of Biotechnology*, vol 158, pp. 248–258.

Zhang, J.K., White, A.K., Kuettner, H.C., Boccazzi, P. and Metcalf, W.W. (2002) 'Directed mutagenesis and plasmid-based complementation in the methanogenic archaeon *Methanosarcina acetivorans* C2A demonstrated by genetic analysis of proline biosynthesis', *Journal of Bacteriology*, vol 184, pp. 1449–1454.

Zhang, Z., Donaldson, A.A. and Ma, X. (2012) 'Advancements and future directions in enzyme technology for biomass conversion', *Biotechnology Advances*, vol 30, pp. 913–919.

Zou, G., Shi, S., Jiang, Y., van den Brink, J., de Vries, R.P., Chen, L., Zhang, J., Ma, L., Wang, C. and Zhou, Z. (2012) 'Construction of a cellulase hyperexpression system in *Trichoderma reesei* by promoter and enzyme engineering', *Microbial Cell Factories*, vol 11, no 21.

Part V

Sustainability in anaerobic digestion

Chapter 16

Life cycle assessment as a tool for assessing biomethane production sustainability

Nicholas E. Korres

26 Grigoroviou str, Patisia, GR–11141, Athens, Greece
Email: nkorres@yahoo.co.uk

Introduction

Energy supply is considered as one of the most important challenges of the future accompanied by many interrelated ecological and economic issues. The energy and transport sectors have been proved to be the main drivers of the greenhouse effect causing global climate changes (WHO, 2006). With the current level of energy consumption, world market energy consumption is projected to increase by 44 per cent from 2006 (497 EJ) to 2030 (715 EJ) (IEO, 2009).

Increased energy consumption, as reported by the Fourth Assessment Report of the IPCC is, in conjunction with the world's growing population, leading to the rapid projected increase in greenhouse gas (GHG) emissions (IPCC, 2007). Carbon dioxide emissions are projected to rise from 29 billion tons in 2006 to 33.1 billion tons in 2015 and 40.4 billion tons in 2030, corresponding to an increase of 39 per cent (IEO, 2009).

Alternative options which could simultaneously mitigate climate change and reduce the dependence on fossil sources are already in development, with the use of biomass for energy production deemed to be one of the most promising (Cherubini and Stromman, 2011). It is usually mentioned that renewable energy sources have a large potential to contribute to sustainable development by providing a wide variety of socioeconomic benefits, including diversification of energy supply, enhanced regional and rural development opportunities, creation of a domestic industry and employment opportunities (del Rio and Burguillo, 2009).

Nevertheless, with increasing use of biomass for energy, questions arise about the validity of bioenergy as a means to reduce greenhouse gas emissions and dependence on fossil fuels (Haas et al., 2001; Gerin et al., 2007; Cherubini et al., 2009). As concluded in the 47th Discussion Forum on Life Cycle Assessment (LCA), in Berne, Switzerland (Emmenegger et al., 2012) there is often a trade-off with other environmental impacts, mainly linked to agricultural production, such as eutrophication or ecotoxicity.

The advantages of biogas production by agricultural biomass and organic wastes are highlighted in the Renewable Directive (EC, 2009) in which it is stated:

> The use of agricultural material such as manure, slurry and other animal and organic waste for biogas production has, in view of the high greenhouse gas emission saving potential, significant environmental advantages in terms of heat and power production and its use [of biogas] as biofuel. Biogas installations can, as a result of their decentralised nature and the regional investment structure, contribute significantly to sustainable development in rural areas and offer farmers new income opportunities.

In the same Directive (Article 17) it is also stated that if a biofuel is to be considered sustainable "the GHG emission saving from the use of biofuels and bioliquids taken into account...shall be at least 35% whereas from 2017 GHG emission savings shall be at least 50% and from 2018 60%".

Under this scenario it has been shown that biogas production has an important potential for the production of biomethane as a transport fuel (Murphy and Power, 2008) in terms of energy inputs (Smyth et al., 2009) and GHG emissions (Korres et al., 2010).

Structure and components of life cycle assessment

Life cycle assessment (LCA) is a structured, comprehensive and internationally standardised method formalised by the International Standards Organisation (ISO) (ISO, 1997) to identify and quantify all relevant emissions and resources consumed together with the related environmental and health impacts including resource depletion issues that are associated with the production of any goods or services, thereby enabling the evaluation and comparison of environmental improvement options (EC, JRC, IES, 2010; Garofalo, 2011). This is particularly important in the case of bioenergy production since consideration of all energy inputs and outputs through the whole production cycle of the product is needed for the determination of energy efficiency of a renewable energy source (Salter and Banks, 2009). In the case of biofuels a full LCA needs to include both direct and indirect emissions (Hitchcock and Lane, 2008). Nitrous oxide (N_2O), lime and biogas production *per se* have been found the main pollutants throughout the biomethane production chain (Korres et al., 2010, 2011; Arnold 2011). For conventional fuels, the direct GHG emissions form the majority of total emissions although indirect emissions can also be significant. In the case of light-duty vehicles, for example, indirect emissions account for around 15 per cent of life cycle emissions (SMMT, 2006).

LCA is considered by the scientific community as one of the best tools for the evaluation of the environmental burdens associated with bioenergy production (Consoli et al., 1993) and resources utilised during the life of the product because it offers a holistic and systemic view of a product through its whole life cycle (Payraudeau et al., 2007). In contrast to other environmental management tools, which tend to focus on specific life stages of a product or process, LCA analyses the entire life cycle, i.e. upstream and downstream supply chain (Garofalo, 2011). Usually, the assessment includes the entire life cycle of the product, process or activity from the extraction of resources and process of raw materials, through production, use and recycling, up to the disposal of remaining waste (Figure 16.1). Additionally, LCA provides a well-established and comprehensive framework to compare the production of renewable energy with fossil-based and nuclear energy technologies (Sathaye et al., 2011).

LCA is a systematic, phased approach and encompasses four components (Figure 16.2) as determined by the ISO series, i.e. ISO 14040 (principles and framework), ISO 14041 (goal, scope and inventory analysis), ISO 14042 (impact assessment), ISO 14043 (life cycle interpretation) accompanied by ISO 14047–14049 (rules of documentation and examples on impact assessment and inventory) (ISO, 1997, 1998a, 1998b, 1999). More particularly:

i goal and scope definition, where the boundaries for the assessment are determined along with the level of detail and the functional basis for comparison;
ii inventory analysis, in which the energy, raw materials and related emissions for each process, usually presented in a flow chart, are quantified;

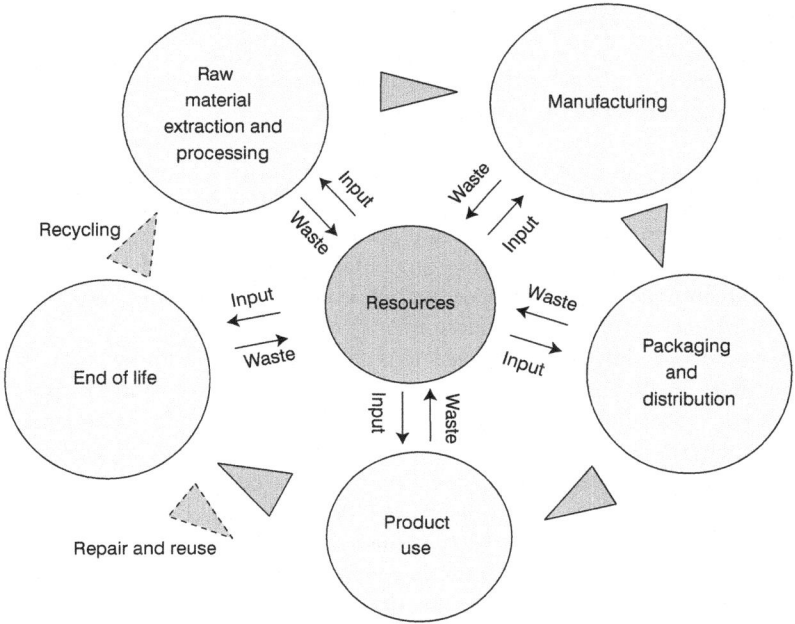

Figure 16.1 Generalised representation of a product's life cycle

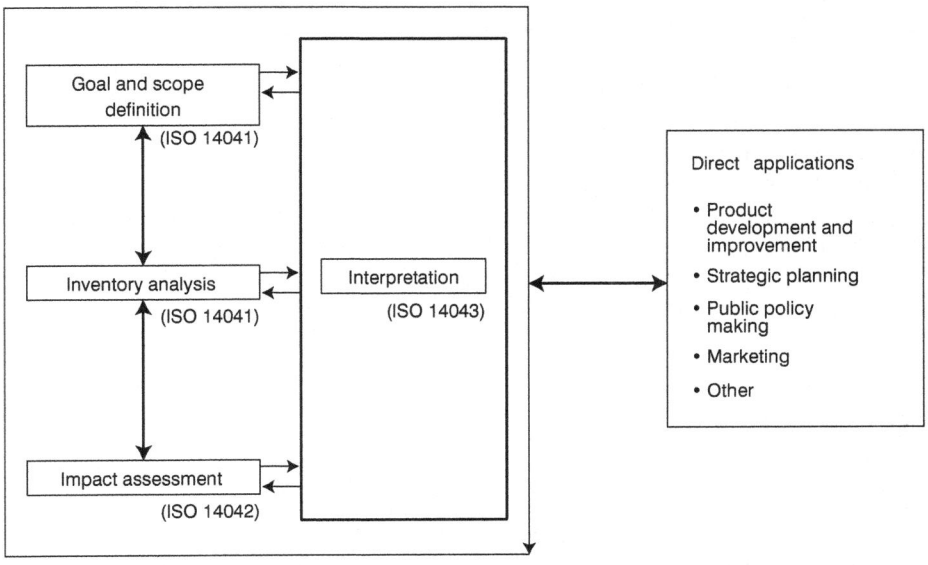

Figure 16.2 Life cycle assessment framework and its direct applications. Based on ISO, 1997; Vikman et al., 2004; Garofalo, 2011

iii impact assessment, which quantifies and groups the effects of the resource use and emissions into a number of environmental impact categories;
iv interpretation, which reports the results and evaluates the opportunities to reduce the environmental impact of the product or service.

The performance of an LCA can assist in i) the identification of opportunities to improve the product's environmental aspects throughout its whole life; ii) decision-making in industry, governmental and non-governmental organisations (e.g. strategy planning, priority setting, etc.); iii) the selection of relevant environmental performance indicators and adequate measurement techniques; and iv) marketing opportunities for "greener" products (e.g. eco-labelling, environmental product declaration, etc.) (ISO, 1997).

The ISO standards have been applied in a large number of assessments of bioenergy systems, but there is still substantial variation among these as to the way in which the standard is implemented. This variability is an inherent characteristic of the ISO standards since they are not constructed as precisely defined tools, but rather as sets of guidelines for good practice. Some parts of the ISO standards, e.g. allocation rules, are continuously under development (Ekvall and Finnveden 2001; Jungmeier et al. 2002; Vikman et al., 2004). Additionally, methodological challenges still exist such as the assessment of direct and indirect land use change emissions and their credits to the bioenergy production, or the influence of heavy metal flows on both human and environmental health or the precise characterisation of pesticidal effects on human health (Korres, 2010). Furthermore, the implementation of a life cycle approach in certification or legislation schemes, as shown in the example of the Renewable Energy Directive of the European Union (Emmenegger et al., 2012) outline another challenge in the application of LCA. These statements are interesting because the quality of LCA methodology used for energy system assessments requires continuous updating of the data and interpretation of the results. These general aspects, as reported by Vikman et al. (2004), are 1) accuracy, which is expressed by the comprehensiveness of the functional unit and system boundaries in time and space and consistency (e.g. consistent treatment of actual and reference system) of the study; 2) transparency in assumptions and calculations, use of flow charts and sensitivity analyses; and 3) efficiency, where an appropriate level of detail must be balanced by ease-of-use and comparable output parameters. Hence, the aim of this chapter is to provide an overview of the objectives, characteristics and components of LCA methodology and to highlight some of the challenging issues for LCA practitioners. This is done in the context of the four phases of LCA as described above for biogas/biomethane production.

Methodological framework

Various approaches have been adopted by LCA practitioners on the type of LCA used. Generally, the categories mostly analysed by the application of LCA are fossil energy consumption and related GHG emissions, i.e. carbon releases into the atmosphere along with other process related emissions (Yu and Chen, 2008). Based on the choice of boundaries various assessments can be applied such as i) a partial LCA from cradle to farm gate (includes production and harvesting of AD feedstock/biomass) or gate to gate (a partial LCA looking at a single added process or material in the product chain); ii) cradle to AD plant gate (includes material acquisition, processing etc. but excludes product end-use); and iii) cradle to grave (includes the entire material/energy cycle of the product, but excludes recycling/

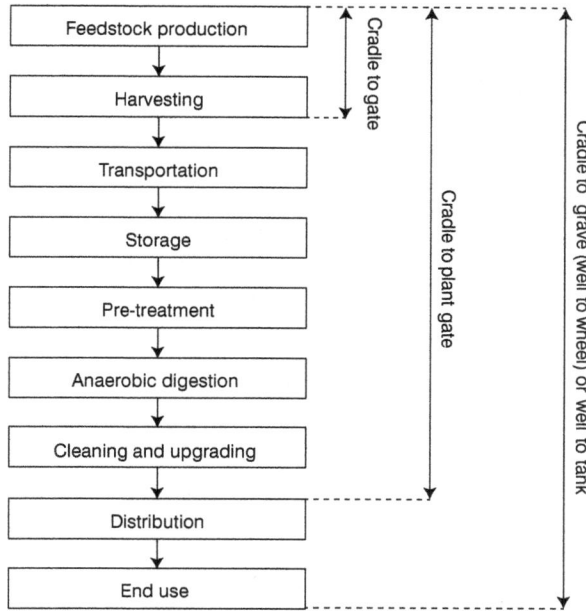

Figure 16.3 LCA of feedstock for biogas production and system boundaries

re-use) or well to wheel (a special type of LCA involving the application of fuel cycles to transportation vehicles that includes the entire material cycle, up to the end use of the biogas or biomethane) or well to tank (includes, in regard to biofuel, the entire life of the biogas/biomethane except its end-use) (Figure 16.3). The first case is focused on the feedstock production and it is widely used for the evaluation of the environmental burden of agricultural feedstock production systems (Kim and Dale, 2005; Bessou et al., 2012), although some modifications including transportation emissions from the farm gate to the AD plant have been adopted (Gasol et al., 2007). The second case includes feedstock and biogas production up to the AD plant gate without taking into account the distribution and the end-use of the product. Some adjustments have been found in the literature that include the injection of the biomethane into the natural gas grid (Arnold, 2011). The last case, which includes the utilisation of the final product, determines a start and an end point, i.e. "well-to-wheels" assessment incorporates "well", i.e. farming and "wheel", i.e. fuel combustion (Sanchez et al., 2012).

Many researchers used a "well to tank" or "well to wheel" LCA approach to compare the environmental impact of biofuels with fossil fuels (McLaughlin et al., 2002; Elsayed et al., 2003; Power and Murphy, 2009; Monti et al., 2009; Korres et al., 2010). According to Singh et al. (2010) the "well to tank" approach is sufficient only for comparing various production technologies for biofuel production from lignocellulosic biomass while the "well to wheel" is suited for comparisons between different biofuels or between biofuels and fossil fuels. In addition to the approaches mentioned above, two other LCA approaches, attributional vs. consequential (also called the change-oriented approach), have also been identified. Multifunctional processes which inevitably occur throughout the biofuels production chain necessitate choices on how to treat co-products in the LCA model. These choices have a strong effect on the performance of the LCA and the distinction between attributable and consequential LCA was developed in the process of resolving the methodological debates over allocation issues and data choices (Thomassen et al., 2008). This distinction is particularly relevant when defining system boundaries in the life cycle inventory (LCI) (Finnveden et al., 2009). Resource flow and related pollution, in the former, is attributed to the unit of analysis of the product under examination by linking together attributable processes along its life cycle (ISO, 2006a). In other words, resource flows and related pollution within the system under study are attributed to the delivery of a specified amount of the functional unit (Rebitzer et al., 2004). Therefore, attributional LCA methodology accounts for immediate physical flows (i.e., resources, material, energy and emissions) involved across the life cycle of a product and typically utilises average data for each unit process within the life cycle (Weidema, 2003; Earles and Halog, 2011). Consequential LCA on the other hand includes processes which are expected to change as a consequence of a change in demand for the unit analysis (Sonnemann and Vigon, 2011). In other words, the consequential approach estimates how pollution and resource flows within a system change in response to a change in output of the functional unit (Ekvall and Weidema, 2004; Rebitzer et al., 2004).

The consequential approach makes use of data that is not constrained and can respond to changes in demand that occurs as a result of changes in production volumes, production technologies, public policies and consumer behaviours (Arnold, 2011). This approach can provide valuable insight in certain applications such as evaluating reduction projects or making public policy decisions (Bhatia et al., 2011). Wenzel and Petersen (2009) reported that the consequential approach requires a comparative LCA, i.e. that alternatives are compared since they are equivalent and provide the same services (primary and secondary). The primary service is the "main function" of the system (e.g. biomethane production as a transport fuel) whereas secondary services are defined as products/services arising from processes in the system under investigation (e.g. the nutrient value of the digestate that can replace mineral fertilisers or the electrical energy of the biogas produced from slurry that can replace equal amount of electricity from the grid).

The core differences between consequential and attributional LCA are that the former includes the processes which are actually affected by a change in demand, instead of the averages used in attributional LCA (Weidema, 2003; Earles and Halog, 2011). Based on this difference, Schmidt (2010) reported that the main argument for the application of the consequential approach is that only the actual affected processes are included whereas those that are not likely to respond to a change in demand should be excluded from the assessment since this will not reflect the actual change in environmental impact. It is obvious that application of the consequential approach can reduce the cost of LCA. Additionally, attributional LCA uses average data reflecting the actual physical flows instead of marginal data used in the consequential approach (Finnveden et al., 2009).

An obstacle for the application of the consequential approach is the difficulty of collecting appropriate and accurate data. Ekvall and Andrae (2006), Lesage et al. (2007) and Thomassen et al. (2008) noted that the adoption of consequential LCA can reveal unique environmental insights beyond attributional LCA. However, Vieira and Horvath (2008) found little difference between the two approaches.

Goal and scope definition

The goal and scope definition phase of an LCA includes several decisions that are of relevance for all subsequent steps (Frischknecht and Jungbluth, 2007). They include the exact formulation of what is to be investigated and how this investigation is to be carried out, along with data requirements (Udo de Haes and van Rooijen, 2005). The scope usually addresses i) the product system to be studied; ii) the function of the product system; iii) the functional unit of the product system; iv) product system boundaries; v) allocation methods; vi) types of impact and methodology of impact assessment; and vii) data requirements, assumptions and limitations (Rebitzer et al., 2004; Udo de Haes and van Rooijen, 2005; Labutong et al., 2012). Nevertheless, the LCA process is iterative and the scope may be revised over the course of the inventory analysis, impact assessment, and interpretation. The goal of an LCA study should unambiguously state the intended application to the intended audience of the study whereas the scope should be adequately defined so as to ensure compatibility with the goal (Singh et al., 2010). Data quality is defined by time, place, technology, and registration method, e.g. measured data or calculated data (Hartmann, 2006). The scope and goal definition of the LCA varies according to socio-economic, environmental and legislative issues but also according to the technicalities on which the framework for the development of the LCA is administered. The technical factors might include the aim of the study, the product and function of the product system, allocation methods (in case of the attributional LCA approach as discussed previously), assumptions and other factors.

It is also worth mentioning the effects imposed by legislation upon the scope and goal definition. For example, in the case of biofuels sustainability, as stated by Renewable Directive 2009/28/EC (EC, 2009), is defined by emissions savings when compared with the fossil fuel it replaces on a whole life cycle analysis. In addition, according to Directive 2009/28/EC, Annex V, C–13, "emissions from the fuel in use shall be taken to be zero for biofuels and bioliquids". Hence emissions from combustion of biomethane in vehicles should not be taken into consideration, although there is a debate going on in the scientific community concerning this scenario. Such legislative impacts obviously determine not only the functional unit of the system assessed but also the inputs and related environmental burdens that should be considered in the life cycle inventory (LCI) and impact assessment development.

Furthermore, an energy balance is not directly related to a greenhouse gas balance because a biofuel system that generates significant quantities of fuel per hectare (GJ/ha) may not necessarily be sustainable. Emissions associated with agricultural production of biomass, source of fertiliser, sources of parasitic or consumed energy demand, and efficiency of vehicle may deem a product unsustainable.

In many LCA studies the scope can be driven by policy issues (Deasy and Power, 2011) or the goal may be multifaceted as when driven by different feedstocks (Thyo and Wenzel, 2007; Stucki et al., 2011) or different technologies (Arnold, 2011). The integration of simple rules into the formulation of goal and scope definition using, for example, root definition or the CATWOE model, which both originate from soft system methodology (SSM), could clarify and enhance the understanding of the whole method. The first step in SSM is to formulate the root definition of the system under study, assessment or design. A properly structured root definition comprises three elements (i.e. what, how, why) and is of the form "a system to do X, by (means of) Y, in order to achieve Z". In other words, the "what" component is the immediate aim of the system, the "how" component is the means of achieving that aim and the "why" component is the longer term aim of the purposeful activity (Korres, 2004). In terms of LCA scope and goal components a possible root definition could be "to evaluate the sustainability of biogas production by the employment of the LCA technique, as described by ISO 14040 standards, in order to improve part of or the entire biogas production chain in terms of human health and/or natural environment and/or natural resources and/or …". Improvements of root definition can be achieved by the employment of CATWOE methodology which supports the identification and categorisation of all elements (e.g. stakeholders, processes, environment etc.) of the system under analysis (Korres, 2004). More particularly, "C" stands for the Customers of the system, those who would be the victims or beneficiaries of the purposeful activity, those who are on the receiving end of whatever it is that the system does (e.g. rural community, stakeholders, policy makers, investors in renewable energy, scholars etc.); "A" stands for the Actors of the system, those who would do the activities, those who transform inputs to outputs (e.g. AD operators, engineers, agronomists etc.); "T" stands for the Transformation Process, the activity which changes a defined input into a defined output (e.g. anaerobic digestion and the break down of lignocellulosic material); "W" stands for Weltanshauung, the view of the world that makes this definition meaningful (e.g. cheap clean energy, mitigation of climate change, etc.); "O" stands for the Owner(s) of the system, those who could stop the activity; "E" stands for the Environmental constraints, the constraints in the environment that the system takes as given (e.g. it includes any social, technical or economic factors, rather than describing the natural environment alone) (Elghali, 2002).

Functional unit

The functional unit is the quantified measure of performance of a product system. It describes the function of the product and it is the basis for the calculations in LCA assessments (ISO, 1997; Bligny et al., 2012). In all bioenergy assessment systems the choice of an appropriate functional unit, as the basis for comparisons, is of major importance (Ekvall and Finnveden, 2001). This is particularly so in studies of systems with more than one output, e.g. energy systems where a combination of heat and electricity is co-generated (CHP plants) (Vikman et al., 2004). All material and energy flows and the effects resulting from these flows are related to the functional unit. This makes the functional unit a base for a variety of comparisons (Hartmann, 2006). The preferred functional unit must be defined and measurable. In practice the functional unit consists of a qualitatively defined function or property (e.g. environmental impact) and a quantified unit (e.g. 1 m^3 or 1 MJ of fuel). There is significant diversity in relation to the functional unit used in LCA, particularly in the case of biofuels. Korres et al. (2010) defined the functional unit as m^3 biomethane yr^{-1} whereas, based on the 2009 EU Renewable Directive (EC, 2009) for the evaluation of grass-biomethane sustainability as a transport fuel, the environmental impacts were expressed as g CO_2 equivalent (CO_2e) MJ^{-1} energy replaced. Börjesson et al. (2011) in their LCA of biofuels in Sweden including biogas from waste (food industry and household), biogas from manure and biogas from crops (i.e. sugar beets, ley crops and maize) used "environmental impact per MJ fuel" as the functional unit. They argued that other options such as "environmental impact per kilometre of transport service" could increase uncertainty in the results when improvements in the fuel efficiency of different vehicles, for example, are implemented rapidly and new technologies such as electric hybrid technology are introduced. They also stated that the functional unit expressed as per "MJ of fuel" can be easily converted into per kilometre of transport service for the specific vehicles in question. Nevertheless, according to Bergsma et al. (2006) the functional unit when biofuels are compared with conventional fuels should be "1 km driving of a standard car on gasoline or diesel". This way, efficiency differences between biofuels and conventional fuels can be included. According to Cherubini et al. (2009) and Kim and Dale (2009) the functional unit "per hectare and year" should be used for fuels based on crops in parallel with "per MJ fuel" (and if possible per km transport service). This would allow the area efficiency to be assessed given the increased competition of cropland for food, feed and energy. Gasol et al. (2007) used the "per ha per year cultivation " of *Brassica carinata* as the system function whereas the energy stored in the crop from one year (83.69 GJ) was used as an energy reference value in order to compare the natural gas and biomass systems. Arnold (2011) used "Nm^3 of biogas/ton fresh matter (FM)" as the functional unit for the production of biogas from various substrates, i.e. maize, rye, sorghum, whole crop triticale and barley silage and the grass *Landsberger Gemenge*, a mixture of hairy vetch (*Vicia villosa*), crimson clover (*Trifolium incarnátum*) and Italian ryegrass (*Lolium multiflorum*). The basis for comparing bio-electricity and fossil electricity is "1 kWhe delivered to the customer" (Bergsma et al., 2006) whereas in other studies the functional unit of electric energy produced by the biogas plant is "1 TJ electricity fed into the public electricity network" (Hartmann, 2006). Nevertheless, in attributional life cycle inventory (LCI), as Rebitzer et al. (2004) reported, results describe the environmental exchanges of the average electricity production in a geographic area and/or an electricity supplier. According to some authors the results could be presented as the emissions per MWh produced. The magnitude of the functional unit (megawatt hour, smaller, or larger) does not affect the conclusions since the average emissions of the electricity system scale linearly with the functional unit. In some LCA studies it can be helpful to use several functional units. If this is the case their use must be explained in detail while the goal definition is developed, especially if the comparison between the production of various kinds of biofuels or the cultivation of land is the object of the study (Bergsma et al., 2006). According to Lindfors et al. (1999) the definition of the functional unit of a product system must take the following aspects into consideration: i) the efficiency of the product; ii) the durability of the life span of the product; and iii) the performance quality standard (if any).

Boundaries of the system

The choices and assumptions made during system modelling, especially those concerning the system boundaries and the processes that should be included within these boundaries, are often decisive for the result of an LCA study (Lundin et al., 2000; Rebitzer et al., 2004). As mentioned by Tillman et al. (1994), the system must be delimited by boundaries, which encapsulate all processes and activities that will be included in the study. It is therefore understood that the selection of appropriate boundary conditions is a fundamental point in a sustainability analysis as the use of different boundaries in fact means using different inputs or outputs of the system under study, resulting in the generation of incomparable results.

Assessment of the GHG benefits requires particular knowledge of bioenergy production and conversion technology *per se* (Horne and Matthews, 2004). Energy, CO_2 and parameters affecting GHG emissions such as emission factors are the main LCA parameters of interest in GHG accounting and can be further defined as primary energy inputs, net CO_2 savings and GHG emissions, compared with current fossil fuel equivalents. Nevertheless, the literature on biofuel LCAs contains conflicting studies that often employ differing units and system boundaries, making comparisons across studies difficult (Kammen et al., 2008)]. Different

LCA practitioners treat system boundaries in a different way which can reduce accuracy and increase uncertainty between comparisons of various studies. For example, fuels comparisons in a LCA study was made on the basis of the energy content of the fuel, excluding end-use conversion (Elsayed et al. 2003) whereas in other studies the system boundaries have been expanded to include the transportation work in the functional unit (Beer et al. 2002; Jungmeier and Hausberger, 2002). The system boundary should as far as possible include all relevant life cycle stages and processes (EC, JRC, IES, 2010) although this would result in a complicated analysis (Tillman et al., 1994) because of the difficulty of handling large life-cycle inventories (Hartmann, 2006).

Dimensions within system boundaries, as Tillman et al. (1994) and Guinee et al. (2002) reported, are:

i   Boundaries between the technological system and environment. The LCA usually begins at the extraction point of raw materials and covers the entire life cycle of the final product up to final stages which normally include waste generation in any stage, i.e. gaseous, liquid or solid (Figure 16.1). Nevertheless, partial LCAs as mentioned earlier are included in this category (e.g. cradle to gate).
ii Boundaries between current life cycle and related life cycles of other products or between the technological system under study and other technological systems. In this case most activities are interrelated and for that reason must be isolated from each other for further study. It is obvious that the system boundary between the technological system under study and other technological systems is affected in various other ways by the LCA approach as for example when a consequential LCA is applied.
iii Boundaries between significant and insignificant processes. The identification of system boundaries between "significant" and "insignificant" processes is difficult because often it is not known, at the beginning of a study, which processes based on data collected/measured/estimated are significant or not (Finnveden et al., 2009).
iv Spatial (geographical) boundaries. Geography holds an important role in most LCA studies since many processes such as electricity production, waste management, feedstock production, transport systems along with ecosystem sensitivity etc. differ from one area to another.
v   Temporal boundaries. LCAs are conducted to evaluate present impacts but also to predict future scenarios although limitations due to technologies involved, pollutants lifespan etc. entail high uncertainty. Incorporation of stochastic modelling and/or simulation techniques, including the time dimension, could benefit LCA performance.
vi Boundaries which are based on the production of capital goods in which the evaluation of the economic feasibility of new and more eco-friendly processes in comparison with current technologies can be achieved.

Finnveden et al. (2009) stated that spatial and temporal limits as LCA boundaries could be received as special technological and environmental boundaries cases. It is recommended, whenever reasonable, to use expanded system boundaries, since this will give the most accurate representation of the real system functions under study. This may be the best approach when the functions provided are strongly integrated, and thus extremely difficult to separate. In co-product situations, where some form of causality might be used for allocation, then allocation may be the most feasible approach in order to avoid systems which are too large (Lindfors et al., 1999).

Reference system

A typical objective of LCA is to discover essential differences in potential environmental impacts between two alternative systems fulfilling the same functions (Lindfors et al., 1999). The intended use of the results of such a study might be to: i) compare different systems, i.e. identify major differences in potential environmental impacts between the systems, ii) select types of impacts caused by a system, that are potential candidates for improvements when compared with a reference system, iii) evaluate the potential environmental options given by the choice between different alternatives, i.e. identifying potential environmental benefits of an alternative, iv) identify impact categories that are not significantly affected by the choice between potential alternatives (Lindfors et al., 1999). Therefore the choice of the reference system with which the bioenergy system is compared is critical because the estimated benefits of bioenergy based on the replacement of the assumed energy system can differ significantly depending on the reference system. Fossil-derived electricity, for example, might be produced from oil, natural gas or coal, all of which have different GHG emissions per kWh of electricity generated. It would not be meaningful for example to compare the GHG emissions caused by the AD of maize silage for the production of electricity with these caused by a fossil energy system that would not be replaced (e.g. it could be a coal-derived electricity system or natural-gas-derived electricity system or other) by the bioenergy system (i.e. in our case that produces electricity through the AD of maize silage). A possible solution is to estimate the GHG emission savings of the bioenergy system and compare them with the average GHG emissions from fossil energy systems. Alternatively, a conservative evaluation of the bioenergy system GHG emissions and their comparison with the GHG emissions of the best available fossil energy technology could offer a feasible solution. For example, it could be assumed that electricity in the fossil fuel reference system is produced by the lowest emission fossil technology such as natural gas, rather than coal. Since natural-gas-generated electricity has a GHG emission factor of around 400 g CO_2 eq/kWh (110 g CO_2 eq/MJ) compared with 990 g CO_2 eq/kWh (240 g CO_2 eq/MJ) for coal-based electricity (Staple and Swisher, 2011), assuming that natural gas is displaced would give a conservative estimate of emission reduction (Figure 16.4).

Jungmeier and Spitzer (2001) studied various bioenergy systems from agriculture for heat supply and combined electricity and heat supply and compared them with the fossil energy systems such as those mentioned above (i.e. oil, natural gas, and coal). Results were expressed as emissions of CO_2 equivalents per kWh in comparison with fossil fuel systems and as a percentage of CO_2 equivalent reduction. They reported lower GHG emissions from bioenergy systems compared with those from fossil fuel systems. Additionally, Broch et al. (2012) showed that LCA modelling results of the N_2O emissions impact on total GHG varied widely among different crops and land reference systems chosen. A land or agricultural

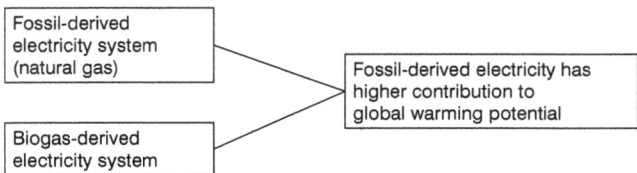

Figure 16.4 Schematic illustration after LCA application between biogas and a fossil fuel (a reference system) derived electricity

reference system defines what the cultivated land area would be used for if the investigated product were not to be produced (Jungk et al., 2002). When a comparison is being made between a bioenergy and a fossil energy carrier, it is always necessary to define an alternative way in which the required land might be used if not for the production of energy (Jungk et al., 2002). Therefore, the choice of proven reference system can be of great significance for the policy makers, utilities providers and industry by allowing the identification of effective agricultural biomass options in order to reach emission reduction targets. Omission of alternative land use for the production system under concern would not adequately assess impact and would put any claim of sustainability in question.

Life cycle inventory

The life cycle inventory analysis (LCIA) specifies the processes that occur during the life cycle of a product. In LCI, an inventory is made of all the inputs and outputs of processes that occur during the life cycle of a product (Udo de Haes and van Rooijen, 2005).

Procedures

Life cycle inventory (LCI) is one of the most critical stages of the LCA study. A generic step-by-step procedure for LCI development is presented in Figure 16.5. The data collection sheet and data collection process, data validation, its relation to unit process (i.e. the smaller part in the production chain, for example the production of fertilisers or the production of feedstock or biogas upgrading etc.) and functional unit along with data aggregation are discussed below.

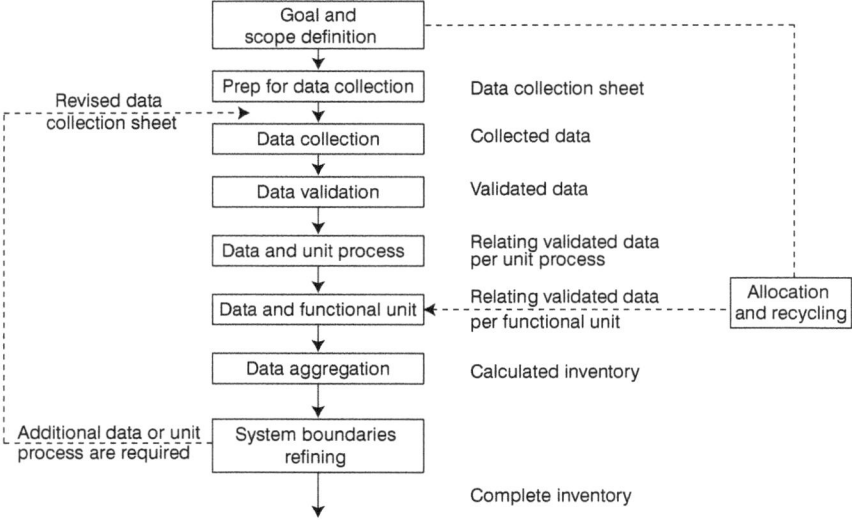

Figure 16.5 A generic step-by-step LCI development plan (based on ISO, 2006b)

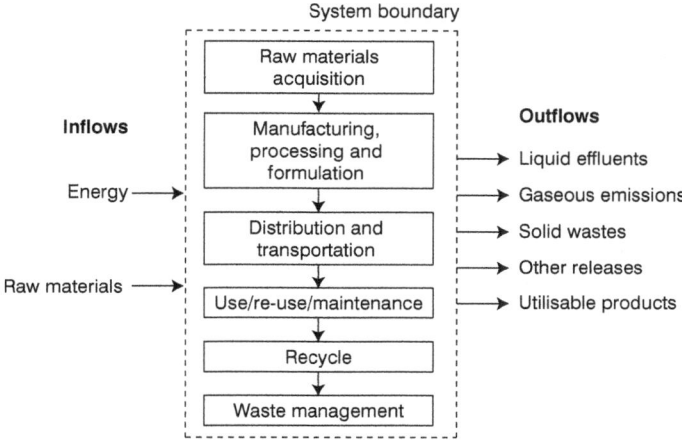

Figure 16.6 Life cycle inventory and determination of system boundaries (based on Fava et al., 1994)

DATA COLLECTION AND DATA COLLECTION SHEET

Data collection is any process of preparing and collecting data as part of a process improvement. The purpose of data collection is to obtain recorded information which after its validation can be used in decision-making procedures or to be passed on to various stakeholders. Data are usually recorded on special data sheets (Table 16.1), which can be either a separate form for each process unit as these were identified during the determination of system boundaries (Figure 16.6) or a summary form on which the data are combined as accurately as possible to ensure that subsequent decisions based on arguments embodied in the findings are valid (Sapsford and Jupp, 2006).

During LCI development every piece of information/data which is either entering the system as input (e.g. raw materials, fuels, agrochemicals and other natural inputs) or is exported to the environment as output (products, gaseous emissions, liquid effluents, solid wastes by-products or wastes) is recorded and quantified.

According to Fava et al. (1994) the major LCA inventory stages to be considered while collecting data for LCA development and which are generally applicable to biogas production, include activities such as raw materials (feedstock) acquisition and energy consumption; manufacturing, processing and formulation; distribution and transportation; use/re-use/maintenance; recycling and waste management (Figure 16.6). Each stage receives inputs of materials and energy and produces outputs of materials to a subsequent stage, and wastes which may or may not be discharged to the output of materials.

DATA VALIDATION

It has been stated by Jensen et al. (1997) that LCA is very data-intensive hence mistakes can easily be made but are hard to find and correct. It is therefore imperative to systematically validate the data during the data collection process. Validity checks such as mass/energy balances (i.e. comparisons between inputs with outputs of mass and energy according to the laws of mass and energy conservation) are useful for checking the general consistency of the collected data and identifying whether an input or output may be missing. Comparative

Table 16.1 Example of a renewable/biofuel data collection sheet.

General info, e.g. name of researcher, name of the site, date of visit (if applicable), stakeholders, reviewer group etc.			
Question(s)	Response(s)	Units	Source and data type E = Estimated M = Measured
Type of renewable energy produced			
Feedstock			
What kind of feedstock is used in the process?			
Feedstock origin			
Transportation to the site			
Pre-treated? (If yes, where and how?)			
Quantity (per time period or per unit of energy produce)			
Final produce (per unit of time or per unit of feedstock weight/volume)			
Energy used in the production process (per unit of time period or per unit of renewable produce)			
Electricity used			
Diesel used			
Gas used			
Other fuel used in the production process (e.g. heat/energy recovery)			
Raw materials used in the process			
Water			
Chemicals			
Any other material used (e.g. lime)			

Secondary outputs			
Co-products, by-products, wastes & residues			
Disposal or recycling of secondary outputs			
Energy or raw material used or produced as a result of secondary outputs			
Environmental burden			
Air			
Water			
Soil			
Use of renewable energy			
Use on or off site?			
If on site, how is the renewable energy used?			
If off site, how is the renewable energy used? How is it transported/distributed? Estimated or measured losses?			
Equipment used for the production of the renewable energy (for the determination of the embodied environmental impact in the machinery used)			
Equipment used to produce the fuel			
Special building? Specifications			
Is it available off the shelf? If yes, details, if possible, are needed			
Manual/Procedures if available			
Flow chart			
Other important information			

analysis of emission factors from similar processes for assessing data plausibility can be conducted (Palsson and Riise, 2011). Any anomalies in the data can be replaced by (alternative) data values complying with the data quality requirements (Jensen et al., 1997). During the collection of data for various processes in the product system, it may be necessary to allocate the inputs and outputs between different products. This may be necessary for processes that produce more than one product. In this case, the use of raw materials and energy, and the releases to air, water and land will need to be divided between the products. The remaining stages of the LCI development, i.e. relating data to unit process and the functional unit, data aggregation and when necessary, refining system boundaries, form the compilation of the inventory result with the collected data (ISO, 1998; Palsson and Riise, 2011).

DATA AGGREGATION AND RELATING PROCESSES

Structuring and preparing the collected data for the unit processes included in the production system and then normalising them to the functional unit (Bird et al., 2011) is the first step before data aggregation. Table 16.2 shows data requirements for the GHG and energy balances of a biogas production system.

The reference flow of each unit process, which was determined during the construction of the flow chart (Figures 16.7, 16.8 and 16.9) relates the collected data for inputs (raw materials and energy) and outputs to this particular flow and consequently to the corresponding unit process. The normalisation of data to the functional unit involves the scaling of the inputs and outputs in each unit process to the functional unit. In other words, normalisation of the data to the functional unit assigns the corresponding inputs and outputs (i.e. energy and environmental burdens) from each unit process to the functional unit.

The actions with which normalisation of the data can be achieved include: i) calculation of the scaling factors of each unit process by using the flow chart and the corresponding inputs and outputs data for each unit process; attention should be given in situations where, for example, electricity or other products are used in several included unit processes; ii) scaling factors for transport are usually derived by multiplying the amount of material transported with the distance the material is transported; iii) scaling each unit process to the functional unit. The scaling factors are used to scale all inputs and outputs for each included unit process. This is achieved by multiplying the inputs and outputs for each unit process with the corresponding scaling factor. The scaled unit process delivers the amount of material needed to produce the functional unit. The next step in the inventory development is the aggregation of the inputs and outputs for all unit processes included in the system. For example, CO_2 emissions to air from all unit processes are added together into the total CO_2 emission to air for the biogas production chain. This is the inventory result for the entire production system.

Data categories

Primary data are clearly preferred to secondary data during the LCI development. As Bhatia et al. (2011) stated, primary data are collected from specific processes of a product's life cycle and include: i) activity data (the quantitative measures of a level of activity or a unit process that results in GHG emissions or removals) including process activity data and financial activity data (these are always secondary types of data); ii) direct emissions data

Table 16.2 Relating biogas data to unit process.

Unit process	Data to be collected or estimated	Variable calculated
Land management change	Carbon stocks in soil and vegetation affected by the bioenergy system	Carbon stock change due to land use change
Cultivation and harvest of biomass	Biomass yield; residue amount and use (co-products and type); fertiliser amount and type; herbicides and pesticides use; fuel use by machines during field operations; indirect emissions for fertiliser, herbicide and pesticide production	Energy input (direct* and/or indirect**) and related GHG emissions (direct* and/or indirect**) from production of biomass or collection of organic/industrial wastes
Transportation of feedstock	Transport distance and mode; fuel per km	Energy input and GHG emissions from transportation
Conversion/digestion	Auxiliary materials input (e.g. enzymes in case or feedstock pre-treatment); co-products amount and type; digestion energy and material efficiency; energy (direct and/or indirect) demand of digestion facility; indirect emissions from auxiliary materials	Energy input (direct and/or indirect) and GHG emissions (direct and/or indirect) from anaerobic digestion plant
Distribution	Distribution distance and mode; distribution losses; energy demand of distribution system (e.g. fuels if biogas is compressed and transported by vehicle); fugitive emissions from the distribution system (e.g. natural gas grid)	Energy input and related GHG emissions from distribution
Use	Energy efficiency; auxiliary energy demand; auxiliary materials input	Energy input and related emissions from use
Co- and/or by-product/waste management	Quantity of digestate	Energy and related GHG emissions from transport and application as substitute of manufactured fertiliser. Fertilisation management emissions balance

Based on Bird et al., 2011.
*Direct energy and the corresponding direct GHG emissions are defined as the energy and related emissions from agronomic operations, e.g. ploughing, harrowing, rolling, application of agrochemicals, harvesting etc., and from energy consumed in anaerobic digestion plant, e.g. heating of digesters (Korres et al., 2010; Korres et al., 2011).
**Indirect energy and corresponding indirect greenhouse gas emissions are defined as the energy and related emissions from the manufacture of biomass production inputs, e.g. fertilisers, herbicides, lime etc. and from the pre-treatment of the feedstock (e.g. macerating) or other activities, e.g. mixing and water pumping activities in anaerobic digestion plant (Korres et al., 2010, 2011).

(i.e. determined through direct monitoring, stoichiometry, mass balance, or similar methods) from a specific site; iii) allocated data; iv) emission factors; v) data that is averaged across all sites that contain the specific process; and vi) secondary data (e.g. financial activity data). Examples of primary data include kWh consumed by a process from an individual site or an average across sites; litres of fuel consumed by a process in the product's life cycle, either from a specific site or an average across all production sites; kg of material added to a process or GHG emissions from the chemical reaction of a process.

ACTIVITY DATA

More specifically, process activity data measure the physical inputs/outputs or other metrics of a process and when combined with a process emission factor result in the calculation of GHG emissions. They include energy data (e.g. KJ of energy consumed), mass data (e.g. kg of a material), volume (e.g. volume of chemicals or fuels used), area (e.g. cultivated area), distance (e.g. distance feedstock travelled to AD plant), and time (e.g. retention time). In contrast, financial activity data consist of monetary measures of a process that results in GHG emissions. They measure the financial transaction associated with a process. These data, when combined with a financial emission factor (e.g., environmentally extended input–output emission factor, EEIOEF) result in the calculation of GHG emissions. An example taken from Saunders et al. (2006) demonstrates how financial activity data can be used for the calculation of GHG emissions in barley production (UK) used as feedstock. A similar analysis requires information on the production system in the UK regarding yield of barley, inputs used by type and associated energy and emission coefficients. Such data are provided by the *John Nix Farm Management Pocket Book* (Nix, 2004 cited in Saunders et al., 2006). The data listed in Table 16.3 are general and not necessarily accurate but will be used to demonstrate the following example on financial data activity.

The calculation of the energy and the emission component from the information provided in Table 16.3 requires the conversion of the inputs into their physical quantities and in some cases breaking them down further. An assumption made, for reasons of simplicity, was that machinery repairs, seed costs and fixed costs were excluded for calculations of energy and associated emissions in barley production. The cost of fuel and repairs for barley equals £100/ha. A more detailed breakdown shows that the cost of fuel, electricity and oil equals £35/ha and the cost of machinery repairs equals £40/ha in a medium-sized cereal production system (100–200 ha). Therefore 46.67 per cent of fuel and repairs is billed to fuel. Consequently, given the £100 price reported in Table 16.3 for fuel and repairs, the input of fuel, electricity and oil is £46.67/ha. The transformation of the fuel value into the physical amount of litres consumed in the process (i.e. diesel equivalent), knowing the cost of the fuel consumed in a process and the cost of the fuel per litre, can be easily found. The price of "red diesel" or gas oil, which is only available to farmers and has very small rates of excise duty attached, is assumed to be 24p per litre, hence the diesel used in barley production equals 194.5 litres/ha. The energy coefficient of diesel, as reported by Saunders et al. (2006), equals 41.2 MJ/L hence 8013 MJ/ha or 1233 MJ/t of barley. The CO_2 emissions associated with barley production can be calculated by multiplying the energy component per tonne of barley found above with the emission factor of 65.1 g CO_2/MJ (e.g. 65.1/1000 × 1233 = 80.2 kg CO_2/t of barley).

Table 16.3 Financial activity data in winter barley production.

Inputs and outputs	Item input/output per ha
Barley yield (average)	6.5 tonnes
Fuel and repairs	£100.00
Fertiliser	£87.50
Sprays	£85.00
Seed	£37.50

Source: Nix, 2004 cited in Saunders et al., 2006

DIRECT EMISSIONS DATA

These data are derived from emission releases and are determined through direct monitoring, stoichiometry, mass balance, or similar methods. They include emissions from an incinerator measured through a continuous emissions monitoring system, a chemical reaction's emissions determined using stochiometric equation balancing or fugitive refrigerant emissions determined using a mass balance approach (Bhatia et al., 2011).

ALLOCATED DATA

Allocated data are considered primary data as long as the data meets the other primary data requirements (i.e. are based on process-specific information and they provide sufficient supporting information to enable users to understand how the data were gathered, what calculation methodologies were used and the quality of inventory). Collected primary data which do not require allocation to other process outputs are preferable to data that require allocation. For example, with other data quality indicators being roughly equal, data gathered at the process level that does not need to be allocated is preferable to facility-level data that needs to be allocated between the studied product and other facility outputs (Bhatia et al., 2011).

EMISSION FACTORS

Emission factors (EF) are the GHG emissions per unit of activity data, and they are multiplied by activity data to calculate GHG emissions. Emission factors may cover one type of GHG (for example, CH_4/litre of fuel) or they may include many gases in units of CO_2 equivalents (CO_2e). Emission factors can include a single process in a product's life cycle, or they can include multiple processes aggregated together. Life cycle emission factors that include emissions from all attributable upstream processes of a product are often called cradle-to-gate emission factors (Bhatia et al., 2011).

SECONDARY DATA

According to Bhatia et al. (2011), secondary data are those that are not from specific processes in the product's life cycle. As such, financial activity data cannot be used to meet the primary data requirement and therefore are always classified as secondary. Examples of secondary data include i) average number of litres, obtained from a life cycle database, of a fuel consumed for the completion of a process; ii) average weight of material input into a process; iii) kWh consumed by another similar process used as an alternative in a product's LCA; iv) average GHG emissions from a process's chemical reaction; and v) amount spent on process inputs, either specific to the process or a company/industry average. Usually, secondary data originate from external sources (e.g. life cycle databases, industry associations, etc.) or can be data from another process or activity in the reporting company's or supplier's control that is used as a proxy for a process in the inventory product's life cycle. According to EC, JRC, IES (2010) the selection criteria for the inclusion of the secondary data into LCI should consider: i) use of consistent secondary data sets; ii) the selection should be quality oriented; iii) preference should be given to pre-verified data sets; and iv) preference should be given to well-documented data sets. In addition the rules for data quality described in the following section are applied to primary and secondary data.

Data quality

As already mentioned, a comprenhensive LCI involves the collection and integration of hundreds upon thousands of pieces of data regarding the product, process or activity under study (Fava and Pomper, 1997). As such, it is essential that the manegement of data quality is an integral part of the overall process. Data quality was defined by SETAC (cited in Fava and Pomper, 1997) as: "...the degree of confidence one has in the individual data input from a source and in the data set as a whole and ultimately in the decisions based on the life cycle study using such data as input." Many authors (Fava et al., 1994; Fava and Pomper, 1997; Jensen et al., 1997; Bhatia et al., 2011; Labutong et al., 2012) have agreed about the type and number of quality indicators that can be used in assessing data quality. These include an assessment of the necessary level of the following. i) Precision, i.e. a measure of the spread of the data set values about their mean. Precision measures such as mean and standard deviation of reported values can be calculated and reported for each process unit along the entire life cycle). ii) Completeness, i.e. the degree to which the data are statistically representative of the process sites or the percentage of locations reporting primary data from the potential number in existence for each data category in a unit process. iii) Consistency, i.e. one of the most important qualitative measures for the assessment of the methodology is uniformity of application to the various components of the analysis. iv) Representativeness, i.e. a qualitative assessment of the degree to which the data set used in the analysis is a true and accuarate measurement of the average processes under examination. In many cases more specified indicators such as technological (i.e. representation of the actual technology used in the assessment), geographical (i.e. representation of the actual geograpical location of the processes within the inventory boundaries) and temporal (i.e. representation of actual time or process age) can be used to assess the reprsentativeness of the data. v) Reproducibility, i.e. a qualitative measurement that assesses the possibility of performing the calculations and reproducing the results reported in the study. vi) Comparability, i.e. a measure by which the boundary of the system under examination, data categories, assumptions, sampling methodology and quality assurance are clearly documented and so permit comparisons to be made on the results and conclusions obtained for different components of an analysis. vii) Accesibility or availability, i.e. whether information concerning the study is available to either internal or external reviewers for examination of the methodology and data values. viii) Reliability, i.e. the degree to which the sources, data collection methods and validation/verification procedures used to obtain the data are dependable. Finally the identification of outliers (i.e. extreme values in the data set which deviate markedly from the main body of data and strongly influence the outcome of descriptive measures such as the mean) through statistical tests (e.g. Dixon test, Grubb's test for outliers), or other data anomalies or missing values, is of major importance for the quality of LCI. The application of data warehousing and data mining techniques as described in Chapter 13 has proved an invaluable tool towards smoothing data sets, particularly when these originate from a continuous process control.

Cut-off criterion (a)

The cut-off criterion (a), according to ISO 14044, is the specification of the amount of material or energy flow or the level of environmental significance associated with unit processes or product systems to be excluded from LCI. According to Yu (2009), the cut-off criterion is a key point in the system boundary determination. In principle, an LCA should track all the processes in the life cycle of the product system, but in practice, due to the lack of readily accessible data, this may not be feasible. The cut-off criteria used in the LCA should be described clearly (Singh et al., 2010) and define an appropriate balance between result representativeness and data collection effort by users (Chomkhamsri and Pelletier, 2011). It is important to clarify and describe rules for omitting inventory data which are negligible from the point of view of being relevant in the study (IEC, 2008). Using cut-off rules should not give the perception of "hiding" information but rather to facilitate the data collection for practitioners. Cut-off criteria are quantified in relation to the percentage of environmental impacts estimated to be excluded because of the cut-off (e.g. it is usually suggested that the percentage of the overall environmental impact or that of a selected impact category related to cutting off should equal no more than 5 per cent of the total environmental impact) (EC, JRC, IES, 2010). Nevertheless, as stated by IEC (2008) a default value for a cut-off rule of 1 per cent regarding energy, mass and environmental relevance should apply to upstream and core processes. This means that for the overall LCI result of the product 99 per cent of the elementary flows by energy content, mass and environmental relevance are included.

Flow diagrams and calculations

Before data collection, all sub-systems included in the production system under study need to be defined at the level of detail intended to be used in the study. Flow diagrams (input–output models) at the maximum possible resolution (level of details) are quite often used to fulfil this task and it is recommended that the inventory data be given in relation to those flow diagrams (Lindfors et al., 1999).

The flow chart in Figure 16.7 summarises the fundamental data which characterise most bioenergy production technologies for the calculation of the GHG emissions throughout the entire production chain. As such, it represents major inter-linked processes which comprise a biomass technology where each major process is identified and related data are specified for the inputs and outputs associated with it. Consequently, in terms of biomass technologies, the main materials are the initial input resources (seeds, fertiliser, land etc.) and the main products are forms of delivered energy (solid, liquid and gaseous fuels, heat, electricity etc. which are purchased by the consumers for operating equipment, appliances, and other devices). Intermediate products exchanged between processes are included up to the level of detail at which clarity is not jeopardised. However, it is essential for the flow chart to indicate any co-products, by-products or wastes which occur at any stage of the production chain.

In addition to the flow chart in Figure 16.7, a supplementary, more detailed, biomethane production flow chart can be used, as shown in Figure 16.8. The production of feedstock is represented in detail along with direct and indirect energy flows and related emissions.

Figure 16.9 adds more detail into the overall biogas/biomethane production chain since it depicts an overall view of the energies (and related GHG emissions) that are involved in AD *per se* (named as "Anaerobic Digestion" in Figure 16.8).

CALCULATION METHOD FOR DIRECT AND INDIRECT ENERGY FOR BIOGAS/BIOMETHANE PRODUCTION FROM AGRICULTURAL BIOMASS

When performing a sustainability analysis of a technology, great care should be taken in the evaluation of both direct and indirect energy, under the defined system boundaries, by converting all the material flows into energy units. In brief, the analysis of energy balance for biogas/biomethane production (as depicted in Figures 16.8 and 16.9) may be calculated based on the following equation:

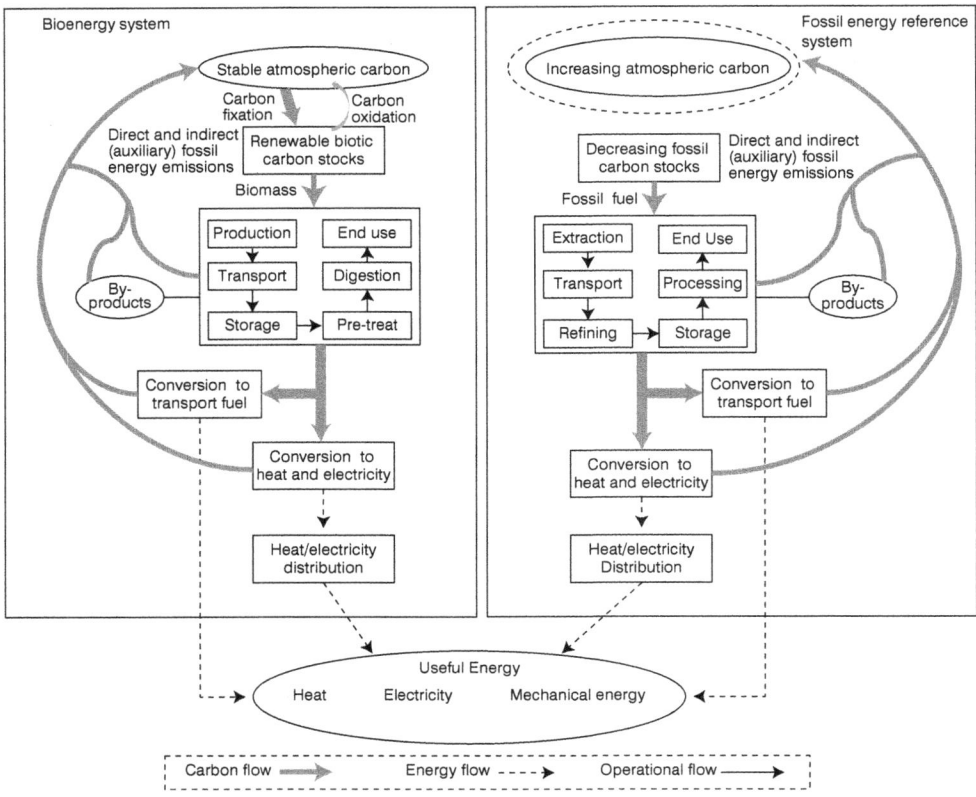

Figure 16.7 Flow chart for the calculation of GHG emissions for bioenergy and fossil fuels (reference system for comparative LCA) production chains (based on IEA Task 38). Note: Auxiliary fossil energy inputs and other GHG emissions have been excluded from the figure for reasons of simplicity (see Figures 16.8 and 16.9).

$$E_{\text{Net}} = E_{\text{CH}_4} - \left[E_{\text{Feedstock}} + E_{\text{Trans}} + E_{\text{AD}} \right] \quad \text{(Eq. 16.1)}$$

where E_{Net} = net energy; E_{CH_4} = energy produced (gross energy); $E_{\text{Feedstock}}$ = energy consumed for feedstock production; E_{Trans} = fuel energy consumed for the transportation of inputs and outputs throughout the entire biogas/biomethane production chain; EAD = energy consumed in AD plant.

The energy consumed for the production of the feedstock as shown in Figure 16.8 can be calculated according to the following equation:

$$E_{\text{Feedstock}} = E_{\text{direct}} + E_{\text{indirect}} \quad \text{(Eq. 16.2)}$$

where E_{direct} = energy consumed in field operations and E_{indirect} = energy consumed for the production of biomass production inputs. More particularly, as already mentioned, direct energy is the fuel energy consumed during field operations (e.g. ground preparation, establishment and maintenance etc.) According to Korres et al. (2010) the direct fuel energy consumed during field operations can be estimated based on the following equation:

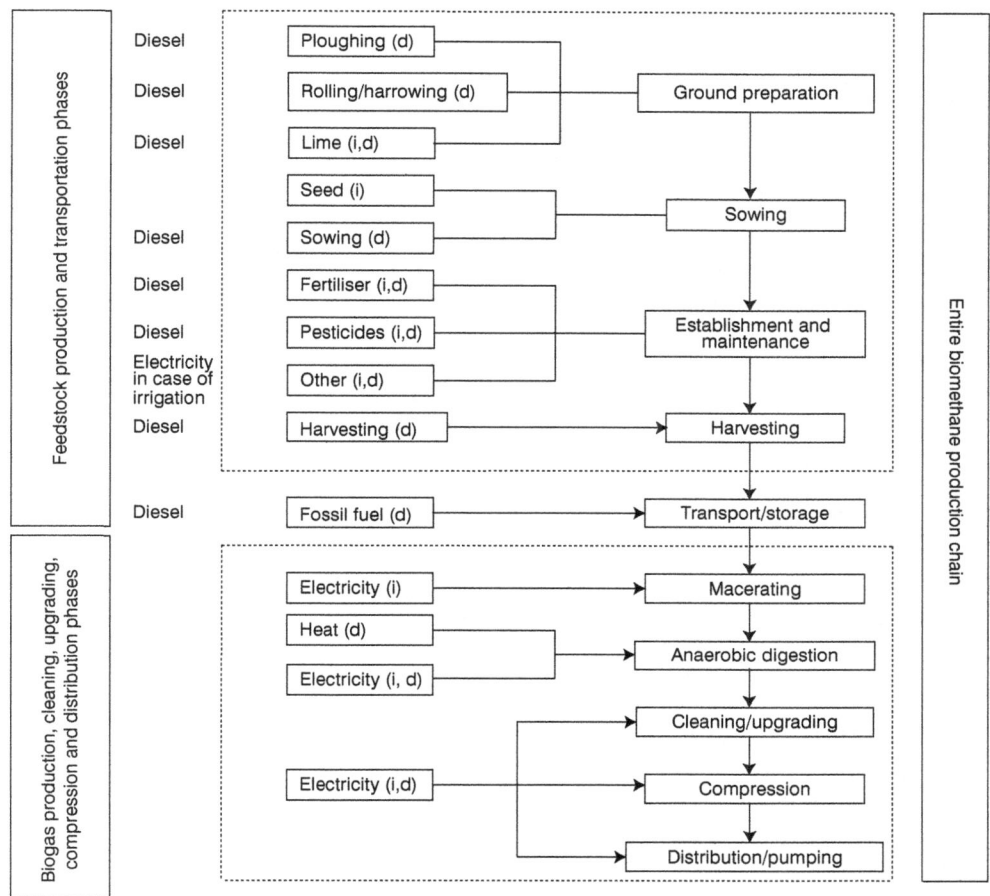

Figure 16.8 A supplementary flow chart to that in Figure 16.7 in which unit processes of the agricultural biomass production and AD is represented in detail (based on Smyth et al., 2009 and Korres et al., 2010).

d=direct; i=indirect energy and related greenhouse gas emissions; Note that digestate is not included in the system. Field operations, i.e. ploughing, harrowing, rolling, lime application, sowing, fertiliser and pesticides application consume diesel for their completion (direct energy). Heat for digester heat up and electricity are used for other operations (e.g. irrigation purposes), feedstock pre-treatment (e.g. maceration) and biogas upgrading, compression etc.

$$FE = \sum_{1}^{i} (F_{ci} \times f_c) / O_{ci} \qquad \text{(Eq. 16.3)}$$

where FE = fuel energy consumed (MJ/ha), F_{ci} = fuel consumption (L/h) for i field operation, f_c = heating value of the fuel and O_{ci} = work capacity for i operation (ha/hr). Indirect energy is the energy consumed in the manufacture of fertilisers and other crop inputs.

Maceration of the silage is carried out before insertion of feedstock into the digester since it reduces particle size to prevent physical obstruction of pipes and pumps by the fibres, and it also increases the surface area available for microbial attack and thus speeds up the

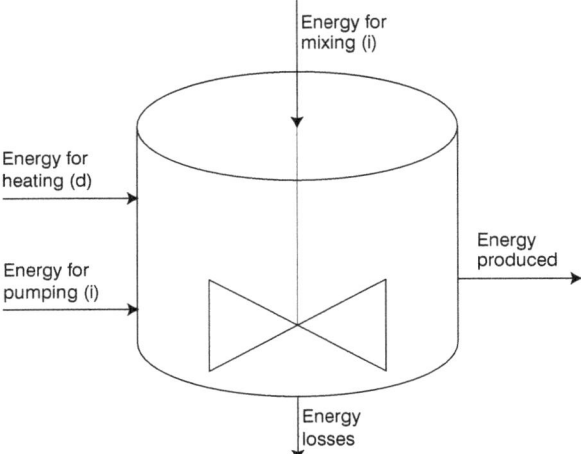

Figure 16.9 Energy flows (and related greenhouse gas emissions) in AD processing *per se*. i = indirect energy consumption and emissions; d = direct energy consumption and emissions.

digestion process (Korres et al., 2010). Mixing may allow better digestion of grass silage by keeping the material homogenous and hindering the settling of silage particles (Korres et al., 2010). It is worth noting that the energy required to heat the digester does not consider the contribution from metabolic heat generation, also called the self-heating effect (the amount of heat provided by metabolic generation is uncertain and it is usually neglected) (Smyth et al., 2009). The energy consumed in the AD plant for the production of biogas (Figure 16.9) can be calculated based on the following equation:

$$E_{AD} = E_{Direct} + E_{Indirect} = E_H + E_{Mc} + E_{Mx} + E_P + E_L \qquad \text{(Eq. 16.4)}$$

where E_H = energy (direct) required to heat the digester up to an operational temperature; E_{Mc} = energy (indirect) for macerating the feedstock (an example of a pre-treatment measure); E_{Mx} = electrical energy (indirect) necessary to mix the feedstock during the digestion process; E_P = electrical energy (indirect) necessary for water pumping or for recirculation of liquid from one digester to the other as in the case of a continuously stirred tank reactor; E_L = energy loss (thermal).

As mentioned in Chapter 12, the biogas produced during anaerobic digestion is composed of CH_4 (55–75 per cent), CO_2 (25–45 per cent) and trace elements. It must therefore be upgraded or scrubbed (Chapter 10) to natural gas standard (methane content of 97 per cent) before being used as transport fuel or injected in the natural gas grid for other use. Moreover, its calorific value is raised and potentially damaging components, e.g. hydrogen sulphide, are removed.

In case of biomethane use as a transport fuel, filling stations have to compress the gas from mains pressure to about 250 bar in order to fully charge vehicle storage tanks where natural gas is stored at about 200 bar. Two types of filling operations are used to achieve this: slow-fill and fast-fill. Slow-fill stations have the simpler design, with the dispensing lines connected directly to the compressor, but have longer filling times, typically from 20 min to a number of hours. A fast-fill operation is more complex, but gives typical filling times of only 3–5 min, and is generally used on a traditional service station forecourt (Smyth et al., 2009). It is understood that energy is consumed during all these operations where biogas is upgraded, compressed and distributed or transported to gas grid or filling stations.

Detailed analysis and calculations of the net energy (and related emissions) in bioreactors which involve the thermal and the electrical energy necessary to run the bioreactor, are reported in Anonymous (2007); Smyth et al. (2009); Ruggeri et al. (2010) and Korres et al. (2010).

The following equation should be considered as well for the evaluation of the total indirect energy:

$$E_{Ind} = E_{Chem} + E_{Mat} + E_{Constr} + E_{Main} + E_{Decom} + E_{Amort} \qquad (Eq.\ 16.5)$$

where E_{Chem} = energy required for chemicals production; E_{Mat} = energy required for construction of materials; E_{Constr} = energy required for plant building; E_{Main} = energy required for maintenance; E_{Lab} = energy required for labour; E_{Decom} = energy required for decommissioning; E_{Amort} = energy required for amortisation.

In many LCA studies, particularly for biofuels, the indirect energy consumed in the manufacture of machinery and equipment, also known as embodied energy (i.e. a cradle-to-gate analysis of the life cycle energy of a product, inclusive of the latent energy in the materials, the energy used during material acquisition and the energy used in manufacturing intermediate and final products) is not considered. This is in line with the "Proposal for a Directive on the promotion of energy from renewable sources" from the Commission of the European Communities (EC, 2009), which states that the manufacture of machinery and equipment is not taken into account when calculating the greenhouse gas impact of biofuels.

The energy consumption in labour during the entire biogas/biomethane production chain deserves particular attention. This can be separated into three components, namely i) the energy of food for the biological support expressed as caloric value; ii) the direct and indirect energy consumption for the entire food LCA (i.e. production, transportation, conservation and preparation); and iii) all the other direct and indirect energy forms linked to daily activities such as energy consumed in clothing, appliances, fuel for transportation etc. It is understood that the evaluation of the energy consumed on labour is inherently difficult (Brown and Herendeen, 1996; Cleveland and Costanza, 2010) and it is often disregarded, although it could be of major significance when comparing different labour vs. capital intensive technologies (e.g. gasification vs. energy crop cultivation). A simplified energy balance of biogas production and usage in a Cambodian family farm is shown in the Appendix (pp. 427–431).

CALCULATION METHOD FOR GHG EMISSIONS FROM BIOGAS/BIOMETHANE SYSTEM

The EU Renewable Energy Directive, Annex V Part C, provides a methodology (Equation 16.6) to calculate the greenhouse gas emissions from a biofuel system (EC, 2009), although, as the author of this chapter believes, its application is suitable for any bioenergy use and type. The functional unit is specified in gCO_{2equiv}/MJ_{fuel}.

$$E_B = E_{ec} + E_L + E_P + E_{TD} + E_U - E_{SCA} - E_{CCS} - E_{CCR} - E_{EE} \qquad (Eq.\ 16.6)$$

where E_B = total emissions from biofuel; E_{EC} = emissions from cultivation/acquisition of raw materials; E_L = emissions from land use change; E_P = emissions from processing;

E_{TD} = emissions from transport and distribution; E_U = emissions from the biofuel in use; E_{SCA} = emission savings via improved agricultural management; E_{CCS} = emission savings from carbon capture and geological storage; E_{CCR} = emission savings from carbon capture and replacement; and E_{EE} = emission savings from excess electricity from cogeneration burning of agricultural crop residues.

Exceptions exist as in the case of the emissions from biofuels in use. A debate has been going on for some time as to whether these emissions should be taken into account in the overall biofuel environmental performance or not. Furthermore, as stated by Ahlgren et al. (2009) the Directive contains very little information on the methodology for calculating the emissions of greenhouse gases from cultivation of crops for biofuel production. More specifically, in Article 19, Annex V, Chapter C, Item 6, it is stated that:

> Emissions from the extraction or cultivation of raw materials, E_{EC}, shall include emissions from the extraction or cultivation process itself; from the collection of raw materials; from waste and leakages; and from the production of chemicals or products used in extraction or cultivation. Capture of CO_2 in the cultivation of raw materials shall be excluded. Certified reductions of greenhouse gas emissions from flaring at oil production sites anywhere in the world shall be deducted. Estimates of emissions from cultivation may be derived from the use of averages calculated for smaller geographical areas than those used in the calculation of the default values, as an alternative to using actual values.

However, the Directive is clear on the choice of allocation method; co-products shall be allocated a share of the greenhouse gas emissions proportional to the lower heating value of the products. Agricultural crop residues are not assumed to have any value and are not burdened with any of the emissions from the cultivation (Article 19, Annex V, Chapter C, Items 17 and 18): "Where a fuel production process produces, in combination, the fuel for which emissions are being calculated and one or more other products ('co-products'), greenhouse gas emissions shall be divided between the fuel or its intermediate product and the co-products in proportion to their energy content (determined by lower heating value in the case of co-products other than electricity)."

In support of the above, Thamsiriroj and Murphy (2010) stated that the highlight of the methodology is that it allows for both the allocation by energy content and the substitution approach to be included. Parameters including E_{EC}, E_P, and E_{TD} are allocated to the main product, co-products, and by-products based on their lower heating values. The substitution approach is allowed through the parameter E_{CCR} when applied to the use of co-products, by-products, and residues from the biofuel system.

The equation above allows for the global warming potential (GWP) of CO_2, CH_4, and N_2O. CO_2 with GWP of one for 1 kg CO_2 is used as a reference gas for the measurement of the emissions. The GWP of 1 kg of N_2O and CH_4 is 296 and 23 respectively. The volume of GHG emissions in terms of CO_2e can be calculated using equation (16.7).

$$\text{GHG}(\text{kg of } CO_2e) = CO_2(\text{kg}) + 23 \times CH_4(\text{kg}) + 296 \times N_2O(\text{kg}) \qquad (\text{Eq. 16.7})$$

Assessment of biogas/biomethane sustainability can be achieved by the calculation of the converted GHG savings compared with fossil fuels based on equation (8).

$$\text{GHG Savings} = (E_F - E_B)/E_F \qquad (\text{Eq. 16.8})$$

where E_B = total emissions from the biofuel and E_F = total emissions from the fossil fuel comparator.

Calculations of nitrous oxide (including these from fertilisation due to its importance in GHG emissions) (Broch and Hoekman, 2012) and methane due to their high GWP as suggested by IPCC (1996) are shown below:

$$N_2O \text{ Direct} = (\text{fertiliser-N} + \text{manure-N} + \text{fixed-N} + \text{crop residue-N}) \qquad \text{(Eq. 16.9)}$$
$$\times EF_1 + \text{histosol area} \times EF_2$$

where fixed-N = fraction of nitrogen in N-fixing crops; crop residue-N = fraction of nitrogen in crop residue. The default emission factor EF_1 is 0.0125 kg N_2O-N/kg N input and the default emission factor EF_2 is 5 kg N_2O-N/ha yr for histosol soils in temperate regions, and 10 kg N_2O-N/ha yr for histosol soils in tropical regions.

According to Bouwman (1996) direct N_2O emissions from agricultural soils are calculated as follows:

$$ER_{N_2O} = 1 + 0.0125 \times N_{applied} \qquad \text{(Eq. 16.10)}$$

where ER_{N_2O} is the emission rate (kg N_2O-N ha^{-1} y^{-1}), the value of 1 kg N ha^{-1} is the background emission rate and $N_{applied}$ is the fertiliser application rate (kg N ha^{-1} y^{-1}). According to the same author and Hamelin et al. (2010), Equation (16.10) may not be adequate to estimate emissions for specified local conditions or specific crops. However, the IPCC methodology (IPCC, 2006) for estimating of direct N_2O emission from synthetic fertiliser applied to agricultural soils is based on Bouwman's (1996) work. It assumes the emission to be a fixed percentage, 1.25 ± 1%, of the N applied (Smith et al., 1997).

Elevated concentrations of dissolved N_2O in groundwater, drainage ditches, rivers and estuaries contaminated by agricultural and sewage nitrogen have been observed, suggesting that indirect N_2O emissions associated with agriculture may be just as important as direct emissions (Minami and Ohsawa, 1990; Seitzinger and Kroeze, 1998). Consequently both short- and long-term N_2O inventories considering the fate of indirect agricultural nitrogen have been developed (Nevison, 1999). Indirect N_2O emissions can be calculated based on equation (16.11).

$$N_2O \text{ Indirect} = N_2O(G) + N_2O(L) + N_2O(S) \qquad \text{(Eq. 16.11)}$$

where $N_2O(G)$ = volatilisation and subsequent atmospheric deposition of NH_3 and NO_x; $N_2O(L)$ = nitrogen leaching and runoff; $N_2O(S)$ = human consumption of crops followed by municipal sewage treatment. $N_2O(L)$ accounts for over 75 per cent of estimated indirect emissions. $N_2O(G)$ and $N_2O(L)$ are closely linked to and dependent on assumptions about the fate of fertiliser and manure, while $N_2O(S)$ is calculated from independent activity data (Nevison, 1999).

In case of a livestock production system (dairy or beef) inclusion into the biogas production system (expansion of system boundaries) nutrient losses/emissions (direct and indirect N_2O and NH_3 emissions) from animal waste management systems (i.e. during waste storage and waste handling before and after spread to the soil) but also emissions from urine and dung deposited on the pasture by livestock (Edwards-Jones et al., 2009) should also be considered in LCI (Figure 16.10).

The example depicted in Figure 16.11 represents a situation where slurry is deposited outside (outside storage with natural crust cover) by outdoor fed dry cattle and daily

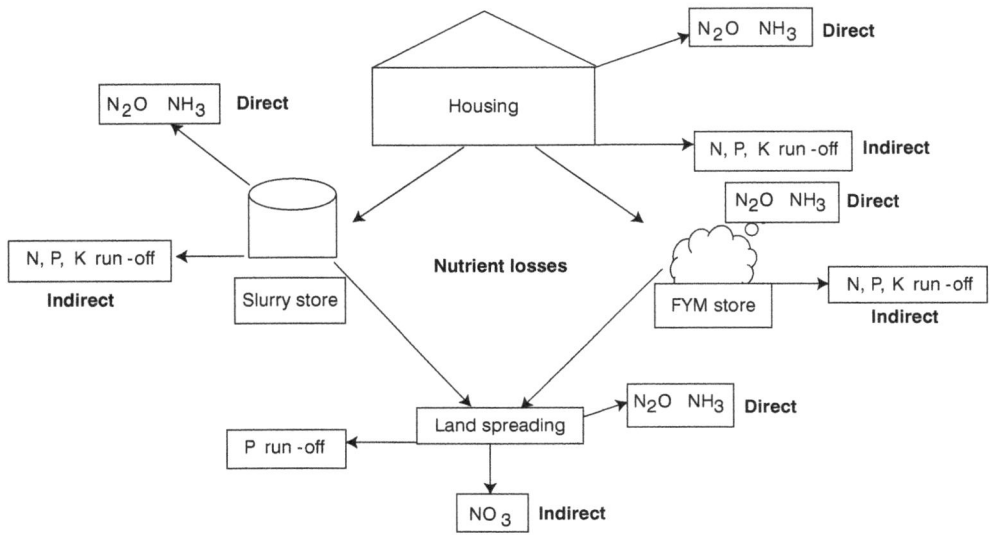

Figure 16.10 Possible nutrient losses from manure management

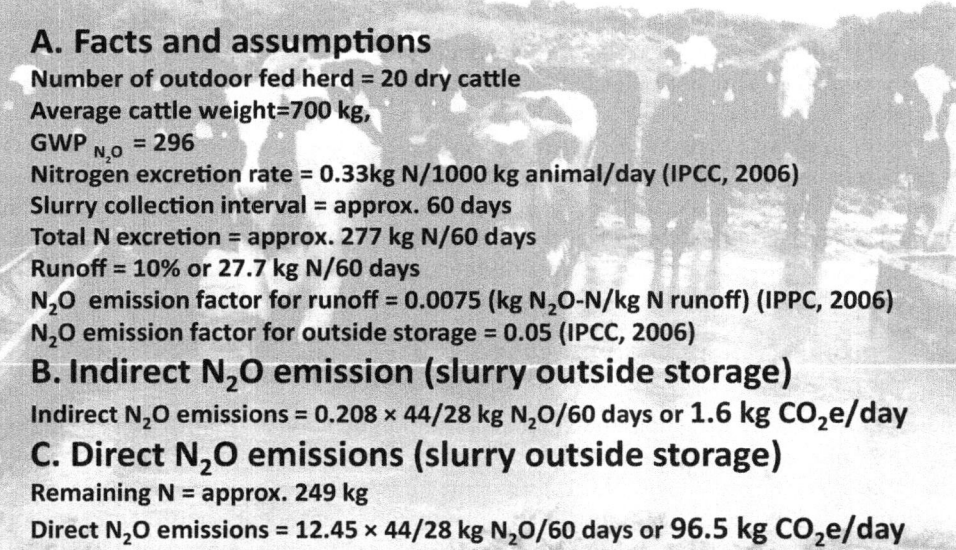

Figure 16.11 Estimation of direct and indirect N_2O emissions due to run-off of the slurry produced by outdoor fed dry cattle (author's unpublished data from visits to Irish beef farms).

emissions due to run-off (direct and indirect), expressed as $kgCO_2e$ (IPCC, 2006; Hamelin et al., 2010), are estimated.

The complexity of the system is obvious and caution needs to be exercised in accepting the results of analyses, particularly of multifunctional systems such as this mixed system of livestock-biogas/biomethane production.

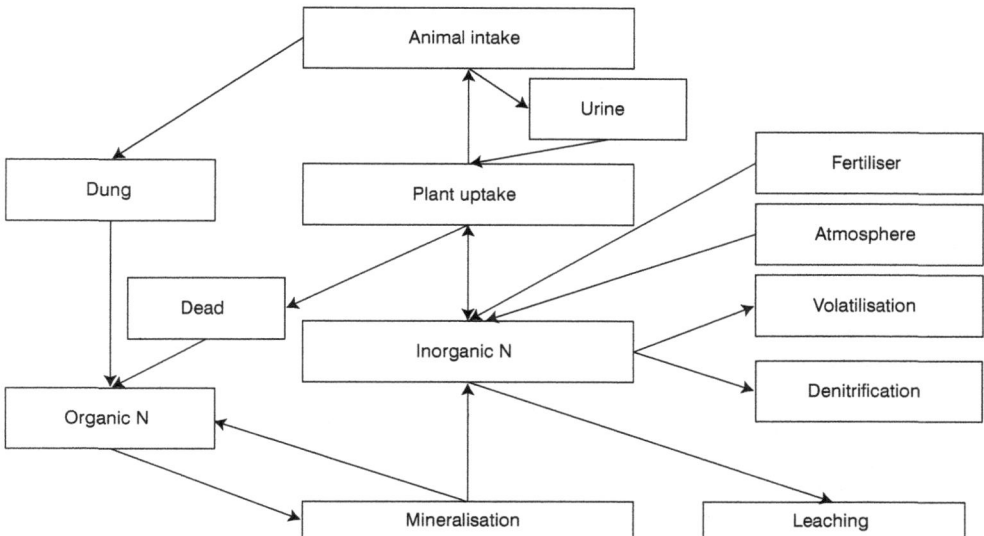

Figure 16.12 Integration of biotic, abiotic and factors influencing soil nitrogen dynamics

Modelling nitrogen dynamics in the soil is based on numerous processes for litter input (Figure 16.12), including N deposition, microbial decomposition, N mineralisation, and other processes (Korres et al., 2010) which are beyond the scope of this chapter.

Methane emissions for the estimation of biogas GWP were not considered in this chapter due to space limitation. Nevertheless, many authors have included methane emissions in their LCA of biogas (Börjesson and Berglund, 2006, 2007; Ahlgren et al., 2009; Cherubini and Jungmeier, 2010; Jury et al., 2010; Fruergaard and Astrup, 2011) as well as other environmental impact assessments besides the GWP used in this chapter.

Multifunctionality and allocation

If a process provides more than one function, i.e. delivering several goods (and/or services) often also called simplified "co-products", it is known as "multifunctional" (EC, JRC, IES, 2010). An example is the production of compost/biosolids and energy (electricity and heat) through the anaerobic digestion of sewage sludge and possible other feedstock since various inputs and services such as feedstock(s) (e.g. sewage sludge, organic fraction of municipal solid waste – OFMSW, slurry), transportation etc. along with fossil fuels, electrical energy and heat are blended together to deliver compost or biosolids (product A) and bioenergy (electricity and/or heat) (product B) and digestate (an AD residue which is used as a substitute for manufactured fertilisers) (Figure 16.13). If carefully applied, co-digestion of sewage sludge with OFMSW can deliver beneficial synergies for the water industry and authorities responsible for food waste management (Iacovidou et al., 2012). Therefore, conducting an LCA properly is vital because the outcome of the study can be proved significant in decision making about waste management.

In most LCI/LCA studies of simple goods and services, one is interested in the specific life cycle inventory of only one of the co-functions (e.g. only of the electricity or the heat of the above example). To achieve this, only the appropriate inputs and outputs of the

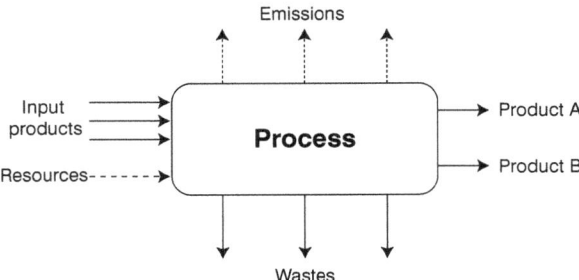

Figure 16.13 Multifunctional process with several input products and resources consumed accompanied by various wastes and emissions along with two co-products as output.

process (i.e. consumed materials, energy carriers, resource flows, emissions, wastes, etc.) are to be counted for the analysed function, i.e. the inventory of the specific function is to be isolated. Different approaches are used for solving multifunctionality. The choice of the most appropriate approach depends among others on the goal and scope of the study, data availability and the features of the multifunctional process or product. Under these conditions, it might be necessary, to assign to each of the functions of a multiple-function system only those environmental burdens associated with that particular function, a process known as allocation (Azapagic and Clift, 1999). Nevertheless, it is recommended that allocation, when possible, should be avoided (ISO, 2006a, b) either by subdivision of multifunctional processes or by system expansion and its variant substitution (Jensen et al., 1997; EC, JRC, IES, 2010). Subdivision refers to the collection of data individually for those of the monofunctional processes that relate to the analysed system and that are contained in the multifunctional process. In addition, as Bhatia et al. (2011) mentioned, process subdivision, to avoid allocation, can be used when it is possible to divide the common process into two or more distinct processes but only to the point at which the studied product and its function are isolated, not to the point that every co-product has a unique and distinct process. Under attributional modelling subdivision is the most appropriate approach but is also applicable under consequential modelling (EC, JRC, IES, 2010).

Another way to avoid allocation, as Weidema (2003) and Vikman et al. (2004) stated, is by expanding system boundaries to include both main- and by-products. System expansion and substitution are the corresponding methodological approaches under consequential modelling for solving multifunctionality. Substitution is also applicable for attributional modelling that is interested in including existing interactions with other systems. Substitution means to subtract the inventory of another system from the analysed system (EC and JR, 2010).

In general, Bhatia et al. (2011) suggested a decision process for selecting the best method for avoiding or performing allocation for a given common process in various situations. To complete this process they also mentioned that if the output is a waste then no allocation is needed and all emissions are allocated to the studied product whereas the waste treatment is also included as an attributable process (waste without value is not subsequently used). In the situation where waste is subsequently used, that output would have some economic value and is no longer classified as "waste". According to the Clean Development Mechanism (CDM, 2007, cited in Singh et al., 2010) co-products are defined as products with similar revenues to the main product, by-products are products with lower revenues than that of the main product, whereas wastes have little or no revenue. According to SAIC (2006)

Table 16.4 Decision-making process for selecting an allocation method.

	Response	Action
Allocation can be avoided		
Is the process output a waste (of no economic value)?	Yes	No allocation
Can the common process be divided and evaluated as separate processes?	Yes	Process subdivision
Can the studied product and co-product(s) be combined into one single functional unit?	Yes	Redefine functional unit*
Is it possible to model co-product's emissions using a similar process or product? Does direct knowledge about the eventual use of the co-product(s) exist?	Yes	System expansion
Allocation cannot be avoided		
Is there an underlying physical relationship between the studied product, co-product(s) and their emissions contributions?	Yes	Physical allocation
Neither a physical relationship can be established nor is applicable		
Are the market values of the product and co-product(s) free of significant market effects on their valuation (e.g. brand value, constrained supply, etc.)?	Yes	Economic allocation
Are there other relationships between the studied product and co-product(s) that can be established?	Yes	Use other relationship

Based on Bhatia et al., 2011.
*Another method to avoid allocation is to redefine the unit of analysis to include the functions of both the studied product and the co-product.

co-products are all output streams other than the primary product that are not waste and are not used as raw materials elsewhere in the system examined in the inventory. In brief, Table 16.4 describes a decision-making process of selecting the most appropriate allocation method.

According to ISO (2006b) allocation should be performed in accordance with the underlying causal physical relationship (implicitly also covering chemical or biological relationships) between the different products or functions. This should reflect the way in which the individual inputs and outputs are quantitatively changed by quantitative changes in the multiple functions delivered by the process or system. More particularly, mass-based allocation (a physical property that is easy to interpret) is a usual approach although some researchers argue that it cannot be an accurate measure of energy functions (Malca and Freire, 2006; Shapouri et al., 2002). When a clear common physical causal relationship between the co-functions cannot be established, ISO (2006) suggests performing the allocation according to another relationship between them. This may be an economic relationship or a relationship between some other (e.g. non-causal physical) properties of the co-functions such as energy content. The latter is often used in the allocation between different fuels co-produced in a CHP plant or in a refinery. However, energy-based allocation methodology has been questioned when co-products are not meant for energy purposes (Gnansounou et al., 2009). The economic allocation is based on the economic values since allocations of the inputs and emissions to the product and co-product(s) are based on the market value of each when they exit the common process (Bhatia et al., 2011).

However, price variation, subsidies, and market interferences could introduce difficulties in its implementation (Wang, 2005; Luo et al., 2009).

Life cycle impact assessment

In the life cycle impact assessment (LCIA), which comprise the material and energy flow analysis of the studied system within defined system boundaries, the results (the quantified inputs and outputs) of the inventory analysis are interpreted in terms of the impacts they have on the environment (Lindfors et al., 1999). The (environmental) impact assessment is a technical, quantitative and/or qualitative procedure which classifies, characterises and partitions the emissions of a product, as identified in the LCI analysis stage, into different impact categories. Usually, the impacts of the systems under study focus on the areas of ecological health, human health and resource depletion (Figure 16.14).

Economic considerations or social effects are not usually addressed during this phase although social and socio-economic assessments could add extra dimensions of impact analysis and valuable information for those who seek to produce or purchase responsibly

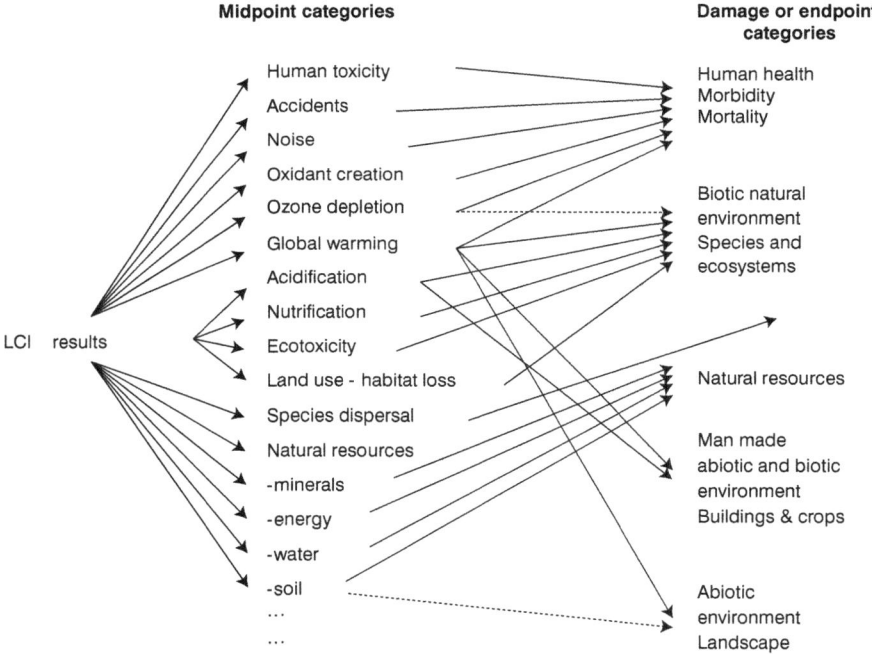

Figure 16.14 Representation of midpoint* and damage or endpoint** impact categories (dotted lines represent results of LCI analysis that are still unclear) (based on EC, JRC, IES, 2010).

"*Class representing environmental issues of concern to which LCI results may be assigned, consequently midpoint indicator is a quantifiable representation of a midpoint impact category; **Class representing damage on an ultimate areas protection (i.e. operational group of items of direct value to human society) to which state/midpoint categories may be assigned, consequently damage indicator is a quantifiable representation of a damage category" (Udo de Haes and van Rooijen, 2005). A midpoint indicator might communicate the amount of emissions being released into an environment while an endpoint indicator would communicate the extent of biodiversity loss or human health impact as a result of those emissions.

(Benoit and Mazijn, 2009). The impact assessment is used for the comparison of different systems and products based on selected indicators or for the localisation of environmental issues. ISO 14042 (ISO, 1998a), setting the rules for LCA impact assessment, clearly stated that three steps are mandatory for the determination of impact categories: first, selection and definition of impact categories (i.e. identifying relevant environmental impact categories such as those of global warming, acidification, terrestrial toxicity); second, classification (i.e. assigning LCI results to the impact categories, e.g., classifying carbon dioxide emissions to global warming); third, characterisation (i.e. modelling LCI impacts within impact categories using science-based conversion factors, e.g. modelling the potential impact of carbon dioxide and methane on global warming). The remaining steps of the impact assessment procedure concern the normalisation of impacts (i.e. in this step potential impacts expressed in such a manner that can be compared, e.g. global warming impact of carbon dioxide and methane for the two options), the grouping of impact indicators (i.e. arrangement or ranking the indicators by location, e.g. local, regional and global as in Table 16.5), weighting (i.e. emphasising the most important potential impacts) and finally evaluating and reporting LCI analysis results.

Interpretation of LCA

Life cycle assessment interpretation is a systematic process to evaluate the results of the inventory analysis and impact assessment, to select the preferred product, process or service with a clear understanding of the uncertainty and the assumptions used to generate the results (SAIC, 2006). Life cycle assessment interpretation is performed in interaction with the three other phases of the LCA. If the results of the inventory analysis or the impact assessment are found not to fulfil the requirements defined in the goal and scoping phase, the inventory analysis must be improved, e.g. by revising the system boundaries, further data collection etc. followed by an improved impact assessment (Jensen et al., 1997). The main issues (ISO, 1997) relating to the interpretation are i) the identification of significant environmental issues; ii) evaluation of the methodology and results based on quality criteria described earlier in this chapter (e.g. completeness, consistency etc.); and iii) checking that the conclusions are consistent with the requirements of the goal and scope of the study. If these conditions are satisfied the conclusions of the study can be reported and recommendation for action based on these can be suggested.

Conclusions

LCA has be proved an invaluable tool for the examination of the sustainability of biogas production but caution is necessary for the proper application of working protocols and rules. More particularly, the functional unit(s) of the studied system(s) should properly describe the services provided and be clearly defined and reported. Where consistent with the goal of the study, the reference alternatives, particularly in comparative studies, should be chosen so as to be representative of the purpose and to maximise the data quality of the compared alternatives. Legislative issues and recent policies in the bioenergy production arena require a clear and concise conclusion to be drawn upon the completion of LCA based on clear evidence. The development of the life cycle inventory, one of the core phases of the technique, should follow a strict protocol so comparability, representativeness and completeness are fulfilled. To assist decision-makers, recommendations should be based on

Table 16.5 Various LCA impact categories.

Impact category	Scale	Relevant LCI data (classification)	Characterisation factor	Units	Description of characterisation factor
Global warming	Global	Carbon dioxide (CO_2); nitrous oxide (N_2O); methane (CH_4); chlorofluorocarbons (CFCs); hydrochlorofluorocarbons (HCFCs); methyl bromide (MHBr)	Global warming potential	$kgCO_2$e eq.	Converts LCI data to carbon dioxide equivalents
Stratospheric ozone depletion	Global	Chlorofluorocarbons (CFCs); halons; hydrochlorofluorocarbons (HCFCs); methyl bromide (MHBr)	Ozone depleting potential	CFC 11 eq.	Converts LCI data to trichlorofluromethane equivalents
Acidification	Regional, local	Sulphur oxide (SOx); nitrogen oxides (NOx); hydrochloric acid (HCl); hydrofluoric acid (HF); ammonia (NH_3)	Acidification potential	kg SO_4 eq. Moles H^+ eq.	Converts LCI data to hydrogen ion equivalents
Eutrophication	Local	Phosphate (PO); nitric oxide (NO); nitrogen dioxide (NO_2); nitrates (NO_3); ammonia (NH_3)	Eutrophication potential	kg PO_4^{3-} eq.	Converts LCI data to phosphate equivalents
Photochemical smog	Local	Nitrogen oxides (NOx); volatile organic compounds; airborne particles (aldehydes, nitrogen oxides, peroxyacetyl nitrate $CH_3CO-OO-NO_2$)	Photochemical oxident creation potential	Kg C_2H_4 eq.	Converts LCI data to ethane equivalents
Terrestrial toxicity	Local	Toxic chemicals with a reported lethal concentration to rodents	LC_{50}	—	Converts LC_{50} data to equivalents; uses multi-media modelling, exposure pathways
Aquatic toxicity	Local	Toxic chemicals with a reported lethal concentration to fish	LC_{50}	—	Converts LC_{50} data to equivalents; uses multi-media modelling, exposure pathways
Human health	Global Regional Local	Total releases to air, water, and soil	LC_{50}	—	Converts LC_{50} data to equivalents; uses multi-media modelling, exposure pathways

Resource depletion	Global Regional Local	Quantity of minerals used Quantity of fossil fuels used	Resource depletion potential	—	Converts LCI data to a ratio of quantity of resource used versus quantity of resource left in reserve
Land use	Global Regional Local	Quantity disposed of in a landfill or other land modifications	Land availability	ha yr^{-1}	Converts mass of solid waste into volume using an estimated density
Water use	Regional Local	Water used or consumed	Water shortage potential	—	Converts LCI data to a ratio of quantity of water used versus quantity of resource left in reserve

Based on Goedkoop and Spriemsma, 2000; Guinee et al., 2002; SAIC, 2006.
Note: LC_{50} denotes the lethal concentration 50 which is an estimated concentration of a toxicant that kills 50 per cent of the organisms being exposed to this toxicant. It is usually expressed as parts per million (ppm) (Korres, 2005)

solid evidence which in turn should be based on proper analysis. Life cycle assessment can be highly misleading if used in an undisciplined way. However, it can be an invaluable tool if practised properly. Furthermore, social and socio-economic assessments could add extra dimensions to impact analysis and valuable information for those who seek to produce or purchase responsibly. The establishment of interactions with other scientific approaches such as data warehouse and data mining techniques, statistics and/or even simulation or soft system methodology approaches are likely to improve the LCA process. Although nothing was mentioned here about sensitivity analysis or uncertainty, major issues in LCA, the detailed step-by-step instructions on LCI and LCIA along with the numerous and explicit graphical information which accompany the examples should provoke and intrigue the curiosity of the conscientious scholar to find her/his way forward.

References

Ahlgren, S., Hansson, P.A., Kimming, M., Aronsson, P., and Lundkvist, H. (2009) 'Greenhouse gas emissions from cultivation of agricultural crops for biofuels and production of biogas from manure', Implementation of the Directive of the European Parliament and of the Council on the promotion of the use of energy from renewable sources. Revised according to instructions for interpretation of the Directive from the European Commission 30 July 2009, Dnr SLU ua 12-4067/08.

Anonymous (2007) 'Energy balance optimisation for an integrated arable/livestock farm unit', Cropgen, Renewable energy from crops and agrowastes, SES6-CT-2004-502824, University of Vienna (BOKU-IFA).

Arnold, K. (2011) 'Greenhouse gas balance of biomethane – which substrates are suitable?', *Energy Science and Technology*, vol. 1, no 2, pp. 67–75.

Azapagic, A. and Clift, R. (1999) 'Allocation of environmental burdens in co-product systems: product-related burdens', *International Journal Life Cycle Assessment*, vol. 4, no 6, pp. 357–369.

Beer, T., Grant, T., Morgan, G., Lapszewicz, J., Anyon, P., Edwards, J., Nelson, P., Watson, H. and Williams, D. (2002) *Comparison of transport fuels – life cycle emissions analysis of alternative fuels for heavy vehicles*, Aspendale: Commonwealth Scientific and Industrial Research Organisation.

Benoit, C. and Mazijn, B. (2009) 'Guidelines for social life cycle assessment of products', United Nations Environment Programme.

Bergsma, G., Vroonhof, J. and Dornburg, V. (2006) 'The greenhouse gas calculation methodology for biomass-based electricity, heat and fuels', Project Group 'Sustainable Production of Biomass' Report from Working Group CO_2 Methodology.

Bessou, C., Basset-Mens, C., Tran, T. and Benoist, A. (2012) 'LCA applied to perennial cropping systems: a review focused on the farm stage', *International Journal Life Cycle Assessment*, vol. 18, no 2, pp. 340–361.

Bhatia, P., Cummis, C., Brown, A., Draucker, L., Rich, D. and Lahd, H. (2011) 'Product life cycle accounting and reporting standard', Greenhouse Gas Protocol, World Resources Institute and World Business Council for Sustainable Development.

Bird, N., Cowie, A., Cherubini, F., and Jungmeier, G. (2011) 'Using a life cycle assessment approach to estimate the net greenhouse gas emissions of bioenergy', Bioenergy:ExCo:2011:03. International Energy Agency.

Bligny, J.C., Pennington, D., De Camillis, C., Palyi, B., Bauer, C., Schenker, U., King, H., Martin, N. et al. (2012) 'ENVIFOOD Protocol Environmental Assessment of Food and Drink Protocol', Draft Version 0.1, Pilot Testing, November 2012, European Food Sustainable Consumption & Production Round Table.

Börjesson, P. and Berglund, M. (2006) 'Environmental systems analysis of biogas systems – Part I: Fuel-cycle emissions', *Biomass and Bioenergy*, vol. 30, no. 5, pp. 469–485.

Börjesson, P. and Berglund, M. (2007) 'Environmental systems analysis of biogas systems – Part II: The environmental impact of replacing various reference systems', *Biomass and Bioenergy*, vol. 31, no. 5, pp. 326–344.

Börjesson, P., Tufvesson, L. and Lantz, M. (2011) 'Life cycle assessment of biofuels in Sweden', Report No. 70, Lund University, Dept. of Technology and Society, Environmental and Energy Systems Studies.

Bouwman, A.F. (1996) 'Direct emissions of nitrous oxide from agricultural soils', *Nutrient Cycling in Agroecosystems*, vol. 46, pp. 53–70.

Broch, A. and Hoekman, S.K. (2012) 'Transportation fuel life cycle analysis. A review of indirect land use change and agricultural N_2O emissions', CRC Project No. E-88-2, Final Report, Desert Research Institute, Life Cycle Associates.

Broch, A., Hoekman, S.K. and Unnasch, S. (2012) 'A review of indirect land use change and agricultural N_2O emissions. Transportation fuel life cycle analysis', CRC Project No. E-88-2, Final Report.

Brown, M.T. and Herendeen, R.A. (1996) 'Embodied energy analysis and EMERGY analysis: a comparative view', *Ecological Economics*, vol.19, no 3, pp. 219–235.

Cherubini, F. and Jungmeier, G. (2010) 'LCA of a biorefinery concept producing bioethanol, bioenergy, and chemicals from switchgrass', *International Journal Life Cycle Assessment*, vol. 15, pp. 53–66.

Cherubini, F. and Stromman, A.H. (2011) 'Life cycle assessment of bioenergy systems: State of the art and future challenges', *Bioresource Technology*, vol. 102, pp. 437–451.

Cherubini, F., Bird, N.D., Cowie, A., Jungmeier, G., Schlamadinger, B. and Woess-Gallasch, S. (2009) 'Energy-and greenhouse gas-based LCA of biofuel and bioenergy systems: Key issues, ranges and recommendations', *Resources, Conservation and Recycling*, vol. 53, pp. 434–447.

Chomkhamsri, K. and Pelletier, N. (2011) 'Analysis of existing environmental footprint methodologies for products and organizations: recommendations, rationale, and alignment', Deliverable 1 to the Administrative Arrangement between DG Environment and Joint Research Centre No. N 070307/2009/552517, including Amendment No 1 from December 2010, European Commission (EC), Joint Research Centre (JRC), Institute for Environment and Sustainability (IES).

Cleveland, C.J. and Costanza, R. (2010) 'Net energy analysis', in *Encyclopedia of Earth*, National Council for Science and Environment, http://www.eoearth.org/article/Net_energy_analysis [Accessed July 2012].

Consoli, F., Allen, D., Boustead, I., Fava, J., Franklin, W., Jensen, A.A., de Oude, N., Parrish, R., Perriman, R., Postlethwaite, D., Quay, B., Séguin, J. and Vigon, B. (eds). (1993) 'Guidelines for life-cycle assessment: a code of practice', Society of Environmental Toxicology and Chemistry (SETAC); SETAC Workshop, Sesimbra, Portugal, 31 March – 3 April 1993.

Deasy, E. and Power N. (2011) 'A life-cycle technical assessment of biofuels options for Ireland', in *Proceedings of the ITRN2011, Biofuel options for Ireland*, 31 August – 1 September, University College Cork, Cork, Ireland.

del Rio, P. and Burguillo, M. (2009) 'An empirical analysis of the impact of renewable energy deployment on local sustainability', *Renewable and Sustainable Energy Reviews*, vol. 13, pp. 1314–1325.

Earles, J.M. and Halog, A. (2011) 'Consequential life cycle assessment: a review', *International Journal Life Cycle Assessment*, vol. 16, pp. 445–453.

EC (2009) 'Directive 2009/28/EC of The European Parliament and of The Council of 23 April 2009 on the promotion of the use of energy from renewable sources and amending and subsequently repealing Directives 2001/77/EC and 2003/30/EC', *Official Journal of the European Union* 2009, pp. 16–62.

EC, JRC, IES (2010) *International Reference Life Cycle Data System (ILCD) Handbook – General guide for life cycle assessment – detailed guidance*, EUR 24708 EN, 1st edn March 2010, European Commission – Joint Research Centre – Institute for Environment and Sustainability, Luxembourg, Publications Office of the European Union.

Edwards-Jones, G., Plassmn, K. and Harris, I.M. (2009) 'Carbon footprint of lamb and beef production systems: insights from an empirical analysis of farms in Wales, UK', *Journal of Agricultural Science*, vol. 147, pp. 707–719.

Ekvall, T. and Andrae, A. (2006) 'Attributional and consequential environmental assessment of the shift to lead-free solders', *International Journal Life Cycle Assessment*, vol. 11, no 5, pp. 344–353.

Ekvall, T. and Finnveden, G. (2001) 'Allocation in ISO 14041 – a critical review', *Journal of Cleaner Production*, vol.9, pp. 197–208.

Ekvall, T. and Weidema, B. (2004) 'System boundaries and input data in consequential life cycle inventory analysis', *International Journal Life Cycle Assessment*, vol. 9, no.3, pp. 161–171.

Elghali, L. (2002) 'Decision support tools for environmental policy decisions and their relevance to life cycle assessment', CES Working Paper 2/02, Augus, TRL Limited.

Elsayed, M.A., Matthews, R. and Mortimer, N.D. (2003) 'Carbon and energy balances for a range of biofuels options', Final report, Prepared for the Department of Trade and Industry, Renewable Energy Programme Unit of Sheffield Hallam University and Forest Research.

Emmenegger, M.F., Stucki, M. and Hermle, S. (2012) 'LCA of energetic biomass utilization: actual projects and new developments – April 23, 2012, Berne, Switzerland', *International Journal Life Cycle Assessment*, vol. 17, pp. 1142–1147.

Fava, J. and Pomper, S. (1997) 'Life cycle critical review! Does it work? Implementing a critical review process as a key element of the aluminium beverage container LCA', *International Journal Life Cycle Assessment*, vol. 2, no.3 pp. 145–153.

Fava, J.A., Denison, R., Jones, B., Curran, M.A., Vigon, B., Selke, S and Barnum J. (eds) (1994) 'A technical framework for life cycle assessment', Society of Environmental Toxicology and Chemistry (SETAC) and SETAC Foundation for Environmental Education.

Finnveden, G., Hauschild, M.Z., Ekvall, T., Guinee, J., Reinout Heijungs, R., Hellweg, S. Koehler, A., Pennington, D. and Suh, S. (2009) 'Recent developments in life cycle assessment', *Journal of Environmental Management*, vol. 91, no 1, pp. 1–21.

Frischknecht, R. and Jungbluth, N. (eds) (2007) 'Overview and methodology. Data v2.0 (2007)', Ecoinvent report No. 1, Dubendorf, December.

Fruergaard, T. and Astrup, T. (2011) 'Optimal utilization of waste-to-energy in an LCA perspective', *Waste Management*, vol. 31, pp. 572–582.

Garofalo, R. (2011) 'Lifecycle assessment and environmental assessment', Aquafuel FP7-241301-2, Algae and aquatic biomass for a sustainable production of 2nd generation biofuels, Coordination Action FP7-ENERGY-2009-1.

Gasol, C.M., Gabarrell, X., Anton, A., Rigola, M., Carrasco, J., Ciria, P., Solano, M.L. and Rieradevall, J. (2007) 'Life cycle assessment of a *Brassica carinata* bioenergy cropping system in southern Europe', *Biomass and Bioenergy*, vol. 31, pp. 543–555.

Gerin, P.A., Vliegen, F. and Jossart, J.M. (2007) 'Energy and CO_2 balance of maize and grass as energy crops for anaerobic digestion', *Bioresource Technology*, vol. 99, pp. 2620–2627.

Gnansounou, E., Dauriat, A., Villegas, J. and Panichelli, L. (2009) 'Life cycle assessment of biofuels: energy and greenhouse gas balances', *Bioresource Technology*, vol. 100, no 21, pp. 4919–4930.

Goedkoop, M.J. and Spriemsma, R. (2000) *The eco-indicator 99. A damage oriented method for life cycle impact assessment.* Methodology report and methodology annex, 2nd edn, available at www.pre.nl

Guinee, J.B., Gorree, M., Heijungs, R., Huppes, G., Kleijn, R., de Koning, A., van Oers, L., Wegener Sleeswijk, A., Suh, S., Udo de Haes, H.A., de Bruijn, J.A., van Duin, R. and Huijbregts, M.A.J. (2002) *Handbook on life cycle assessment: operational guide to the ISO standards*, Eco-efficiency in Industry and Science, Dordrecht: Kluwer Academic Publishers.

Haas, G., Wetterich, F. and Kopke, U. (2001) 'Comparing intensive, extensified and organic grassland farming in southern Germany by process life cycle assessment', *Agriculture, Ecosystems and Environment*, vol. 83, pp. 43–53.

Hamelin, L., Wesnaes, M., Wenzel, H. and Petersen, B.M. (2010) 'Life cycle assessment of biogas from separated slurry', Environmental Project No. 1329 2010 Miljoprojekt, Environmental Protection Agency, Danish Ministry of the Environment.

Hartmann, J.K. (2006) 'Life-cycle-assessment of industrial scale biogas plants', PhD dissertation, Fakultät für Agrarwissenschaften, Georg-August-Universität Göttingen, Göttingen.

Hitchcock, G. and Lane, B. (2008) 'LCA road transport biofuels', Final Report v7, Sustainable Transport Solutions.

Horne, R.E. and Matthews, R. (2004) 'BIOMITRE technical manual', Biomass-based climate change mitigation through renewable energy (BIOMITRE) Project Number: NNE5-00069-2002, November.

Iacovidou, E., Ohandja, D.G. and Voulvounis, N. (2012) 'Food waste co-digestion with sewage sludge – realising its potential in the UK', *Journal of Environmental Management*, vol. 112, pp. 267–274.

IEC (2008) 'Supporting annexes for environmental product declarations. Annex A: Application of LCA methodology', International EPD Cooperation (IEC).

IEO (2009) 'International energy outlook', Energy Information Administration, Office of Integrated Analysis and Forecasting, US Department of Energy, Washington DC.

IPCC (1996) 'Revised 1996 IPCC guidelines for national greenhouse gas inventories: reference manual'.

IPCC (2006) 'Guidelines for national greenhouse gas inventories', Prepared by the National Greenhouse Gas Inventories Programme, Eggleston, H.S., Buendia, L., Miwa, K., Ngara, T. and Tanabe, K. (eds). IGES, Japan, http://www.ipccnggip.iges.or.jp/public/2006gl/index.html [Accessed June 2012].

IPCC (2007) 'Climate change 2007: mitigation', in Metz, B. et al. (Eds.) *Contribution of Working Group 3 to the Fourth Assessment Report of the Intergovernmental Panel on Climate Change*, Cambridge, UK and New York.

ISO (1997) 'Environmental management – life cycle assessment – principles and framework' (ISO/FDIS 14040). ISO TC 207. International Organization of Standardization.

ISO (1998a) 'Environmental Management – life cycle assessment – impact assessment – ISO 14042:2000', ISO TC 207/SC5/WG4, International Standards Organization (ISO), Geneva, Switzerland, http://www.iso.org/iso/catalogue_detail?csnumber=37456.

ISO (1998b) 'Environmental management – life cycle assessment – life cycle interpretation – ISO 14043:2000', ISO TC 207, International Standards Organization (ISO), Geneva, Switzerland, http://www.iso.org/iso/catalogue_detail?csnumber=37456.

ISO (1999), 'Environmental management – life cycle assessment – goal, scope and inventory analysis – ISO 14041:1999', International Organization for Standardization (ISO), Geneva, Switzerland, http://www.iso.org/iso/catalogue_detail?csnumber=37456.

ISO (2006a) 'ISO Norm 14044:2006 international standard', in *Environmental management – life cycle assessment – requirements and guidelines*, International Organisation for Standardisation, Geneva, Switzerland.

ISO (2006b) 'ISO Norm 14040:2006. Life cycle assessment: principles and framework. Environmental Management', International Organisation for Standardisation, Geneva.

Jensen, A.A., Hoffman, L., Møller, B.T., Schmidt, A., Christiansen, K., Elkington, J. and van Dijk, F. (1997) 'Life cycle assessment. A guide to approaches, experiences and information sources', Environmental Issues Series, no.6, European Environmental Agency.

Jungk, N.C., Reinhardt, G.A. and Gärtner, S.O. (2002) 'Agricultural reference systems in life cycle assessments', in van Ierland, E.C. and Oude Lansink, A. (eds), *Economics of sustainable energy in agriculture*, pp. 105–119. Dordrecht: Kluwer Academic Publishers.

Jungmeier, G., and Hausberger, S. (2002) 'Greenhouse gas emissions of cars with biofuels in Austria – A comparison to cars with conventional fuel', in *Biomass for Energy, Industry and Climate Protection*, 12th European Biomass Conference, ETA-Florence, Amsterdam, The Netherlands, pp. 1128–1131.

Jungmeier, G. and Spitzer, J. (2001) 'Greenhouse gas emissions of bioenergy from agriculture compared to fossil energy for heat and electricity supply', *Nutrient Cycling in Agroecosystems*, vol. 60, no 1–3, pp 267–273.

Jungmeier, G., Werner, F., Jarnehammar, A., Hohenthal, C. and Richter, K. (2002) 'Allocation in LCA of wood-based products – Experiences of cost action E9 Part I. Methodology'. *International Journal of Life Cycle Assessment*, 7, pp. 290–294

Jury, C., Benetto, E., Koster, D., Schmitt, B. and Welfring, J. (2010) 'Life cycle assessment of biogas production by monofermentation of energy crops and injection into the natural gas grid', *Biomass and Bioenergy*, vol. 34, no 1, pp. 54–66.

Kammen, D.M., Farrell, A.E., Plevin, R.J., Jones, A.D., Nemet, G.F. and Delucchi, M.A. (2008) 'Energy and greenhouse impacts of biofuels: A framework for analysis', OECD Research Roundtable-Energy & Greenhouse Gas Impacts of Biofuels, Research Report, UCB-ITS-TSRC-RR-2008-1, eScholarship, University of California, http://escholarship.org/uc/item/7zg2 x 23t [accessed, August 2012].

Kim, S. and Dale, B.E. (2005) 'Life cycle assessment of various cropping systems utilized for producing biofuels: Bioethanol and biodiesel', *Biomass and Bioenergy*, vol. 29, pp. 426–439.

Kim, S. and Dale, B.E (2009) 'Regional variations in greenhouse gas emissions of biobased products in the United States – corn-based ethanol and soybean oil', *International Journal of Life Cycle Assessment*, vol. 14, pp. 540–546.

Korres, N.E. (2004) 'Introduction to soft system methodology', Personal notes from the postgraduate course in Operational Research and Applied Statistics at Salford University, UK.

Korres, N.E. (2005) *Encyclopaedic dictionary of weed science. Theory and digest*, Andover: Intercept and Paris: Lavoisier SAS.

Korres, N.E. (2010) Food Climate Research Network (FCRN). Interview Series 2010 Cycle, November 2010, http://www.fcrn.org.uk/interviews/biogas-grasses [Accessed August 2012].

Korres, N.E., Singh, A., Nizami, A.S. and Murphy, J.D. (2010) 'Is grass biomethane a sustainable transport biofuel?', *Biofuels, Bioproducts and Biorefining*, vol. 4, no 3, pp. 310–325.

Korres, N.E., Thamsiriroj, T., Smyth, B., Nizami A.S., Singh, A. and Murphy J.D. (2011). 'Grass biomethane for agriculture and energy', in Lichtfouse, E. (Ed.), 'Genetics, biofuels and local farming systems', *Sustainable Agriculture Reviews*, vol. 7, pp. 5–50.

Labutong, N., Mosley, J., Smith, R. and Willard, J. (2012) 'Life-cycle modeling and environmental impact assessment of commercial scale biogas production', MSc Thesis, University of Michigan, School of Natural Resources and Environment.

Lesage, P., Ekvall, T., Deschenes, L. and Samson, R.J. (2007) 'Environmental assessment of brown-field rehabilitation using two different life cycle inventory models: part 2', *International Journal of Life Cycle Assessment*, vol. 12, no. 7, pp. 497–513.

Lindfors, L.G., Christiansen, K., Hoffman, L., Kruger, I., Virtanen, Y., Juntilla, V., Hanssen, O.J. and Ronning, A. (1999) 'Nordic guidelines on life-cycle assessment: Nord', 1995:20, 3rd edn, Nordic Council of Ministers, Copenhagen.

Lundin, M., Bengtsson, M. and Molander, S. (2000) 'Life cycle assessment of wastewater systems – Influence of system boundaries and scale on calculated environmental loads', *Environmental Science and Technology*, vol.34, pp. 180–186.

Luo, L., van der Voet, E., Huppes, G. and Udo de Haes, H.A. (2009) 'Allocation issues in LCA methodology: a case study of corn stover-based fuel ethanol', *International Journal of Life Cycle Assessment*, vol. 14, pp. 529–539.

Malca, J. and Freire, F. (2006) 'Renewability and life-cycle energy efficiency of bioethanol and bio-ethyl tertiary butyl ether (bioETBE): assessing the implications of allocation', *Energy*, vol. 31, pp. 3362–3380.

McLaughlin, S.B., Delatorreugarte, D.G., Garten Jr., C.T., Lynd, L.R., Sanderson, M.A., Tolbert, V.R., and Wolf, D.D. (2002) 'High-value renewable energy from prairie grasses', *Environmental Science and Technology*, vol. 36, pp. 2122–2129.

Minami, K. and Ohsawa, A. (1990) 'Emission of nitrous oxide dissolved in drainage water from agricultural land', in Bouwman, A.F. (ed.). *Soils and the greenhouse effect* pp. 503–509. New York: John Wiley & Sons.

Monti, A., Fazio, S., and Venturi, G. (2009) 'Cradle-to-farm gate life cycle assessment in perennial energy crops', *European Journal of Agronomy*, vol.31, pp. 77–84.

Murphy, J.D. and Power, N.M. (2008) 'An argument for using biomethane generated from grass as a biofuel in Ireland', *Biomass and Bioenergy*, vol. 33, no 3, pp. 504–512.

Nevison, C. (1999) 'Indirect N_2O emissions from agriculture', *Good practice guidance and uncertainty management in national greenhouse gas inventories*, pp. 381–397, Geneva: IGES. http://www.ipcc-nggip.iges.or.jp/public [Accessed November 2012].

Palsson, A.C. and Riise, E. (2011) 'Performing the inventory in a LCA study', *TOSCA Sustainability Framework*, http://www.tosca-life.info/ [Accessed September 2012].

Payraudeau, S., van der Werf, H.M.G. and Vertes, F. (2007) 'Analysis of the uncertainty associated with the estimation of nitrogen losses from farming systems', *Agricultural Systems*, vol. 94, pp. 416–430.

Power, N. and Murphy, J.D. (2009) 'Which is the preferable transport fuel on a greenhouse gas basis: biomethane or ethanol?', *Biomass and Bioenergy*, vol. 33, pp. 1403–1412.

Rebitzer, G., Ekvall, T., Frischknecht, R., Hunkeler, D., Norris, G., Rydberg, T., Schmidt, W.P., Suh, S., Weidema, B.P., and Pennington, D.W. (2004) 'Life cycle assessment. Part 1: framework, goal and scope definition, inventory analysis, and applications', *Environmental International*, vol.30, no 5, pp. 701–720.

Ruggeri B., Tommasi T. and Sassi G. (2010). 'Energy balance of dark anaerobic fermentation as a tool for sustainability analysis', *International Journal of Hydrogen Energy*, vol. 35, pp. 10202–10211.

SAIC (2006) 'Life cycle assessment: principles and practice', Scientific Applications International Corporation (SAIC), Report No. EPA/600/R-06/060. Cincinnati, OH.: National Risk Management Research Laboratory, Office of Research and Development, US Environmental Protection Agency.

Salter, A. and Banks, C.J. (2009) 'Establishing an energy balance for crop-based digestion', *Water Science & Technology*, vol. 59, no 6, pp. 1053–1060.

Sanchez, S.T., Woods, J., Akhurst, M., Brander, M., O'Hare, M., Dawson, T.P., Edwards, R., Liska, A.J and Malpas, R. (2012) 'Accounting for indirect land-use change in the life cycle assessment of biofuel supply chains', *Journal of the Royal Society Interface*, vol. 9 no. 71 pp. 1105–1119.

Sapsford, R. and Jupp, V. (2006) *Data collection and analysis*, London: The Open University, SAGE.

Sathaye, J., Lucon, O., Rahman, A., Christensen, J., Denton, F., Fujino, J., Heath, G., Kadner, S., Mirza, M., Rudnick, H., Schlaepfer, and Shmakin, A. (2011) 'Renewable energy in the context of sustainable development', in Edenhofer, O., Pichs-Madruga, R., Sokona, Y., Seyboth, K., Matschoss, P., Kadner, S., Zwickel, T., Eickemeier, P., Hansen, G., Schlomer, S. and von Stechow, C. (eds), *IPCC special report on renewable energy sources and climate change mitigation*, Cambridge: Cambridge University Press.

Saunders, C., Barber, A. and Taylor. G. (2006) 'Food miles-comparative energy/emissions performance of New Zealand's agriculture industry', Research Report 285, Lincoln University New Zealand, http://www.lincoln. ac.nz/documents/2328_rr285_s13389.pdf [Accessed March 2012].

Schmidt, J.H. (2010) 'Comparative life cycle assessment of rapeseed oil and palm oil', *International Journal Life Cycle Assessment*, vol. 15, pp. 183–197.

Seitzinger, S.P. and Kroeze, C. (1998) 'Global distribution of nitrous oxide production and N inputs in freshwater and coastal marine ecosystems', *Global Biogeochemical Cycles*, vol. 12, pp. 93–113.

Shapouri, H., Duffield, J. and Wang, M. (2002) 'The energy balance of corn ethanol: an update', Agricultural Economic Report No. 813. US Department of Agriculture.

Singh, A., Pant, D., Korres N.E., Nizami, A.S., Prasad, S. and Murphy, J.D. (2010) 'Key issues in life cycle assessment of ethanol production from lignocellulosic biomass: A review', *Bioresource Technology*, vol.101, pp. 5003–5012.

Smith, K., Bouwman, L., and Braatz, B. (1997) 'N_2O: direct emissions from agricultural soils', *Good practice guidance and uncertainty management in national greenhouse gas inventories*, pp. 361–380, http://www.ipcc-nggip.iges.or.jp/public [Accessed November 2012].

SMMT (2006) 'Annual CO_2 Report: 2006', Society of Motor Manufacturers and Traders.

Smyth, B.M., Murphy, J.D. and O'Brien, C.M. (2009) 'What is the energy balance of grass biomethane in Ireland and other temperate northern European climates?' *Renewable and Sustainable Energy Reviews*, vol. 13, no 9, pp. 2349–2360.

Sonnemann, G. and Vigon, B. (2011) 'Global guidance principles for life cycle assessment databases. A basis for greener processes and products', Shonan Guidance Principles, UNEP/SETAC Life Cycle Initiative.

Staple, G.C. and Swisher, J.N (2011) 'The climate impact of natural gas and coal-fired electricity: A review of fuel chain emissions based on updated EPA national inventory data', American Clearn Skies Foundation, April 2011, www.cleanskies.org [Accessed July 2012].

Stucki, M., Jungbluth, N. and Leuenberger, M. (2011) 'Life cycle assessment of biogas production from different substrates', Version 103298_SB_154346_LCA Biogas_ESU-services_v1.0, ESU-services, December 2011, www.esu-services.ch [accessed September 2012].

Thamsiriroj, T. and Murphy, J.D. (2010) 'Can rape seed biodiesel meet the European Union sustainability criteria for biofuels?', *Energy Fuels*, vol. 24, no3, pp. 1720–1730.

Thomassen, M.A., Dalgaard, R., Heijungs, R. and de Boer, I. (2008) 'Attributional and consequential LCA of milk production', *International Journal of Life Cycle Assessment*, vol. 13, pp. 339–349.

Thyo, K.A. and Wenzel, H. (2007) 'Life cycle assessment of biogas from maize silage and from manure – for transport and for heat and power production under displacement of natural gas based heat works and marginal electricity in northern Germany', 2nd draft, June 21, 2007, Report for Xergi A/S, Institute for Product Development.

Tillman, A.M., Ekvall, T., Baumann, H. and Rydberg, T. (1994) 'Choice of system boundaries in life cycle assessment', *Journal of Cleaner Production*, vol. 2, no1, pp. 21–29.

Udo de Haes, H.A. and van Rooijen, M. (2005) 'Life cycle approaches. The road from analysis to practice', UNEP/SETAC-Life Cycle Initiative, United Nations Environment Programme.

Vieira, P.S. and Horvath, A. (2008) 'Assessing the end-of-life impacts of buildings', *Environmental Science and Technology*, vol.42, no.13, pp. 4663–4669.

Vikman, P., Gustavsson, L. and Klang, A. (2004) 'Evaluating greenhouse gas balances and mitigation costs of bioenergy systems – A review of methodologies', Biomass-based Climate Change Mitigation through Renewable Energy (BIOMITRE)-Work-package 1, June 2004.

Wang, G.R. (2005) 'Sugarcane', in *Taiwan agriculture encyclopedia*, [Taiwan nong-jia yao-lan- zeng-siou-ding san-ban] 3rd edn. pp. 171–180. Taiwan: Council of Agriculture, Executive Yuan.

Weidema, B.P. (2003) 'Market information in life cycle assessment', Environmental Project no. 863, Danish Environmental Protection Agency, Copenhagen.

Wenzel, M.H.W. and Petersen, B.M. (2009) 'Life cycle assessment of slurry management technologies', Environmental Project No. 1298 2009. Environmental Protection Agency. Danish Ministry of the Environment.

WHO (World Health Organisation) (2006) 'Energy, sustainable development and health', http://www.euro.who.int/globalchange/topics/20030310_7 [Accessed August 2012].

Yu, B. (2009) 'Life cycle environmental and economic impact of using switchgrass derived bioethanol as transport fuel'. Masters thesis. Leiden University, the Netherlands.

Chapter 17

The use of digestate as a substitute for manufactured fertilizer

Brian J. Chambers[1]* and Matthew Taylor[1]

[1]ADAS UK Ltd, Gleadthorpe, Meden Vale, Mansfield, Notts, NG20 9PD
*Corresponding author email: Brian.Chambers@adas.co.uk

Introduction

Increasing pressure is being placed both on industry and government to decrease the amounts of waste being sent to landfill in Europe. The European Community Landfill Directive (EC, 1999) sets strict limits on the amount of biodegradable municipal waste that can be disposed of via landfill; the amounts must be reduced by 65 per cent in 2020 compared with 1995 levels (EC, 1999). As a result, there are strong legislative and government drivers to recycle and re-use materials that in the past would have been destined for landfill as the cheapest way of disposal.

Anaerobic digestion (AD) is one of the best ways to recover value from organic 'wastes' – primarily because energy in the form of methane (biogas) is produced naturally as part of the digestion process. This biogas can be used as a substitute for natural gas (biogas needs to be upgraded to biomethane for injection into the national gas grid), to produce green electricity or heat, or it can be compressed for use as a transport biofuel. Anaerobic digestion is a key part of many governments' strategies to increase the production of renewable energy and help combat climate change. Digestate is a natural product from AD, through the controlled biological decomposition of biodegradable materials in the absence of oxygen. Suitable input materials include domestic and commercial food 'wastes', animal manures and purpose-grown crops.

The recycling of organic materials, such as digestate, to land is regarded as the best practicable environmental option in most circumstances, completing both natural nutrient and carbon cycles. Digestate is a valuable source of major plant nutrients (i.e. nitrogen – N, phosphate – P_2O_5, potash – K_2O and sulphur – SO_3), which are essential for plant growth and therefore sustainable crop production. Historically, agriculture has relied on the addition of livestock manures etc. to support crop growth. However, rapid population growth in the twentieth century combined with a need to grow more from less land (i.e. increase crop yields) has led to the extensive world-wide use of manufactured fertilizers. As a result, modern agriculture is dependent on manufactured fertilizer inputs to sustain high crop yields, maintain soil fertility and to replenish the nutrients removed in harvested crops and livestock products.

Digestate also supplies organic matter, which increases the capacity of soils to bind nutrients and regulate their supply, and act as a store for carbon. Soil physical properties, such as plant available water supply, aggregate stability and friability are also dependent on soil organic matter (SOM). Notably, organic matter can hold up to 20 times its weight in water and can, therefore, directly affect soil water retention (Dick and Gregorich, 2004), helping to reduce the effect of drought stress in times of reduced rainfall. At an EU level, there is increasing concern that low organic matter levels are compromising the ability of soils to sustain crop production and that soils are becoming increasingly prone to degradation through erosion and compaction (EU, 2006).

There is increasing agricultural demand for organic materials (e.g. digestate, compost etc.) as a source of crop available nutrients. This demand has recently been driven by dramatic (three- to four-fold) increases in the price of manufactured fertilizers, as a consequence of the growing demand for food as the world's population continues to grow. As a result, farmers and growers are becoming increasingly interested in and reliant on organic materials to supply crop nutrient requirements and to enhance soil organic matter levels. Additionally, as the importance placed on carbon footprinting increases, farmers and land managers are being encouraged to reduce the 'carbon footprint' of their products through using organic materials to offset the use of manufactured fertilizers.

The nutrient supply properties of digestate are largely a reflection of the feedstocks used, which include a wide range of biodegradable organic materials, for example, livestock manures, purpose-grown crops such as grass, maize and whole-crop cereals, and food waste etc. In the UK, for example, the quantity of digestate (from source-segregated biodegradable materials) currently recycled to agricultural land is relatively small (around 200,000 tonnes fresh weight), when compared with around 90 million tonnes fresh-weight of livestock manures (Williams et al., 2000), 1.1 million dry solid tonnes of biosolids (Water UK, 2010) and around 2 million tonnes fresh-weight of compost (WRAP, 2011a). However, the quantity of digestate recycled to land is expected to increase to around 5 million tonnes (fresh-weight) by 2020 (Defra/DECC 2011).

Quality of digestate (biofertilizer)

A publicly available specification, PAS 110 (BSI, 2010), for the processing and production of digestate, and an Anaerobic Digestate Quality Protocol (ADQP) (WRAP/EA, 2009) have been produced in Britain. PAS110 requires producers to undertake hazard analysis and critical control point (HACCP) planning to ensure digestate meets minimum requirements for microbial pathogens, heavy metals, stability and physical contaminants, and is fit for purpose (Table 17.1). The ADQP sets out criteria for the recovery/production of quality digestate from source-segregated biodegradable waste (which includes compliance with PAS110). Adherence with the ADQP means that although 'waste' feedstocks are used as an input to the process, the resultant digestate (biofertilizer) can be used as a product, and hence is not subject to Environmental Permitting Regulations (SI, 2010) when applied to land in England, Wales and Northern Ireland, and in Scotland in accordance with a Regulatory Position Statement "The Use of PAS110 Certified Digestate from Anaerobic Digestion" (SEPA, 2011). Notably, the European Commission (JRC/IPTS, 2011) is seeking to develop end-of-waste criteria for digestate (and compost) at an EU level, building upon work in individual EU countries that has developed end-of-waste status for digestate (e.g. BSI, 2010) and compost.

There are three main types of digestate (whole, liquid and fibre), with whole digestate being the most commonly available for land recycling. Some AD plants opt to separate the digestate into liquid and fibre fractions for management reasons. The fibre fraction typically has a dry matter content of between 20 and 40 per cent and the separated liquid fraction between 1 and 4 per cent, although these proportions will vary depending upon the separation process (or processes) employed.

Table 17.1 BSI PAS100 quality standards.

Pathogens (human and animal indicator species)		
E.coli	cfu/g fw[a]	1000
Salmonella spp	25g fw	Absent
Potentially toxic elements		
Cadmium (Cd)	mg/kg dm[b]	1.5
Chromium (Cr)	mg/kg dm[b]	100
Copper (Cu)	mg/kg dm[b]	200
Lead (Pb)	mg/kg dm[b]	200
Mercury (Hg)	mg/kg dm[b]	1.0
Nickel (Ni)	mg/kg dm[b]	50
Zinc (Zn)	mg/kg dm[b]	400
Stability		
Volatile fatty acids (screening value)	g COD/gVS[c]	0.43
Residual biogas potential	l/gVS[d]	0.25
Physical contaminants		
Total glass, metal, plastic, and any 'other' non-stone man-made fragments > 2 mm	% m/m dm[e]	0.5 (of which none are 'sharps')
Stones > 5 mm	% m/m dm[e]	8

[a] colony forming units/gram fresh weight; [b] milligrams/kilogram dry matter; [c] grams chemical oxygen demand/gram volatile solids; [d] litre/gram volatile solids; [e] percentage mass/mass dry matter.

The whole and liquid fractions have similar management properties to livestock slurries (see below) whereas the fibre fraction has properties more akin to solid livestock manures (e.g. farmyard manure, poultry manure). Whole digestates (and the separated liquid fraction) are a very good source of readily available nitrogen and fibre digestate is a valuable source of stable organic matter.

Digestate properties

Nutrient content

The 'typical' total N content of food-based digestate (i.e. digestate produced from food 'waste') is around 5 kg/m^3 (WRAP, 2011b); by way of context 'typical' total N concentrations in pig and cattle slurry are 3.6 and 2.6 kg/m^3, respectively (Table 17.2). As a general rule, digestate total nitrogen and phosphate concentrations increase with dry matter content.

The readily available nitrogen – RAN (i.e. ammonium-N) content of digestate is the nitrogen that is potentially available for rapid crop uptake. Livestock slurries and poultry manures are 'high' in RAN (typically in the range 35–70 per cent of total N) compared with farmyard manure that is 'low' in RAN (typically in the range 10–25 per cent of total N). In contrast, organic N is the nitrogen contained in organic forms, which will be slowly released and become potentially available for crop uptake over a period of months/years. Food-based digestate typically has c. 80 per cent of its total N content present as RAN; compared with

Table 17.2 Mean nutrient concentrations in food-based digestate and livestock slurries.

Nutrient (kg/m³ fresh weight)	Food-based digestate+ (4% dry matter)	Pig slurry* (4% dry matter)	Cattle slurry* (6% dry matter)
Total N	5.0	3.6	2.6
Readily available (NH_4) N	4.0	2.5	1.2
Total phosphate (P_2O_5)	0.5	1.8	1.2
Total potash (K_2O)	2.0	2.4	3.2
Total magnesium (MgO)	0.05	0.7	0.6
Total sulphur (SO_3)	0.4	1.0	0.7

+ data from Taylor et al., (2010) and WRAP (2011b).
* typical slurry values taken from Defra (2010).

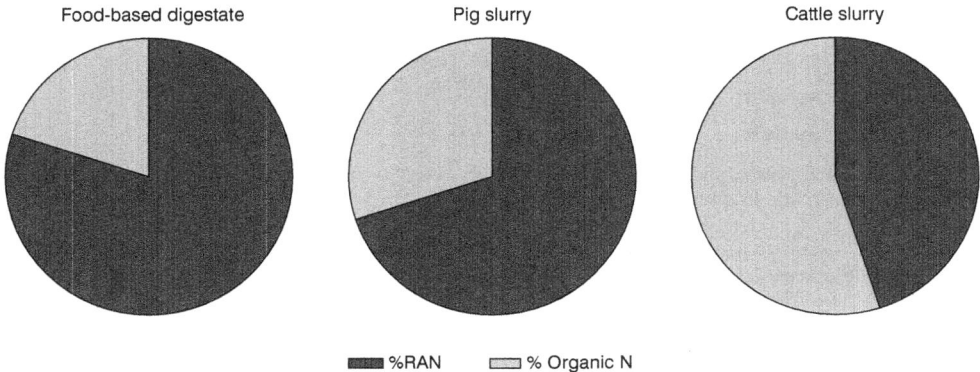

Figure 17.1 Readily available N (RAN) contents of food-based digestate in comparison with typical values for pig and cattle slurries

typical RAN contents for pig slurry of 70 per cent and cattle slurry 45 per cent of total N (Defra, 2010) (Figure 17.1). Typically the digestion of livestock slurries will increase RAN by 10 per cent of the total N content and decrease dry matter content by 2 per cent (ADAS and SAC, 2007; Smith et al., 2010). RAN is the most important fraction controlling crop available N supply from organic materials (Chambers et al., 1999; Birkmose, 2009; Defra, 2010; Lukehurst et al., 2010).

In addition to laboratory analysis for ammonium (readily available) N, it is possible to undertake *on-site* 'rapid' N measurements (using a Quantofix or Argos N meter, Figure 17.2). Measurements made with rapid N meters have been shown to be in good agreement with laboratory analysis data (Williams et al., 1999). These data can then be input directly into decision support systems such as MANNER-NPK (Nicholson et al., 2010a) to calculate the crop available N supply from digestate applications to agricultural land. MANNER-NPK predicts the fertilizer N value of field applied organic materials, taking into account organic material analysis data (total N, RAN and organic N), soil type, application timing and technique, ammonia volatilization, nitrate leaching, denitrification losses and the mineralization of organic N.

In order to determine the crop available N supply from digestate, field experiments have been undertaken in the UK (WRAP, 2012; Taylor et al., 2012). Yield and N offtake measurements at

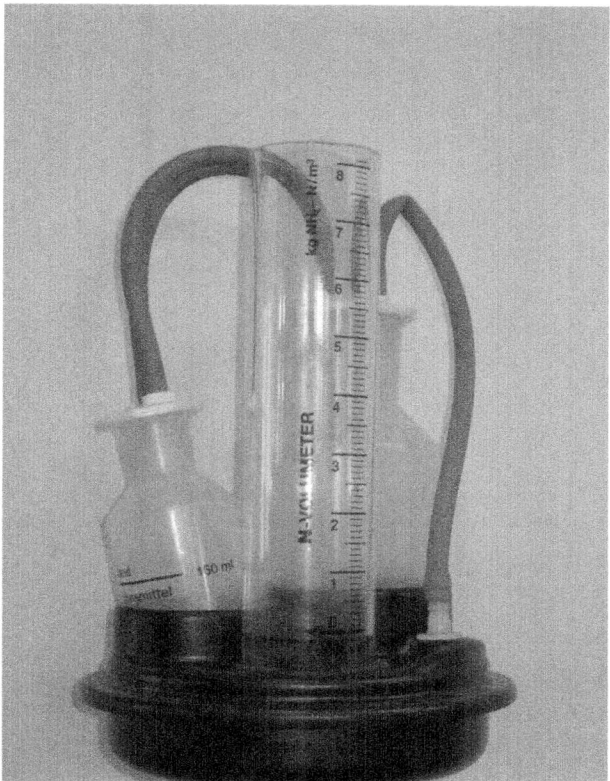

Figure 17.2 Quantofix meter for rapid on-site ammonium-N measurement

harvest, and direct comparison with different rates of manufactured N applied, enable fertilizer N replacement values and the N use efficiency of digestate to be quantified. At a sandy soil site in Nottinghamshire, growing potatoes (variety Harmony) during the 2011 harvest year, food-based digestate (applied in early and late spring using a band spreader) had a mean N use efficiency of 64 per cent (RAN represented 76 per cent of total N), manure-based digestate a mean N use efficiency of 63 per cent (RAN represented 59 per cent of total N) and cattle slurry a mean N use efficiency of 47 per cent (RAN represented 47 per cent of total N) compared with manufactured fertilizer N (Figure 17.3). The N use efficiency of the applied organic materials reflected their RAN content and low ammonia losses to air as a result of rapid soil infiltration; there was no effect of application timing on N utilization. In Denmark, Birkmose (2009) reported an N use efficiency for spring bandspread digestate applications to winter wheat of 80 per cent of total N applied (RAN represented 83 per cent of total N).

The 'typical' phosphate content of food-based digestate is around 0.5 kg/m³ compared with 'typical' concentrations of 1.8 and 1.2 kg/m³ in pig and cattle slurry, respectively. Similarly, the 'typical' potash content of food-based digestate is around 1.8 kg/m³ compared with 'typical' concentrations of 2.4 and 3.2 kg/m³ in pig and cattle slurry, respectively (Table 17.2). Notably, the extractable (plant available) potash content of food-based digestate is around 80 per cent of the total potash content, which is similar to livestock slurry where 90 per cent of the total potash is plant available (Defra, 2010).

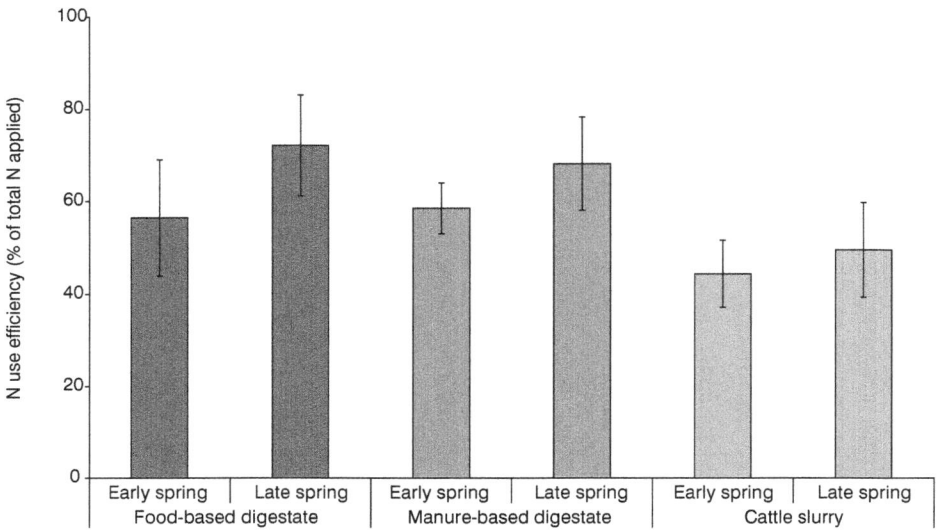

Figure 17.3 Digestate N use efficiency (percentage of total N applied)

Organic matter

Based on an application rate of 250 kg total N/ha, the maximum field N rate permitted in nitrate vulnerable zones – NVZs in Britain as part of the EU Nitrates Directive (EEC, 1991), 'typical' organic matter loadings from food-based digestate would be c. 1 t/ha; by way of context 'typical' organic matter loadings from cattle slurry would be c. 3 t/ha (Bhogal et al., 2008). However, given that the organic matter contained in digestate has a higher proportion in the lignin (i.e. recalcitrant) fraction than cattle slurry (WRAP, 2011b), the long-term benefits per unit of organic matter applied are likely to be greater.

Heavy metal concentrations

Mean heavy metal concentration in food-based digestate in comparison with PAS110 (BSI, 2010) limit values are presented in Table 17.3 (Nicholson et al., 2010b); by way of context 'typical' heavy metal concentrations in livestock slurries are also included in the table.

Residual biogas potential

Digestate stability in PAS110 (BSI, 2010) is assessed by measuring the residual biogas potential (RBP). For digestate to be compliant with PAS110 the RBP must be below 0.25 l/g of volatile solids; at this point the rate of biological activity is considered to have slowed to an acceptably low level and the material can be considered sufficiently treated and stable for landspreading. The mean RBP of food-based digestates analysed by Taylor et al. (2010) was 0.22 l/g volatile solids. Notably, the draft EU end-of-waste document (JRC/IPTS, 2011) proposes a difference methodology for assessing digestate stability through the measurement of organic acid concentrations.

Table 17.3 Mean digestate and livestock slurry heavy metal concentrations (mg/kg).

Metal concentration (mg/kg dry matter)	Food-based digestate*	PAS110 upper limit values	Pig slurry	Cattle slurry
Total zinc	104	400	870	196
Total copper	21.5	200	279	137
Total cadmium	0.9	1.5	0.3	0.1
Total nickel	19.7	50	3.9	3.4
Total lead	6.1	200	3.5	4.8
Total chromium	10	100	2.3	2.9
Total mercury	< 0.05	1	n.d	n.d

n.d. = not determined
* Data from Taylor et al., (2010)
Based on an application rate of 250 kg total N/ha 'typical' zinc additions from food-based digestate would be c. 0.14 kg/ha and copper 0.03 kg/ha; by way of context 'typical' zinc additions from pig slurry would be c. 2.5 kg/ha and copper c. 0.8 kg/ha (Nicholson et al., 2010b). All other metal addition rates from food-based digestate were < 0.05 kg/ha.

Biochemical oxygen demand

Biochemical oxygen demand (BOD) is a measure of the oxygen used by microorganisms to decompose organic materials; the mean BOD of food-based digestates analysed by Taylor et al. (2010) was c. 9000 mg/l. By way of comparison, pig slurry typically has a BOD level of 20,000–30,000 mg/l, cattle slurry 10,000–20,000 mg/l and dirty water 1000–5000 mg/l (MAFF, 1998). Notably, treated sewage effluent must typically have a BOD of ≤ 20 mg/l before discharge to surface water systems. As with livestock slurry management, care is needed when applying digestate to land to minimize water pollution risks.

Physical contaminants

The PAS110 (BSI, 2010) and proposed EU end-of-waste (JRC/IPTS, 2011) limit for total physical contaminants (> 2 mm) in digestate is 0.5 per cent mass/mass dry matter. All of the digestate samples analysed by Taylor et al. (2010) complied with the limit, indeed no glass, metal or stone contamination was found in any of the samples.

Microbial pathogens

To comply with PAS110 (BSI, 2010) and the proposed EU end-of-waste (JRC/IPTS, 2011) limit *Salmonella* spps. must be absent in 25 g of fresh matter; *Salmonella* spps. were absent from the food-based digestates analysed by Taylor et al. (2010). By way of comparison, around 5 per cent of pig and cattle slurry samples in the UK contain *Salmonella* (Hutchison et al., 2002). *E. coli* numbers in the food-based digestates (< 10 colony forming units – CFU/g fresh weight – fw) analysed by Taylor et al. (2010) were below the PAS110 and proposed EU end-of-waste (JRC/IPTS, 2011) limit of 1000 cfu/g fresh weight. By way of comparison, pig and cattle slurries typically contain c. 7 $\log_{10}$ CFU/g fresh weight.

A major benefit from the digestion of biodegradable organic materials is the reduction and in many cases the elimination of human and animal pathogens; for example, *E coli*, *Camplylobacter*, *Listeria* etc. (Bohm et al., 1999; Bendixen, 1994, 1999; Martens et al., 1998; Termorshuizen et al., 2003) and plant pathogens, for example, nematodes, ring/brown rots, *Phytophthora*, *Fusarium* etc. (Engeli et al., 1993, Ryckeboer, 2002; Mikkelson et al., 2006; Spaull et al., 1989; Ryckeboer et al., 2002; Termorshuizen et al., 2003).

Financial value of digestate

Digestate is a valuable source of major plant nutrients, especially nitrogen, which can be used to replace manufactured fertilizer additions. When applied at 250 kg total N/ha, digestate has a value of around £230 per hectare, based on the crop available N and total phosphate and potash supplied (Table 17.4).

Carbon footprint benefits

In addition to the financial value of digestate, use can also help to reduce a farm's carbon footprint by replacing the need to apply manufactured fertilizers, which are either manufactured using fossil fuels (e.g. ammonium nitrate) or quarried and refined for transport/processing using fossil fuels (e.g. phosphate and potash) (Table 17.5). Hence, replacing the use of manufactured fertilizer with food-based digestate will reduce a farm's carbon footprint by around 20 kg CO_2-e/tonne of digestate applied or around 1 tonne CO_2-e per hectare (if applied at the maximum field limit of 250 kg total N/ha in NVZs). However, there will also be nitrous oxide emissions from the digestate application, which will to a lesser or greater extent negate the carbon reduction benefits of reduced manufactured fertilizer use.

Table 17.4 Financial value of food-based digestate.

Nutrient	kg/ha[a]	£/ha[b]
Crop available N[c]	150	£150
Total phosphate	25	£20
Total potash	100	£60
Total		£230

[a] Based on a 250 kg total N/ha application of food-based digestate.
[b] Assuming N = 100 p/kg, P_2O_5 = 80 p/kg, K_2O = 60 p/kg.
[c] Assuming crop available N = 60% of total N applied.

Table 17.5 Displaced carbon footprint benefit of food-based digestate.

Nutrient	CO_2-e (kg CO_2-e/kg nutrient)[a]	Food-based digestate nutrient content	CO_2-e (kg/t) saving
Crop available N	6.2	3.0	18.6
Total phosphate P_2O_5	0.7	0.5	0.4
Total potash K_2O	0.5	2.0	1.0
Total			20.0

[a] Taken from Brentrup and Paliére (2008).
[b] Assuming crop available N = 60% of total N applied.

Land application controls

Animal by-products regulations

Where relevant, digestate applications to agricultural land must also comply with the EU Animal By-Products Regulations (EC, 2005). Notably, pasture land cannot be used for grazing within three weeks (or two months for pigs) of applying digestate (SI, 2005; SRNI, 2003; SSI, 2003; WSI, 2006).

Nitrate vulnerable zones

In NVZs in Britain, established under the EU Nitrates Directive (EEC, 1991), the total quantity of N applied in organic materials (including digestate) must not exceed 250 kg total N/ha in any 12-month period (i.e. the field N limit). And in some situations, lower application rates may be appropriate, for example, where the amount of crop available N supplied would exceed the crop requirement. As the readily available N (RAN) content of liquid digestate exceeds 30 per cent of its total N content, digestate (like cattle and pig slurry) applications are subject to mandatory closed spreading periods during autumn/winter (SI, 2008; SSI, 2008; WSI, 2008).

Good agricultural and environmental condition

Farmers who receive payments under the Single Payment scheme in the EU must maintain their land in good agricultural and environmental condition (GAEC). GAEC measures relating to soils, environmental protection and the land application of organic materials highlight the important of managing applications to maintain and enhance soil organic matter levels, avoiding causing soil compaction and minimizing the risks of soil erosion (RPA/Defra, 2012).

How and when to apply digestate

To make optimum use of the N content of digestate, it should be applied at times of maximum crop growth – generally during the late winter to summer period (Defra, 2010; SAC, 2010). Software decision support systems, such as MANNER-NPK (Nicholson et al., 2010a) and PLANET which incorporates MANNER-NPK (Dampney and Sagoo, 2008) are available in the UK to help farmers and growers integrate digestate (and other organic material) nutrient supply in their farm nutrient management plans.

To make best use of digestate RAN, the ADQP recommends that 'low emission application equipment' is used, namely a bandspreader (trailing hose/trailing shoe, Figure 17.4) or shallow injector (Figure 17.5), to reduce ammonia losses (and odour nuisance) compared with surface broadcast application, and thereby increase crop available N supply (Figure 17.6).

Additionally, bandspreading and shallow injection application techniques increase the number of spreading days and cause less sward contamination than surface broadcast application. Bandspreading equipment is available that enables the accurate topdressing of arable crops across full tramline widths, without causing crop damage and contamination.

Ammonia loss (and odour nuisance) can also be reduced by ensuring that organic materials are rapidly incorporated into soils. On uncropped land, soil incorporation following application (within 24 hours) is a mandatory requirement in NVZs (SI, 2008; SSI, 2008; WSI, 2008).

Figure 17.4 Bandspread application to arable land

Figure 17.5 Shallow injection application to grassland

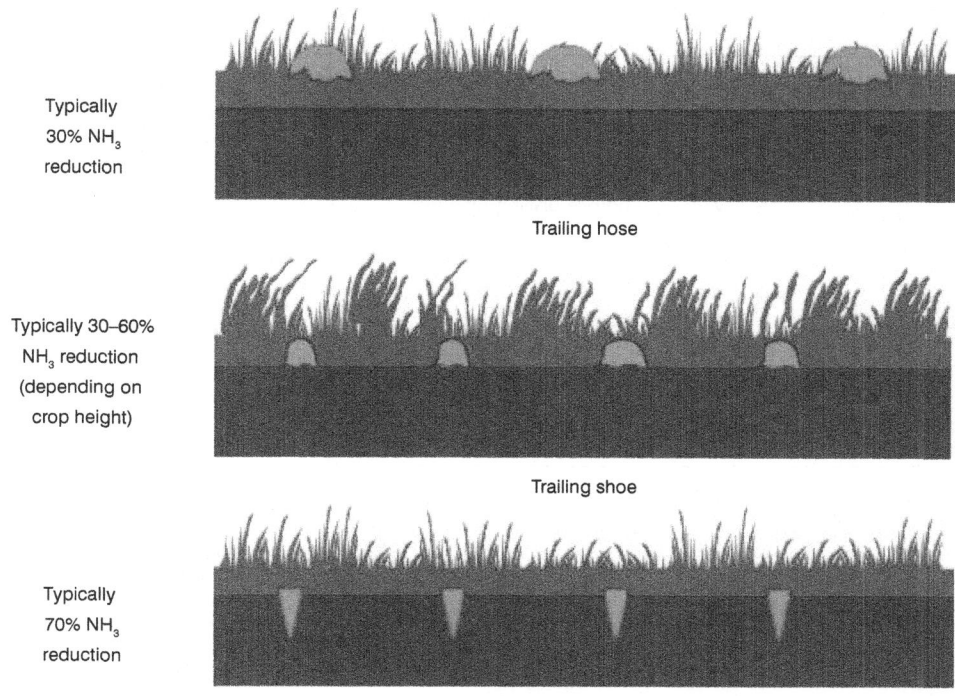

Figure 17.6 Application type and effect on ammonia (NH3) losses

To minimize water pollution risks, digestate (and other organic materials) applications should not be made when:

- the soil is waterlogged; or
- the soil is frozen hard; or
- the field is snow covered; or
- the soil is cracked down to field drains or backfill; or
- the field has been pipe or mole drained or subsoiled over drains in the last 12 months; or
- heavy rain is forecast within the next 48 hours.

Also, applications should not be made:

- within 10 metres of any ditch, pond or surface water; or
- within 50 metres of any spring, well, borehole or reservoir that supplies water for human consumption or for farm dairies; or
- on very steep slopes where run-off is a high risk throughout the year.

In addition to Environmental Regulations and Codes of Good Agricultural Practice (Defra, 2009; Scottish Executive, 2005; WAG, 2011), farmers who use digestate need to comply with the requirements of farm assurance schemes (FAS) in the UK. To ensure that digestate recycling to agricultural land is viewed by the FAS as a safe and sustainable operation, WRAP has proposed a Biofertiliser Matrix (Table 17.6 and Table 17.7).

Table 17.6 Proposed Biofertiliser Matrix (PAS110/ADQP input materials): agriculture and field horticulture.

Crop type		Pasteurized[+]		Not pasteurized	
Fresh produce	Category 1	✓	before drilling/planting	✗	12 month harvest and 6 month no drilling interval applies
	Category 2	✓	before drilling/planting	✗	12 month harvest and 6 month no drilling interval applies
	Category 3	✓		✓	
Grassland and forage		✓	[a] 3 weeks no grazing period and harvest interval applies	✓	3 weeks no grazing period and harvest interval applies
Combinable and animal feed crops etc.		✓		✓	[b]

[+] Pasteurization process compliant with animal by-products regulations (SI 2005; SRNI 2003; SSI 2003; WSI 2006).
[a] 2 months no grazing or harvest period for pigs.
[b] If feedstocks contain maize, biofertilizer applications should be ploughed into the soil ahead of following crops.

Table 17.7 Biofertiliser Matrix: crop categories.

Fresh produce			Combinable and animal feed crops etc	Grassland and forage
Category 1	Category 2	Category 3		
Wholehead lettuce, leafy salads (including any vegetable leaf you can eat raw), celery, salad, onions, radish, fresh and frozen herbs etc.	Apple, beetroot, blackcurrant, broad bean, broccoli, cabbage, carrot, capsicum, cauliflower, celeriac, cherry, courgette, cucumber, garlic, green beans (other than runner beans), melon, mushroom, onion (red and white), pea, pear, peach, plum, raspberry, strawberry, sugar snap peas, sweetcorn, tomato and tree nuts etc.	Artichoke, runner beans, leeks, marrow, parsnip, potato, pumpkin, squash, swede, turnip etc.	Wheat, barley, oats, rye, triticale, field peas, fields beans, linseed/flax, oilseed rape, sugarbeet, sunflower, borage, nursery stock, bulbs etc.	Grass maize hay haylage swede/turnip fodder mangold/beet/kale forage rye and triticale herbage seeds turf production etc.

The proposed Biofertiliser Matrix differentiates between pasteurized (e.g. batch heated at 70°C for 1 hour) and un-pasteurized digestates, with the latter having more restricted recycling options on land growing fresh produce (i.e. ready-to-eat crops). Notably, both pasteurized and unpasteurized applications need to comply with the Animal By-products Regulations (EC, 2005) which stipulate no graze/harvest intervals (SI, 2005; SRNI, 2003; SSI, 2003; WSI, 2006).

Integrating digestate with manufactured fertilizer – a good practice guide

An integrated digestate and manufactured fertilizer use policy should aim to maximize as far as practically possible the nutrients supplied by digestate. Failure to adequately allow for digestible nutrient supply, particularly nitrogen, not only wastes money because of unnecessary fertilizer use, but can also reduce crop yields and quality (e.g. lodging in cereals, poor fermentation in grass silage and low sugar levels in beet) and cause environmental pollution. The key steps for farmers/land managers are:

1   Calculate the quantity of crop available nutrients (equivalent to fertilizer) supplied.
2   Identify the fields that are available and will benefit most from digestate application. Take into account accessibility and likely soil conditions at the time of application, and the application equipment available. Crops with a high nitrogen demand should be targeted first.
3   Plan the application rate for each field ensuring that no more than 250 kg/ha total N is applied in any 12-month period. As far as practically possible, apply digestate in the late winter to summer period – this will make best use of the RAN content.
4   Aim for digestate application to supply no more that 50–60 per cent of the total N requirement of the crop, with manufactured fertilizer used to make up the difference. This approach will minimize the potential impact of (any) variations in digestate N supply on crop yields and quality.
5   Make sure that digestate application equipment is well maintained and suitable for applying digestate evenly and at the target rate. For whole digestate (or separated liquor), use precision application equipment (e.g. trailing hose, trailing shoe or shallow injector). All equipment should be routinely calibrated for the type of digestate being applied.
6   Following application, use a nutrient management recommendation system (e.g. *The Fertilizer Manual (RB209)* (Defra, 2010) or the MANNER-*NPK*/PLANET decision support systems) to calculate the amount of crop available nitrogen and phosphate/potash etc. supplied from each digestate application in each field.
7   Calculate the nutrient requirement of the crop and then deduct the nutrients supplied from the digestate. This will give the balance that needs to be supplied by manufactured fertilizer.

References

ADAS and SAC. (2007) *Nutrient value of digestate from farm based biogas plants in Scotland* ADAS and Scottish Agricultural College report for Scottish Executive Environment and Rural Affairs Department – ADA/009/06.

Bendixen, H.J. (1994) 'Safeguards against pathogens in Danish biogas plants', *Water Science and Technology*, vol. 2, pp. 171–180.

Bendixen, H.J. (1999) 'Hygienic safety – results of scientific investigations in Denmark, in *Hygienic and Environmental Aspects of Anaerobic Digestion: Legislation and Experiments in Europe;* vol. 2, pp. 27–47. Stuttgart, Hohenheim: International Energy Association Bioenergy Workshop.

Bhogal, A., Chambers, B.J., Whitmore, A.P. and Powlson, D.S. (2008) 'The effects of reduced tillage practices and organic material additions on the carbon content of arable soils', Report for Defra Project SP0561.

Birkmose, T. (2009) 'Nitrogen recovery from organic manures: improved slurry application techniques and treatment – the Danish scenario', *Proceedings of the International Fertiliser Society* No 656, York, UK.

Bohm, R., Martens, W. and Philipp, W. (1999) 'Regulations in Germany and results of investigations concerning hygienic safety of processing biowastes in biogas plants', in *Hygienic and environmental aspects of anaerobic digestion: legislation and experience in Europe'*, Proceedings of IEA Bioenergy Workshop, Task 34, volume 2, pp. 48–68.

Brentrup, F. and Paliére, C. (2008) 'GHG emissions and energy efficiency in european nitrogen fertiliser production and use', *Proceedings of the International Fertiliser Society* No. 639, York, UK.

BSI (2010) 'Specification for whole digestate, separated liquor and separated fibre derived from the anaerobic digestion of source-segregated biodegradable materials', Publicly Available Specification No 110. BSI, London.

Chambers, B.J., Lord, E.I., Nicholson, F.A. and Smith, K.A. (1999) 'Predicting nitrogen availability and losses following application of organic manures to arable land: MANNER', *Soil Use and Management*, vol.15, pp. 137–143.

Dampney, P. and Sagoo, E. (2008) 'PLANET – The national standard decision support and record keeping system for nutrient management on farms in England', in *Agriculture and the environment vii – land management in a changing environment*. Crighton, K. and Audsley, R. (Eds), pp. 239–245. Proceedings of the SAC and SEPA Biennial Conference, 26–27 March 2008, Edinburgh.

Defra (2009) *Protecting our water, soil and air: a code of good agricultural practice for farmers, growers and land managers*, Norwich: The Stationery Office.

Defra (2010) *The fertiliser manual (RB209)*, 8th edition. Norwich: The Stationery Office.

Defra/DECC (2011) *Anaerobic digestion strategy and action plan: a commitment to increasing energy from waste through anaerobic digestion*, London: Department for Environment Food and Rural Affairs/ Department of Energy and Climate Change.

Dick, W.A. and Gregorich, E.G. (2004) 'Developing and maintaining soil organic matter levels', in *Managing soil quality challenges in modern agriculture*, Schjonning, P., Elmholt, S. and Christensen, B.T. (eds), pp. 103–120. Wallingford: CABI Publishing.

EC (1999). European Union Council Directive 1999/31/EC. The Landfill of Waste.

EC (2005) Commission Regulations (EC) No. 93/2005 amending Regulation (EC) No. 1774/2002 of the European Parliament and of the Council as regards processing of animal by-products of fish origin and commercial documents for the transportation of animal by-products.

EEC (1991) Council Directive of 12 December 1991. Concerning the protection of waters against pollution caused by nitrates from agricultural sources (91/676/EEC).

Engeli, H., Edelmann, W. and Fuchs, J.K. (1993) 'Survival of plant pathogens and weed seeds during anaerobic digestion', *Water Science and Technology*, vol.27, pp. 69–76.

EU (2006) Thematic Strategy for Soil Protection. COM (2006) 231 final http://ec.europa.eu/environment/soil/pdf/com_2006_0231_en.pdf.

Hutchison, M.L., Ashmore, A.K., Crookes, K.M., Wilson, D., Groves, S.J., Keevil, W.C. and Moore, A. (2002) 'Enumeration of pathogens in livestock manures and factors affecting their survival', in *Proceedings of the Joint CIWEM and Aqua Enviro Technology Transfer 7th European Biosolids and Organic Residuals Conference*, Session 3, Paper 15.

JRC/IPTS (2011) 'Technical report for end-of-waste criteria on biodegradable waste subject to biological treatment'. Second Working Document. Joint Research Centre of European Commission and Institute for Prospective Technological Studies, 11 October.

Lukehurst, C.T., Frost, P. and Al Seadi, T. (2010) 'Utilisation of digestate from biogas plants as biofertiliser', IEA Bioenergy, Task 37, 22pp.

MAFF (1998). *A code of good agricultural practice for the protection of water*. London: Ministry of Agriculture Fisheries and Food.

Martens, W., Fink, A., Philipps, W., Weber, A., Winter, D. and Bohm, R. (1998). 'Inactivation of viral and bacterial pathogens in large scale slurry treatment plants'. Proceedings of RAMIRAN conference http://www.ramiran-net./98/FIN-ORAL/Martens.pdf [accessed September 2012].

Mikkelson, L., Elphistone, J., and Jensen, D.F. (2006). 'Literature review of the detection and eradication of plant pathogens in sludge, soils and treated biowaste'. http://www.ecn.nl/docs/society/horizontal/Hor_30_plant_pathogens.critical%20.review%20DJ.2.pdf [accessed September 2012].

Nicholson, F.A., Rollett, A.J., Bhogal, A., Lord, E., Thorman, R.E., Williams, J.R., Smith, K.A., Misselbrook, T.H., Chadwick, D.R. and Chambers, B.J. (2010a) 'MANNER NPK', In *Agriculture and the Environment VIII. Climate, Water and Soil: Science, Policy and Practice*. Crighton, K. and Audsley, R. (Eds), pp. 328–333. Proceedings of the SAC and SEPA Biennial Conference, Edinburgh,

Nicholson, F.A., Rollett, A.J. and Chambers, B.J. (2010b) 'The Defra agricultural soil heavy metal inventory – for 2008', Report 3 for Defra Project SP0569.

RPA/Defra (2012). *The guide to cross compliance in england: 2012 Edition*. Sheffield: Rural Payments Agency.

Ryckeboer, J. (2002) 'The fate of plant pathogens during anaerobic and composting', *Biocycle*, vol. 43, pp. 50–53.

Ryckeboer, J., Cops, S and Coosemans, J. (2002) 'The fate of plant pathogens and seeds during anaerobic digestion and aerobic composting of source separated household wastes', *Compost Science and Utilization*, vol.10, pp. 204–216.

SAC (2010) 'Optimising the application of bulky organic fertilisers', Technical Note TN622, Edinburgh: Scottish Government.

Scottish Executive (2005) *Prevention of environmental pollution from agricultural activity – a code of good practice*, Edinburhg: Scottish Executive.

SEPA (2011) 'Classification of outputs from anaerobic digestion processes' http://www.sepa.org.uk/waste/waste_regulation/idoc.ashx?docid=e2cfd0f4-6fc5-4cab-af13-aa94e506af0b&version=-1

SI (2008) 'Nitrate Pollution Prevention Regulations 2008', Statutory Instrument 2008/2349.

SI (2005) 'The Animal By-Products Regulations 2005', Statutory Instrument 2347.

SI (2010) 'The Environmental Permitting (England and Wales) Regulations 2010, SI No 675'. www.opsi.gov.uk [accessed August 2012].

Smith, K.A., Jeffrey, W.A. Metcalfe, J.P. Sinclair, A.H. and Williams, J.R. (2010) 'Nutrient value of digestate from farm-based biogas plants', *Proceedings of RAMIRAN conference* http://www.ramiran.net/ramiran 2010/docs/Ramiran2010_0171_final.pdf [accessed September 2012].

Spaull, A.M., McCormack, D.M. and Pike, E.B. (1989) 'Effects of various sewage sludge treatment processes on the survival of potato cyst-nematodes (*Globodera* spp.) and the implications for disposal', *Water Science and Technology*, vol.21, pp. 909–916.

SRNI. (2003) 'Animal By-Products Regulations 2003', Statutory Rules of Northern Ireland 495.

SSI (2003) 'The Animal By-Products (Scotland) Regulations 2003'. Scottish Statutory Instrument 411.

SSI (2008) 'The Action Programme for Nitrate Vulnerable Zones (Scotland) Regulations 2008', Scottish Statutory Instrument 2008/298.

Taylor, M.J., Rollett, A.J., Tompkins, D. and Chambers, B.J. (2010) 'Digestate quality and fertiliser value', *Proceedings of the 15th European Biosolids and Organic Resources Conference*, CD-rom.

Taylor, M.J., Rollett, A.J., Williams, D.J. and Chambers, B.J. (2012). 'Digestate – a low carbon nitrogen fertiliser', In: *8th World Potato Congress*, Edinburgh, 27–30 May 2012, p. 37 [abstract].

Termorshuizen, A.J., Volker, D., Blok, W.J., Ten Brummeler, E., Hortog, B.J., Janse, S.D., Knol, W. and Wenneker, M. (2003) 'Survival of human and plant pathogens during anaerobic mesospheric digestion of vegetable, fruit and garden waste', *European Journal of Soil Biology*, vol.39, pp. 165–171.

WAG (2011) 'The Code of Good Agricultural Practice for the Protection of Water, Soil and Air for Wales'. Welsh Assembly Government.

Water UK (2010) *Recycling of biosolids to agricultural land*. London: Water UK.

Williams, J.R., Hunt, C.L., Chambers, B.J., Brookman, S. and Chadwick, D. (1999) 'Rapid methods for the analysis of readily available nitrogen in manure', in *Accounting for nutrients: A challenge for grassland farmers in the 21st century*. Cornall, A.J. (Ed.), British Grassland Society Occasional Symposium No. 33, pp. 171–172.

Williams, J.R., Chambers, B.J., Smith, K.A. and Ellis, S. (2000) 'Farm manure land application strategies to conserve nitrogen within farming systems', in *Agriculture and waste management for a sustainable future*. Petchey, T., D'Arcy, B. and Frost, A. (Eds), pp. 167–179. SAC/SEPA Biennial Conference, Edinburgh.

WRAP (2011a) *A study of the UK organics recycling industry in 2009*. Banbury: WRAP.

WRAP (2011b) Digestate and compost in agriculture, Bulletin 2: Beat rising costs of fertiliser and extreme weather by using digestate and compost', Available from: http://www.wrap.org.uk/content/digestate-compost-agriculture [Accessed on 21 May 2012].

WRAP (2012) 'Digestate and compost in agriculture, Bulletin 3: Field Experiments focus on crop available nitrogen supply from digestate', Available from: http://www.wrap.org.uk/content/digestate-compost-agriculture [Assessed on 21 May 2012].

WRAP/EA (2009) *Quality protocol anaerobic digestate. End of waste criteria for the production and use of quality outputs from anaerobic digestion of source-segregated biodegradable*, Banbury: WRAP. http://www.environment-agency.gov.uk/static/documents/Business/W524AnaerobicDigestatev4(1).pdf

WSI (2006) The Animal By-Products Regulations 2006. Welsh Statutory Instrument 1293.

WSI (2006) 'The Animal By-Products Regulations 2006', Welsh Statutory 540 Instrument 1293.

WSI (2008) 'The Nitrate Pollution Prevention (Wales) Regulations 2008', Welsh Statutory Instrument 2008/3134.

Chapter 18

The sustainability of small-scale anaerobic digesters at farm scale

Phillip Hobbs[1] and Allan Butler[2]

[1]Anaerobic Analytics Ltd., 2 East Street, Okehampton EX20 1AS, www.bioenergy.tumblr.com. [2]Farm Management Researcher, SRUC, SRUC Edinburgh Campus, Peter Wilson Building, King's Buildings, West Mains Road, Edinburgh EH9 3JG
Corresponding authors email: philip.hobbs@btinternet.com; Allan.Butler@sruc.ac.uk

Introduction

Mention small-scale anaerobic digesters (SSADs) and eyebrows will be raised. Some identify this with the 'low' technology approach adopted by millions in south-east Asia for heating food with biogas generated from inverted barrels or underground vessels with gas holding capacity. Conversely, small-scale digesters are thought of as a mixed bag of different technologies in the more financially driven economies where the large scale tends to dominate. However, the possibilities of what could be achieved by SSADs are substantial. From a sustainability perspective, what if every farm or community could have electricity and heating, or indeed cooling and transport fuel, without the dependency on global markets with a rapid payback of 3–5 years? With SSADs all of the electricity and heat could be used within a system that reduces greenhouse gas (GHG) emissions (only about 22 per cent of energy generated at a power station is received by the consumer – 7 per cent transmission losses for 30 per cent electricity generation from coal). The impact of anaerobic digestion (AD) on anthropomorphic GHG emissions would contribute about 4–7 per cent of heat and power for the UK to help meet national and European obligations by 2020 (Lukehurst, 2009).

To examine such possibilities, this chapter first develops what small-scale means in terms of size and the criteria for sustainability. Furthermore we will focus on the all-important economic and environmental sustainability that form the basis of a pragmatic approach. After exploring size and sustainability, the effects of policy and the regulatory framework that currently encourage SSAD development are discussed. The UK for example, depends upon legislative drivers for economic performance of large-scale AD and SSADs. The drivers stem from international and national protocols and legislation to reduce GHG emissions, especially from agriculture. For example, dairy cows can emit 500 to 600 litres methane of per day from feedlots in the USA (Kaharabata and Schuepp, 2000), 14 per cent of which is emitted from manure. Furthermore, by processing the manure using AD the resulting digestate when applied to land as a fertiliser contains less available carbon leading to substantially reduced nitrous oxide (N_2O) emissions. N_2O has 296 times the global warming potential of carbon dioxide (CO_2) and about 13 times that of methane (CH_4) (Steinfeld et al., 2006). SSADs may enable economic access to excreta particularly if the costs of GHG emissions are included.

If the outcomes of installing an SSAD look so promising in reducing GHG emissions, what issues are preventing uptake? There are numerous texts that describe such hurdles. As the technology develops, overcoming such hurdles may improve SSAD performance with time. As a consequence, new configurations and better process control will provide more efficiencies at the small-scale. Maximising biogas potential from farm resources will be challenging and there are many factors that can influence both the quality and quantity of biogas and the concomitant economic returns. For example, ensuring the quality of feedstock, the degree of process control, and the efficiency of the fermentation process and the quality of the digestate all potentially impact upon the cost effectiveness and usefulness of SSADs. These and other factors are therefore discussed.

The concluding sections of this chapter focuses on the contribution that SSADs can make to society. Costing GHGs emissions using social cost benefit analysis illustrates that on-farm SSADs looks favourable from a sustainability perspective.

Size of sustainable small-scale anaerobic digestion

There are about 40 million small digesters worldwide that function mostly in warmer areas of Asia, which are simple in design and provide energy at the very small scale. Typical benefits in for example Costa Rica include (1) a reduction in deforestation associated with firewood collection; (2) fewer hours devoted to firewood collection; (3) eliminating the need to purchase propane for cooking; and (4) a reduction in GHG emissions to the atmosphere (Lansing et al., 2007). In colder climates, the effectiveness of these systems are less conducive since heating is required for reasonable kinetic rates of biogas output. The main concern from an economic perspective is whether SSADs are efficient to first sustain fermentation and second to provide sufficient energy for an economic return. The size of SSADs also undoubtedly impacts upon their economic sustainability, whether they function as standalone economic units or receive supportive subsidies from allocated funds or grants. The argument for subsidised biogas production is more convincing when recognising their environmental benefits.

Defining small-scale is problematic and may vary from country to country depending on the size and structure of the industry. We can define sizes of digesters in terms of total energy output from the biomethane. This approach can give an indication of the payback but not necessarily the cost of the plants. Generally, large-scale can be considered as greater than 500 kW, small-scale below 250 kW with very small below 25 kW. In terms of farming, a dairy herd of about 100 cows will produce sufficient slurry to produce over 30 kW over the winter or housed period. As might be expected, small-scale biogas production becomes less economically sustainable as the energy output of the plant reduces, thus decreasing the return of revenue. Whether SSAD is integrated into a business or a standalone operation, reduced revenue may necessitate subsidised production as a way forward to sustain a greater uptake of the technology at this scale.

Small-scale, particularly very small-scale, AD plants include a number of distinct advantages over their larger scale cousins. First, mobile SSADs mean that the portability of the digester allows it to travel to the waste source, rather than the waste being brought to the digester. Second, both the operators of SSADs and small-scale waste producers may gain greater control over the feedstock. Third, mobile SSADs have the advantage over developers of static off-the-shelf (or turnkey) systems as these do not always suit local or changing circumstances. Fourth, a small unit may not require planning permission or may only need to meet minimal planning regulations. Fifth, capital purchase may be superseded since the option of renting or leasing of the plant is possible, making the uptake of biogas production more attractive and a cheaper option. However, these advantages are peripheral to economic sustainability and SSADs at the smallest scale (Table 18.1) may only be viable within a niche market.

How does small-scale anaerobic digestion contribute to sustainability?

Debates on 'sustainability' and 'sustainable development' reconcile human social and economic activities alongside the long-term resilience, vulnerability and regenerative capacity of the local–global continuum of ecological systems (Sneddon, 2000). One of the greatest challenges to sustainability is climate change. Indeed, global GHG emissions resulting from human activities have increased since pre-industrial times, with a rise of 70 per cent between 1970 and 2004 (IPCC, 2007). The impacts of climate change threaten long-term resilience and regenerative capacity of physical and ecological systems through potentially irreversible and catastrophic feedbacks (Adger et al., 2005). Considering sustainability as three long-term constraints (eco-centric concerns, socio-centric concerns and techno-centric concerns) helps conceptualise the potential contribution of SSADs. Therefore, SSADs viewed as a mitigation technology can not only ameliorate some of the impacts of GHG emissions but also contribute to solutions for other environmental problems; namely, eutrophication, acidification and air pollution (Börjesson and Berglund, 2007; Deublein and Steinhauser, 2008). Clift (2007) describes eco-centric concerns as representing the constraints in that (in thermodynamic terms) the earth is a closed system with a finite biospheric capacity to absorb or adapt to the emissions from human activities. Techno-centric concerns characterise the deployment of technology and the economic system within which the technology operates, whereas socio-centric concerns represent human expectations of providing a better quality of life as well as a capacity for future generations to be as well off as we are (Guest, 2010). Adapted from Clift (2007), these three aspects of sustainability are illustrated in Figure 18.1 alongside the potential contributions of SSADs.

The most compelling reason to encourage SSAD technology is its potency as a mitigation technology to reduce GHG emissions, which is particularly relevant to livestock farming. In terms of eco-centric concerns, the livestock sector accounts for 18 per cent of GHG emissions as measured by CO_2 equivalents; but larger shares for some gases with the sector emitting 37 per cent of methane (which has 23 times the global warming potential (GWP) of CO_2) and 65 per cent of nitrous oxide (with 296 times the GWP of CO_2) (Steinfeld et al., 2006). Since the mean decay time of methane (CH_4) in the atmosphere is much shorter than that of CO_2 – 8 years compared with 50–200 years – decreasing methane emissions at source is an effective strategy to reduce the anthropogenic greenhouse effect (Deublein and Steinhauser, 2008). Using technologies such as SSADs is therefore, an important strategy in maintaining the resilience and regenerative capacity of physical and ecological systems since it captures biomethane from livestock manure that would otherwise escape into the atmosphere. Furthermore, from a techno-centric perspective, SSADs can convert biomethane into heat and, if appropriate, electrical energy using combined heat and power (CHP) units; this can further displace energy derived from alternative sources, such as fossil fuels.

Additional ecological benefits from using SSADs are derived from replacing undigested liquid manure by digested manure as a fertiliser; recovering crop residues as a feedstock;

Table 18.1 Some very small digesters available commercially.

Digester name	Supplier	Vessel output/size	Notes
Puxin	Shenzhen Puxin Science & Technology Co., Ltd. (Guangdong, China)	2.5 m³ glass fibre reinforced plastic	A single-stage AD plus a gas storage bag, a gas pump, a solar charger and appliances puxinbiogas.en.alibaba.com/productgrouplist-210340983/Portable_solar_digester_projects.html#products
Evergreen Gas	Evergreen Gas Ltd, Barrett's Mill, Ludlow, Shropshire SY8 4AH	20 to 250 kWe	evergreengas.co.uk
Muckbuster	SEaB Energy Southampton Science Park, Southampton SO16 7NP	0.5 m³	A self-contained AD, built inside a shipping container seabenergy.com/products/anaerobic-digesters/muckbuster-faqs/
Marches Biogas	Unit 4 Lower Barns Business Park, Ludlow Shropshire SY8 4DS	12–500 kWe continuously stirred tank AD	www.marchesbiogas.com
Methanogen	The Nurton, Linley, Shropshire SY9 5HW	200–500 litre 30,000 l. for larger community schemes	www.biogastronomy.co.uk The largest food digester can produce about 60 m³ of gas per day on a tonne of food waste.
Nethy energy	5th floor, 14–16 Regent Street, London SW1Y 4PH	50–150 kWe	www.nethyenergy.co.uk/?gclid=COnQtjX-trACFVMetAod-COU8Q
ITDI Portable Biogas Digester	The DOST Compound Gen. Santos Ave., Bicutan, Taguig, Metro Manila	0.2–0.5 m³	The waste-to-fuel process involves mixing 1.5 kilos of animal manure with 4.5 kilos of kitchen waste. The manure starts the fermentation. (no website)
Mini Biogas Holder	Arjun Energy Corporation Arjun Towers, 16 Rajaji Road, Salem, Tamil Nadu – 636 007, India	40 m³	Food waste based biogas plant – The gas thus generated is being used in their own canteen for cooking, which replaces the LPG cylinders to a large extent.

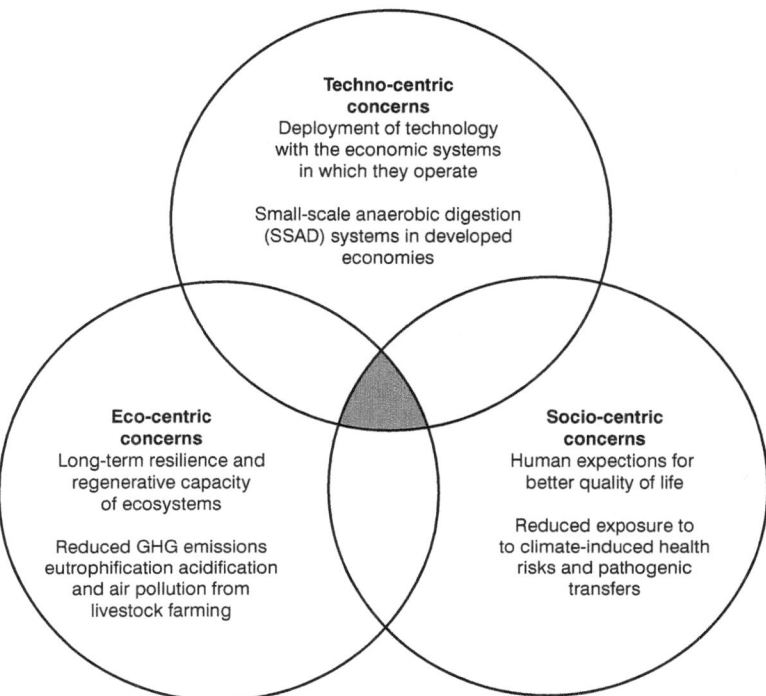

Figure 18.1 A conceptual framework for the sustainability of small-scale anaerobic digestion. Adapted from Clift (2007).

and changes in land use cropping patterns (Börjesson and Berglund, 2007). In particular, untreated or poorly managed livestock manures can be a major source of air and water pollution through nutrient leaching, mainly nitrogen and phosphorous, ammonia evaporation and pathogen transfers (Holm-Nielsen et al., 2009). For example, liquid manure in the form of pig or dairy cattle slurry, after fermentation in a biogas plant, penetrates into the soil better thus decreasing loss of N-containing gases (Deublein and Steinhauser, 2008). However, while SSADs can ameliorate externalities from human activity on ecosystems, its ultimate impact will depend on the raw materials digested, the types of energy provided and the systems which it replaces (Börjesson and Berglund, 2007).

To a large extent, the contribution that SSADs makes to social-centric concerns is determined by ecological impacts that impinge upon the rights of individuals within society and its ability to provide a better quality of life. Ecological impacts from raw manure in agriculture include GHG emissions (particularly methane), nitrous oxide emissions, and potential contamination of watercourses by pathogenic organisms. Each of these is briefly explored. First, there is growing evidence that climate–health relationships pose increasing health risks and uncertainties (Patz et al., 2005; Manzi et al., 2010). Examining studies from both developing and developed countries, Cifuentes et al. (2001) commented that mitigating against GHG emissions can have powerful and immediate benefits to public health by reducing the adverse effects of local air pollution. Next, the lower rate of nitrous oxide emissions from digestate has benefits to society, although there are also potential hazards. Evidence suggests (see Börjesson and Berglund, 2007; Holm-Nielsen et al., 2009) that compared with raw manure, the lower carbon content of digestate reduces nitrous oxide emissions during its application to land. In addition to its role in global warming, nitrous oxide is a contributor to ozone depletion, and ozone protects the biosphere from the harmful effects of solar generated ultraviolet radiation (Bolin et al., 1981). Consequently, nitrous oxide emissions, either at the source of the pollution or many kilometres downwind, can be detrimental to human health (World Health Organisation, 2003). While widespread adoption of SSADs may lessen nitrous oxide emissions, in particular circumstances inadequate storage facilities or inappropriate applications that do not match crop needs can increase both nitrous oxide and ammonia emissions (Holm-Nielsen et al., 2009; Defra, 2007). Finally, pathogenic organisms in untreated raw manure transferred from farms to watercourses also risks human exposure via bathing waters or through the consumption of contaminated shellfish (Chadwick et al., 2008). A SSAD has the potential to reduce a number of different pathogens that could otherwise transfer to water courses although the degree of this reduction depends on a number of factors including the processing temperature, the number of stages of digestion, days in digestion and diligence to prevent recontamination (Horan et al., 2004; Wagner et al., 2008). Given these three illustrations, using SSAD as a mitigation technology to ameliorate GHG and nitrous oxide emissions as well as to reduce pathogenic transfers has the potential to reduce socio-centric concerns centred around public health.

Up to now, SSADs have been proposed as a technological application to address particular eco-centric and socio-centric concerns without considering economic systems in which SSADs operate. Arguably, the convergence of SSADs with economic sustainability is not straightforward, particularly in advanced economies. For example, appraisals are often based on narrow economic definitions which assert that projects should sustain beneficial ratios for liquidity, profitability and growth (Charter and Tischner, 2001). Small-scale biomethane production becomes less economically sustainable as the energy output of the plant reduces, thus decreasing revenue. Accordingly, measuring the economic sustainability of SSADs by purely financial indicators often renders them uneconomic without governmental support. Moving away from this narrow definition of economic sustainability to one that encompasses the notion of justice embodied in sustainability, between and across generations, some of the wider eco-centric and socio-centric benefits of the technology are calculable. In particular, costing the benefits of reduced GHG emissions that SSADs potentially have on the environment and society creates a much stronger form of economic sustainability. Later in this chapter, this more inclusive form of economic sustainability is demonstrated by modelling SSADs accounting for their amelioration benefits to eco-centric systems through reduced the GHG emission from agriculture. However, for SSADs to practically marry techno-centric, socio-centric and eco-centric concerns that illustrate the technology's sustainability credentials, conducive policies and appropriate incentives are necessary, which are now explored.

Policy and incentives

At both national and international levels, climate change is a prominent political issue that requires the search for regulatory 'solutions' (Splash, 2010). The Kyoto Protocol included legal binding emission limits for industrial countries and specifically identified the promotion of renewable energy as a key strategy for reducing GHG emissions (Anandarajah and Strachan, 2010). Developing policies that reduce GHG emissions and renewable energy production to reduce fossil fuel use has fostered the development of AD at all scales.

If developed appropriately, policy can aid development and, as with many 'developing' technologies, government support is essential until an infrastructure or critical mass of manufacturing of biomethane output volume or both are achieved.[1] Incentives to support AD often reflect national and international policy agendas. For instance, developing AD, particular on farms, satisfies EU polices and directives such as the Climate Change package that commits to a highly energy-efficient, low-carbon economy (European Commission, 2010); the directive to promote renewable energy (2009/28/EC); the landfill directive (1999/31/EC); and the water framework directive (2000/60/EC). At a national level in the UK, it is only recently that AD has received government support, recognising its potential contribution in meeting the UK Government's own climate change obligations. The Climate Change Act 2008, and specific policy documents including 'Developing an Implementation Plan for AD' (Defra, 2009) are designed to meet the objectives from the EU directives and promote biogas as a tool to reduce GHG emissions and water pollution, as well as improve waste management.

Different countries and sometimes states or provinces within those countries have adopted different economic incentives to support AD, including SSADs. Principle incentives are detailed in Table 18.2 although many of these are also generic to other renewable technologies. Tradable green certificate schemes, such as Renewable Obligations Certificates (ROCs) in the UK, and Renewable Energy Certificates (RECs) in the USA and Australia, tend to promote biogas schemes that produce a minimum of 1 MWh. These schemes, while intended for larger scale development, align more closely with a market economy whereas feed-in tariffs are often regarded as more bureaucratic. Other mechanisms that assist the development of biogas include low interest loans, subsidies and grants. For example, the initial development of biogas in Denmark was assisted by long-term, low-interest, and

Table 18.2 Principal incentives that are used to promote energy generation from biogas

Principal incentives	Description	Examples of use*
Investment subsidies	Subsidies and grants to assist development	For climate mitigation projects and agricultural development in Sweden, USA and Canada (varies in different states/provinces/territories)
Low interest/soft loans	Loans with below or at market rates	USA (varies in different states)
Feed-in tariffs	A fixed price is paid for each generated kWh that is fed into the electric grid	UK, Germany, France, Italy, Ireland, Malaysia
Fixed price premiums	Government fixes a price premium that is added to the electricity price.	Denmark, Luxembourg, Spain
Green certificates	Tradable systems oblige electricity suppliers, users or retailers to hold a proportion of green certificates and demonstrate compliance at the end of a specific period. Fines may be imposed for non-compliance.	Renewable Obligation Certificates (ROCs) in the UK; Renewable Energy Certificates (RECs) in the USA and Australia; Green certificates in Belgium, Italy and Poland

Sources: Haas et al., (2011); Wilkinson (2011); Ofgem (2011), Resch et al., (2005), Biogas Regions (2008), European Renewable Energy Council (2009).
*These are illustrative examples rather than an extensive list of countries that adopt particular incentives.

indexed-linked loans (Raven and Gregersen, 2007). In the USA, similar financial incentives such as grants, tax incentives, low-interest loans are available through state government programmes (Environment Protection Agency, 2011).

Many use feed-in tariffs (FITs) as an economic incentive for biogas production. However, the targeting of these towards SSADs varies widely and in some countries is non-existent. For instance, in Malaysia, the lowest FIT band for biogas is for plants up to 4 MW, which discourages development in SSADs. Contrast this to the German feed-in tariff system that has a lowest band for plants up to 75 kW, which is designed to promote the expansion of SSADs on dairy farms with about 150 cows using 80 per cent of manure and 60 per cent of its heat (EBA, 2012). Furthermore, the German example illustrates that targeting the use of heat is increasingly important which is often paid as a bonus to the FIT payment. In the UK, the Renewable Heat Initiative (RHI) is argued to be the first of its type offering businesses and public sector organisations the opportunity to receive a reward for eligible installations (DECC, 2011). Its tariff not only provides for biomethane injection into the national gas grid, but also for heat produced by the combustion of biomethane from AD at a reasonable small-scale, under 200 kW of thermal energy. SSADs on farms, provided they can utilise the heat, may benefit economically from such incentives or FIT bonuses. For example, a pig farmer might use an SSAD to heat sheds housing livestock, or a dairy farm might use the heat for another business on the farm, such as a farm shop or bed and breakfast facility, while horticultural producers may heat greenhouses. While FITs and other incentives are designed to stimulate markets, if the scale of the lowest FIT band is set too high, this may destabilise markets for small-scale development rather than boost its development.

In addition to economic incentives the market for biogas is often delineated by national, federal and local government policies, and environmental and planning regulations. Planning can be an issue for large-scale systems but is much less onerous for SSADs, or non-existent for portable systems. Furthermore, while SSADs may circumvent many regulations that are statutory for their larger cousins, the type of feedstock often determines which regulations are obligatory. In the USA, farms seeking to install manure-fed biogas plants must coordinate with federal and/or state environmental agencies. However, as on-farm AD installations tend to be smaller (in US terms), a 'fast track process' is possible through the Federal Energy Regulatory Commission (FERC) saving both time and money rather than the standard route through the Small Generator Interconnection Process (SGIP) (Bramley et al., 2011). In the EU, complications can occur if SSADs wish to use food or abattoir wastes since additional waste licensing permission may be required to meet EU regulations on the use of animal by-products (EC 1069/2009 and EC 142/2011). However, since many SSADs are located on farms using only manure and energy crops, exemption from this may occur. For instance, in England, exemptions exist for low-risk small-scale operations providing the quantity of waste treated is less than 1250 cubic metres of waste and storage does not exceed 50 cubic metres (Environment Agency, 2010).

Some factors influencing the development of small-scale anaerobic digestion

In agriculture, as highlighted in the introduction, numerous SSADs can contribute significantly to reductions in GHGs emissions, particular from dairy and beef herds, but what is the scale of this effect? In the UK, there are about 12,000 dairy farms (DairyCo 2012) which are primary sources of GHG emissions from the agricultural industry. This is second to the transport industry. Addressing GHG emissions from dairy farms tackles the direct reduction of GHG emissions (1) from slurry and the indirect emissions of the GHG nitrous oxide after land spreading; and (2) by directly replacing fossil fuels from the production of biomethane from cattle slurry. As already noted, the SSAD is a useful technology to tackle both eco-centric and socio-centric concerns that are conspicuous in the dairy industry. Our calculations estimate that the CO_2 equivalent (CO_2e) savings per year from methane and nitrous oxide emissions from one SSAD plant treating the waste from a dairy herd of 100 cows will be 65 tonnes and from a pig unit with 300 breeding sows, 55 tonnes. These figures include housing of the livestock, manure management and emissions after land spreading. These figures do not include the replacement of fossil fuel by biomethane or artificial fertiliser by digestate. Including these, as is examined later, can create even greater savings. Assuming sales of 200 dairy SSAD plants and 100 pig SSAD plants the GHG savings will be 18,500 tonnes of CO_2e annually. Should all dairy farms use an SSAD then about 555,000 tons of CO_2e annually could be saved.

In addition to GHG reductions, there are also advantages in using digestate instead of spreading unprocessed slurry to the land. Digestate has less dry matter and hence less viscosity than the original slurry and requires less energy to be pumped. Once spread on land, the digestate will infiltrate the soil more quickly, reducing ammonia emission losses, and P and N are more readily available for crop utilisation. Mineralisation of feedstock and manure in particular provides a more predictable crop nutrient delivery and application with less run-off and leaching losses of nutrients.

Despite the potential that on-farm SSADs can contribute to sustainability by ameliorating GHG emissions and improving manure management, there are many factors that influence the efficiency and effectiveness of these systems. For example, we explore the quality of feedstock, the degree of monitoring during the AD of slurry, the temperatures that are used in the fermentation process and the fermentation process itself. While discussing each of these, their impact on sustainability is noted, although these links are elaborated more thoroughly at the end of this section.

Quality of feedstock on revenue flow

Many would like a quick answer to the question as to which is the smallest most economically viable (in a narrow economic sense) SSAD system. However, there are no quick answers given the complex interrelationships between dependencies on feedstock availability, the resulting biogas yield, and how well the SSAD system can be integrated into the existing business environment. For example, if we consider feedstock, we have to also consider combinations of feedstock that provide the highest output per unit volume while not overloading fermentation. High-performance feedstock has to provide high biogas yield per unit mass and be readily degradable. Such feedstock includes biodiesel residues, fats and greases and bakers' waste as identified in Table 18.3. If we compare the range of economic outputs from the highest and lowest feedstock sources, then the reader may have an answer to the 'burning' economic question, although this may not be as expected. A further complication is the balance of C:N trace nutrients and retention times for each feedstock as this is needed to assess pragmatic biomethane productivity.

To illustrate the effects of including high energy feedstock such as biodiesel residue, the AD analytical model (ADAM) was developed (Butler et al., 2011). The program calculates

Table 18.3 Range in values assumed for DM, ODM and potential biogas yield.

Feedstock	Dry matter (DM) $(g\ kg^{-1})$	Organic dry matter (ODM) $(g\ kg^{-1})$	Biogas yield $m^3\ kg^{-1}$ ODM
Bakers' waste	870 – 880	930 – 950	0.50 – 0.60
Biodiesel residue (glycerine)	> 980	900 – 930	1.00 – 1.10
Brewers' grains	200 – 260	750 – 950	0.50 – 1.10
Dairy slurry	60 – 110	650 – 850	0.10 – 0.80
Food waste	90 – 370	750 – 980	0.40 – 1.00
Grass silage	210 – 400	760 – 900	0.60 – 0.70
Maize silage	200 – 400	940 – 970	0.60 – 0.70
Meat waste	80 – 250	900 – 900	0.80 – 1.20
Pig slurry	30 – 100	770 – 850	0.30 – 0.80
Potato waste	60 – 180	850 – 960	0.30 – 0.90

Sources: Defra (2001, 2010); Deublein and Steinhauser (2008).

Table 18.4 Improving biogas production by adding high energy feedstocks.

	Biogas yield m^3	kW (elec)	kW (heat)	N ha^{-1}	P ha^{-1}	Biogas profit/loss	Farm profit	Total profit
UK dairy$_0$	447 073	104	130	80	30	£0	£225,869	£225,869
UK dairy$_{BDR}$	541 226	126	153	81	36	–£12,401	£226,986	£214,585
UK dairy$_{BG}$	462 713	108	134	94	32	£815	£229,179	£229,994

Source: Calculated from ADAM (Butler et al. 2011).

the increase in profit when for example 100 tonnes of biodiesel waste (glycerine) is combined with feedstock from a dairy farm. Table 18.4 presents three scenarios. The first, UK dairy$_0$, acts as a control, in that it represents the structure of UK dairy farming as determined by data from the Farm Business Survey (FBS) and the Scottish equivalent, the Farm Accounts Survey (FAS). These data are scaled to enable biogas production to break even in terms of profit (cost = revenue). As such, the required farm size is 299 hectares, of which 167 hectares is allocated to dairy production enabling a dairy herd of 334 cows, while the remainder of land is used to grow grass silage and maize silage as energy crop feedstock. To reflect the geographical limitations of growing of maize in certain parts of the UK, it is assumed that 65 per cent of the energy crop feedstock is grass silage while the remainder is maize silage. Using these assumptions, a dairy farm of this size would produce enough biomethane to produce 104 kW of electricity and 131 kW of heat. In terms of the impact on the farm, there would be savings to fertiliser costs and energy costs, as well as reducing the farm's carbon footprint. The second scenario (UK dairy$_{BDR}$) includes the 100 tonnes of biodiesel residue in the AD process, while the third scenario (UK dairy$_{BG}$) includes the addition of 100 tonnes of brewers' grains (brewery waste).

The results of this exercise demonstrate that the former produces greater biogas output (21 per cent extra as compared with 3.5 per cent for brewers' grains). However, while biodiesel residue can greatly boost biogas production, this comes at a cost. In both scenarios, costs of purchasing additional feedstock were calculated on the basis of sellers' prices. Therefore, with biodiesel residue at a cost of £200 per tonne and brewers' grain at £30 per tonne the economics of additional feedstock is mixed. Both feedstock types marginally increase farm profit but only brewers' grains increase profit from producing biogas since the revenue from the additional biogas is greater than its costs, unlike biodiesel residue.

Co-digestion with another feedstock can yield more than the sum of the two parts in some cases. So planning the feedstock intake is of great benefit to the SSAD owner. Finally, while the type and mix of feedstock used in SSADs undoubtedly influences economic output, the quality of feedstock can mean the difference between profit and loss. As such, poor harvests of energy crops, poor storage conditions or mishandling of feedstocks can all reduce biogas produced.

Monitoring and process control

Monitoring and process control technology implemented in SSADs is important since they differ in their requirements to the larger scale; but in what way? Often all types of digesters are operated below their capacity to buffer any unknown or contaminated feedstock that may sour the fermentation. However, and importantly, optimum economic and hence fermentation performance of the biogas plant is more necessary for small-scale systems and therefore some means of fundamental measurements are essential. These include: volume; temperature, as methanogens are sensitive to $\pm$ 1°C in the mesophilic or thermophilic range; and pH, although pH changes are mostly catastrophic rather than slow and need to be predicted. Volume measurements are necessary to prevent the digester from having too high or too low liquid content; and measuring the feed rate of the digester prevents overloading that can cause acidification whereas underloading results in poor performance.

In SSADs, monitoring approaches are more necessary as small volumes are more sensitive to change. Often there are a range of different parameters that can be measured to assess the fermentation, such as volatile fatty acid (VFA) composition, alkalinity, chemical oxygen demand (COD) and possibly dissolved hydrogen. In the gas phase, methane, CO_2, VFAs, H_2S and NH_3 can be of use. Despite the need for monitoring, implementation of automatic monitoring and control is considered an added expense to AD that has little economic return. However, if process control strategies are introduced then the SSAD performance may be improved. Various researchers have recently suggested measuring alkalinity (Boe et al., 2010; Punal et al., 1999; Ward et al., 2008, 2011). Additional means of monitoring can include spectrophotometric approaches which are described in numerous publications (Nordberg et al., 2000; Hansson et al., 2002; Holm-Nielsen et al., 2007; Luck et al., 2010; Ward et al., 2011). These primarily concern measurements by near infrared red spectroscopy that has reduced spectral interference from the water bands compared with the mid-infrared region. However, the ruby-tipped near-infrared sensor is one of the few that is robust enough to withstand the effect of the abrasive fermentation liquid. Moreover, these multi-optical fibre systems are expensive.

The digester needs to be characterised for process control to maximise digester performance. Manual control may work well on a daily basis but automatic process control offers more advantages, such as adjustment to variations of incoming feedstock and recognition of impending digester failure. There are a range of numeric means of process control that have been proven in the laboratory, and these include neural networks (Holubar et al., 2000), fuzzy logic of difficult feedstock (Scherer et al., 2009), and variable structure model (Tartakovsky et al., 2005) as well as nonlinear approaches (Aguilar-Garnica et al., 2009) that have yet to be proven for uptake by the industry. Such approaches are dependent on reliable sensor data and have only recently become available as a consequence of the wastewater industry's efforts. Finally, little is known of how the dynamics of the SSAD may impact on performance. Here further research is required.

Use of thermophilic temperatures

Some potentially interesting environmental benefits associated with SSADs can hang in the balance because of economics. One is these is the use of thermophilic temperatures, optimally around 49–57°C rather than mesophilic temperatures of around 30–38°C. Thermophilic temperatures require more process control with small-scale systems and there is the need for effective insulation. However, the benefits of achieving higher temperatures are improved kinetic rates of biogas formation, better pathogen kill and the removal of weed seeds from feedstock. Spore formers, such as *Bacillus* spp, generally are unaffected whereas *Clostridium* spp, also a spore former, is affected according to species and generally reduces by about 1 $\log_{10}$ order of magnitude. Both weed seeds and pathogens are reduced substantially at thermophilic temperatures. *Ascaris* eggs are reduced at 55°C. At 50°C, *Salmonella* spp., *Campylobacter* spp., *Listeria* spp. and *E. coli* were reduced below their detection limit within 24 h (Wagner et al., 2008). These are all useful benefits for farmers in reducing cross-contamination from the spreading of digestate.

Effectiveness of fermentation

In SSADs, as with larger-scale digesters, the effectiveness of the fermentation is important. There are several easy approaches to having a high-performing digester, for example, using a feedstock with a high biological methane potential such as bakers' waste with about 700 m^3 of biogas per tonne. However, with feedstock that is not readily decomposed, unlike biodiesel residues, energy crops do not hydrolyse quickly because of structural interference due to the relatively high lignocellulose content that prevents a high rate of hydrolysis prior to methanogenesis. However, biomass throughput can be enhanced by: (1) producing more efficient fermentation with high flow through digesters that would enable a higher gas production per capital expenditure; (2) enabling process control to operate nearer to the maximum loading rate; (3) pre-treating the feedstock for more rapid decay, thus producing more biogas per unit feedstock mass (Ward et al., 2008); and (4) the use of appropriate mixtures of different feedstock types. There are additional approaches that involve heat, mechanical or ultrasound maceration and the addition of chemicals, normally acids or bases. Biological treatment can involve the addition of organisms or direct application of enzymes to improve the degradation of the feedstock. These include rumen fluid that contain the bacterium *Fibrobacter succinogenes*, which play a major role in degradation of fibre (Lissens et al., 2004). Often mineral supplements that are co-enzyme factors are necessary for continued fermentation performance. These and other factors are discussed in more detail in Chapter 12.

Improving performance is not exclusively limited to biogas production, but also to appropriate processing of the biogas for use. Paramount to successful operation of SSADs is the need to ensure an integrated approach that may provide advantages. For example, a microbrewery may require large amounts of heat and has the resulting feedstock of mash from the hops for the digester. These circumstances may require heat and not need a relatively expensive CHP unit. In other situations, heat produced from the CHP system can dry materials from other industries or the CO_2 and digestate could be used to enhance tomato growth within a glasshouse production system. Other possibilities for SSADs include the need to reduce pathogen content in the waste stream.

Sustainability, complexity and factors influencing small-scale anaerobic digestion

What the four factors discussed offer is a glimpse at the complexity that these contribute towards different sustainability concerns. These four factors are only a few of the numerous other technological approaches available but discussion of these are outside of the scope of this chapter. For instance, inexpensive pre-treatment of feedstock requires an extensive investigation of the most appropriate technologies. Currently most pre-treatments probably involve unused heat from the CHP. Incentives to practically use surplus heat are encouraged by the Renewable Heat Initiative in the UK, while other countries, such as Germany, pay bonus tariffs. Pre-treatment approaches therefore need to be investigated, particularly if the heat is utilised for income or as a resource including enhancing or extracting more methane from the feedstock.

The type and quality of feedstock is likely to have both positive and negative impacts on all three aspects of sustainability. Feedstock producing higher quantities of biogas, providing their cost of production or purchase price are not too expensive or difficult to integrate into an SSAD, can result in positive economic gains, reduced GHG emissions and with the associated societal benefits. However, poor-quality feedstock, whether this is the result of a poor harvest, poor handling or inappropriate storage facilities, curtails environmental benefits and potentially increases GHG emissions over and above what is expected, and delivers poor economic returns.

The sustainability benefits of monitoring and processing feedstock at thermophilic temperatures in an SSAD can address both eco-centric and socio-centric concerns through ensuring optimal biogas production while minimising GHG emissions. However, in the short-term these technological solutions do not necessarily increase economic returns if the marginal cost of producing an additional unit of biogas does not cover the cost of implementing the technology. Indeed, it is likely to require a far greater diffusion of these technologies before costs are reduced, making their application economically sustainable. Improving the fermentation of feedstock in the digester has the potential to meet all three sustainability concerns. More efficient fermentation, operating near maximum loading rates, pre-treatment and appropriate feedstock mixes, improve biogas production and revenue as well as reducing GHGs emissions. Furthermore, for SSADs these need not be expensive solutions.

Finally, trying to balance the three sustainability concerns illustrates that there are no simple trade-offs. Instead, potentially sustainable solutions that SSADs provide are found on a complex multi-dimensional trade-off frontier. Whereas using SSADs to reduce eco-centric and socio-centric concerns tend to complement each other, improving the technology to provide greater impacts will often struggle to deliver worthwhile economic returns (Figure 18.2). This does not meant that this technology should not be pursued; indeed it should since the long-term benefits that SSADs offer, particular if widespread uptake occurs, could provide solutions to eco-centric, socio-centric and techno-centric concerns.

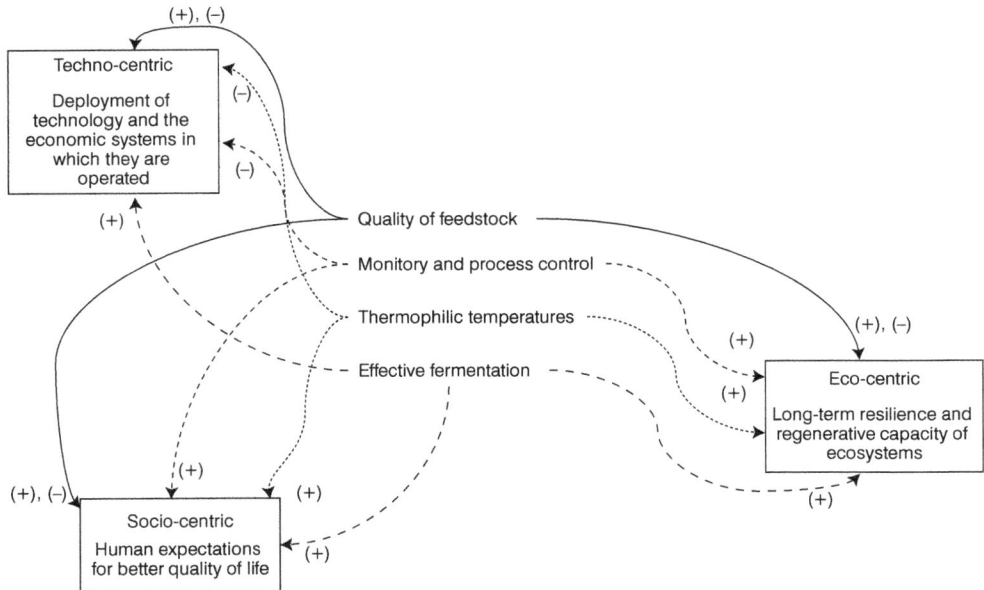

Figure 18.2 Complex interplay in developing sustainable small-scale anaerobic digestion

Valuing the sustainability of SSAD

Minimising GHG emissions, as already noted, benefits both eco-centric and socio-centric concerns. But what is the value of the reductions that SSADs can make to society? Using the shadow price of carbon (SPC) in a social cost–benefit analysis (CBA) indicates the potential value of carbon savings.[2] Two sets of carbon price estimates are used (Figure 18.3): tradable values for the EU Emissions Trading Scheme; and non-tradable emissions including those from agriculture.

On the revenue side of the CBA, carbon savings achieved by SSADs on farms include savings in CO_2e from energy (heat and electricity) production; reduced GHG emissions from manure management and spreading slurry direct to land; and lower fertiliser needs because of better nutrient uptake by plants from applied digestate. On the cost side, the costs captures the construction of a plant; the steel used in the CHP generator;[3] the production and use of fertiliser; greenhouse gas emissions from livestock slurry systems and land spreading of manures (CO_2, N_2O and CH_4); transport emissions from moving feedstocks and digestate; and the production of farm-based feedstocks.[4]

Modelling an average dairy farm (122 cows) using ADAM, with an SSAD plant (38 kW electric and 60 kW thermal), is likely to produce £0.25 million of CO_2e savings (7000 tonnes over 20 years) for only a modest cost (£0.02 million). Also modelled are the costs and benefits for a larger SSAD (104 kW electric and 131 kW heat) previously illustrated in this chapter. The marginal costs per kW in both the smaller and larger SSADs in Table 18.5 illustrate that SSAD on the small dairy farm provides the best value in terms of NPV. The higher value of benefits associated with the small dairy farm are related to GHG savings associated with reduced CH_4 and N_2O emissions from slurry stores and manure spreading operations. Since the large dairy farm uses more energy crops as a feedstock this increases kW but reduces the

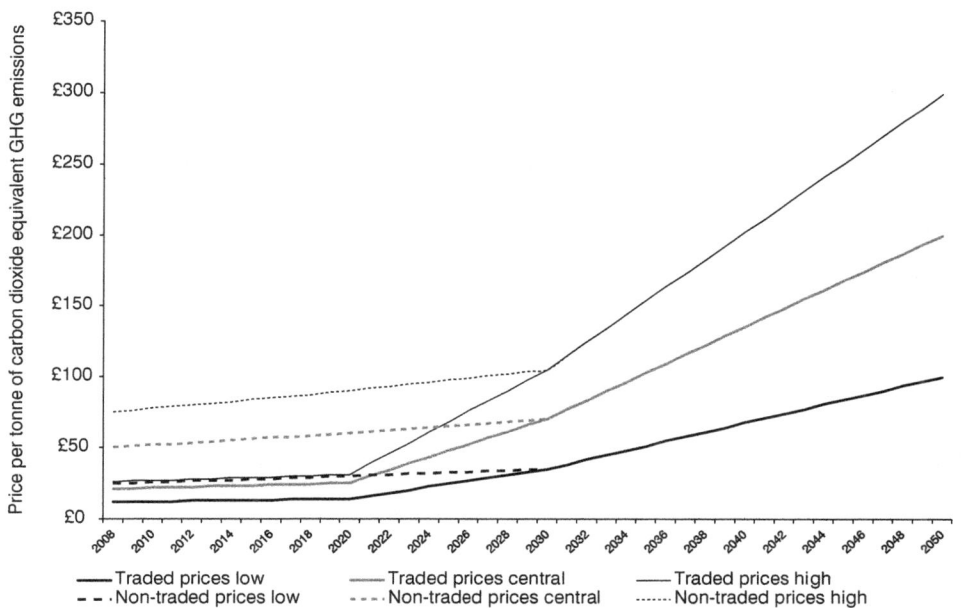

Figure 18.3 Shadow price of carbon based on the EU ETS between 2008 and 2050

Table 18.5 Discounting the shadow price of carbon (SPC) for SSAD plant scenarios.

	SSAD farm scenarios	
	Dairy farm (small with SSAD 38kW$_e$)	Dairy farm (large with SSAD 104 kW$_e$)
SPC benefits (£m)	£0.25	£1.16
SPC costs (£m)	£0.02	£0.08
SPC NPV (£m)	£0.23	£1.08
SPC benefits per kW installed	£6,665	£3,455
SPC costs per kW installed	£476	£249
SPC NPV per kW installed	£6,189	£3,205

Source: Calculated from ADAM (Butler et al., 2011).

relative benefits from reduced GHG emissions, since the production of energy crops, such as maize and grass silage, are not carbon neutral.

The marginal carbon cost curves for the small dairy farm with an SSAD is illustrated in Figure 18.4. SSADs on farms may not be economically viable due to their size yet it is important to identify how carbon savings can occur. The first observation is that while there are some carbon costs in the construction of the SSAD plant and in the supply of feedstock, the benefits are much greater with 42.1 per cent of total tonnes of CO_2e saved on dairy farms being attributable to producing electricity from renewable sources instead of fossil fuels. The second observation is the reduction in GHG emissions from livestock slurry systems and manure spreading. On the small dairy farm with an SSAD these save 23.6 per cent from livestock systems and 20.7 per cent from better manure management of total tonnes of CO_2e.

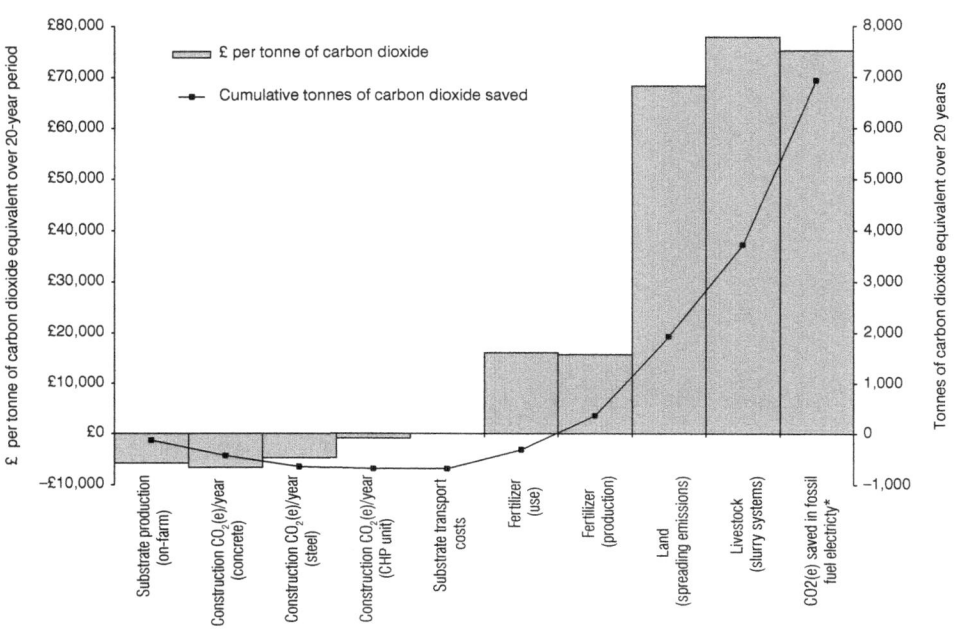

Figure 18.4 Marginal carbon cost curve per kW electricity produced on a dairy farm

When costed these carbon savings account for 30.8 per cent by changing livestock slurry systems and 27.0 per cent from managing manures. It is evident from Figure 18.4 that even if the electricity generated is additional rather than substitutive for energy produced by fossil fuels, the reduction in GHG emissions associated with livestock slurry systems and spreading of manures outweigh the carbon costs associated with the construction, maintenance and supply of feedstock for the SSAD plant on the dairy farm. Costing SSAD plants in terms of carbon clearly demonstrates the benefits, even at the small-scale. Indeed, further carbon savings could be made if heat is also utilised within the farming system.

Returning to the question about the economic sustainability of SSADs, this chapter opines that SSADs are unlikely to be economically sustainable at the smallest scale. This is particularly pertinent at the lower range when using narrow definitions of economic sustainability. However, a more comprehensive definition incorporating the value of environmental and social benefits, through the proxy measure of GHG emission savings, demonstrates how SSADs provide economic sustainability.

Concluding remarks and future developments for the SSAD

As 75 per cent of the population in India live in villages, SSADs at the smallest scale have considerable potential. Under such circumstances, the total amount of gas that can be produced from manure and sewage is estimated to be equivalent to 930 million tonnes of coal (Chiranjivi, 1978). Not surprisingly, results from an analytical hierarchy process of ranking of barriers for very small SSADs at the smallest scale in Thailand show that the three most important barriers in the adoption of SSADs are (i) high investment cost, (ii) lack of financial resources, and (iii) lack of experts and skilled manpower (Limmeechokchai and Chawana, 2005).

This is probably not so dissimilar to the situation in many countries. SSADs and those at larger scales are generally considered appropriate for developed or first-world countries where incentives or policies are necessary. Developing policies that reduce GHG emissions and increase renewable energy production has fostered the development of AD at all scales. In the broader context, there should be a place for SSADs both as low-technology and high-technology options. More consideration should be given and aimed at how SSADs can integrated into agriculture more effectively to produce biogas, with particular emphasis placed on transferring knowledge and practices from the laboratory into real-life conditions – for example, an efficient utilisation of different feedstock at advantageous ratios as well as pre-treatment options.

With the correct infrastructure and financial support, the sustainability credentials of SSADs are evident. SSADs can effectively contribute to a reduction in GHG emissions from agriculture with complementary benefits to both ecosystems and society. For SSAD plants at the smallest scale, financial incentives through FITs and/or the RHI scheme as in the UK, can provide economic stability and replace fossil fuels both directly as biomethane injections and indirectly as heat and electricity. So in answering the question of what would be the current smallest scale of digester that would give a reasonable rate of return, say less than a 10-year payback, Table 18.6 demonstrates a rule-of-thumb response, based on the following assumptions for the UK, as the authors are more familiar with these circumstances: (1) the FIT is at 14p per kWh of electricity and that heat is utilised and 5p RHI is received for each kWh of heat; (2) standard biogas production rates are 25, 190 and 210 m^3 for cattle slurry, grass silage and maize silage respectively; (3) cattle slurry is assumed to be available through a 200-day period of the year; (4) all the heat generated is available for the Renewable Heat Initiative. Within Table 18.6 there are clearly highlighted issues such as the difference in higher retention time (HRT) for different feedstock and the cost of dealing with dryer feedstock. Integrating a small digester into an existing farming system may make the farm more viable or provide other business possibilities. For instance, diversifying to use and treat food wastes not only provides extra income from biogas production but also additional NPK and trace elements that will reduce the cost of imported fertiliser. However, these benefits need weighing against the costs of additional permits and equipment for pre-treatment. Furthermore, the heat produced from a CHP unit may be used to dry crops or locally produced products. Thus, the green credibility of the business is improved with financial reward. If carbon savings were also rewarded, these credentials would be further improved. Nevertheless, SSADs have disadvantages of high capital costs with enabling technologies such as gas clean-up for grid injection or for use as transport fuel. Yet these approaches ensure more effective energy or carbon utilisation, which may be available once these technologies improve or the volume of large-scale gas purification reduces unit costs. A further high cost may be incurred if electricity is exported from the farm, since an SSAD may experience potentially high connection charges.

Future research for SSADs, as you may suspect, will be driven by the need for economic return in developed economies. The development for improved SSADs requires new and improved materials for buildings and performance. Process control and improved fermentation need to be better for SSADs due to instability and therefore require new understanding. Major research targets for biogas development are difficult to meet because the experimental period and inoculation (or start-up) require considerable periods of time. New approaches have

Table 18.6 Viable SSAD according to 2012 costs.

Feedstock	Digester size m³	Digester capital costs £,000	Realistic HRT days	Biogas yield m³ day⁻¹	Annual income £s	Annual costs @ 5% pa over 10 yrs	Payback years	Notes
Cattle slurry	500	400	20	625	98,094	200,000	6	No additions of waste silage
Cattle slurry	100	150	15	167	26,158	75,000	9	Improved technology lower HRT
Grass silage	150	600	40	712.5	111,827	480,000	10	Dry digester 8% running costs
Maize silage	120	500	40	630	98,879	400,000	9	Dry digester 8% running costs
Maize silage	150	600	40	787.5	123,598	480,000	9	Dry digester 8% running costs

Source: Calculated from ADAM (Butler et al., 2011).

been adopted to reduce resources and time and include, for example, the use of shortened factorial design for experiments to predict which contributions of variables are significant to biogas production (Rao et al., 2008). Finally, the progress of SSAD technology needs to be periodically assessed to determine the effects on management, economic and environmental impacts through benchmarking to achieve best practice.

Notes

1 SSADs is technically not a new technology since it has been around since the late nineteenth century, with the first anaerobic digester being credited to India, in 1897, at the Matunga Leper Asylum in Bombay (Mumbai) (Khanal, 2008). However, in the context of this chapter, SSADs is a developing technology.
2 SSAD is assumed to be a project over 20 years using a discount rate of 3.5 per cent, as used in the UK Treasury Green Book (see Defra, 2007).
3 While CHP are constructed from many components made from different materials, steel and alloy steel are frequently used and is therefore used as a proxy.
4 Carbon equivalent figures are calculated from IPCC (2007).

References

Adger, W.N., Arnell, N.W., and Tompkins, E.L. (2005) 'Successful adaptation to climate change across scales', *Global Environmental Change*, vol. 15, pp. 77–86.

Aguilar-Garnica, E., Dochain, D., Alcaraz-Gonzalez, V., and Gonzalez-Alvarez, V. (2009) 'A multivariable control scheme in a two-stage anaerobic digestion system described by partial differential equations', *Journal of Process Control*, vol. 19, p1324.

Anandarajah, G. and Strachan, N. (2010) 'Interactions and implications of renewable and climate change policy on UK energy scenarios', *Energy Policy*, vol. 38, no 11, pp. 6724–6735.

Biogas Regions (2008) 'Biogas production and market development through local and regional partnerships', www.biogasregions.org/doc/newsletter/newsletter1.pdf (accessed 20 September 2011).

Boe, K., Batstone, D.J., Steyer, J.P., and Angelidaki, I. (2010) 'State indicators for monitoring the anaerobic digestion process', *Water Research*, vol. 44, no 20, pp. 5973–5980.

Bolin, B., Crutzen, P.J., Vitousek, P.M., Woodmansee, R.G., Goldberg, E.D. and Cook, R.B. (1981) 'An overview of contributions and discussions at the SCOPE workshop on the interaction of biogeochemical cycles', Örsundsbro, Sweden, 25–30 May, www.icsu-scope.org/downloadpubs/scope21/chapter01.html (accessed 31 March 2012).

Börjesson, P. and Berglund, M. (2007) 'Environmental systems analysis of biogas systems—Part II: The environmental impact of replacing various reference systems', *Biomass and Bioenergy*, vol. 31, pp. 326–344.

Bramley, J., Fobi, L., Peterson, C., Rainville, L., Cheng-Hao Shih, J., Teferra, A., and Wang, R.Y. (2011) 'Agricultural biogas in the United States: A market assessment', Department of Urban & Environmental Policy & Planning, Tufts University.

Butler, A., Hobbs, P., and Winter, M. (2011) 'Expanding biogas on UK dairy farms: a question of scale', 85th Annual Conference of the Agricultural Economics Society, University of Warwick, 18–20 April.

Chadwick, D., Fish, R., Oliver, D.M., Heathwaite, L., Hodgson, C., and Winter, M. (2008) 'Management of livestock and their manure to reduce the risk of microbial transfers to water–the case for an interdisciplinary approach', *Trends in Food Science*, vol. 19, no 5, pp. 240–247.

Charter, M. and Tischner, U. (2001) *Sustainable Solutions*, Sheffield: Greenleaf.

Chiranjivi, C. (1978) 'Design analysis of small-scale anaerobic digesters in India', Energy from biomass and wastes: symposium, August 14–18, Washington, DC.

Cifuentes, L., Borja-Aburto, V.H., Gouveia, N., Thurston, G. and Davis, D.L. (2001) 'Hidden health benefits of greenhouse gas mitigation', *Science*, vol. 293, pp. 1257–1259.

Clift, R. (2007) 'Climate change and energy policy: The importance of sustainability arguments', *Energy*, vol. 32, pp. 262–268.

DairyCo (2012) 'Farm data – registered dairy production holdings' www.dairyco.org.uk/media/89690/uk_producer_numbers.xls (accessed 18 July 2012).

DECC (2011) *Renewable Heat Initiative*. London: Department of Energy and Climate Change.

Defra (2001) *Managing livestock manures: Making better use of livestock manures on grassland*, London: Defra.

Defra (2007) *The social cost of carbon and the shadow price of carbon: What they are, and how to use them in economic appraisal in the UK*, London: Defra.

Defra (2009) 'Developing an implementation plan for anaerobic digestion', Report of the Anaerobic Digestion Task Group, July 2009, Defra, London.

Defra (2010) *Fertiliser recommendations for agricultural and horticultural crops (RB209)*, London: Defra.

Deublein, D. and Steinhauser, A. (2008) *Biogas from waste and renewable resources: An introduction*, Weinheim: Wiley-VCH.

EBA (2012) 'Newsletter February 2012. European Biogas Association', www.european-biogas.eu/eba/images/stories/eba%20newsletter%2002.2012.pdf (accessed 24 May 2012).

Environment Agency (2010) 'Anaerobic digestion and environmental permitting', www.environment-agency.gov.uk/static/documents/Business/Anaerobic_Digestion_and_Environmental_Permitting.pdf (accessed 26 May 2012).

Environment Protection Agency (2011) 'Funding resources', www.epa.gov/chp/funding/bio.html#tabnav (accessed 20 September 2011).

European Commission (2010) 'The EU climate and energy package', ec.europa.eu/environment/climat/climate_action.htm (accessed 29 June 2011).

European Renewable Energy Council (EREC) (2009) 'Renewable energy policy review – Italy', www.erec.org/fileadmin/erec_docs/Projcet_Documents/RES2020/ITALY_RES_Policy_Review_09_Final.pdf (accessed 20 September 2011).

Guest, R. (2010) 'The economics of sustainability in the context of climate change: An overview', *Journal of World Business*, vol. 45, no 4, pp. 326–335.

Haas, R., Panzer, C., Resch, G., Ragwitz, M., Reece, G., and Held, A. (2011) 'A historical review of promotion strategies for electricity from renewable energy sources in EU countries', *Renewable and Sustainable Energy Reviews*, vol. 15, no 2, pp. 1003–1034.

Hansson, M., Nordberg, Å., Sundh, I., and Mathisen, B. (2002) 'Early warning of disturbances in a laboratory-scale MSW biogas process', *Water Science Technology*, vol. 45, no 10, pp. 255–260.

Holm-Nielsen, J.B., Andree, H., Lindorfer, H. and Esbensen, K.H. (2007) 'Transflexive embedded near infrared monitoring for key process intermediates in anaerobic digestion/biogas production', *Journal of Near Infrared Spectroscopy*, vol. 15, no 2, pp. 123–135.

Holm-Nielsen, J.B., Al Seadi, T., and Oleskowicz-Popiel, P. (2009) 'The future of anaerobic digestion and biogas utilization', *Bioresource Technology*, vol. 100, pp. 5478–5484.

Holubar, P., Zani, L., Hager, M., Froschl, W., Radak, Z., and Braun, R. (2000) 'Modelling of anaerobic digestion using self-organizing maps and artificial neural networks', *Water Science and Technology*, vol. 41, p149.

Horan, N.J., Fletcher, L., Betmal, S.M., Wilks, S.A. and Kevil, C.W. (2004) 'Die-off of enteric bacterial pathogens during mesophilic anaerobic digestion', *Water Research*, vol. 38, pp. 1113–1120.

IPCC (2007) 'Contribution of Working Groups I, II and III to the Fourth Assessment Report of the Intergovernmental Panel on Climate Change', Geneva: Intergovernmental Panel On Climate Change.

Kaharabata, S.K. and Schuepp, P.H. (2000) 'Estimating methane emissions from dairy cattle housed in a barn and feedlot using an atmospheric tracer', *Environmental Science Technology*, vol. 34 no 15, pp 3296–3302.

Khanal, S.K. (2008) *Anaerobic biotechnology for bioenergy production: principles and applications*, Ames, IA: Wiley-Blackwell.

Lansing, S., Botero, R.B. and Martin, J. (2007) 'Small-scale digesters in Costa Rica', *BioCycle*, vol. 48, no 2, pp. 48–51.

Limmeechokchai, B. and Chawana, S. (2005) 'Sustainable energy development strategies in the rural Thailand: The case of the improved cooking stove and the small biogas digester', *Renewable and Sustainable Energy Reviews*, vol. 11, no 5, pp. 818–837.

Lissens, G., Verstraete, W., Albrecht, T., Brunner, G., Creuly, C., Seon, J., Dussap, G. and Lasseur, C. (2004) 'Advanced anaerobic bioconversion of lignocellulosic waste for bioregenerative life support following thermal water treatment and biodegradation by *Fibrobacter succinogenes*', *Biodegradation*, vol. 15, pp. 173–183.

Luck, S., Buge, G., Plettenberg, H. and Hoffmann, M. (2010) 'Near infrared spectroscopy for process control and optimisation of biogas plants', *Engineering Life Science*, vol. 10, no 6, 537–543.

Lukehurst, C. (2009) 'International Energy Agency Bioenergy Task 37', University of Reading Rural Research and Strategy Partnership, March 25, 2009.

Manzi, T., Lucas, K., Lloyd, T. and Allen, J. (2010), *Social sustainability in urban areas: Communities, connectivity and the urban fabric*, London: Earthscan.

Nordberg Å., Hansson, M., Sundh, I., Carlsson, H. and Mathisen, B. (2000) 'Monitoring of a bio-gas process using electronic gas sensors and near-infrared spectroscopy (NIR)', *Water Science Technology*, vol. 41, pp. 1–8.

Ofgem (2011) 'Renewables obligation', www.ofgem.gov.uk/Sustainability /Environment/RenewablObl/Pages/RenewablObl.aspx (accessed 20 September 2011).

Patz, J.A., Campbell-Lendrum, D. and Holloway Foley, J.A. (2005) 'Impact of regional climate change on human health', *Nature*, vol. 436, pp. 310–317.

Punal, A., Lorenzo, A., Roca, E., Hernandez, C. and Lema, J.M. (1999) 'Advanced monitoring of an anaerobic pilot plant treating high strength wastewaters', *Water Science and Technology*, vol. 40, no 8, pp. 237–244.

Rao, R.S., Kumar, C.G., Prakasham, R.S and Hobbs, P.J. (2008) 'The Taguchi methodology as a statistical tool for biotechnological applications: A critical appraisal'. *Biotechnology Journal* vol. 3 no 4, pp. 510–523.

Raven, R.P.J.M. and Gregersen, K.H. (2007) 'Biogas plants in Denmark: successes and setbacks', *Renewable and Sustainable Energy Reviews*, vol. 11, no 1, pp. 116–132.

Resch, G., Lopez-Polo, M., Auer, H. and Haas, R. (2005) 'Electricity from renewable energy sources in EU-15 countries – A review of promotion strategies', Energy Economics Group (EEG), Institute of Power Systems and Energy Economics, Vienna University of Technology, Austria.

Scherer, P., Lehmann, K., Schmidt, O. and Demirel, B. (2009) 'Application of a fuzzy logic control system for continuous anaerobic digestion of low buffered, acidic energy crops as mono-substrate', *Biotechnology and Bioengineering*, vol. 102, p. 736.

Sneddon, C.S. (2000) 'Sustainability in ecological economics, ecology and livelihoods: a review', *Progress in Human Geography*, vol. 24, no 4, 521–549.

Splash, C.L. (2010) 'The brave new world of carbon trading', *New Political Economy*, vol. 15, no 2, pp. 169–195.

Steinfeld, H., Gerber, P., Wassenaar, T., Castel, V., Rosales, M. and Haan, C. de (2006) *Livestock's long shadow: Environmental issues and options*, Rome: FAO.

Tartakovsky, B., Morel, E. and Guiot, S.R. (2005) 'Application of a VSM-based process control to a bench-scale anaerobic digestor', *Industrial & Engineering Chemistry Research*, vol. 44, p106.

Wagner, A.O., Gstraunthaler, G. and Illmer, P. (2008) 'Survival of bacterial pathogens during the thermophilic anaerobic digestion of biowaste: laboratory experiments and in situ validation', *ANAEROBE*, vol. 14, no 3, pp. 181–183.

Ward, A.J., Hobbs, P.J., Holliman, P.J. and Jones, D.L. (2008) 'Optimisation of the anaerobic digestion of agricultural resources', *Bioresource Technology*, vol. 99, pp. 7928–7940.

Ward A.J., Hobbs, P.J., Holliman, P. and Jones, D.L. (2011) 'Evaluation of near infrared spectroscopy and software sensor methods for determination of total alkalinity in anaerobic digesters', *Bioresource Technology*, vol. 102, no 5, pp. 4083–4090.

Wilkinson, K.G. (2011) 'A comparison of the drivers influencing adoption of on-farm anaerobic digestion in Germany and Australia', *Biomass and Bioenergy*, vol. 35, no 5, pp. 1613–1622.

World Health Organisation (2003), *Health aspects of air pollution with particulate matter, ozone and nitrogen dioxide*', Bonn: WHO.

Chapter 19

Biogas technology for developing countries

An approach to sustainable development

M.S. Dhanya,[1,2] S. Prasad[1] and Anoop Singh[3*]

[1]Division of Environmental Sciences, Indian Agricultural Research Institute, New Delhi, India; [2]Present address: Centre for Environmental Sciences & Technology, Central University of Punjab, Bathinda, India; [3]Department of Scientific and Industrial Research (DSIR), Ministry of Science and Technology, Technology Bhawan, New Mehrauli Road, New Delhi 110016 India
* Corresponding author email: apsinghenv@gmail.com

Introduction

Non-conventional, easily available energy sources like waste from animals and agricultural residues has attracted the attention of developing and underdeveloped countries due to the energy crisis, which prompted a steep rise in fuel prices. The current disposal practices for agricultural residues have caused widespread environmental concern as they represent hindrance to sustainable development in rural areas as well as to national economies (Sheehan, 2009). Environmental contamination has also necessitated identification of environmentally sound and economically feasible technologies for waste management (Prasad et al., 2007a,b). Anaerobic digestion is a versatile, effective and established method that is being used world-wide for the digestion of different organic wastes and the production of energy in the form of biogas (Verstraete et al., 2005). Organic waste materials of agricultural, industrial and municipal origin can be converted anaerobically into biogas by the action of rumen-derived microorganisms. This technology has the potential to reduce greenhouse gas emissions, because methane as a main constituent of biogas would otherwise be released into the atmosphere, provoking a greenhouse effect that is 21-fold that of the same volume of carbon dioxide. Apart from this, biogas being a cost-effective and environmentally sound energy source helps in improvement of livelihood and promotes sanitation and a healthier household environment in rural areas. Biogas technology also improves rural livelihoods as it is a cleaner fuel than traditional fuels like wood and fossil fuel and can be used for cooking, lighting and heating.

Energy scenario in developing countries

Traditional fuels like wood, charcoal, agricultural residues and animal wastes are major contributors to household energy supply in many of the developing countries having agrarian economies (FAO, 2005). Thornton et al. (2002) reported that the livestock sector directly supports the livelihoods of 600 million poor smallholder farmers in the developing world. Many tonnes of manure from this sector (cattle, buffalo herds, etc.) are produced per day.

This was often used as fertilizer but, due to shortage of firewood, is increasingly dried and used for cooking fuel. About 2 billion people globally use biomass for cooking and heating (Anon, 2005).

Li et al. (2005) compared the total energy usage in 15 least developed countries (LDC) and developed countries and found that 30–95 per cent of energy use was for household consumption, especially for cooking in LDC, which is 25–30 per cent more than that of developed countries. India uses half of its total energy consumption for cooking, of which 87 per cent was from non-commercial fuels such as firewood (Bhatt and Sachan, 2004). Similarly in Yunnan province of China, household energy accounts for 64 per cent of the province's total energy consumption and 55 per cent is from traditional biomass fuels such as firewood and straw (Li et al., 2005). Similarly, an average household in Pakistan depends on biomass for their energy needs, typically using 23–25 kg of firewood, 14–80 kg of dung or 11–60 kg of crop residues (Mirza et al., 2008). But this energy use is often associated with many environmental, health and social problems.

Negative impacts from usage of conventional energy sources

Methane contribution to GHG

Methane is a potent GHG, contributing 17 per cent of total GHG emissions. With a global warming potential of 21 over carbon dioxide, its concentration has increased by 150 per cent to 1800 ppm since the 1750s and accounts for 20 per cent of total radiative forcing (IPCC, 2001, 2007). 590–880 million tons of methane is released worldwide into the atmosphere through microbial activities from the decomposition of biomass (GATE and GTZ, 2007a). On a regional basis, Asia was reported to be the major methane emitter in 2005; China, India, the United States, the European Union and Brazil were the top five methane-emitting economies (Climate Analysis Indicators Tool (CAIT) Version 6.0, 2009). China and India contributed 853 Tg (13 per cent of the world total) and 548 Tg (9 per cent of world total) CO_2e, respectively. Agriculture is the leading source of GHG emissions in the world, contributing around 14 per cent of the total. The primary sources of agricultural GHGs emissions are livestock (particularly cattle, poultry and pigs, which contribute 37 per cent of human-induced methane emissions) and the cultivation of rice paddy (FAO, 2006). Methane emission from enteric fermentation accounts for 16 per cent of the world's annual methane emissions (Tyler, 2007). GHG emissions by India currently comprise 55 per cent CO_2, 23 per cent methane and 22 per cent N_2O. Though the agricultural sector of India contributes only 1 per cent to the country's total CO_2 emissions, it dominates in methane (5 MMT, i.e. 50 per cent) and N_2O emissions (0.31 MMT). The country has a livestock population of around 500 million (with an expected growth rate of 1.23 per cent), accounting for 15 per cent of the global livestock population on 2 per cent of the world's geographical area (Steinfeld et al., 2006).

Methane emissions from livestock wastes

As stated above, livestock contributes 17 per cent of global GHG emissions and 37 per cent of anthropogenic methane. Two factors affecting methane emission from livestock manure are the amount of manure produced (i.e. average amount of manure produced per animal and the number of animals) and the portion of the manure that decomposes anaerobically (which depends on the manure management system used and the climate – primarily temperature).

More than 230 million metric tons of carbon dioxide equivalent (MMTCO$_2$e) of methane emissions, roughly 4 per cent of total anthropogenic methane emissions, were contributed by livestock manure in 2005, of which 40 per cent was derived from swine, 20 per cent from non-dairy cattle, and 20 per cent from dairy cattle. Poultry is also a significant source of methane emissions in certain countries. The total estimated methane emissions (including enteric fermentation and manure management) from Indian livestock were 11.75 Tg in 2003 (reported in India's Initial National Communication (IINC) to the United Nations Framework Convention on Climate Change). Enteric fermentation constitutes a major part of the total methane emissions, accounting for approximately 91 per cent, or 10.65 Tg of the total, while manure management of livestock accounts for only 9 per cent, or 1.09 Tg. Cattle and buffalo are the major source of methane emissions (10.9 Tg) compared with emissions from other livestock (0.86 Tg) (Chhabra et al., 2009).

Methane emissions from agro-industrial wastes

Agro-wastes, such as crop debris, and spoilt or discarded produce and agro-industrial wastes, such as seed meal residue after oil extraction, can emit methane from its biodegradable organic fraction, which may even be more readily biodegradable than the organic fraction of manure (Doorn et al., 1997). Anaerobic digestion technology for agro-industrial waste management provides enhanced environmental and financial performance than traditional waste management systems (e.g. manure storages and lagoons) and is very effective in reducing methane emissions. Additionally, it also controls air and water pollution due to the controlled retention of the effluent.

Deforestation

Deforestation is a major source of greenhouse gases, responsible for 17–25 per cent of all anthropogenic GHG emissions worldwide (Strassburg et al., 2009). A large part of the deforestation in developing countries is due to over-consumption of firewood for primary energy and accounts for 54 per cent of the world's deforestation. Apart from its impact on climate change, deforestation is also an important factor in soil erosion and land degradation (Osei, 1993; Gautam et al., 2009). The tribal communities of the North Eastern Himalayan region of India have left more than 50 per cent of the region as wasteland as a result of deforestation for firewood. Biogas is therefore a technology that could help to reduce dependence on wood as a fuel.

Due to deforestation and burning of biomass and dung cake, some essential plant nutrients are lost, leading to the loss of those soil nutrients and soil degradation (Dendukurit and Mittal, 1993). In Taktse province in Tibet, 41.6 per cent of the dung from grazing animals is collected and this puts a heavy load on the already overgrazed fields (Liua et al., 2008).

Health risks

Burning of firewood, dung cakes, straw and agricultural residue without the correct chimney or ventilation systems creates many hazardous particles, which can cause severe health problems. The WHO (2002) reported that indoor air pollution from biomass burning increases the risk of acute lower respiratory infections in children, chronic obstructive pulmonary disease in adults, tuberculosis, low birth weight, asthma, ear infections, and even cataracts (Bajgain and Shakya, 2005). In sub-Saharan Africa, according to *Renewables 2005*, burning solid fuels caused 1–2 million deaths, comprising 3–4 per cent of total global mortality in 2000 (REN 21, 2005).

The collection of firewood and dung is a time-consuming process (often taking several hours every day with over 5 km travelled) and is generally carried out by women and children (Topa et al., 2004). Lighting provided directly from biomass fuel is a poor light source, particularly for children to study after dark. Time spent collecting firewood also takes up time that could be used for education (Gautam et al., 2009). Biogas can be used as a replacement for these conventional fuels and can help to solve many of the problems associated with biomass fuels.

Properties of biogas

Methane is a colourless and odourless gas. The composition and properties of biogas are given in Table 19.1. Methane is a very light gas and is the only combustible constituent of biogas. The specific gravity of methane is less than petrol, LPG and air, hence biogas will rise and dissipate from the site of a leak, which makes it safer than other fuels. In terms of calorific value, one cubic metre of biogas is equivalent to 5–7.5 kWh energy, 1 lb of LPG, 0.54 L of petrol, 0.52 L of diesel and can produce about 1.63 kWh of electricity (with 32 per cent production efficiency) (Horst, 2000; Asankulova and Obozov, 2007); calculated considering 60 per cent biomethane content (with LHV of biomethane: 9.5–10 kWh/Nm^3).

Biogas production technology

As described in previous chapters, biogas is produced by anaerobic fermentation, which decomposes complex organic material (biomass), by four groups of microorganisms under humid conditions. The hydrolysis takes place at the initial stage and the hydrolytic enzymes create a suitable environment for acid-forming bacteria, while intermediate metabolites, i.e. soluble and insoluble monomers, are synthesized. The monomers are then converted into fatty acids with a small amount of hydrogen. The most frequently detected organic acids, propionic and butyric acids, are produced with small quantities of valeric acid. This stage is called acetogenesis (acid formation). Biogas production is a multiple-stage process, shown in Figure 19.1.

Bacteria of genera *Bacteroides*, *Lactobacillus*, *Propionibacterium*, *Sphingomonas*, *Sporobacterium*, *Megasphaera*, and *Bifidobacterium* are most commonly responsible for hydrolysis, including both facultative and obligatory anaerobes. Acid-forming bacteria (acidogens) include both facultative and obligate anaerobic fermentative bacteria, including *Clostridium* spp., *Peptococcus anaerobus*, *Bifidobacterium* spp., *Desulphovibrio* spp., *Corynebacterium* spp., *Lactobacillus* spp., *Actinomyces* spp., *Staphylococcus* spp., and *Esherichia coli*. Acetotrophic methanogens, including bacteria from the genera *Methanosarcina* and *Methanosaeta*, perform the conversion and generate methane and carbon dioxide (Sharma, 2008). The various environmental factors influencing anaerobic fermentation of organic substrates are pH, alkalinity, volatile acids concentration, temperature, nutrient availability and toxic materials. The operational factors include composition of organic substrate, retention time, concentration of the substrate, carbon: nitrogen ratio, organic loading rate and degree of mixing (Khanal, 2008).

Table 19.1 Composition and properties of biogas (Bedoya et al., 2009; Gadde, 2006; Paul and Kemnitz, 2006; Horst, 2000).

Composition of biogas	
CH_4	50–70%
CO_2	25–50%
N_2	1–5%
H_2	1–3%
H_2S	0.1–0.5%
Properties of biogas	
Specific gravity	1.21 kg/m³
Heating value	20–26 MJ/m³ (4713–6126 kcal/m³)
Density	0.72 kg/l

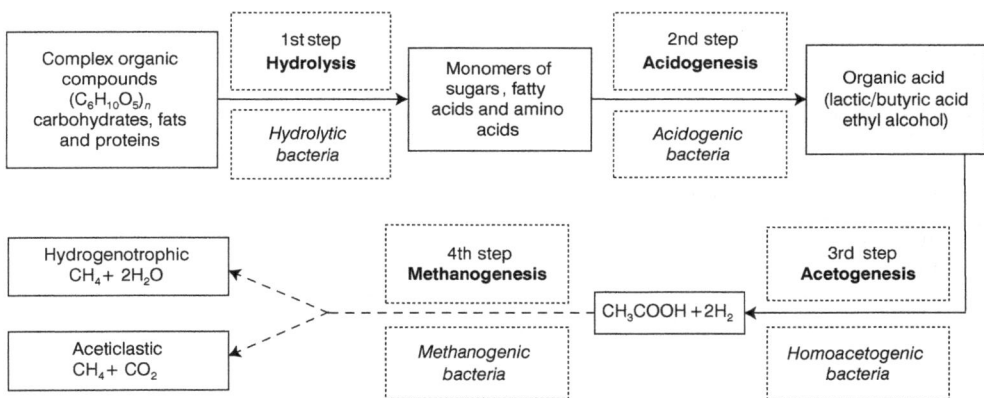

Figure 19.1 Biogas production process

Scope of biogas in developing countries

Biogas can be produced from the organic fraction of biomass like firewood, agricultural wastes and animal wastes which are readily available in developing countries.

Biogas from manures

The advantage of biogas over other fuels is that it requires limited feedstock preparation regardless of the composition and moisture content. The choice of feedstock for biogas depends on its availability. Manures are a widely used feedstock globally (Methane to Markets, 2008). The biogas production potential of commonly used manures is given in Table 19.2. The period for efficient anaerobic fermentation varies with feedstock and for cattle manure, pig manure and poultry manure is 20–30 days, 15–25 days and 20–40 days respectively, considering mesophilic conditions.

Table 19.2 Potential biogas production from different feedstocks (Korres et al., 2010; Singh et al., 2010; Navickas, 2007; Murphy et al., 2006; Murphy and Power, 2006; GATE and GTZ, 2007b; Khendelwal and Mahdi, 1986).

Feedstock	Biogas yield ($m^3 t^{-1}$)
Cattle waste	31–36
Buffalo waste	54
Piggery waste	18–55
Poultry waste	11–35
Human excreta	20–28
Slaughter waste	80–156
Organic fractions of municipal solid waste	125
Used fats	800
Fatty wastes	400
Vegetable oil	350
Distillery waste	80
Dairy waste	55
Fruit and vegetables	35

Alternative feedstocks for biogas

The inadequate availability of dung, either due to changes in the size of the cattle holdings of families or its traditional use as farmyard manure and cakes for fuel to meet direct energy requirements, has been a major constraint in the use of biogas. Khendelwal (1990) reported that the availability of cattle waste is capable of supporting only 12–30 million family-size biogas plants against the requirement of 100 million plants. Alternative feedstocks, like herbaceous biomass, used either alone or along with animal wastes for biogas production, are a promising solution to meet the energy requirements of rural areas. A significant portion of 70–88 million biogas plants can be run efficiently with fresh/dry biomass residues. Some reports state 1150 billion tons of biomass is available in India, a fifth of which would be sufficient to meet the Indian energy demand (Jagadeesh, 1996). The potential of many biomass substrates for biogas production were studied by various researchers. The biogas production potential of water hyacinth, *Lantana camera*, castor, neem, groundnut, and coconut were reported (Mallick et al., 1990; Dar and Tandon, 1987; Nagamani et al., 1992). Coir mixed with cattle dung in a ratio of 3:2 gave the best gas output and 80–85 per cent methane content was reported by Radhika et al. (1983). The retention time taken for co-digestion of manure and plant biomass ranged from 50 to 80 days. Co-digestion of biomass often improves the digestibility of the materials, leading to faster and greater biogas yield compared with cattle dung alone. The most important parameters in the selection of particular plant feedstocks are the economic considerations and the yield of methane for fermentation of that specific feedstock (Smith et al., 1992). The biogas yield (m^3/kg VS) for leaves, straw and garden wastes were 0.1–0.3, 0.35–0.45 and 0.20–0.50, respectively (Brachtl, 1998; Thomé-Kozmiensky, 1995; Nordberg and Edström, 1997).

The organic fraction of municipal solid waste (MSW), which consists of food waste, paper waste, grass and branches or leaves, is also a good source for biogas generation. Holliger

Table 19.3 Composition of biogas from alternative feedstocks (Naskeo Environnement, 2009).

Component	Household waste	Waste water treatment plants sludge	Agricultural waste
Methane (%)	50–60	60–75	40–70
Carbon dioxide (%)	38–34	33–19	30–60
Nitrogen (%)	0–5	0–1	0–1
Hydrogen sulphide (%)	0.01–0.09	0.1–0.4	0.3–1

(2008) reported the methane production potential of those various organic fraction of MSW were 500, 330, 310 and 110 m^3/t, respectively.

Agro-industrial wastes, including hotel wastes, bio-sludge from paper and pulp industry, rejected animal food, fishery by-products, harvest residues and ley crops are other potential source for biogas (Amon et al., 2001). Food processing wastewaters, e.g. from citrus processing, dairy processing, vegetable canning, potato processing, breweries, and sugar production, can also be used for biogas production. Estimations from Global Methane Initiatives (2011) on the total biogas generated from the distilleries, tapioca and corn starch production sectors in India for a period of 7 years (2001–2007), average 1507 million, 187,958 and 635,058 m^3/day, respectively. Biogas can even be made from the leftover organic material from other biofuel production like biodiesel (from Jatropha seed cake), bioethanol (bagasse, spent wash), etc. Biogas from various alternative sources is characterized in Table 19.3 based on its chemical composition and the physical characteristics.

Domestic biogas plants convert livestock manure and night soil into biogas and slurry (the fermented manure). As discussed in previous chapters, the optimum temperature for biogas production is in the mesophilic range around 36 °C, which makes the technology suitable for most developing countries with (sub)tropical climates (FAO, 1996). The simplicity in implementation and use of cheap raw materials – most easily available in villages – means that biogas technology is one of the most environmentally sound energy sources for rural needs.

Benefits of biogas technology

- **Cooking:** Biogas is used as cooking fuel (see Appendix) in a specially designed burner. Biogas stoves are safe for indoor use and do not produce unhealthy smoke; they thereby reduce healthcare costs and increase productivity.
- **Lighting:** Biogas is used in silk mantle lamps for lighting purposes (see Appendix). The requirement of gas for powering a 100 candle lamp (60 W) is 0.13 cubic metres per hour. Biogas lamps allow households without electricity to light their homes after dark, enabling children to study and family members to increase their working hours. Biogas lamps are brighter than kerosene lanterns and safer for domestic use (Bedoya et al., 2009).
- **Power Generation:** Biogas can be used to generate electricity, especially useful in remote areas that are not connected to the power grid (Urmee et al., 2009).
- **Transport Fuel:** After removal of CO_2, H_2S and water vapour, biogas can be converted to natural gas quality for use in vehicles. Biogas can either be used to operate a dual fuel engine to replace up to 80 per cent of diesel oil, and diesel engines, petrol and CNG engines can be easily modified to run 100 per cent on biogas (Singh et al., 2010, 2011; Korres et al., 2010).

Table 19.4 Energy efficiency of cooking fuels (Barnes and Floor, 1996; Veena, 1988).

Fuel	Energy (MJ/kg fuel) [a]
Wood	3
Wood, with stove	5
Cow dung	8.76
Crop residue	15.2
Charcoal, with stove	10
Kerosene	12
Biogas	20–26
Liquified petroleum gas (LPG)	25–30

[a] Energy delivered to the cooking pot. Figures are approximate and rounded.

The energy efficiency of biogas as cooking fuel is shown in Table 19.4, derived from a combination of a fuel's energy content and the efficiency with which the fuels are typically burned for cooking in developing countries.

The anaerobic treatment of animal waste through controlled capture of methane and its energetic use means that about 13.24 MMT CH_4/year can be eliminated worldwide. This is mainly acquired by replacing dung with biogas for cooking purposes. In total about 4 per cent of the global anthropogenic methane emissions could be reduced by biogas technology.

The combustion of fossil fuels leads to emission of air pollutants such as CO, NOx, SO_2, volatile organic compounds and particulates (Parashar et al., 2005). Burning of kerosene, firewood and cattle dung cake as fuels emits 0.8 to 2.2, 0.7 to 4.0 g kg^{-1} NOx, and SO_2, respectively along with varying amounts of CO, volatile organic compounds, particulate matters, organic matter, black carbon and organic carbon. Biogas is a smokeless fuel offering an excellent substitute for kerosene, cattle dung cake, agricultural residues and firewood which are used as fuel in most developing countries (MNES, 2006). Burning biogas reduces greenhouse gas (GHG) emissions; it reduces the net CO_2 release and prevents CH_4 release to the atmosphere, a potential means to satisfy various legislative and ecological constraints (Jahangirian et al., 2009). A family size biogas plant substitutes 316 L of kerosene, 5535 kg firewood and 4400 kg cattle dung cake per annum as fuels (Pathak et al., 2009). Substitution reduces emissions of NOx, SO_2 and CO as given in Table 19.5.

Compared with fossil fuel, combustion of biogas reduces the flame temperature, which reduces NOx emissions since the main pathway for NOx formation is thermal (Lafay et al., 2007). N_2O-reduction potential through anaerobic treatment is about 10 per cent, i.e. 49,000 t N_2O/year or 15.7 MMT CO_2-equivalents could be reduced on average (ISAT & GTZ, 1999).

The digester reduces emissions of methane, carbon dioxide and ammonia from manure while in the enclosed vessel. Combustion of the biogas releases some carbon dioxide and sulphur compounds back into the atmosphere. The net saving of GHG for an average-sized biogas plant has been estimated as 4.6 tonnes of CO_2 equivalents per year (Bajgain and Shakya, 2005).

Energy content of biogas compared with other fuels

The amount or volume of biogas is normally expressed in 'normal cubic metres' (Nm^3). This is the volume of gas at 0°C and atmospheric pressure. Pure methane has an energy

Table 19.5 Pollution reductions due to use of biogas plant (Prasad and Dhanya, 2011).

Pollutant	Pollution reduction due to a biogas plant (kg year^{-1})			
	Kerosene	Firewood	Dung cake	Total
Oxides of N (NOx)	0.7	12.2	3.5	16.4
Oxides of S (SOx)	1.3	3.9	6.2	11.4
Carbon monoxide	0.6	549.6	436.9	987.1
VOCs	0.2	38.7	30.8	69.7
Particulate matter^{10}	0.1	16.6	13.2	29.9
Particulate matter$^{<2.5}$	0.1	11.6	28.6	40.3
Organic matter	0.4	7.2	17.6	25.2
Black carbon	0.1	3.3	11.0	14.4
Organic carbon	0.1	19.4	55.4	74.9

Table 19.6 Energy content of biogas compared with other fuels (Asankulova and Obozov, 2007, www.preem.se, www.swedegas.se).

Fuel	Energy value (kWh)
1 Nm^3 biogas (97% methane)	9.67
1 Nm^3 natural gas	11.0
1 litre petrol	9.06
1 litre diesel	9.8
1 litre E85	6.6

1 Nm^3 biogas ≈ 1.1 litres of petrol, 1 Nm^3 natural gas ≈ 1.2 litres petrol

value of 9.81 kWh/Nm^3. The energy value of biogas varies between 4.5 and 9.1 kWh/Nm^3, depending on the relative amounts of methane (50–70 per cent), carbon dioxide and other gases present. Thus, if biogas comprises 60 per cent methane, the energy content is ~6.0 kWh/Nm^3. Energy content of biogas compared with other fuels is given in Table 19.6.

Biogas has very high octane number, approximately 130 in comparison to gasoline (90 to 94) and alcohol (105) at best (Paul and Kemnitz, 2006). This means that a higher compression ratio engine can be used with biogas than petrol. The cylinder head of the engine is faced so that the clearance volume will be reduced and the compression ratio can sufficiently increase, thus volumetric efficiency and power output are increased (Biogas Digest, 2010).

Biogas as transportation fuel

Motive power can be generated by using biogas in a dual fuel internal combustion (IC) engine. Air mixed with biogas is aspirated into the engine and the mixture is then compressed, raising its temperature to about 350°C, the self-ignition temperature of diesel. Biogas has a high (600°C) ignition temperature, therefore, in order to initiate combustion of the charge, a small quantity of diesel is injected into the cylinder just before the end of compression. The charge is thus ignited and the process is continued smoothly. Converting a spark-ignition engine for biogas fuelling requires replacement of the gasoline carburetor with a mixing

Table 19.7 Pollution reductions due to biogas used as vehicle fuel (Traffic and Public Transport Authority, 2000).

Pollutant	Emission (g/km)		
	Diesel	Natural gas	Biogas
Particulate matter	0.1	0.022	0.015
NOx	9.73	1.1	5.44
Carbon monoxide (CO)	0.2	0.4	0.08
Unburned hydrocarbons (HC)	0.4	0.6	0.35
CO_2	1053	524	223

valve (pressure-controlled venturi type or with throttle). A spark ignition engine (gasoline engine) draws a mixture of fuel (gasoline or gas) and the required amount of combustion air. The charge is ignited by a spark plug at a comparably low compression ratio of between 8:1 and 12:1. Power control is affected by varying the mixture intake via a throttle (Biogas Digest, 2010). The biogas used as vehicle fuel presents better characteristics than natural gas (Table 19.7). Some disturbance still appears for the NOx emissions which are below the EU norms. Gasoline produces 2.44 CO_2e kg/l and ethanol 1.94 CO_2e kg/l (Popa, 2010). Biogas is far better than the natural gas used for vehicles (NGV) in terms of the CO_2, hydrocarbons and CO emissions (Traffic & Public Transport Authority, 2000).

Borjesson and Berglund (2006) analysed fuel-cycle emissions of CO_2, CO, NOx, SO_2, hydrocarbons (HC), CH_4 and particles from a life-cycle perspective for biogas systems based on different digestion technologies and raw materials and suggested that the overall environmental impact of biogas depends largely on the status of uncontrolled losses of CH_4, the end-use technology that is used, the raw material digested and the energy efficiency in the biogas production chain.

Better fuel efficiency

The efficiency of fuel usage is one of the major criteria in selecting the best solution to the unsustainable use of biomass as energy. Burning of biomass in a traditional stove and dried dung's heating efficiency can be raised from 10–15 per cent to about 25–30 per cent by using better designed stoves which can also save fuel (Li et al., 2005). The heating efficiency of biogas from dung would be raised to 60 per cent (Mirza et al., 2008). The indoor climate will also be dramatically improved as a result of using clean biogas stoves instead of burning firewood, straw and dung cakes.

Health benefits

Biogas can have significant health benefits, especially in rural areas. According to the Integrated Environmental Impact Analysis carried out by the Biogas Support Programme for 600 biogas users and 600 non-users, 4 per cent more non-biogas users have respiratory diseases (Table 19.8) than those who own biogas plants (BSP, 2000). The qualitative information from various household surveys carried out by BSP has revealed that problems like respiratory illness, eye infection, asthma and lung problems have decreased after installation of biogas plants.

Table 19.8 Health benefits of biogas over traditional fuel wood usage (BSP, 2000).

Disease	Problems in the past (households)		Present status of households	
	Yes	No	Improved	Remained same
Eye infections	72	18	69	3
Cases of burning	29	71	28	1
Lung problems	38	62	33	5
Respiratory problems	42	58	34	8
Asthma	11	89	9	2
Dizziness/headaches	27	93	16	11
Intestinal/diarrhoea	58	42	14	44

Pathogen removal

Human and animal waste is loaded with pathogens like *Salmonella*, *E. coli* O157:H7, *Campylobacter jejuni*, *Yersinia enterocolitica*, *Giardia lamblia*, several types of *Cryptosporidium*, etc. Biogas technology helps in improvement of hygienic conditions through reduction of pathogens, worm eggs and flies, compared with other methods of waste disposal. *Salmonella*, *Shigella*, and *Vibrio cholera*, common pathogens in human waste, were completely eradicated by the digestion process. High temperatures and long retention times kill most pathogens. The time period taken for killing principal pathogenic organisms in biogas plants are one or two weeks for typhoid, paratyphoid, cholera and dysentery bacteria (three weeks for hookworm) and bilharzia, tapeworm and roundworm die completely when the fermented slurry is dried in the sun. Thus by linking biogas plants to toilets, families in rural areas and people living in slums in and around cities have been provided with good sanitation facilities and help in securing hygienic conditions (Brown, 2006; Kunte et al., 1998).

Biofertilizer

The open exposure of manure heaps resulted in loss of nutrients, mainly nitrogen, potassium and some phosphorus, through volatilization and leaching (Matsumoto et al., 1997). Biogas technology is effective in manure management, which reduces nutrient losses from the manure. Moreover, biogas spent sludge (BSS), obtained after anaerobic digestion, is richer in valuable nutrients than the animal manure, which has environmental and economic advantages. BSS is a very relevant input to sustainable agriculture which surpasses the negative problems of energy intensive chemical input based agriculture, like global warming, soil erosion, degradation of ground water, biomagnification of pesticides, etc. The recovery of plant nutrients from the biogas plant, via BSS, represents an additional economic return to the farmer using the technology. The use of biogas slurry in proper combination with chemical fertilizer is one of the major steps in integrated nutrient management for sustainable agriculture (Jain, 1993). The concentration of nutrient elements in the residue after digestion are reported to be higher on a percentage dry weight basis than in the undigested material due to conservation of nutrient elements during digestion and gasification. These residues are readily available for growing plants, as well as residual carbon, phosphorous and trace nutrients, and they can thus be returned to the agricultural land as fertilizer and soil-improvement medium (BORDA, 1997). Samuil et al. (2009) suggested that organic fertilizers, if applied rationally to grasslands, can entirely replace chemical fertilizers. Slurry from 1 kg of digested dung can yield extra nitrogen up to 0.5 kg compared with fresh manure (Sasse et al., 1991). Anaerobically treated biosolids are often more easy to handle and the digestion process is known to inactivate weed seeds, plant pathogens and pests and decrease the amount of phytotoxic compounds in manure (Gunaseelan, 1997). The volume of sludge produced by anaerobic treatment is 80 per cent less than that produced by aerobic treatment processes. By anaerobic digestion technology, plant nutrients, nitrogen, potassium and phosphorus are not removed but rather enrichment occurs, which is an advantage as the effluent can be applied to agricultural fields in place of chemical fertilizer. This can solve problems of soil degradation in areas where earlier dung has been used as a burning fuel and save costs in purchasing chemical fertilizers, creating revenue for the household (Li et al., 2005).

Employment opportunity

Many local jobs are created around biogas projects. By 2000, almost 200,000 permanent jobs were created by more than 2 million biogas plants in India. During 2000–2001 alone, 164 thousand biogas plants were constructed, generating employment to the tune of 5 million man-days (Patel, 2001).

Improvement in standard of living

Apart from these benefits, biogas plants have provided many indirect social benefits, such as reduction in the drudgery of rural people involved in the collection of fuel materials from long distances, reduction in the incidences of lung and eye diseases from cooking in smoky kitchens and an overall improvement in the standard of living. The daily time spent in feeding a small biogas digester could be as little as 15 minutes compared with several hours in biomass collection. Time consumed cleaning pots and other kitchen equipment can also be lowered since biogas does not create as much soot as biomass. Hence people, particularly women and children, would have more time for education when they do not have to spend as much time collecting firewood and other biomass fuels (Bajgain and Shakya, 2005).

Biogas development in developing countries

The anaerobic fermentation of waste products, human excreta and cattle manure, etc. and widespread use of biogas in many developing countries resulted from the problems associated with traditional use of biomass and lack of alternatives such as fossil fuels. Domestic biogas technology is a proven and established technology in many parts of the world, especially Asia. Biogas plants can be made for individual households or for small communities as they are relatively cheap to build, can use existing waste products such as household waste, human excreta and cattle manure, and are a cheap way of generating energy and handling waste products (Gautam et al., 2009). Small biogas plants started to emerge in India in the beginning of the 1950s and have since then spread over many other countries.

The Netherlands Development Organization (SNV) supports national programmes on domestic biogas that aim to establish commercially viable domestic biogas sectors in which local companies market, install and service biogas plants for households. In Asia, SNV is working in Nepal, Vietnam, Bangladesh, Bhutan, Cambodia, Lao PDR, Pakistan and Indonesia; and in Africa in Rwanda, Senegal, Burkina Faso, Ethiopia, Tanzania, Uganda, Kenya, Benin and

Cameroon. Several Asian countries have embarked on large-scale programmes on domestic biogas, such as China and India, Nepal, Pakistan and Bangladesh, due to the thriving population of livestock and agriculture. China, India, and Nepal have conducted the main biogas programmes; all three countries now have large manufacturing industries for biogas plants.

China

China leads the world with 7.5 million household biogas digesters installed and another 750 large- and medium-scale industrial biogas plants. China's extensive biogas programmes began in the 1950s and reached peaks in both 1960 and 1979. Inadequate education and training of households led to technical failures and declining use subsequent to the biogas programme. Since the mid-1980s, a network of rural biogas service centres has been established to provide the infrastructure necessary to support dissemination, financing, and maintenance. The Guangxi project has become a catalyst for other initiatives in the region; 2.73 million biogas tanks have been built in villages, benefiting about 34.2 per cent of the rural households in Guangxi. It is estimated that 7.65 million tons of standard coal and 13.40 million tons of firewood are saved annually in Guangxi because of the use of biogas. In the 1990s, China's biogas strategy was extended to remote communities in west Guangxi, where wood for fuel was in short supply and rural electricity was not available. In 2002, the strategy was a key component of a six-year IFAD-funded project to improve and sustain the livelihoods of poor rural people while rebuilding and conserving natural resources. By 2006, the project had exceeded its target by providing more than 22,600 biogas tanks and helping almost 30,000 households in more than 3100 villages. As a result, 56,600 tons of firewood can be saved in the project area every year, which is equivalent to the recovery of 7470 hectares of forest.

India

The first digestion plant was built at a leper colony in Bombay, India in 1859. Biogas plants in India were experimentally introduced in the 1930s, and research was principally focused around the Sewage Purification Station at Dadar in Bombay. The early plants developed were very expensive and were not cost effective in terms of the gas output. Over the next twenty years, Jashbhai Patel designed and made several small-scale biogas digesters, envisaging farm labourers as the users. Although other individuals and institutions were also designing biogas plants, in 1961 the Khadi and Village Industry Commission chose to promote Patel's design, which, although more costly than other models, was more productive, had a longer life, and required minimal maintenance (KVIC, 1993).

India had a large programme, with about 3 million household plants installed over its potential of 12 million biogas plants. Initial efforts focused on technology development and increased user awareness (MNES, 2002). The Deenabandhu model is a fixed dome biogas-production model popular in India, with a capacity of 2 to 3 cubic metres, constructed using bricks or by a ferrocement mixture. In India, the Ministry of New and Renewable Energy offers a subsidy per model constructed.

The Deenbandhu model is an improved fixed dome model. The entire plant is built with brick in cement mortar. The digester, gas storage chamber and the empty space above the slurry are all provided in the spherical shell. All the slurry displaced out of the gas storage chamber is stored in the outlet displacement chamber as there is no displacement space on the inlet side. The inlet is in the form of a pipe which connects the digester with the slurry mixing tank. The hydraulic retention time for this design is 40 and 50 days for northern plains and hilly locations, respectively.

Nepal

In Nepal, around 80–85 per cent of the population make their living through agriculture in rural areas and the agricultural sector accounts for 39 per cent of Nepal's GDP. Of the energy used in Nepal, 88 per cent comes from biomass such as firewood and agricultural residues. The household usage of energy accounts for 90 per cent of the total energy. The biogas technology in Nepal has reached the country from India where biogas digesters have been used since the 1950s. The most common type of biogas digester used in Nepal is a small household scale fixed dome digester (Gautam et al., 2009). The potential for biogas plants in Nepal is estimated as 1.3 million, calculated from the cattle population of the country (Bajgain and Shakya, 2005). An independent non-profit organization, the Nepal Biogas Support Programme (BSP) which has been financed mainly by the Netherlands, has been active in Nepal since 1992 and through this programme over 208,000 biogas plants had been installed by 2010, of these 97 per cent are operational. On average, each biogas plant in Nepal replaces 2 tonnes of firewood annually. With over 208,000 biogas plants, around 420,000 tonnes of firewood is replaced by the 103 million m^3 of biogas produced annually. The slurry gives 354 thousand tonnes of fertilizer which replaces artificial fertilizer; 6.75 million L of fossil kerosene is also replaced by the biogas. It has been estimated that these biogas plants save around 12,742 ha of forest which corresponds to around 16.86 million trees. On average, each biogas plant reduces CO_2 emissions by 4.6 tonnes annually and together 111 thousand biogas plants reduces GHG by 510,000–600,000 tonnes CO_2e each year as reported by Bajgain and Shakya (2005). With the help of the Biogas Sector Programme, 208,000 biogas plants were installed by 2010, benefitting 1.25 million people across the country and meaning that 420,000 tonnes/year less fuelwood was burnt, resulting in a CO_2 saving of 630,000 tonnes/year (BSP-Nepal, 2002; BSP, 2011).

Pakistan

Pakistan is situated in the south Asian region, covering a total land area of 888,000 square kilometres with a population of 140 million. Pakistan is an agricultural country, more than 70 per cent of the population is involved in agriculture and per capita income is about US$480. Agriculture accounts for more than 35 per cent of the GDP (Ilyas, 2006). In Pakistan, the company, PAK-Energy Solution, is aiming to install 70,000 biogas plants and have designed and developed the Uetians Hybrid Model, in which they have combined a fixed dome and floating drums; and the Uetians Triplex Model. Both of these designs have been innovated for the first time in the world. Moreover, the Pakistan Dairy Development Company has also taken an initiative to develop this kind of alternative source of energy for Pakistani farmers. Biogas is now running diesel engines, gas generators, kitchen ovens, geysers, and other utilities in Pakistan.

Sri Lanka

In Sri Lanka, biomass accounts for 45 per cent of the country's energy needs and the economy is still largely based on agriculture. Although biogas digesters have been introduced in Sri Lanka since the 1970s, poor design, lack of maintenance skills and insufficient capacity to deal with problems meant that only a third of the 5,000 units installed functioned properly.

The Intermediate Technology Development Group (ITDG) started a project in 1996 to improve the success rate of the units on a national level by setting up demonstration units to help to spread information, restoring abandoned units and training users to operate and maintain them. In addition, individual farmers get help to install biogas units on their farms to make use of the manure from their cows.

Sub-Saharan African countries

In sub-Saharan Africa, most of the existing 2,400 biogas units were installed through donor and demonstration projects. But due to inadequate feedstocks, intensive labour demand, high capital costs, poor technical performance, and lack of water, the technology was not completely successful (Gitonga, 1997; Karekesi and Ranja, 1997). A discussion paper prepared by Winrock International for the Dutch Ministry of Foreign Affairs, reports that as a result of biogas, sanitation could yield benefit-cost ratios (BCRs) ranging from 1.22 to 1.35 and financial internal rates of return (FIRRs) from 7.5 per cent to 10.3 per cent for selected African countries (Renwick et al., 2007).

Other developing countries

In Vietnam, Colombia, Ethiopia, Tanzania, Cambodia (see Appendix) and Bangladesh the polyethylene tubular digester was promoted to reduce production costs by using local materials and simplifying installation and operation of biogas plants. The biogas sanitation systems are in operation in many other countries like Barbados, Bolivia, Burundi, Butan, Cameroon, Democratic People's Republic of Korea, Ethiopia, Georgia, Indonesia, Jamaica, Kenya, Laos, Lesotho, Morocco, Mozambique, Nicaragua, Philippines, Republic of Korea, Rwanda, South Africa, Tanzania, Thailand, Uganda, Vietnam, and Zambia.

Policies for adoption of biogas technology

Renewable energy technologies (RETs)

RETs are energy-providing technologies that utilize energy sources in ways that do not deplete the Earth's natural resources and are as environmentally benign as possible (Renewable Energy Association, 2009). A significant effort has been made to mobilize the resources to realize this potential to power rural development (Bhattacharyya, 2006; Boyle et al., 2006). Reducing rural poverty through rural development is the key requirement to achieving the Millennium Development Goals (MDGs) of the United Nations General Assembly by the year 2015, the connection between clean sources of energy and rural energy access. MDG 7 – ensuring environmental sustainability – promotes RETs as a way of expanding access to these services (World Bank, 2004; United Nations Public–Private Alliance for Rural Development, 2009; United Nations, 2009). The United Nations Development Programme (UNDP) 1997 Report, *Energy after Rio: Prospects and Challenges* identified community biogas plants as one of the most useful decentralized sources of energy supply. Unlike the centralized energy supply technologies, such as power plants based on hydroelectricity, coal, oil or natural gas, that have hitherto been the only choices open to rural communities, biogas plants do not require big capital to set up, and do not pose environmental problems that excite public opposition. They also offer solutions to existing environmental problems, and many unexpected benefits (United Nations, 1984).

Clean Development Mechanism (CDM)

The CDM of the United Nations Framework Convention on Climate Change (UNFCCC) can be used for biogas projects. This flexible mechanism makes it possible for Annex 1 (developed) countries to displace emission reductions to developing countries. The avoided GHG emissions from the CDM projects will generate CERs (Certified Emission Reductions) that can be bought by Annex 1 countries. This can help finance further biogas growth in developing countries (Bajgain and Shakya, 2005). The UNFCCC has set up a Clean Development Fund, and the World Bank has put together a Carbon Finance Unit to allow rich countries, which are pumping more carbon into the atmosphere than is allowed under the Kyoto Protocol, to buy emissions that poor countries prevent through conserving forests or promoting renewable energy (Grubb, 2003). The production of biogas will also reduce the use of fossil fuels, thereby reducing the CO_2 emission. This is in line with the Kyoto Summit Agreement as a carbon-reducing technology (Mata-Alvarez et al., 2000). Nepal's successful biogas programme make the rest of the world pay hard cash for not burning firewood to release carbon dioxide into the atmosphere (BSP, 2010). Nepal's biogas programme is internationally regarded as a model for successful use of alternative energy for the rural Third World. Nepal has now overtaken China and India in the number of biogas plants per capita. As stated previously, each of its 125,000 functioning digesters prevents 5 tonnes of carbon dioxide equivalents from being pumped into the atmosphere every year. This 'saved' greenhouse gas is what rich countries are buying to offset their own emissions, and is worth US$5 million. This money can be invested back into clean energy that would make Nepal eligible to trade even more carbon offset to rich polluters.

Other policies or programmes implemented by government of India

Policies from the government have a crucial role in the successful adoption of any technology. In developing countries, many biogas support programmes have focused on rural families using animal manure and human faeces as feedstock in order to reduce the use of firewood by providing people with biogas, to improve soil fertility and to reduce indoor air pollution (Voegeli and Zrubruegg, 2008).

National Project on Biogas Development (NPBD)

This project was launched by the Government of India in 1981, leading to about 3.4 million family-size biogas plants installed all over India by December 2002. This is only 28.3 per cent of the total potential (12 million) of family size biogas plants that can be put up in India (Annual report of MNES, 2002). Also, more than 3,380 Community Biogas Plants (CBP), Institutional Biogas Plants (IBP) and Night-soil based Biogas Plants (NBP) have been installed all over the country.

Biogas based distributed/grid power generation programme

In India, with the purpose of promoting biogas-based power generation, especially in the small capacity range based on the availability of a large quantity of wastes (e.g., animal, forestry, agro/food processing, kitchen), the programme was implemented through nodal departments/agencies of the states/Union Territories, Khadi and Village Industries Commission (KVIC), institutions, and non-governmental organizations in January 2006. The government of India provides subsidies and financial assistance to waste-to-energy projects, especially for households.

National Biogas and Manure Management Programme (NBMMP)

The Central Sector Scheme of the National Biogas Programme in India, which mainly focuses on setting up family-type biogas plants, was implemented in 1981–1982. NBMMP provides subsidies; turn-key job fees linked with three years of free maintenance; financial support for repairing old, non-functional plants; training for users, masons, entrepreneurs, and others; publicity and extension; service charges or staff support; state-level Biogas Development and Training Centres (BDTC); fixed amount of support to institutional biogas plants; and financial support to institutions for cattle dung-based power generation plants. A total of 3.93 million family-type biogas plants have been established in the country, with an estimated potential for 12 million plants.

Problems and challenges

Bioconversion of organic domestic and farm residues has become attractive as its technology has been successfully tested through experience on both small- and large-scale projects. Nevertheless, there are still several problems that impede the efficient working of biogas generating systems.

Technical constraints

It is a very complex technology, requiring the combination of a variety of classical fields of engineering, i.e. of professionals. Highly skilled personnel are required for the installation and maintenance of biogas plants. Improper preparation of influent solids leading to blockage and scum formation, temperature fluctuations, maintenance of pH for optimal growth of methanogenic bacteria, C/N ratio, dilution ratio of influent solids content, corrosion of gas holder, pin-hole leakages (digester tank, holder, inlet, outlet), etc. are some of the technical problems in biogas production and usage. Other technical issues are mostly related to bad maintenance. A common problem is that pipes get blocked due to lack of service and leakage is also a problem that is not unusual with fixed dome biogas plants (Korres et al., 2010; Woods et al., 2006; Bajgain and Shakya, 2005; Han et al., 2008). These problems can be reduced by regular maintenance of the biogas plant.

Financial constraints

Lack of financial capabilities to invest in biogas plants among poor farmers in rural areas remain one of the biggest challenges. Construction costs could be reduced for only slightly higher planning costs. Progress could be enhanced if public funds were available to support small but flexible engineering firms, specializing in anaerobic digestion, rather than large offices, which often stick to known but outdated techniques, because of their high overhead costs.

Social constraints

Social constraints and psychological prejudice against the use of raw materials like night soil also prevents biogas from its wide use (Bajgain and Shakya, 2005). The negative image of biogas production, which was created by the early installations, has considerably reduced the interest of farmers to build biogas plants. In particular, the malfunction of the widely announced, large-scale demonstration units, such as the village biogas plants in Denmark, or the installations erected in agricultural schools in Switzerland, had a detrimental effect on the local diffusion, even though a considerable number of problems arose only due to poor maintenance. The social pressure to build individually-owned installations is lacking (Marchaim, 1992).

Environmental constraints

Most of the digesters in India operate normally at ambient conditions. Northern India records a shortfall in biogas output during winters and in some other parts of the country, especially in dry tracts, which also affect the digester performance due to higher temperature. Cold temperatures will make the biogas plants more inefficient and if the temperature drops below 10°C the whole process will stop. Biogas plants built in areas where temperatures drop below 10°C will need extra insulation and possibly water heating to maintain operation which will add to the construction costs (Bajgain and Shakya, 2005). If there are heavy rains, biogas digesters that are below ground can get flooded and as a result have to be drained, which adds to the maintenance time and costs.

Other constraints

Constraints like cost of cleaning, upgrading (to remove CO_2) and transportation of biomass, limit the use of biogas (Jahangirian et al., 2009). The carbon dioxide and water present in the digester dilutes the energy content of biogas, lowering its calorific value. Hydrogen sulphide is an extremely reactive biogas constituent that forms sulphuric acid in the presence of moisture. The acids formed can corrode engine parts in the combustion chamber, exhaust system, pipelines, storage tank and in various bearings. Therefore, purification to reduce all impurities is a prerequisite for effective biogas utilization as a vehicle fuel.

The lack of effective and clear policies is a major hurdle to overcome in the dissemination of biogas technology. Government has to set policies that promote biogas usage and encourage collaboration with governmental organizations (GO) and nongovernmental organizations (NGO) (Han et al.,2008; Bajgain and Shakya, 2005).

Technical problems can easily be avoided with better site design planning when the biogas plant is built. Public support is very important in the promotion of biogas. It is important to spread knowledge and education about the biogas technology to the people. The government can play an important role to implement pilot biogas projects in rural agricultural areas to showcase the benefits of biogas technology (Ilori et al., 2000).

Use of additives

As discussed in previous chapters, the performance of biogas plant and gas production rate can be improved by stimulating the microbial activity using various biological and chemical additives under different operating conditions. Agricultural crop residues, e.g. maize stalks, rice straw, cotton stalks, wheat straw, onion storage waste and water hyacinth, each enriched with partially digested cattle dung, enhanced gas production in the range of 10–80 per cent. Powdered leaves of some tree spp. (like Gulmohar, *Leucacena leucocephala*, *Acacia auriculiformis*, *Dalbergia sisoo* and *Eucalyptus tereticornis*) and legumes have been found to stimulate biogas production between 18–40 per cent (SPOBD, China, 1979; Chowdhry et al., 1994). The additives also help to maintain favourable conditions in the reactor (such as pH, inhibition/promotion of acetogenesis and methanogenesis for the best yield, etc.), which helps in rapid and higher gas production (Yadvika et al., 2004). Several inorganic additives that improve gas production have also been reported. Dar and Tandon (1987) reported that alkali treated (1 per cent NaOH for 7 days) plant residues (lantana, wheat straw, apple leaf litter and peach leaf litter) when used as a supplement to cattle dung, resulted in almost two-fold increase in biogas and CH_4 production. Improvement in biomethanation of mango processing wastes by the addition of seed extracts of Nirmali, common bean, black gram, guar and guargum at the rate of 1500 ppm was also reported by Babu et al. (1994).

Strains of some bacteria and fungi have also been found to enhance gas production by stimulating the activity of particular enzymes (Yadvika et al., 2004). Cellulolytic strains of bacteria like actinomycetes and mixed consortia have been found to improve biogas production in the range of 8.4–44 per cent from cattle dung (Tirumale and Nand, 1994; Attar et al., 1998). The addition of iron salts at various concentrations (50 mM-$FeSO_4$, 70 lM-$FeCl_3$) have been found to enhance gas production rate. Shimizu (1992) claimed that high concentrations of bacteria could be retained in the digester by the addition of metal cations since cations increase the density of the bacteria, which are capable of aggregating by themselves. Wong and Cheung (1995) reported that the plant with a higher content of heavy metals (Cr, Cu, Ni and Zn) had a higher CH_4 yield than the control. Certain adsorbents are also reported to improve gas production. Patel et al. (1992), in anaerobic digestion of water-hyacinth cattle dung, found a trend of enhanced gas production with high CH_4 content and lower BOD and COD in effluent with increasing doses of different adsorbents (gelatin, polyvinyl alcohol, powdered activated charcoal, pectin, kaolin, silica gel, aluminum powder, bentonite and talc powder). Using Ca and Mg salts as energy supplements, CH_4 production was enhanced and foaming was avoided (Mathiesen, 1989).

Recycling of digested slurry/slurry filtrate

The recycling of digested slurry along with filtrate back into the reactor has been found to improve gas production. Kanwar and Guleri (1994) reported that about 60–65 per cent more biogas can be produced by simply recycling the digested slurry in 1 m^3 plug flow type pilot plants. Some designs use vermiculture to further enhance the slurry produced by the biogas plant for use as compost.

Economics of biogas production

Depending on size and location, a typical brick made fixed dome biogas plant can be installed in the yard of a rural household with an investment of US$300–500 in Asian countries and up to US$1400 in the African context. A high-quality biogas plant needs minimum maintenance costs and can produce gas for at least 15–20 years without major problems and re-investments. A well-maintained digester can pay for itself in one-fifth of that time. The size of the digester can be calculated on the basis of the following formula:

Size of digester (m^3) = Fresh manure/day × No. of animals × 2 (for cow/buffalo) or × 3 (for pig) × Retention time (60 days)

Conclusion

Developing countries depend upon biomass for energy, which results in negative impacts on environment and society. In contrast, biogas technology provides clean energy, reduces indoor air pollution, and reduces the time needed for traditional biomass collection. The slurry is an additional bonanza as a clean organic fertilizer that increases agricultural productivity. Thus anaerobic digestion is considered an important component of the global strategy to increase energy security and is environmentally safe by providing an alternative to fossil fuels for sustainable development. It is a suitable tool for developing countries rich in agriculture, particularly the livestock sector, for maximizing the use of scarce resources and provides significant benefits to human and ecosystem health.

References

Amon, T., Hackl, E., Jeremic, D., Amon, B. and Boxberger, J. (2001) 'Biogas production from animal wastes, energy plants and organic wastes', in van Velsen, A.F.M. and Verstraete, W.H. (Ed), *Proc. 9th World Congress on Anaerobic Digestion, TechnologischInstituttzw*, Antwerp, pp381–386.

Anon. (2005) *Energy for development: The potential role of renewable energy in meeting the Millennium Development Goals*. Dutch Government Report. http://www.worldwatch.org/system/files/ren21-1.pdf accessed on 27 October 2012.

Asankulova, A. and Obozov, A.D. (2007) 'Biogas in Kyrgyzstan', *Applied Solar Energy*, vol 43, no 4, pp. 262–265.

Attar, Y., Mhetre, S.T. and Shawale, M.D. (1998) 'Biogas production enhancement by cellulytic strains of Actinomycetes'. *Biogas Forum I*, vol. 72, pp. 11–15.

Babu, K.S., Nand, K., Srilatha, H.R., Srinath, K. and Madhukara, K. (1994) 'Improvement in biomethanation of mango processing wastes by addition of plant derived additives'. *Biogas Forum III* vol. 58, pp. 16–19.

Bajgain, S. and Shakya, I. (2005) 'The Nepal Biogas Support Program: A successful model of public private partnership for rural household energy supply', Ministry of Foreign Affairs. The Netherlands.

Barnes, D. and Floor, W. (1996) 'Rural energy in developing countries: A challenge for economic development', *Annual Review of Energy and Environment*, vol. 21, pp. 497–530.

Bedoya, I.D., Arrieta, A.A. and Cadavid, F.J. (2009) 'Effects of mixing system and pilot fuel quality on diesel–biogas dual fuel engine performance', *Bioresource Technology*, vol 100, pp. 6624–6629.

Bhatt, B.P. and Sachan, M.S. (2004) 'Firewood consumption pattern of different tribal communities in Northeast India', *Energy Policy*, vol 32, pp1–6.

Bhattacharyya, S. (2006) 'Renewable energies and the poor: niche or nexus?', *Energy Policy*, vol. 34, no. 6, pp. 659–663.

Biogas Digest. (2010) 'Basic information and advisory service on appropriate technology, Vol I', available from http://www.gtz.de/de/dokumente/enbiogas-volume1.pdf (accessed on 5 January 2012).

BORDA. (1997) '*Biogas: Manual for the realization of biogas programmes*', Bremen: Bremen Overseas Research Development Association

Borjesson, P. and Berglund, M. (2006) 'Environmental systems analysis of biogas systems. Part I: Fuel-cycle emissions'. *Biomass and Bioenergy*, vol 30, no. 5, pp. 469–485.

Boyle, G., Deepchand, K., Hua, L. and Bre La Rovere, E. (2006) 'Renewable energy technologies in developing countries: Lessons from Mauritius, China and Brazil', Yokohama: UNUIAS.

Brachtl, E. (1998) 'Pilotversuchezur Cofermentation von pharmazeutischen Abfällenmit Rindergülle'. Master's thesis. InteruniversitäresForschungsinstitutfürAgrarbiotechnologie, Abt. Umwelt biotechnologie, Austria.

Brown, V.J. (2006) 'Biogas: A bright idea for Africa'. *Environmental Health Perspective*. vol. 114, no. 5, pp. A300–A303.

BSP. (2000). 'Biogas Support Program Nepal', Winrock International, http://practicalaction.org/docs/energy/docs50/bp50-nepalbiogas.Pdf (accessed on 5 January 2012).

BSP. (2010) 'Biogas sector partnership, Nepal', At http://www.ashden.org/winners/bsp (accessed on 5 January 2012).

BSP. (2011) *BSP Database*. Bagdol, Laltipur: Biogas Sector Partnership-Nepal.

BSP-Nepal. (2002) *An integrated environment impact assessment, final report, Biogas Support Programme*, Kathmandu: Biogas Support Programme.

CAIT. (2009) Climate Analysis Indicators Tool (CAIT) Version 6.0, Washington DC: World Resources Institute.

Chhabra, A., Manjunath, K.R., Panigrahy, S. and Parihar, J.S. (2009) 'Spatial pattern of methane emissions from Indian livestock', *Current Science*, vol 96, no 5, pp. 683–689.

Chowdhry, S.D.R., Gupta, S.K, Banergy, S.K., and Roy Chowdhry, S.D. (1994) 'Evaluation of the potentiality of tree leaves for biogas production'. *Indian Forester* vol. 120, no. 8, pp. 720–728.

Dar, H.G. and Tandon, S.M. (1987) 'Biogas production from pre treated wheat straw, lantana residue apple and peach leaf litter with cattle dung', *Biological Wastes*, vol 21, pp. 75–83.

Dendukurit, G. and Mittal, J.P. (1993) 'Household energy needs of a village in the Rayalaseema area of Andhra Pradesh, India', *Energy Conservation Management*, vol. 34, no. 12, pp. 1273–1286.

Doorn, M.R.J., Strait, R., Barnard, W. and Eklund, B. (1997) 'Estimate of global greenhouse gas emissions from industrial and domestic wastewater treatment', EPA-600/R-97-091. Final Report prepared for Research Triangle Park, NC: U.S. EPA.

FAO. (1996) 'A system approach to biogas technology', in *Biogas technology: A training manual for extension*. Kathmandu: Food and Agriculture Organization/Consolidated Management Services.

FAO. (2005) 'Bioenergy'. Rome: Sustainable Development Department, FAO. Available from http://www.fao.org/sd/dim_en2/en2_050402_en.htm (accessed on 5 January 2012).

FAO. (2006) *Livestock's long shadow–environmental issues and options*. Rome: Food and Agriculture Organization of the United Nations. http://www.fao.org/docrep/010/a0701e/a0701e00.HTM (accessed on 5 January 2012).

Gadde, B. (2006) 'Economic utilisation of biogas as a renewable fuel for fuel cell', The 2nd Joint International Conference on 'Sustainable Energy and Environment (SEE 2006)' 21–23 November, Bangkok, Thailand, http://www.jgsee.kmutt.ac.th/see1/cd/file/A-009.pdf (accessed on 15 May 2012).

GATE and GTZ. (2007a) *Biogas digest volume I: Biogas basics*, Frankfurt: German Appropriate Technology Exchange (GATE) and German Agency for Technical Cooperation (GTZ).

GATE and GTZ. (2007b) *Biogas digest volume II: Application and product development*, Frankfurt: German Appropriate Technology Exchange (GATE) and German Agency for Technical Cooperation (GTZ).

Gautam, R., Baralb, S. and Herat, S. (2009) 'Biogas as a sustainable energy source in Nepal: Present status and future challenges', *Renewable and Sustainable Energy Reviews*, vol 13, pp. 248–252.

Gitonga, S. (1997) *Biogas promotion in Kenya: A review of experiences*, Nairobi, Kenya: Intermed. Technolloglical Development Group.

Global Methane Initiatives. (2011) 'The U.S. government's global methane initiative accomplishments' Annual Repor, December. Available at http://www.epa.gov/globalmethane/ pdf/2011-accomplish-report/ GMI_USG2012_Full-Report.pdf.

Grubb, M. (2003) 'The economics of the Kyoto Protocol', *World Economics*, vol 4, no. 3, pp. 143–189.

Gunaseelan, V.N. (1997) 'Anaerobic digestion of biomass for methane production', *Biomass Bioenergy*, vol. 13, pp. 83–113.

Han, J., Mol, A., Lu, Y. and Zhang, L. (2008) 'Smallscale bioenergy projects in rural China: Lessons to be learnt', *Energy Policy*, vol 36, pp. 2154–2162.

Holliger, C. (2008) 'MicrobiologieBiotechnologieenvironmentale'. *Enseignements au 2iE*. Lausanne: Swiss Federal Institute of Technologies Lausanne (EDFL).

Horst, W.D. (2000) 'Biofuel generation, proceedings in form (2000)', *National workshop on Integrated Food Production and Resource Management*: 101–105.

Ilori, M.O., Oyebisi, T.O., Adekoya, L.O. and Adeoti, O. (2000) 'Engineering design and economic evaluation of a family sized biogas project in Nigeria', *Technovation*, vol 20, pp. 103–108.

Ilyas, S. (2006) 'Biogas support program is a reason for its success in Pakistan', *American-Eurasian Journal of Scientific Research*, vol 1, no 1, pp. 42–45.

IPCC. (2001) *Climate change 2001, IPCC third assessment report*. Geneva: IPCC/United Nations Environment Programme.

IPCC. (2007) *IPCC fourth assessment report*, Geneva: IPCC/United Nations Environment Programme.

ISAT & GTZ. (1999) *Biogas digest volume III. Biogas – costs and benefits and biogas – programme implementation*. Frankfurt: Information and Advisory Service on Appropriate Technology (ISAT) and German Agency for Technical Cooperation GmbH (GTZ).

Jagadeesh, K.S. (1996) 'Biogas from leaf biomass: prospects and problems', in *Proceedings of the International Conference on Biogas Energy Systems* held at TERI, New Delhi, 22–23 January.

Jahangirian, S., Engeda, A. and Wichman, I.S. (2009) 'Thermal and chemical structure of biogas counter flow diffusion flames', *Energy and Fuels*, vol.23, pp. 5312–5321.

Jain, M.C. (1993) 'Bioconversion of organic waste to fuel and manure', *Fertilizer News*, vol 38, pp. 55–61.

Kanwar, S.S. and Guleri, R.L. (1994) 'Performance evaluation of a family size rubber balloon biogas plant under hilly conditions', *Bioresource Technology*, vol 50, pp. 119–121.

Karekesi, S. and Ranja, T. (1997) *Renewable energy technologies in Africa*, London: Zed Books.

Khanal, S.K. (2008) *Anaerobic biotechnology for bioenergy production*, New York: Wiley-Blackwell.

Khendelwal, K.C. (1990) 'Biogas technology development and implementation strategies – Indian experience', in *Proceedings of the International Conference on Biogas Technology and Implementation Strategies*, Bremen Overseas Research and Development Association, Eschroborn, Germany.

Khendelwal, K.C. and Mahdi, S.S. (1986) *Biogas technology: A practical technology*. New Dehli: McGraw-Hill.

Korres, N.E., Singh, A., Nizami, A.S. and Murphy, J.D. (2010) 'Is grass biomethane a sustainable transport biofuel?' *Biofuels, Bioproducts and Biorefining*. Vol. 4, no. 3, pp. 310–325.

Kunte, D.P., Yeole, T.Y., Chiplonkar, S.A. and Ranade, D.R. (1998) 'Inactivation of *Salmonella typhi* by high levels of volatile fatty acids during anaerobic digestion'. *Journal of Applied Microbiology*, vol. 84, pp. 138–142.

KVIC. (1993) *Khadi and VI Commission and its non-conventional energy programmes*. Bombay: KVIC.

Lafay, Y., Taupin, B., Martins, G., Cabot, G., Renou, B. and Boukhalfa, A. (2007) 'Experimental study of biogas combustion using a gas turbine configuration', *Experimental Fluids*, vol.43, no.2, p 395.

Li, Z., Tang, R., Xia, C., Luo, H. and Zhong, H. (2005) 'Towards green rural energy in Yunnan, China', *Renewable Energy*, vol 30, pp. 99–108.

Liua, G., Lucas, M. and Shen, L. (2008) 'Rural household energy consumption and its impacts on ecoenvironment in Tibet: Taking Taktse county as an example', *Renewable and Sustainable Energy Reviews*, vol 12, pp. 1890–1908.

Mallick, M.K., Singh, U.K. and Ahmad, N. (1990) 'Batch digester studies on biogas production from *Cannabis sativa*, water hyacinth and crop wastes mixed with dung and poultry litter', *Biological Wastes*, vol 31, pp315–319.

Marchaim, U. (1992) *Biogas processes for sustainable development*. Rome: Food and Agriculture Organization.

Mata-Alvarez, J., Mace, S. and Llabres, P. (2000) 'Anaerobic digestion of organic solid wastes: An overview of research achievements and perspectives', *Bioresource Technology*, vol. 74, pp. 3–16.

Mathiesen, N.L. (1989) 'Ca and/or Mg soap solution in biogas production'. WO Patent 8900548.

Matsumoto, T., Noshiro, M. and Hojito, M. (1997) 'The effect of farm yard manure of different degradation levels on grass production', in Ando, T., Fujita, K., Mae, T., Matsumoto, H., Mori, S. and Sekiya, J. (eds), *Plant nutrition for sustainable food production and environment*, pp. 591–592. Tokyo: Kluwer Academic.

Methane to Markets. (2008) 'Managing animal waste to recover methane, international opportunities for project development', http://www.globalmethane.org/documents/ag_fs_eng.pdf (accessed 2 March 2012).

Mirza, U.K., Ahmad, N. and Majeed, T. (2008) 'An overview of biomass energy utilization in Pakistan', *Renewable and Sustainable Energy Reviews*, vol 12, pp. 1988–1996.

MNES. (2002) *Annual Report 2001–2002*, New Delhi: Ministry of Non-conventional Energy Sources, Government of India.

MNES. (2006) *Renewable energy for rural applications*, New Delhi: Ministry of Non-conventional Energy Source.

Murphy, J.D. and Power, N.M. (2006) 'A technical, economic and environmental comparison of composting and anaerobic digestion of biodegradable municipal waste', *Journal of Environmental Science and Health Part A Toxic/Hazardous Substances & Environmental Engineering*, vol 41, no 5, pp. 865–879.

Murphy, J.D., Power, N. and Poliafico, M. (2006) 'The potential for pig slurry to power trains in Ireland'. in *Conference proceedings of 2nd International Conference of Renewable Energy in Maritime Island Climates*, Dublin, Ireland, April 26–28.

Nagamani, B., Chitra, V. and Ramasamy, K. (1992) 'Biogas production technology – An Indian perspective', 32nd Annual Conference of Association of Microbiologists of India held at Madurai Kamaraj University, Madurai.

Naskeo Environnement. (2009) 'Biogas composition'. http://www.biogas-renewable-energy.info/biogas_composition.html (accessed on 2 February 2012).

Navickas, K. (2007) 'Biogas for farming, energy conversion and environment protection'. Bioplin, Technologija in Okolje, 29 November 2007, Rakičan. http://fk.uni-mb.si/fkweb-datoteke/Biosistemsko_inzenirstvo/Bioplin-Navickas.pdf (accessed on 2 February 2012).

Nordberg, Å. and Edström, M. (1997) 'Co-digestion of ley crop silage, source sorted municipal solid waste and municipal sewage sludge'. *Proceedings from 5th FAO/SREN Workshop, Anaerobic Conversion for Environmental Protection, Sanitation and Re-Use of Residue*; 24–27 March; Gent, Belgium.

Osei, W.Y. (1993) 'Woodfuel and deforestation – answers for a sustainable environment', *Journal of Environmental Management*, vol 37, pp. 5162.

Parashar, D.C., Gadi, R., Mandal, T.K. and Mitra, A.P. (2005) 'Carbonaceous aerosol emissions from India', *Atmospheric Environment*, vol.39, pp. 7861–7871.

Patel, H. (2001) 'The biogas alternatives', *IREDA News*, vol 12, no 3, pp. 93–96.

Patel, V., Patel, A. and Madamwar, D. (1992) 'Effect of adsorbents on anaerobic digestion of water-hyacinth-cattle dung'. *Bioresource Technology* vol. 40, no. 2, pp. 179–181.

Pathak, H., Jain, N., Bhatia A., Mohanty, S. and Gupta, N. (2009) 'Global warming mitigation potential of biogas plants in India', *Environmental Monitoring and Assessment*, vol 157, pp. 407–418.

Paul, N. and Kemnitz, D. (2006) *Biofuels–plants, raw materials, products*, Berlin: Fachagenturnachwachsende Rohstoffe e. v. (FNR).

Popa, B. (2010) 'Emissions: gasoline vs. diesel vs. bioethanol', Available from http://www.autoevolution.com/news/emissions-gasoline-vs-diesel-vsbioethanol-3657.html.

Prasad, S. and Dhanya, M.S. (2011) 'Air quality and biofuels'. in Marco Aurélio dos Santos Bernardes (ed.) *Environmental impact of biofuels*, pp. 227–250. New York: InTech Publishers.

Prasad, S., Singh, A. and Joshi, H.C. (2007a) 'Ethanol as an alternative fuel from agricultural, industrial and urban residues'. *Resources, Conservation and Recycling*, vol. 50, pp. 1–39.

Prasad, S., Singh, A., Jain, N. and Joshi, H.C. (2007b) 'Ethanol production from sweet sorghum syrup for utilization as automotive fuel in India'. *Energy and Fuels.* Vol. 21, no. 4, pp. 2415–2420.

Radhika, L.G., Seshadri, S.K. and Mohandas, P.N. (1983) 'Biogas production from a mixture of coir pith and cattle waste', *Journal of Chemical Technology & Biotechnology,* vol B33, pp. 189–194.

REN 21. (2005) 'Renewable Energy Policy Network 2005', *Renewables 2005: Global Status Report,* Washington DC: Worldwatch Institute.

Renewable Energy Association. (2009) 'Energy and environment.' http://www.r-ea.net/info/energy-info (accessed 20 March 2010).

Renwick M., Subedi, P.S. and Hutton, G. (2007) *A cost-benefit analysis of national and regional integrated biogas and sanitation programs in sub-Saharan Africa,* Brussels: Winrock International.

Samuil, C., Vintu, V., Iacob, T., Saghin, Gh. and Trofin, A. (2009) 'Management of permanent grasslands in north-eastern Romania', *Grassland Science in Europe,* vol. 14, pp. 234–237.

Sasse, L., Kellner, C. and Kimaro, A. (1991) *Improved biogas unit for developing countries,* (GATE) Eschborn: DeutschesZentrum fur Entwicklungstechnologien..

Sharma, K.R. (2008) 'Kinetics and modeling in anaerobic processes', in Khanal, S.K. (ed.) *Anaerobic technology for bioenergy production: principles and applications,* Ames, IA: Wiley-Blackwell.

Sheehan, J. (2009) 'Biofuels and the conundrum of sustainability'. *Current Opinion in Biotechnology,* vol. 20, pp. 318–324.

Shimizu, C. (1992) 'Holding anaerobic bacteria in digestion tank', JP Patent 4341398.

Singh, A., Smyth, B.M. and Murphy, J.D. (2010) 'A biofuel strategy for Ireland with an emphasis on production of biomethane and minimization of land-take'. *Renewable and Sustainable Energy Reviews,* vol. 14, no. 1, pp. 277–288.

Singh, A., Nizami, A.S., Korres, N.E. and Murphy, J.D. (2011) 'The effect of reactor design on the sustainability of grass biomethane'. *Renewable and Sustainable Energy Reviews* vol. 15, no. 3, pp. 1567–1574.

Smith, W.H., Wilkie, A.C. and Smith, P.H. (1992) 'Methane from biomass and waste – a program review' *TIDE (Teri Information Digest on Energy),* vol. 2, no 1, pp. 1–20.

SPOBD. (1979) 'Biogas technology and utilization'. Chengdu Seminar, Sichuan Provincial Office of Biogas Development, Sichuan, P.R. China.

Steinfeld H., Gerber, P, Wassennar, T., Castel, V., Rosales, M. and deHaan, C. (2006) *Livestock's long shadow: Environmental issues and option',* Rome Food and Agriculture Organisation of the United Nations. www.virtualcentre.org/ en/library/key_pub/longshad/ A0701E00.htm (accessed on 15 Jan. 2012).

Strassburg, B., Turner, R.K., Fisher, B., Schaeffer, R. and Lovett, A. (2009) 'Reducing emissions from deforestation—The "combined incentives" mechanism and empirical simulations', *Global Environmental Change,* vol 19, pp. 265–278.

Thomé-Kozmiensky, K.J. (1995) *Biologische Abfallbehandlung.* Berlin: EF-Verlag für Energie- und Umwelttechnik.

Thornton, P.K., Kruska, R.L., Henninger, N., Kristjanson, P.M., Reid, R.S., Atieno, F., Odero, A.N. and Ndegwa, T. (2002) *Mapping poverty and livestock in the developing world.* Nairobi: ILRI (International Livestock Research Institute).

Tirumale, S., and Nand, K. (1994) 'Influence of anaerobic cellulolytic bacterial consortia in the anaerobic digesters on biogas production'. *Biogas Forum III* vol. 58, pp. 12–15.

Topa, N., Mizoue, N., Kaib, S. and Nakao, T. (2004) 'Variation in wood fuel consumption patterns in response to forest availability in Kampong Thom Province, Cambodia', *Biomass and Bioenergy* vol 27, pp. 57–68.

Traffic & Public Transport Authority (2000) *Technology and biogas use in Sweden,* City of Gothenburg, Sweden.

Tyler, M.G. (2007) *Sustaining the Earth: An integrated approach.* New York: Thomson Advantage Books.

United Nations. (1984) *Updated guidebook on biogas development,* Energy Resources Development Series, No. 27. New York: United Nations.

United Nations. (2009) 'Millennium Development Goals', http://www.un.org/millenniumgoals/ bkgd.shtml (accessed on 26 October 2011).

United Nations Development Programme (UNDP). (1997) *Energy after Rio: Prospects and challenges*, New York: United Nations.

United Nations Public-Private Alliance for Rural Development. (2009) 'Information'. http://www.un.org/esa/coordination/Alliance/index.htm (accessed 26 October 2011).

Urmee, T., Harries, D. and Schlapfer, A. (2009) 'Issues related to rural electrification using renewable energy in developing countries of Asia and Pacific', *Renewable Energy*, vol 34, pp. 354–357.

Veena, D.R (1988) *Rural energy: Consumption, problems, and prospects – a replicable model for India*. New Delhi: Ashish.

Verstraete, W., Morgan-Sagastume, F., Aiyuk, S., Waweru, M., Rabaey, K. and Lissens, G. (2005) 'Anaerobic digestion as a core technology in sustainable management of organic matter', *Water Science & Technology*, vol 52, no 1–2, pp59–66.

Voegeli, Y. and Zrubruegg, C. (2008) 'Biogas in cities – A new trend?', *Sandec News* 9.

WHO. (2002) *Addressing the links between indoor air pollution, household energy and human health*. Geneva: World Health Organization.

Wong, M.H., and Cheung, Y.H. (1995) 'Gas production and digestion efficiency of sewage sludge containing elevated toxic metals'. *Bioresource Technology* vol. 54, no. 3, pp. 261–268.

Woods, J., Hemstock, S. and Burnyeat, W. (2006) 'Bioenergy systems at the community level in the South Pacific: Impacts & monitoring', *Mitigation and Adaptation Strategies for Global Change*, vol 11, pp. 469–500.

World Bank. (2004) *'Renewable energy for development: the role of the World Bank Group'*, Washington DC: World Bank.

Yadvika, Santosh, Sreekrishnan, T.R., Kohli, S. and Rana, V. (2004) 'Enhancement of biogas production from solid substrates using different techniques—a review'. *Bioresource Technology* vol. 95, no. 1, pp. 1–10.

Chapter 20

Concluding remarks

The biogas industry is beginning to provide energy to both domestic and industrial sectors at an increasing pace. Given the appropriate legislative, political and economic framework, biogas used either as steam, heat, electricity or as vehicle fuel can be a significant provider of cheap and clean energy.

Much effort has been invested in this book in providing a holistic, multidimensional assessment of the science and technology of biogas production and utilisation. In addition, key areas for improvement and future trends or needs have been identified. It is evident that biogas/biomethane brings a plethora of benefits to many economic sectors including transport, energy, waste management services and agriculture. Biogas is needed to mitigate climate change and respond to fuel shortages or high costs. At the same time it can provide alternative options due to the range and abundance of materials that can serve as feedstock for its production, the advanced state of many technologies used in its production and the extent to which its production and utilisation are in compliance with existing policies.

It seems inappropriate to use potential food crops other than 'waste' products and agricultural 'residues' for biofuel production in order not to compete directly with human food needs. Grassland is an obvious source of feedstock for biofuel production, and the sequestration of carbon under permanent grassland has obvious benefits relating to reducing greenhouse gas emissions. Organic wastes along with industrial residues, particularly from the food, beverages and biofuel industries, are easy to use in addition to grass and maize silage feedstock. The anaerobic digestion of these materials provides solutions to problems of their disposal and recycling management. Finally, algae due to their composition and productivity seem another great potential source of feedstock for biogas/biomethane production.

Improvements in biogas production are also essential through engineering solutions and improved understanding of the microbiology and biochemistry of fermentation. Clearly the desire to optimise feedstocks by pre-treatments or mixing of different feedstock components is tempered by practical considerations such as the need to manage waste material from other industries in a safe and efficient way. Selection of the proper digester design for biomethane production is an important management/design decision. The wet continuous two-stage system, the leach bed system with UASB, the dry continuous system and batch digesters all have potential for biomethanation of agricultural residues. The chosen digester type will lead to different biogas production rates even for the 'same' feedstock due to differences in seasonal growth, harvesting regimes, pre-treatment, digester operating conditions and procedures.

Biogas can be used for the production of heat and/or electricity, or it can be transformed to biofuel or natural gas substitute. However, it has to be cleaned and upgraded in order to reach the quality requirements for vehicle fuel or grid injection. Many factors influence the selection of the most suitable upgrading technique.

Biomethane, as the final product of AD, is less corrosive than biogas and potentially is more valuable as a fuel. For these reasons, it may be both possible and desirable to store biomethane for on- or off-site uses. Compressed biomethane is generally stored in order to increase the energy density, save storage space and facilitate transportation. The provision of sufficient gas storage is an important design consideration which can have a major effect on the efficiency and viability of a biogas plant. For biomethane distribution, the use of biogas pipelines or gas grid injection at larger scale, offers the most efficient mode of distribution from an economic and environmental impact viewpoint. It allows for greater energy efficiency and flexibility of end use.

Monitoring biogas/biomethane production for better control of AD plant performance aims at optimising operational processes. This, in combination with techniques suitable for data storage and subsequent analysis such as data warehousing and data mining have the potential to deliver invaluable tools to secure sustainable biogas/biomethane production.

As with food crops, genetic improvements of non-food biofuel crops is in progress but more research is needed. In particular, the time is now right to transfer results from model plant systems such as *Arabidopsis* into biofuel and food-crop plants.

The ability to detect, identify and monitor species in complex mixtures has demonstrated a far greater diversity of micro-organisms participating in AD processes than had been thought. Their composition is strongly influenced by operating conditions including the feedstocks used. Although some core microbial groups appear to occur frequently, there is dynamic change in their populations over time, with a strong stochastic element. The significant role of hydrogenotrophic methanogens working with syntrophic bacteria in many effective gas producing communities, when acetogens had been thought to be dominant before, has been demonstrated. Reducing the number of micro-organisms involved in the process to a few highly engineered taxa may be the future for some AD processes, but the need for a number of taxa with redundant metabolic performance for robust digester function suggests other strategies are also required. The extent to which surveys of digesters have revealed species that are new to science and whose biology will need to be characterised indicates how much more work is needed.

Molecular techniques have revolutionised the study of the role of micro-organisms in the AD process. First, by providing the means to reliably identify and enumerate which organisms are involved in the process. Second, by helping speed the understanding of their metabolism through gene isolation and characterisation. Genomics has allowed aspects of metabolism to be elucidated even in unculturable organisms. Metagenomics has enabled the targeted search for more efficient new enzymes and new strains. Knowledge of metabolism and its genetic controls can be used to engineer microbes with novel capability to improve efficiency of bioenergy processes. The application to biogas production had to await the challenging development of tools to investigate and engineer the methanogens, critical for biogas production. Genetic engineering promises major breakthroughs but these will take time.

Treating biogas as an energy source alone does not do justice to its overall value. If the substrate is properly selected, the digestate serves as an excellent fertiliser of high quality replacing manufactured fertiliser. Hence, to make optimum use of the N content of digestate, it should be applied at times of maximum crop growth – generally during the late winter to summer period. Additionally, best use of digestate is made when 'low emission application equipment' is used, namely a band-spreader (trailing hose/trailing shoe) or shallow injector, to reduce ammonia losses (and odour nuisance) compared with surface broadcast application, and thereby increase crop available N supply. An integrated digestate and manufactured fertiliser use policy should aim to maximise as far as practically possible the nutrients supplied by digestate.

The constantly increasing costs of both energy and food production have led many enterprises within the food and beverage production sectors, small-scale farms and households, particularly in developing countries, to reconsider practices concerning the utilisation of their organic residues by using integrated AD technology.

With the correct infrastructure and financial support, the sustainability credentials of small-scale anaerobic digesters (SSADs) are evident. SSADs can effectively contribute to a reduction in GHG emissions from agriculture with complementary benefits to both ecosystems and society. It is therefore imperative that more consideration should be given to how SSADs can be integrated into agriculture more effectively to produce biogas, with particular emphasis placed on transferring knowledge and practices from the laboratory into real-life conditions – for example, an efficient utilisation of different feedstock at advantageous ratios as well as pre-treatment options. The most important key areas in biogas/biomethane production are listed below.

Key issues in biogas/biomethane production

- A policy addressing production, distribution and use (e.g. the production of electricity by anaerobic digestion of energy crops will not be economically feasible without subsidies) is a key factor in the development of the biogas industry. Stakeholders should be involved in policy formation. A biomethane for transport industry can reduce GHG emissions and local pollutants, while providing waste treatment and a source of renewable energy.
- The high biomass production, the high biomethane yield along with advances in husbandry techniques for crop establishment, maintenance and storage make grass and maize important feedstocks for biomethane production. This is particularly true when the similar principles of silage production for both biogas and feed production are taken into consideration. In addition, algae are a suitable resource for biogas production in terms of composition and general productivity, but require more work before this resource can be accessed on a large scale.
- Among the industries presented in the feedstock section, the sugar industry and abattoirs (due to lack of alternative utilisation) have the strongest interest in integrating AD technology to their production process.
- There is need to compare the potential of various pre-treatment options to increase the process efficiency. Digester design for biomethane production is an important management/design decision. A high-rate reactor such as UASB, for example, converts volatisable organics into biogas with great efficiency and has a short retention time. This type of reactor is well suited to substrates with solids content between < 4 to 15 per cent and can process a wide range of substrates when attached to a leach bed reactor.
- The selection of the biogas upgrading includes adsorption (pressure swing adsorption), absorption (water scrubbing and chemical amine), membrane separation and cryogenic techniques. The most appropriate technology is dependent on the amount and composition of the raw biogas, the quality of biomethane desired, and economic and environmental issues.
- Biomethane is less corrosive than biogas and also, potentially, is more valuable as a fuel. For these reasons, it may be both possible and desirable to store biomethane for on- or

- off-site uses. Compressed biomethane is generally stored in order to increase the energy density, save storage space and facilitate transportation. Depending on the purpose, biogas can be compressed to high pressure, medium pressure or low pressure prior to storage.
- The rationale for monitoring biogas production is that biogas plants which operate without monitoring may be underperforming. As stated in Chapter 8, the main bottlenecks in the anaerobic digestion of industrial feedstocks are ammonia inhibition and foaming (slaughterhouse waste, bioethanol residues), management of digestate (bioethanol residues), lignocellulose containing compounds (brewers' spent grains) or other inhibiting substances (e.g. polyphenols in olive oil waste). Additionally, the chosen digester type, although it is an important management decision for biogas production, will lead to different biogas production rates even for the 'same' feedstock due to differences in seasonal growth, harvesting regimes, pretreatment, digester operating conditions and procedures. Managing this variation is a key element of AD system operation. Data warehouse and data mining techniques are invaluable tools in knowledge discovery from databases and can greatly assist in understanding and manging the variation met thoughout anaerobic digestion production chains.
- Microbial communities are highly diverse and their composition is strongly influences by temperature, pH, feedstock, the concentration of VFAs and operating conditions.
- Genetics has provided accurate tools to detect and monitor species, even in complex mixtures, and to trace the biological processes occurring in AD. This has demonstrated considerable differences between microbial communities in different fractions of the digestate (e.g. liquid and solid fractions and biofilms) and between digesters fed different substrates or having different environmental circumstances.
- LCA can be an invaluable tool for the examination of the biogas production sustainability as long as its application follows, as far as possible, transparent methodology. This allows critical review following its completion by its commissioner or independent parties.
- Soil fertility can be affected by the removal of biomass from land through increased erosion, losses of nutrients and organic carbon. The application of digestate as a substitute for manufactured fertilisers is a management option that supports biogas/biomethane sustainability significantly. Optimum use of the N content of digestate can be achieved when digestate is applied when crop growth is maximum and when 'low emission application equipment' such as a bandspreader (trailing hose/trailing shoe) or shallow injector is used.
- SSADs can effectively contribute to a reduction in GHG emissions from agriculture with complementary benefits to both ecosystems and society in developing and developed countries. This is especially so nowadays when strategic management for efficiencient energy use and rural developent are recognised as essential elements of sustainability.

Future needs in biogas/biomethane production

- Research towards open ocean aquaculture for increased algae biomass production needs to be addressed in detail.
- Identification of the problems and suggested solutions regarding the use of maize in mono-digestion under various AD production scenarios, e.g. high loading rates, is needed.
- Research towards mono-digestion of large-scale AD plants for breweries and/or olive mills, merits further consideration due to their considerable potential.

- Research towards co-digestion of lignocellulosic feedstock with other substrates and subsequent benefits on biogas/biomethane production is required.
- Research on the effects of nanotechnology applications to monitoring and controlling the inhibitory effects of VFA, COD, pH and temperature deserves further consideration.
- Investigation of the applicability of data warehouse and data mining techniques on monitoring and controlling the entire biogas production chain will be important.
- The relative merits of biogas production strategies which move in the direction of reducing the number of micro-organisms involved in a process to a few highly genetically engineered taxa, or rely on more complex communities with high resilience and redundancy, need to be assessed.
- The integration of LCA with advanced data management approaches such as these of data warehouse and data mining techniques could benefit decision-making processes for increased biogas/biomethane production.
- Investigation of the effects of various types of digestate and manufactured fertilisers on the feedstock production for AD is required.

Appendix A

Anaerobic digestion application in a typical Cambodia family farm – a case study

Andrea Salimbeni and Giuliano Grassi

EUBIA, European Biomass Industry Association
Rue d'Arlon 63–65
B-1040 Brussels, Belgium
Email: eubia@eubia.org
Web: www.eubia.org

Description of the system

The following case study represents the potential of anaerobic digestion to cover the basic needs of a typical Cambodia family farm that cultivates 1.5 ha of rice, and owns two pigs and two buffalo. The family is assumed to consist of five members.

Pigs usually remain in their sty for many days during their life span. Nevertheless, in comparison with intensive breeding units rearing pigs out of doors, as in this case study, this makes the collection of manure they produce a difficult task. Thus, even if they produce effluents and manure every day, it is assumed that the amount of available manure (for feeding the digester) is equal to that produced in 200 days per year. However, to reach this amount of manure and consequently a considerable organic content (for the needs of the anaerobic digestion), the farmer must collect the excreta periodically.

Buffalos are often used in cultivation and sowing activities which require many working hours per day. For this reason, it is almost impossible to consider all the manure they produce as being available for anaerobic digestion. In this case study the manure produced is available only for 150 days per year. The buffalos are fed twice/three times a day. Manure should be collected, at least, every evening and every morning before daily cultivation activities

Rice crop residues (straw, husks and leaves) are used as high value feedstock to feed the digester and produce biogas. Worldwide rice is largely the most important potential feedstock for biogas production. The tropical climate conditions allow for two rotations annually which permit the family to have a continuous biogas production throughout the year.

Facts and assumptions

- In this small AD plant, biogas is not purified and upgraded into biomethane, because of the costs and technical requirements related to upgrading technology.
- Biogas is composed of CO_2 (35 per cent), CH_4 (50–55 per cent) and others (such as N_2, H_2; 10 per cent).
- Power generation is not considered here because it is accompanied by high costs, difficulties in installation and, above all, would involve complicated, frequent and expensive maintenance operations for this small family farm.

Therefore, the raw biogas produced will cover the needs of various devices installed such as gas lamps for lighting, a boiler for heating water and a burner for cooking. Gas lamps will have lower efficiency in comparison with those burning natural gas or propane lamps due to the CO_2 presence in biogas. Nevertheless, three of them should provide sufficient light to cover the daily needs of the family.

Biogas production life cycle

Figure A.1 describes the life cycle of biogas production within the family farm. All the organic material produced by these three sources as mentioned earlier is co-digested in the anaerobic digester to produce light and heat as shown in Figure A.1.

Biogas production

In the following sections, calculations of the potential biogas production based on crop residues, pig and buffalo manure are presented.

Estimated biogas production from rice crop residuals, pig and buffalo manure

Tables A.1, A.2 and A.3 represent the parameters needed for the estimation of methane potential production from rice residues (straw, husk and leaves), pig and buffalo manure.

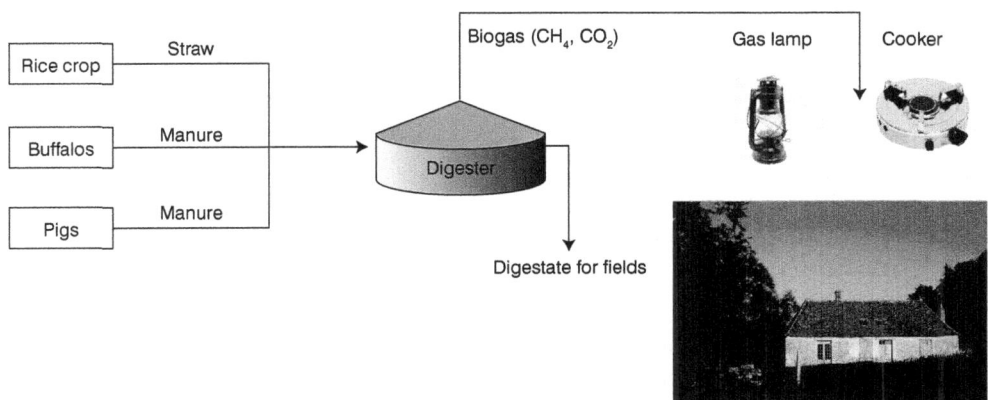

Figure A.1 Life cycle of biogas production

Table A.1 Parameters for methane estimation from rice crop residuals.

Rice crop parameters[1]	Value[1]
Straw (t/ha)	6 fresh tons/ha per cycle
Husk and leaves (0.5 fresh tons/ha per cycle)	0.5
Moisture (%)	50
Biogas (m³/t vs)	430[2]
Methane content (%)	50

[1] EUBIA estimation unless stated otherwise; [2] Contreras et al. (2012)

Table A.2 Parameters for methane estimation from pig manure.

Pig manure parameters[1]	Value[1]
Total manure (kg/day)	12
Dry matter (kg/day)	1.8
Volatile solids produced (kg vs/day)	1.4
Biogas yield (m³/t vs)	400[2]
Methane content (%)	55

[1] EUBIA estimation based on working activities and farm audits; [2] Evaluation based on EUBIA experience

Table A.3 Parameters for methane estimation from buffalo manure.

Buffalo manure parameters	Value
Total manure (kg/day)	25
Dry matter (kg/day)	3
VS produced (kg vs/day)	2.4[1]
Biogas yield (m³/t vs)	400[2]
Methane content (%)	55

[1] Maithel (2009); [2] Evaluation based on EUBIA experience

The annual methane production (Table A.4) was estimated by taking into account the sub-systems the family farm manages (i.e. rice crop, pigs and buffalo).

Energy consumption

Total biomethane consumption is shown in Table A.5.

Conclusions

The potential of anaerobic digestion to provide energy for lighting, heating and cooking needs of a small family farm in a developing country for the whole year is feasible since the biomethane potential production over limits the annual needs in energy of this family farm model.

Table A.4 Total methane yield.

Estimated biomethane production (m^3 per year)	
Rice crop residues	1069.1
Pig manure	123.2
Buffalo manure	158.7
Total	1350.7

Table A.5 Annual energy consumed for the satisfaction of the family farm basic needs.

Energy consumed devices	No of devices	Methane consumption (m^3/hr)[1]	Hours in operation (hrs/yr)[1]	Total CH_4 consumption (m^3/yr)[1]
Lighting devices	3 silk mantle lamps	0.13	1200	468
Heating devices	1 boiler	5.00	150	68
Cooking devices	1 burner	2.16	182	394
Total				930

[1]Maithel (2009)

References

Contreras, L.M., Schelle, H., Sebrango, C.R. and Pereda, I. (2012) 'Methane potential and biodegradability of rice straw, rice husk and rice residues from the drying process', *Water Science Technology*, vol. 65, no 6, pp. 1142–1149.

Maithel, S. (2009) *Biomass energy. resource assessment handbook*, Prepared for APCTT (Asian and Pacific Centre for Transfer of Technology) of the United Nations – Economic and Social Commission for Asia and the Pacific (ESCAP).

Index

abattoirs 15–16, 100, 111–17, 126, 130, 205, 424
Abbotsford, B.C. 147
Abonyi, J. 231
absorption 156–8, 160–70
acetate 275–7, 294–6
acetic acid 249, 276, 294–5
aceticlastic bacteria 116, 213, 274–5, 280
acetogenesis 194–5, 209, 237, 239, 262–5, 294–5, 400
acetyl-CoA pathway 295
acidification 59, 73–4
Acidobacteria 274
acidogenesis 32–3, 294, 400; data mining 237, 239; microbial dynamics 262–4; pH 209–10; process monitoring 194–5
Actinobacteria 205, 274, 400
activity data 332–5
additives 414–15 *see also* pre-treatments
ADM1 280–1
adsorption 156–8, 160, 415; pressure swing 165–7, 170–6, 178, 190
aeration 275
Africa 191, 400, 408, 411, 415
agricultural residues 15–16, 24, 342; developing countries 397, 399, 403, 415; land-use change 33–7, 39–40; organic wastes 100, 106–8
agriculture 398, 410; carbon emissions 4–5; digestate as fertiliser 359–70; policy 14–16, 24 *see also* small-scale anaerobic digesters
Ahlgren, S. 342
Åhman, M. 9
Ahring, B.K. 208
Alcaligeaceae 274
algae 82–4, 92–6, 422, 424; macro- 88–92; micro- 84–8
alkalinity 50, 56–7, 207–8, 210–11, 217–18
allocation 326, 342, 346–7
Amer, M. 22
amine scrubbing 164–5, 170–7
ammonia 75, 160–1, 270; fertiliser 367, 369; industrial residues 116, 120; process monitoring 210–11, 213–14
Amon, T. 209

amplified fragment length polymorphism (AFLP) 298, 300
Anaerobaculum 206
ANAEROBECHIP 281
anaerobic digestate quality protocol (ADQP) 360, 367
anaerobic digestion (AD) 34, 67–9, 139–44; biochemistry 290–6; design/process 72–3; future research 148–9; industry challenges 144–7; modelling 279–81 *see also* digester types
anaerobic digestion analytic model (ADAM) 383, 388
Andrae, A. 322
Angelidaki, I. 208
animal by-products (ABP) 16, 112–13, 382 *see also* slaughterhouse wastes
Animal By-Products Regulations 16, 367, 371
animal wastes 100, 199, 270, 343, 402, 404, 407–8 *see also* manures; slurry
aquaculture 85, 88, 94–5, 425
arable crops 4–6, 14, 368
archaea 264, 267, 270–6, 280, 294, 301–3, 307–9
Argentina 9
Arnold, K. 325
Ascaris 386
Ascomycota 265
Asia 139, 191, 375–6, 398, 408–10, 415
Aspergillus 206
Astals, S. 105
ATRES 123
attributional modelling 321–2, 346
Australia 381
Austria: energy/agricultural policy 9, 11, 25; industrial residues 114, 122–4, 130; maize 67, 73; technology 139, 147, 191

Babu, K.S. 415
Bacillus 206, 307–8, 386
bacteria 116, 212, 400, 415; algae 91–2; grass 58–9; inoculants 205–6; maize 69–70, 74; molecular biology 301, 303, 307–9; population dynamics 265–7, 270–5, 278
Bacteroides 400

Bacteroidetes 265, 270
Bajgain, S. 410
bandspreading 367–8
Bangladesh 408–9, 411
Banks, C. 209, 237
Barbados 411
barley 4, 334
Baserga, U. 59
Basidiomycota 265
Batstone, D.J. 253, 280
Bavaria 123
Bayesian theory 244–6
Beddington, J. 6
Belgium 18, 92, 177
Benefield, D.L. 253
Benin 408
Berglund, M. 406
Bergsma, G. 324–5
Berlin 10
Berne 317
Berry, P.M. 4–5
Bhatia, P. 332, 335, 346–7
Bhutan 408
bicarbonate concentrations 210–11
Bifidobacterium 400
biochemical methane potential (BMP) 119, 130, 140, 144
biochemical oxygen demand (BOD) 216–17, 365
biochemistry 279–82, 290–7
bioCNG 9, 11, 17–18, 20, 22–4, 192
biodegradable municipal waste (BMW) 100–1, 105–7 see also organic fraction of municipal solid waste
biodiesel 4, 86–7, 308, 383–5
biodiversity 32, 41–2
Biodiversity Action Plan 17
bioethanol 3, 86–7, 89, 117–20
Biofertiliser Matrix 369–71 see also digestate
biofilters 158
biofuels 3–4, 422–3; energy/agricultural policy 9–15, 17–18, 20–1, 24; land-use change 31, 34, 39, 41–2; life cycle assessment 318, 321, 323–5, 341–2; molecular biology 307–8
biogas: benefits 403–4; composition/production process 401–3; compression/storage 177–8; contaminants/treatment 155–61
Biogas Development and Training Centres (BDTC) 413
biogas spent sludge (BSS) 407 see also digestate
Biogas Support Programme (BSP) 406, 410
biological enrichment 170
biological oxygen demand (BOD) 216–17, 365
biological pre-treatment 205–6
biorefinery 33–4, 66, 92–4
Birkmose, T. 363
Bischofsberger, W. 130
Blanc, F.C. 213
blended fuels 18
blood 16, 100, 113–16, 130
Bochmann, G. 123

boilers 184–5
Bolivia 411
Bomb, C. 10
Bombay 409
Börjesson, P. 324, 406
Bothast, R.J. 118
bottlenecks 131, 425; bioethanol 118–20; olive mill waste 125–7; organic waste 105–8; slaughterhouse waste 116–17
boundaries 235–6
Bouwman, A.F. 343
Brassica carinata 325
Braun, R. 130
Brazil 3, 11, 33, 121, 398
Brennan, L. 87
Brereton, A.J. 55
breweries 111–12, 121–4, 384–6
British Columbia 140, 147
British Sugar 128
Broch, A. 327
Brooks, L. 130
Bruni, E. 72, 204
BSE 111–12
Buekens, A. 212
buffalo 428–31
buffering capacity (BC) 50, 56–7, 207–8, 210–11, 217–18
Bulgaria 9
Burkina Faso 408
Burundi 411
Buswell, E. 155

C_4 plants 50–1, 54, 56, 66
Caldicellulosiruptor bescii 308
California 147
Cambodia 408, 411, 427–31
Cameroon 409, 411
Campylobacter 366, 386, 407
Canada 11, 147
carbon 3–5; land dynamics 31–2; savings 366, 388–91; sequestration 5–6, 88, 422
carbon dioxide 37–8, 55–6, 86, 148, 152, 155; absorption techniques 161–70; comparison of upgrading techniques 173–8
Carbon Finance Unit 412
carbon monoxide 295
carbon/nitrogen (C/N) ratio 73, 103, 141, 383; industrial residues 126, 129; process monitoring 199–200, 207–8, 210–11
Casella, E. 55
Catalyst Power 147
cations 103, 116, 415
cattle 100, 212, 268; developing countries 399, 401–2, 404, 408, 415; digestate/fertiliser 361–5; life cycle assessment 343–4; slaughterhouse wastes 112–15; small-scale digesters 375–6, 382–4, 391–2
CATWOE model 323–4
cell wall 52 see also hemicellulose; lignin

cellulose 69, 144; industrial residues 122–3; molecular biology 291–2, 305–6, 308; process monitoring 194–6, 199, 204–5
cellusomes 291
centralised anaerobic digestion (CAD) 15, 21
Ceriporiopsis subvermispora 206
Certified Emission Reductions (CERs) 412
Chae, K.J. 209–10
chemical absorption 158, 160–1
chemical characteristics 197–200, 202–3
chemical oxygen demand (COD): data warehouse/mining 236–7, 239, 249; industrial residues 123, 125–6, 130; process monitoring 199, 204, 216–18
chemical pre-treatment 205
chemical scrubbing 164–5, 170–7
chemo-autotrophic upgrading 170
Chen, L.Z. 249
Cheng, J. 206
Cherubini, F. 325
Cheung, Y.H. 103, 415
China 11, 105, 121, 139, 196, 398, 409, 412
Chloroflexi 272, 274
chopping 72
chromatography 216
Chynoweth, D.P. 92, 96
Cifuentes, L. 379
classification 244–50, 254
Clean Development Fund 412
Clean Development Mechanism (CDM) 346
Clemens, J. 212
Clift, R. 377–9
Climate Change Act 381
cloning 299
Clostridium 70, 265, 303, 386, 400
Clostridium acetobutylicum 306–7
Clostridium hastiforme 273
clustering 253–4
co-digestion 72, 91, 104–5, 126, 148, 385, 402
co-products 321, 342, 345–6
cobalt 73–4
cocksfoot 50–2, 202
Collins, D.P. 57
Colombia 411
combined heat and power (CHP) 155, 183–4; algae 92; industrial residues 111–12, 115–16; maize 67, 70; small-scale 387, 393n3
combustion 123, 154, 156, 158–60, 185, 404–6
Common Agricultural Policy (CAP) 14
Common Market Organisation (CMO) 127
Community Biogas Plants (CBP) 412
complexity 387–8
compressed natural gas (CNG) 9, 19–21, 31, 191–2
compression 177–9, 184–5, 190–1, 425; ratio 405–6
condensate 119–21
consequential modelling 321–2, 326, 346
contaminants 155–61, 365
continuously stirred tank reactors (CSTRs) 89–90, 114, 143, 207, 212, 216

cooking 397–8, 403–4, 429–31
Coprothermobacter 206
corn cob mix (CCM) 67
corn connection 15
Corradini, M.G. 278
corrosion 156–7, 161, 185
Corynebacterium 400
cost benefit analysis (CBA) 388
cost-effectiveness 145–6, 173–4, 376, 411
Costa Rica 376
cows *see* cattle
Crenarchaeota 270
crop residues *see* agricultural residues
crop rotation 76
crude protein (CP) 56
cryogenic condensation 160, 168–72, 174–5
Cryptosporidium 407
cut-off criteria 336–7

Dadar Sewage Purification Station 409
Daim, T.U. 22
dairy 375–6, 382–4, 388–9
Dale, B.E. 325
Dalton's law 162
Daniel, R. 301
Dar, H.G. 415
data 214–16, 423, 425–6; life cycle assessment 323, 328–37
data banks 297, 300
data cleansing 234
data cubes 236–8
data mining (DMN) 231–2, 238–43, 254, 425–6; classification 244–50; clustering 253; regression 250–3
data warehouse (DWH) 231–8, 254, 425–6; data mining and 240–1
DDGS 118–19
DeBaere, L.D. 141
decision support system (DSS) 232
decision trees 246–7
Deenabandhu model 409–10
deforestation 399
Dehue, B. 37
Delgenes, J.P. 204
Delucchi, M.A. 42
demand pull 19
demonstration projects 18–19, 24, 411, 414
denaturing gradient gel electrophoresis (DGGE) 277, 299–300
dendrogram 203, 254
Denmark 15, 21, 363, 381, 414
Desai, M. 208
desugared molasses (DMo) 104–5
Desulphovibrio 400
Deublein, D. 21, 194
developing countries 390–1, 397–404, 414–16; biogas development 408–11; policies 411–13; problems/challenges 413–14
digestate 56, 359–66, 383, 423–5; agricultural policy 9, 16–17, 21; developing countries 407–8; how/

when to apply 367–71; industrial residues 112, 117–20, 124; maize 74–6; with manufactured fertilizer 371
digester types 101–2, 141–3, 206–7, 275–6, 422
dilution rates 274
dimensional data 234–6
dimethylsulphide (DMS) 157
Dinenympha 303
directed evolution 306–7
disposal costs 111–12, 130
distribution 183–5, 191–2 *see also* gas grid
DNA 264, 297–301 *see also* genetics
Donoso-Bravo, A. 279
Drake, H.L. 295
Drosg, B. 75
dry digesters 142, 146, 207
dry matter (DM) 50–9
dry solids (DS) 141, 144
dry-grind 117–18
drying methods 156–7
Dubreuil, A. 42
Dugat-Bony, E. 301
dust 155–6

economics 38, 199, 415–16; industrial residues 117, 124; maize 76–7; of small-scale anaerobic digesters 380; of upgrading technologies 173–5
ecosystem damage potential (EDP) 42
efficiency 387, 406
effluents 57–9
Egger, K. 59
Eichhorinia crassipes 205
Ekvall, T. 322
electricity 146, 325, 327, 388–91, 403, 412–13 *see also* combined heat and power
electrochemistry 216
Ellenrieder, J. 72
Embden–Meyerhof–Parnas (EMP) 32
emission factors (EF) 335, 343
emissions *see* greenhouse gases
Emissions Trading Scheme 388
employment 408
Enbasys 123
End-of-Waste 360, 364–5
energetic performance 170–1
energy 3, 111, 416; developing countries 397–400; life cycle assessment 317–18, 322–6, 332, 334, 337–41, 347–8 *see also* renewable energy
energy content 404–5
energy crops 14–15, 36–7, 40, 75 *see also* grass; maize
energy demand 121, 124
energy policy 10–14, 24
Energy Policy Act 147
energy recovery 120–1
energy value 154, 342
England 128, 360
enteric fermentation 398–9
Enterobacteria 70

environment 17, 320–7, 336–7, 411; life cycle assessment 317–27, 336–7, 411; upgrading processes 175–7
Environmental Permitting Regulations 360, 369
environmental variables 268–9, 276–7, 400
enzymes 73, 123, 205, 291, 294–6, 302–9
equilibrium selectivity 166
erasable programmable read-only memory (EPROM) 218
Escherichia coli. 307, 365–6, 386, 400, 407
ethanol 4, 11 *see also* bioethanol
Ethiopia 408, 411
ETL (extract, transform and load) 232–3
EU 4, 6, 39, 191; digestate/fertiliser 360, 364–5, 367, 371; energy/agricultural policy 10–11, 14, 16–18; industrial residues 112, 118, 127; life cycle assessment 317–18, 320, 323–4, 341–2; process monitoring 194, 197; sustainable development 381–2, 388, 398, 406
Eucarya 301
Eukaryotes 301
European Commission 111, 360
European Community Landfill Directive 359
European Environment Agency (EEA) 9
Euryarchaeota 280
eutrophication 82
expression profiling 299

Fang, C. 104
Farm Assurance Schemes (FAS) 369
farm surveys 384
farming *see* agriculture
fats 194–6, 270, 274
fatty acids *see* volatile fatty acids
Fava, J. 329
Federal Energy Regulatory Commission (FERC) 382
feed-in-tariff (FIT) 67, 77, 140, 381–2, 391
feedstocks 39, 139, 141–6, 148, 247, 422; energy/agricultural policy 8, 11–13, 16; life cycle assessment 321, 337–40; microbial communities 270–1; molecular biology 290–3; pre-treatment 203–6; process monitoring 194–203; sustainable development 383–7, 401–3 *see also* algae; grass; industrial residues; maize; organic wastes
Feil, B. 231
fermentation 32–3, 66, 261–2; algae 89–92; effectiveness 386–7; grass 57–60; molecular biology 290–2, 307; sugar processing 120–1
fertiliser 4–5, 56–7, 74–6, 99, 343, 423–4 *see also* digestate
Fibrobacter succinogenes 386
Fick's law 167
filling stations 18, 20, 192, 340
filtration 158
Finland 17, 54
Finnveden, G. 326
Firmicutes 265, 270, 272, 274
Fischer, G. 40

fleet vehicles 18–19
flexible membrane covers 188–90
floating covers 186–7
flow diagrams 337–9
fluorescent in-situ hybridization (FISH) 261, 299–300, 303
foaming 112, 114, 116, 131, 156
Fonknechten, N. 303
food 3–6, 341, 422–3; algae 94–5; industry 130, 403; land-use change 34–6 *see also* agriculture
France 191
Franke-Whittle, I.H. 301
Fuchs, W. 75
fuels 397–8 *see also* biofuels
functional unit 324–5, 332, 349
fungi 206, 267, 274, 301, 305, 307–9, 415
Fusarium 366
fuzzy logic 249, 280

Garba, B. 208
gas bags 188
gas engines 159–60
gas grid 11, 177, 191–2
Gasol, C.M. 325
gasoline 9, 11, 17, 400, 405–6
Gemmeke, B. 73
genetic algorithms 249
genetics 290, 296–304, 425; biomethane production 304–7; engineering 281–2, 307–9
Genomes OnLine Database (GOLD) 300
Genosorb 164
Georgia 411
Germany 139–40, 147; energy/agricultural policy 9–11, 14, 16–17, 20–1, 25; industrial residues 121, 123; maize 67, 72–3, 75; storage/distribution 191–2; sustainability 382, 387; upgrading/compression 173, 175, 178
Giardia lamblia 407
Global Methane Initiatives 403
global warming potential (GWP) 175, 178, 342–5 *see also* greenhouse gases
glucose 32, 59, 66, 268–9, 294
Golueke, C.G. 84
Gompertz model 253
Gösser brewery 123
government *see* policies
Gracilobacter thermotolerans 273
granulation 275
grass 5–6, 14, 56–60, 368, 422; environmental factors 54–6; growth stage at harvest 52–4; organic wastes 101–2; process monitoring 201–3, 207; species 49–52
Greece 124
Green Electricity Act 67
Green Energy and Green Economy Act 140
greenhouse gases (GHGs) 41, 57, 75, 148, 220; algae 82; developing countries 397–9, 404, 412; energy/agricultural policy 4–6, 11–16; industrial residues 111–12, 127; land-use change 30–1, 36–8; life cycle assessment 317–18, 323, 325, 327, 332–5, 337–45; small-scale digesters 375–7, 379–80, 383, 387–91
Grieder, C. 251
Grilc, V. 208–9, 213
grinding 204
Gronauer, A. 51
Grover, M. 238
Guangxi project 409
Guinee, J. 326
Guleri, R.L. 415
Gunnarsson, C. 54
Guwy, A.J. 247

halogenated compounds 158
Hamburg University of Applied Sciences 280
Hamelin, L. 343
harvest date 52–4
Hazard Analysis and Critical Control Point (HACCP) 360
health 16–17, 348, 379–80; benefits 406–7, 416; risks 399–400
heating value 154, 342
heavy metals 364–5
Helsinki 10
hemicellulose 104, 123, 291–2
Henry's law 161–2
Herrmann, C. 69, 72
high-pressure water scrubbing (HPWS) 190–1
Himalayas 399
Hodotermopsis sjoestedti 303
Holliger, C. 402–3
Holubar, P. 248
Horn, S.J. 89
Horvath, A. 322
household consumption 397–8, 408–10, 427–31
Hughes, D.J. 4
human waste 100, 407–8
Human–Machine Interface (HMI) 215, 218
Hungary 127
Hutnan, M. 129–30
hydraulic retention times (HRT) 72, 103, 141, 274; industrial residues 126–7, 129–30; process monitoring 196–7, 208, 211–12
hydrogen 126, 294–5
hydrogen peroxide 205
hydrogen sulphide 157–8, 162–4, 166–70, 174, 176, 185–6, 414
hydrogenotrophs 116, 213, 270–1, 274–7, 281, 295, 308, 423
hydrolysis 32–3, 148, 400; data mining 236–7, 239; microbial communities 262–4, 271; molecular biology 291, 294; process monitoring 194–5, 197, 199, 204–5, 209

Idler, C. 205
ignition 405–6

impact assessment 318–20, 323, 348–52
incentives 24, 146, 148, 391; small-scale digesters 380–2
India 196, 212, 390, 398–9, 402, 408–14
indirect emissions 318
indirect energy 337, 339–41
indirect land use change (ILUC) 14–15, 32, 36–7, 41
Indonesia 408, 411
industrial residues 111–12, 130–1; from bioethanol 117–21; from breweries 121–4; from olive mills 124–7; from slaughterhouses 112–17; from sugar processing 127–30
information policies 18–19
inhibitory compounds 90–2, 116, 214
inocula 91, 144, 205–6, 272
Institutional Biogas Plants (IBP) 412
instrumentation 215–17
Integrated Environmental Impact Analysis 406
Intergovernmental Panel on Climate Change (IPCC) 317, 343
Intermediate Technology Development Group (ITDG) 411
internal combustion 185, 405
International Energy Agency (IEA) 20
International EPD Cooperation (IEC) 337
International Fund for Agricultural Development (IFAD) 409
International Standards Organisation (ISO) 318–20, 323–4, 336, 347
investment costs 124, 173–4, 183
Iran 9
Ireland 9–10, 16, 21, 54–5
Israel 87
Italian ryegrass (IRG) 50, 53
Italy 9, 124, 127

Jacobi, F.H. 252
Jamaica 411
Japan 88
Jarvis, A. 214
Jensen, A.A. 329
Jerez 87
Jungmeier, G. 327

Kaiser, F. 51
Kampmann, K. 270
Kanwar, S.S. 415
Kaposvár (Hungary) 127
Keane, T. 55
Kenya 408, 411
Khadi and Village Industry Commission (KVIC) 409, 412–13
Khendelwal, K.C. 402
Kim, S. 325
kinetic selectivity 166
kinetics 277–81
knowledge discovery in databases (KDD) 231, 238
Köchling, T. 300

Koellner, T. 42
Korea 411
Korres, N.E. 253, 324, 338, 341
Krakat, N. 270
Kumar, R. 306
Kurakake, M. 205
Kyoto Protocol 67, 380, 412

L-ornithine 303
labour 341
lactate dehydrogenase (*ldh*) 308
lactic acid 57–9, 69–70
Lactobacillus 69, 400
Laholm Bay 15
Lal, R. 106
land-use change (LUC) 31–2, 36–7, 127; agricultural policy 14–15; biomethane production 34–5; carbon sequestration 5–6; energy crops 14–15; life cycle assessment 327–8; maize 76
land-use intensity 39, 41–2
landfill 40, 99, 152–3, 155, 157–9, 161, 359
Landsberger Gemenge 325
Lantana camera 402
Laos 408, 411
Latin America 191
leach bed reactors 73, 218, 422, 424
leachate 99, 262
legumes 50–2, 56
Lehtomäki, A. 103, 207
Lesage, P. 322
Lesotho 411
Li, Z. 398
Life Cycle Assessment (LCA) 317–20, 349–52, 425–6; flow diagrams/calculations 337–45; goal/scope definition 323–8; methodology 320–3; multifunctionality/allocation 345–7
Life Cycle Impact Assessment (LCIA) 348–9, 352
Life Cycle Inventory (LCI) 325, 328–37, 343, 345, 348–9, 352
lighting 403, 429–31
lignin 52, 69, 84, 144; industrial residues 122–3, 129; molecular biology 291, 308; organic wastes 100, 103; process monitoring 194–6, 199, 204–6
Lindeijer, E. 42
Lindfors, L.G. 325
linearity 252
lipids 84, 86–7, 89, 114, 270, 274, 291, 293
liquefied biomethane (LBM) 31
Listeria 366, 386
Liu, J.G. 218
livestock 343–4, 359–65, 377, 388–90, 397–9 *see also* cattle; pigs; poultry
loans 381–2
London 10
Lotka-Voltera model 277
Low Indirect Impact Biofuels initiative 41
Lübken, M. 278
Lurgi GmbH 120

McDonald, P. 58–9
machine learning 242
McHugh, D. 88
macroalgae 82–5, 88–96
Mahmuti, M. 4
Mahnert, P. 51
maintenance costs 173
maize 14, 20, 66–70, 204, 384, 425; anaerobic digestion 72–7; data mining 251–2; storage/pre-treatment 70–2
Malaysia 382
malt 121
Malthus model 253, 277
MANNER-NPK 362, 367, 371
manure 15, 40, 56, 276; developing countries 397–9, 401–3, 407–8; digestate/fertiliser 359–61; family farm case study 428–31; industrial residues 126, 130; life cycle assessment 343–4; maize 72–5; organic wastes 104–5; process modelling 205, 212; small-scale digesters 375, 377, 379, 383, 388–90
Martinez, E. 249
massive parallel sequencing (MPS) 264, 297, 300, 303
Mata, T.M. 87
Mattheeuws, B. 141
measuring techniques 214–18
meat industry *see* abattoirs; animal waste
Mediterranean 124
Megasphaera 400
Meher, K.K. 209
membrane separation 167–8, 171–4, 176, 178
mesophilic digestion 130, 145, 170, 208–9, 244–6, 272–4
metagenomics 299–300, 303, 423
methane 17, 183, 345, 377; biogas upgrading/compression 152–5, 164–70, 175; developing countries 398–400, 404–5; fermentation 32–3; from macroalgae 89–90; grass 49–54, 57, 60; land-use change 30–1, 38, 40; maize 69–70
Methanisarcina 275
Methanobacteria 277
Methanobacteriales 74, 275
Methanobacterium 170, 270, 273, 281
Methanobrevibacter 303
Methanocella paludicola 303
Methanocorpusculum 273, 275
Methanoculleus 270, 273–5
methanogenesis 170, 262–5, 400; data mining 237, 239; kinetics/modelling 277–81; maize 73–4; molecular biology 294–6, 302–3, 308–9; pH 209–10; process monitoring 194–5, 213
Methanomicrobiales 303
Methanosaeta 273–7, 400
Methanosaetaceae 275
Methanosarcina 206, 270, 273–7, 307–9, 400
Methanosarcinaceae 74, 275
Methanospirillum 273–5
Methanothermobacter 270, 273–5

Mexico 218
micro-feed-in-tariff (MFIT) 140
microalgae 82–8, 92–5
microarrays 298, 300–1
microbial communities 261–7, 295; digester environment/operation 267–77; kinetics/modelling 277–81; molecular genetics 296–304 *see also* pathogens
Microjoenia 303
Milan 10
Millennium Development Goals (MDGs) 411
milling 122–3, 204
miscanthus 35, 50–1, 201
mitigation: of biofuel impact 37, 39, 41; of climate change 35–6, 377, 379–80
mixing 54, 90, 141–2, 275, 340
Miyazaki, K. 301
modelling 261; data warehouse/mining 233, 235–6, 240, 242–4, 249–54; life cycle assessment 325–7, 346, 350; methanogenesis 277–81
moisture content 201
molasses 127–30
molecular biology 290, 296–304, 423; biomethane production 304–9
Molof, A.H. 213
monitoring 202–3, 214–19, 261–2, 303–4, 385–6, 425–6
mono-digestion 73–5, 126, 425
mono-fermentation 114, 116
monocultures 15, 75–7
Monod equation 280–1
Montan University 123
Morand, P. 90–1
Morocco 92, 411
Morrison, J. 55
moulds 70
Mozambique 411
multifunctionality 345–7
municipal wastes 8, 100–1, 105–7, 140, 345; developing countries 402–3; process monitoring 204–5, 207
Murphy, J. 101–2, 120, 141, 216, 342
mutagenesis 304–7, 309

NADH 32
Naïve Bayes 244–6
Nand, K. 205
nanotechnology 148, 426
Narihiro, T. 303
National Biogas and Manure Management Programme (NBMMP) 413
National Project on Biogas Development (NPBD) 412
natural degradation potential (NDP) 41–2
natural gas 8–9, 11, 154, 177, 191–2
natural gas vehicles (NGVs) 9–10, 17–19, 21, 146
near-infrared spectroscopy (NIRS) 252, 385
Nepal 105, 139, 408–10, 412
net metering 146

Netherlands 18, 127, 178, 410–11
Netherlands Development Organization (SNV) 408–9
neural networks 247–9
New Zealand 21
Nicaragua 411
nickel 73
Niessner, R. 160
Night-soil based Biogas Plants (NBP) 412
Nirmali 415
nitrate leaching 15, 118, 362
Nitrate Vulnerable Zones (NVZs) 364, 367
Nitrates Directive 17, 118, 364, 367
nitrogen 4–5, 345, 423–4; biogas upgrading 162, 166–7; digestate/fertiliser 361–4, 366–7, 371; grass 50, 55–7; industrial residues 120–1; maize 74–5 see also carbon/nitrogen ratio
nitrous oxide 75, 160, 343–4, 375, 379–80, 404, 406
Nizami, A.S. 101–2, 141, 144, 196, 216
non-linearity 252–3
normalisation 235–6, 332
Northern Ireland 360
Norway 89
Nozhevnikova, A.N. 209

obligation systems 18
octane number 405
oilseed rape 4–5
olive mill waste 124–7
Olsson, B.M. 218
online analytical process (OLAP) 233, 236–7, 240
online transaction processing (OLTP) 232–3
Ontario 140, 147
operational taxonomic units (OTUs) 265
organic fraction of municipal solid waste (OFMSW) 8, 204–5, 207, 345, 402–3 see also biodegradable municipal waste
organic loading rates (OLR) 72, 103, 123, 126–7, 129, 141, 143, 208, 212
organic matter 103, 359–61, 364
organic solvents 164
organic wastes 99–108, 197–8, 207, 397, 422
Oswald, W.J. 84
Owen, W.E. 199
Owende, P. 87
oxidation 155
oxygen 162, 166–7
ozone 380

Pakarinen, O. 58
Pakistan 9, 196, 398, 408–10
Palatsi, J. 206
Parma 127
PAS100 361, 364–5
pasteurization 371
Patel, J. 409
Patel, V. 415
pathogens 365–6, 380, 386, 407

Patterson, T. 10
pectins 291–2
Peleg, M. 278
Peltola, R.J. 204
Peptococcus anaerobus 400
Perendeci, N.A. 205
perennial ryegrass (PRG) 50–3, 55, 67, 201–2
permeability 167
Peterson, B.M. 322
petrol 9, 11, 17, 400, 405–6
pH 385; biogas upgrading 155, 162; data warehouse/mining 237, 239, 244–5; process monitoring 207–11, 213–14, 217–18
Phanerochaete chrysosporium 206
Philippines 411
phosphate 363, 366
phosphorous 55–7, 407–8
phylogeny 301–2
physical absorption 160–1, 164, 171–5
Phytophthora 366
phytoplankton 82–8, 92–5
Pieper, H.J. 118
pigs: digestate/fertiliser 361–3, 365; family farm case study 427, 429–31; organic waste 100, 105–6; slaughterhouse waste 112–15, 130; small-scale digesters 382–4 see also swine
pivot table 238–9
Planctomycetes 265, 274
planning permission 376–7
Pleurotus ostreatus 206
Plöchl, M. 69
policies 10–17, 411–14, 424; biomethane promotion 17–25; small-scale anaerobic digesters 380–2
Polit, M. 249
politics 38
pollution 175–7, 322, 379–80, 399–400, 404–5
polyhydroxyalkanoates (PHA) 114
polymerase chain reaction (PCR) 261, 264, 298, 300–1, 303, 306
polymers 32, 103–4, 167, 290–2
polyphenols 125–6
population growth models 278, 281
potash 363
potassium (K) 55–7
poultry 100, 106, 112–13, 126, 361, 398–9, 401–2
power see combined heat and power; electricity
Power, N.M. 120
pre-treatments 275, 387, 424; of brewery residues 122–4; of macroalgae 91; of maize 69, 72; of olive oil residues 126; of organic waste 103–4; process monitoring 196, 203–6 see also additives
precision 336
prediction 251–2
pressure 162–5, 177, 185–7
pressure swing adsorption (PSA) 165–7, 170–6, 178, 190
primary data 332–3, 335
probes 298–300
process control 214–19, 385–6

process subdivision 346–7
Product Carbon Footprint (PCF) 127
production chain 194–6
programmable logic controller (PLC) 214–18
Propionibacterium 400
propionic acid 203, 249
proteins 70, 94, 270; land use 31–2; molecular biology 291, 293; slaughterhouse waste 111–12, 114, 116
Proteobacteria 265, 274
proteolysis 57–8
Pseudomans mendocina 273
Pseudomonas veronii 308
Pyrococcus furiosus 308

Qi, B.C. 204
Quebec 140

Radhika, L.G. 402
RAPD 298, 300
rational protein design (RD) 307
readily available nitrogen (RAN) 361–3, 367, 371
Rebitzer, G. 325
Reclaim, Recycle and Reuse (3Rs) 99
red diesel 334
redox potential 208, 213
reference systems 327–8
regression 250–3
regulations 16–17, 360, 367, 369
reliability 336
rendering 111–14
renewable energy 3, 10–11, 194, 318, 411
Renewable Energy Act 20
Renewable Energy Certificates (RECs) 381
Renewable Energy Directive 14, 16, 317–18, 320, 323–4, 341–2
Renewable Energy Sources Act 67
Renewable Heat Initiative (RHI) 382, 387, 391
Renewable Obligations Certificates (ROCs) 381
Resch, C. 69
residual biogas potential (RBP) 364
residues 35–7, 102; breweries 121–4; from sugar production 127–30 *see also* agricultural residues; industrial residues
resource flows 322
respiration 57–8
retention times 123, 142, 200 *see also* hydraulic retention times
Reticulitermes speratus 303
RFLP 298, 300
rice 206, 427–30
Richards model 253
Ridla, M. 206
rigid digester cover 188
Ritari, J. 265
RNA 298–301
roadmaps 22–5
Ronteltap, M. 105
root definition 323

RT-PCR 298, 300–1
Ruggeri, B. 341
rumen 40, 114–15
rural development 317, 397, 406–15
Russia 121
Rwanda 408, 411
Ryckebosch, E. 156
ryegrass 50–3, 55, 67, 201–2

Saccharomyces cerevisiae 307
St Martin (Austria) 114–16
Sajko, H. 206
Sala, O.E. 55
Salmonella 365, 386, 407
Samuil, C. 407–8
Sanchez, E.P. 212
sanitation 411
Sanz, J.L. 300
Saunders, C. 334
SCADA 214–15, 218
Schievano, A. 251
Schlicher, M.A. 118
Schmidt, J.H. 322
Schnurer, A. 214
Scholz, R.W. 42
Schweigkofler, M. 160
Scientific Applications International Corporation (SAIC) 346
Scotland 360, 384
scrubbing 162–5, 170–7, 178, 190
Seambiotic 87
seaweeds 82–5, 88–96
secondary data 335
seed inocula 272
Sekiguchi, Y. 303
Selexol 164–5
Senegal 408
Senn, T. 118
Seppala, M. 52
sequencing 297
sequential leach bed reactor with upflow anaerobic sludge blanket (SLBR-UASB) 216
sequestration 5–6, 88, 422
sewage 8–9, 100, 152–3, 159, 268–9, 274, 343, 345
Sezun, M. 124
Shadow Price of Carbon (SPC) 388–9
Shahriari, H. 205
Shakya, I. 410
Shapiro-Piatetsky, G. 231
Shen, B. 308
Shigella 407
Shimizu, C. 103, 415
shocks 274–5
Siegart, I. 209, 237
silage 384; grass 50, 53–4, 56–60; life cycle assessment 339–40; maize 66–76; process monitoring 202–3
silage ripeness index (SRI) 68
silica gel 160

Index 441

siloxanes 158–60, 169
Silvestrini, A. 10
Simon, C. 301
Singh, A. 100, 321
Single Payment Scheme 14, 367
site-directed mutagenesis (SDM) 307
Sjoberg, J. 247
slaughterhouse wastes (SW) 15–16, 100, 111–17, 126, 130, 205, 424 *see also* animal by-products
slurry 15–16, 99–100; developing countries 415–16; digestate/fertiliser 361–5; life cycle assessment 343–4; small-scale digesters 383, 388–91
Small Generator Interconnection Process (SGIP) 382
small-scale anaerobic digesters (SSADs) 375–80, 390–3, 424–5; factors influencing 382–8; policy/incentives 380–2; sustainability 388–90
smart grid technologies 214
Smyth, B.M. 191, 341
snowflake schema 236
soft system methodology (SSM) 323
software 215, 218, 277, 280, 300, 367
soil 35, 40, 55, 367, 369, 399, 425 *see also* fertilizer
soil organic matter (SOM) 359–60
solvents 164
Sopex 92
Soussana, J.F. 55
South Africa 411
South America 15
South East Asia 375
soy 15
Spain 87, 124
spectrometry 216–17
Speece, R.E. 130
Sphingomonas 206, 400
Spitzer, J. 327
spore formers 386
Sporobacterium 400
Sri Lanka 196, 410–11
stage digesters 262–3, 265–6, 276
stakeholder involvement 19, 22, 24–5
Staphylococcus 400
star schema 235–6
starch 32, 66–8, 72
statistics 202, 238, 240–3, 250–4, 336
Steinhauser, A. 194, 213
stillage 117–21
stomachs 40, 114–15
storage 177–9, 183–92, 423–5
Strik, D.P.B.T.B. 247
Stuttgart 86
sub-Saharan Africa 400, 411
subsidies 14–15, 18, 77, 413
substitution approach 342, 346
substrates *see* feedstocks
sugar industry 104–5, 117, 127–30, 424
Suiker Unie 127
sulphur 90, 157–8, 163, 166

Sun, Y. 206
SunChem 87
sunlight 54
supervised learning 242, 247
support vector machine (SVM) 242, 250
sustainability 11–14, 36, 425; agriculture 3, 5–6; developing countries 397, 411, 416; small-scale digesters 375–80, 383, 387–8, 390–1
Sweden 9, 11, 15–16, 20, 25, 147, 178, 191, 324
swine 126, 268, 399 *see also* pigs
Switzerland 11, 87, 127, 191, 317, 414
syntrophic bacteria 116, 206, 277, 281, 423
Syntrophobacter wolinii 295
Syntrophomonas 206
Syria 124
syrup 118–21
system expansion 346–7
Sytrophomonos wolfei 295

Takashima, M. 208
Talbot, G. 300
tall fescue 50, 52
Tandon, S.M. 415
Taniguchi, M. 206
Tanzania 408, 411
targets 10–11, 22–4
tariffs 20–1 *see also* feed-in-tariff
tax 20; exemptions 17–18, 21
Taylor, M.J. 364–5
Tchobanoglous, G. 212
technical availability 170
technical considerations 199, 413–14
temperate grassland 50–1
temperature 54, 73, 103, 162, 207–9, 244–6, 272–4, 386, 414
Tereso navalis 303
Thailand 390, 411
Thamsiriroj, T. 9, 22, 342
thermal gasification 40
thermophilic digestion 92, 103, 126, 130, 145, 170, 208–9, 213, 386
Thermotogae 270
Thomassen, M.A. 322
Thomasson, A.J. 55
Thornton, P.K. 397
Tibet 399
Tillman, A.M. 325–6
time series 246
timothy 50, 52
Tirumale, S. 205
titrimetry 211, 216–17, 253
Tokyo 92–3, 96
Tomei, M.C. 279
total Kjeldahl nutrition (TKN) 115–16, 119–20, 125
total organic carbon (TOC) 216
trace elements 73–4, 116–17
training processes 247–8
Trametes versicolor 206
transcriptomics 299

transport fuel 191–2, 340; algae 93–4; biomethane experience 20–1; developing countries 403, 405–6; energy/agricultural policy 8–11, 15–16, 18–19; policy roadmap 22–5
Trchunian, A. 213
Tredici, M.R. 95
tree diagram 203, 254
trend 246
Trichoderma reesei 291, 305–6
Tunisia 124
Turkey 124

Uchiyama, T. 301
Uetians Models 410
Uganda 408, 411
UK 4, 6, 10; digestate/fertiliser 360, 362, 365, 367, 369; small-scale digesters 375, 381–2, 384, 387, 391
Ulva 88, 90
UN 411–12
uncertainty 146, 231, 326
unit processes 322, 328, 332–3, 339
Unitronics DataXport 218
unsupervised learning 242, 247, 253
upflow anaerobic filter (UAF) 126
upflow anaerobic sludge blanket (UASB) 42, 141–3, 148, 422, 424; industrial residues 123, 126; process monitoring 207, 216
upgrading 178, 190–1, 422–4
Urban, W. 191
US 139, 144–5, 147; energy/agricultural policy 9, 11, 15, 17; industrial residues 117–18, 121; organic wastes 105–6; sustainability 375, 381–2, 398
Us, E. 205
US Environmental Protection Agency (US-EPA) 140, 147

van der Laak, W.W.M. 10
Van Vooren, L. 253
Vandevivere, P. 141
variability 194, 203, 206–7, 219, 231, 241
Vasilian, A. 213
vehicle fuel *see* transport fuel
Vehida, S. 206
Verma, S. 206
vertical flow reactors (VFRs) 90
Vervaeren, H. 69
Vibrio cholera 407
Vieira, P.S. 322
Vietnam 408, 411
Vikman, P. 320, 346
Voigt, J. 123
volatile fatty acids (VFA) 146, 148; data warehouse/mining 237, 239, 244–5; microbial communities 271, 274–5, 277, 281; molecular biology 294–5; process monitoring 203, 209–11, 214, 216–18
volatile solids (VS) 144, 196, 199–200, 202, 247–8
Von Nordenskjöld, R. 123

Wales 360
Walter, A. 276
Ward, A.J. 207, 249
Ward's method 254
waste 40, 67, 346, 359–60; policy 16, 21; streams 38–9, 175–7 *see also* agricultural residues; animal wastes; human waste; industrial residues; municipal wastes; organic wastes
wastewater treatment plants (WWTPs) 140, 146
water 3, 55, 380
Water Framework Directive 17
water scrubbing 162–4, 170–6, 178, 190
water soluble carbohydrates (WSC) 50, 54–5, 57, 67
water vapour 156
weed seeds 386
Weidema, B.P. 346
Weihenstephaner Brewery 123
Weiland, P. 74, 129
well to tank/wheel 321
Weltanshauung 324
Wenzel, M.H.W. 322
West Africa 191
wet cake 118–21
wet continuous digesters 141–2, 422
Wilson, R.K. 57
Winrock International 411
Wisenthal, T. 10
Wissington 128
Wobbe index (WI) 154, 161, 184
Wong, M.H. 103, 415
Wood–Ljungdahl pathway 295
Woolford, M.K. 205
World Bank 412
World Health Organization (WHO) 399–400
WRAP 369

Xie, S. 212

Yadvika, S. 209–10
yeasts 70, 120–3
Yeh, S. 10, 17
Yersinia enterocolitica 407
Yu, B. 336
Yu, H.W. 207
Yunnan 398

Zambia 411
Zeng, M. 204
Zupancic, G.D. 208–9, 213